37 50/1 EC

...RNED

M

D1237195

QUANTUM MECHANICS:
NEW APPROACHES TO SELECTED TOPICS

This book is dedicated to the memory of

AMOS DE-SHALIT

who enjoyed finding new approaches to all problems
and encouraged others to look for them

QUANTUM MECHANICS

NEW APPROACHES TO SELECTED TOPICS

HARRY J. LIPKIN

THE WEIZMANN INSTITUTE OF SCIENCE
REHOVOTH, ISRAEL

1973

NORTH-HOLLAND PUBLISHING COMPANY – AMSTERDAM · LONDON
AMERICAN ELSEVIER PUBLISHING CO., INC. – NEW YORK

© North-Holland Publishing Company – 1973

All rights reserved. No part of this publication may be reproduced, stored in a retrieval system, or transmitted, in any form or by any means, electronic, mechanical, photocopying, recording or otherwise, without the prior permission of the copyright owner.

Library of Congress Catalog Card Number 72–79733
ISBN North-Holland 0 7204 0258 1
ISBN American Elsevier 0 444 10450 X

Publishers:

NORTH-HOLLAND PUBLISHING COMPANY – AMSTERDAM
NORTH-HOLLAND PUBLISHING COMPANY, LTD. – LONDON

Sole distributors for the U.S.A. and Canada:

AMERICAN ELSEVIER PUBLISHING COMPANY, INC.
52 VANDERBILT AVENUE, NEW YORK, N.Y. 10017

PRINTED IN THE NETHERLANDS

764819

PREFACE

This book grew from lecture notes written while teaching a first year graduate course in quantum mechanics over a ten year period, both in the United States and in Israel. Many topics were presented in a manner very different from treatments in the standard texts. These topics are collected in this book which can be used together with a text like Messiah's 'Quantum Mechanics' to present a course which contains both the standard material and the new treatments of special topics. The individual topics contained in the book can also be considered as self-contained 'Monographs for Pedestrians' on the Mössbauer Effect, Many-Body Quantum Mechanics, Kaon Physics, Scattering Theory, Feynman Diagrams and Relativistic Quantum Mechanics. These treatments should be useful to active research physicists as well as students.

The principal unusual feature of this book is the introduction of many interesting physical problems and mathematical techniques at a much earlier level than in conventional texts. The student can thus see the physical implications and useful applications of quantum theory before he has mastered the formalism in detail. He also has new mathematical tools available at an earlier stage for use in subsequent problems. To the active physicist this provides a 'pedestrian' treatment of many topics which is accessible without having to 'go back to school to learn the formalism'.

A historical approach is not used, because more modern developments are often easier for the student to understand than earlier work. Furthermore, the student already has learned about many physical phenomena which were not yet known in the early days of quantum theory, such as the existence of antiparticles, pair production and the motion of particles and holes in solids. The use in many-particle problems of techniques borrowed from field theory is an example of how techniques originally developed for highly sophisticated problems find application in simple problems. These techniques are presented at a very early stage in the course, without reference to field theory. Thus the student is already familiar with second quantization when he studies time-dependent perturbation theory, and can treat processes in which particles are emitted and absorbed. This simplifies the derivation of the 'golden rule'. The student is already familiar with the concept of particles and holes from many-body theory when he studies the Dirac equation, and also obtains a good introduction to field theory from the many-body problem.

The book is divided into several distinct topics as follows:

Chapter I is 'Dirac's Polarized Photons for Pedestrians'. This introduction to basic principles is motivated by the first chapter of Dirac's famous book. The problem of polarized photons shows the basic concepts and mathematical techniques used in quantum theory and is treated with very simple mathematics. It is a useful introduction for a graduate course in which the entering students all have different backgrounds and must be brought to a common level. The material is easily read by a student with some background, and each can fill in the gaps in his own particular background while going through the material. The problems at the end of the chapter can serve as a useful 'entrance examination' to check the student's background.

Chapters 2, 3 and 4 constitute a monograph 'Momentum Transfer to Bound Systems and the Mössbauer Effect'. These chapters provide excellent examples of the use of quantum theory in practical problems. Chapters 2 and 3 can be presented very early in a course, where the student has learned only a minimum of the formalism. Chapter 2 requires only the knowledge of the free particle and the harmonic oscillator in one dimension and the transformation between the two bases. Chapter 3 requires only the knowledge that wave functions exist for systems having many degrees of freedom, but no other knowledge of many-body quantum mechanics. This simple framework is sufficient to allow the introduction and practical use of quite sophisticated mathematical techniques, such as manipulating exponentials of Heisenberg operators. Chapter 4 requires the knowledge of second quantization and time-dependent perturbation theory. A reader unfamiliar with these techniques should read chapters 5–7 before reading chapter 4. More advanced readers will find in chapter 4 a unified treatment of momentum transfer to bound systems, applied to a wide variety of problems not normally considered in the same framework. These include emission, absorption and scattering of photons by atoms and nuclei, nuclear beta decay and electron scattering by complex nuclei.

Chapters 5, 6, 9, 10 and 11 constitute a monograph 'Many-Body Quantum Theory for Pedestrians'. Chapter 5 introduces the notation of second quantization at a level where the student has mastered one-particle quantum mechanics and is just beginning to study systems of several identical particles. Chapter 6 uses the second-quantized formalism to treat composite particles. The deuteron is shown to be almost but not quite like a boson, and Cooper pairs are shown in a very elementary mathematical treatment to have the properties underlying superconductivity and nuclear pairing correlations. Chapters 9, 10 and 11 present more sophisticated theory, still 'for pedestrians', including the Fermi–Thomas, Hartree–Fock, BCS and random-phase approximations, with applications to nuclear, atomic and solid state physics. Differences between the physical bases for these approximations in different areas are discussed, pointing out how the same random-phase approximation formalism is used in nuclear and solid state problems for different reasons. Elegant time-dependent formalisms are discussed which have simple interpretations in many-body theory and provide the student with an intuitive basis for later applications in field theory.

Chapter 7 is a monograph 'Kaon Physics for Pedestrians'. The student should have already read the earlier chapters, in particular the introduction to second quantization in chapter 5. Time-dependent perturbation theory is introduced in the simple Heitler description, and deals immediately with radiation processes since the student has already learned second quantization. It is presented again in the modern formalism in the discussion of the K_1–K_2 mass difference. Many basic concepts of quantum mechanics are illustrated in the simple K_1–K_2 system. The discussion of C, P and T provides an introduction to these discrete symmetries. The introduction of phenomenological Hamiltonians shows the student how many practical problems are attacked.

Chapter 8 is 'Scattering for Pedestrians' and illustrates many features of scattering problems already present in one-dimensional systems where they can be treated with simple mathematics. One-dimensional polar coordinates provide an introduction to partial waves and phase shift analysis. The one-dimensional delta potential is used to present resonances, bound states, analytical properties of scattering amplitudes, inelastic scattering and the relation between particles and fields in quantum theory. Perturbation theory, the Born series, Green's functions and the S- and T-matrices are also introduced.

A special self-contained topic in chapter 8 is the treatment of a one-dimensional periodic potential using the phase shift method to derive the band structure of solids. This derivation may be useful in solid state courses, as it is much simpler and more general than the standard derivations.

Chapter 12 is 'Feynman Diagrams and Field Theory for Pedestrians' and shows in an elementary manner the difference between the new formalism and the old 'intermediate state' perturbation theory. The sections on 'Fieldsmanship' show how mathematical techniques and intuition developed in treating the many-body problem can be used to give an impression of the problems encountered in field theory without going into detail.

Chapters 13, 14 and 15 give 'Symmetries and Relativistic Quantum Mechanics for Pedestrians'. Chapter 13 introduces the Dirac equation from an unorthodox point of view, making use of the student's knowledge of the many-body problem and of the non-conservation of parity. Chapters 14 and 15 discuss symmetries and invariance from a very elementary point of view, presenting the physical interpretations and mathematical techniques used in different cases. Chapter 15 develops the properties of the Lorentz group as a time-dependent symmetry, following the general approach of the author's book 'Lie Groups for Pedestrians'.

ACKNOWLEDGEMENT

The approach presented in this book is the result of 25 years of first studying and then teaching quantum mechanics and shows the influence of many people, too numerous to mention here. The development of quantum mechanics during these years could be loosely classified into three periods: 1. The depressing period, 2. The useful period and 3. The exciting period.

I was a graduate student at Princeton during the dark period just before the dramatic break-through in quantum electrodynamics. Quantum mechanics was presented as a theory which had outlived its usefulness. All physical phenomena which could be described properly by quantum mechanics had already been calculated. All phenomena which had not already been calculated gave nonsensical infinities when quantum mechanics was used to describe them. The general feeling around Princeton was that a drastic modification at the fundamental level was necessary to cure this sick theory, perhaps even as remarkable a revolution as the introduction of relativity and quantum mechanics in the first quarter of the century. David Bohm's book on quantum mechanics, from his lectures which I heard at Princeton, gives a good description of the prevailing mood at the time with the emphasis on studying quantum mechanics in depth in order to find its basic difficulties and inconsistencies. The aim was arriving at a better theory, rather than studying it as a useful tool for later applications. My education at Princeton encouraged an iconoclastic approach and a continuous questioning of accepted beliefs while at the same time stimulating new imaginative and exciting approaches and the following up of crazy ideas. I should like to thank all members of the faculty at Princeton and my contemporaries as graduate students for their part in this phase of my education. The names are too numerous to mention and any attempt will invariably leave someone out.

The useful era of quantum mechanics began in the early 50's with the development of the nuclear shell and collective models and of modern solid state physics where quantum mechanics found extensive use as a basic tool without the necessity to solve the deep problems of ultraviolet divergences. During this period I worked at the Weizmann Institute in Israel and was much influenced by the Israeli School of Physics with its pragmatic use of highly sophisticated but down-to-earth theory to describe the real world of experimental data. I should like to thank my friends and colleagues in

Rehovoth who helped to teach me in this period that quantum mechanics could be useful and that many apparently complicated and sophisticated formal descriptions were in reality based on very simple physics.

The exciting period came when research at the frontiers of physics found new and exciting phenomena which could be described simply with elementary quantum mechanics. Kaon physics, the BCS theory of superconductivity and the Mössbauer effect are a few examples. It was my good fortune to spend a year at the University of Illinois in 1958–59 shortly after the famous BCS paper and Mössbauer's first experiments, when there were still very lively discussions about both subjects and many experts around from whom I could learn a great deal. When I returned to Urbana for another year in 1962–63 and was asked to teach the introductory quantum mechanics course, the idea occurred to me to include many of these exciting developments which could be presented at an elementary level. It is a pleasure to thank my colleagues and my students at the University of Illinois who taught me many things about the exciting new areas in physics and who encouraged me in my teaching of quantum mechanics. I taught the quantum mechanics course three more times in the next several years at the Weizmann Institute and developed the notes which led to this book during that period. I wish to thank all those colleagues at the Weizmann Institute and the graduate students who listened to the courses, helped me with the exercises and gave numerous suggestions on the manuscript.

Particular thanks are due to J.D. Jackson who taught the Illinois Quantum Mechanics course one semester ahead of me in 1962 for much useful advice and encouragement, to J.H.D. Jensen for pointing out to me the particular role of unpolarized light in foreshadowing its quantum nature, to J. Bardeen for helpful comments on my oversimplified treatments in chapter 6, to Y. Zarmi and Y. Frishman for much helpful criticism of the lecture notes and to M.D. Frank who encouraged me to write books.

Finally, I want to thank all the very helpful secretaries who taught me how to use the dictaphone and typed innumerable pages of messy lecture notes, in particular Margaret Runkel, Eva Kinstle, Connie Carter, Ilana Eisen, Georgette Azogue and Corinne Hasdai.

I regret that Amos de-Shalit, whose constant encouragement, stimulation and criticism was of inestimable value in the teaching of the course and the preparation of the book, did not live to see the final results. This book is dedicated to his memory.

CONTENTS

POLARIZED PHOTONS FOR PEDESTRIANS

Photon polarization was used by Dirac as an introduction to quantum mechanics. Chapter 1 presents an expanded treatment of this simple case which introduces many basic concepts, phenomena and paradoxes of quantum mechanics at a level which requires only the simple mathematics of 2×2 matrices. This chapter is easily read by a student familiar with elementary matrix algebra and the Dirac notation for vector spaces and matrix elements.

All the mathematical formalism necessary for the quantum-mechanical description of photon polarization is introduced naturally at the classical level: the description states of polarization by complex vectors in a two-dimensional space, of apparatus which change the polarization by matrices, of polarization measurements by expectation values of Hermitean operators which also describe changes of polarization produced by measuring apparatus and whose eigenvectors are the particular states which pass through the measuring apparatus unaffected (e.g. those polarized in the direction of the principal axes of a nicol prism) and the use of unitary transformations between different bases.

Unpolarized and partially polarized light introduce the quantum nature of light already at the classical level, since they cannot be described as continuous classical waves but only as series of discrete short pulses or quanta with no correlation between the polarizations of successive pulses. The density matrix formalism is introduced to describe partially polarized light in the two-dimensional vector space.

Sections 1.5–1.9 show how this classical description must be modified in view of the quantum nature of electromagnetic radiation; i.e. that it is composed of discrete quanta. The basic features of the quantum description are obtained by considering how a single photon must behave in the various experimental situations discussed previously for the classical case, with the aid of the correspondence principle which requires that experiments performed with very large numbers of photons should give the classical results. The results of certain experiments are shown to be unpredictable even if all the available information about the initial conditions of the experiment are known. This leads to a discussion of the statistical description, the interaction of the observer with the system, the incompatibility of different measurements and wave–particle duality.

CHAPTER 1

POLARIZED PHOTONS AND QUANTUM THEORY

1.1 Polarization measurement for a classical beam

Consider a polarization measurement on a classical beam of polarized light. The measuring apparatus is a Nicol prism which splits the beam into two components, one horizontally polarized and one vertically polarized as shown in fig. 1.1.

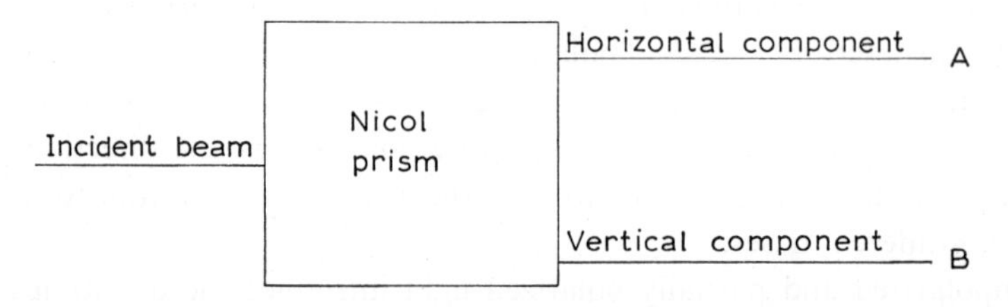

Fig. 1.1

The relative magnitudes of the horizontally and vertically polarized components depend upon the properties of the incident beam. Measuring the intensities of these two components gives information about the incident beam, but does not generally supply sufficient information to determine its polarization. For example, if the intensities of the horizontal and vertical components are found to be equal, the incident beam might be circularly polarized left-handed or right-handed, plane polarized at an angle of $\pm 45°$ with respect to the horizontal, or elliptically polarized in any one of a continuum of states. To specify the polarization of the incident beam completely, we need to measure not only the relative magnitudes of the horizontal and vertical components, but also the relative phase.

More information about the incident beam is obtained from two successive measurements of the type shown in fig. 1.1, with the Nicol prism in different orientations. The second measurement could be, for example, with the Nicol prism rotated by an angle of 45°, thus separating the beam into two plane polarized components with the planes of polarization at angles of $+45°$ and $-45°$ with respect to the horizontal. In the case considered above, where the horizontally and the vertically polarized components were found to be equal, the second measurement would enable a choice to be made between circular polarization, plane polarization at an angle of 45°, and elliptical polarization. However, even two measurements do not completely specify the polarization of the incident beam. There remains one ambiguity. Suppose, for example, that the intensities of the two 45° components were also found to be equal, as well as the horizontal and vertical components. This would definitely establish the beam to be circularly polarized, but would not distinguish between right-circular and left-circular polarization.

The number of measurements required to specify the initial polarization is seen by the following analysis. The beam is completely specified by giving the values of the magnitudes and the phases of its two components; i.e., by the values of four parameters. If we are not interested in the absolute intensity of the beam or its absolute phase, but only in the state of polarization, we require only two parameters: the ratio of the amplitudes and the relative phase. Two measurements give us two equations from which we can determine these two parameters. However, measurements of *intensities* which are *quadratic* functions of the amplitudes give quadratic equations which have two solutions. These two solutions produce the ambiguity which in the above example fails to distinguish between the right and left circular polarization.

Thus more than a single measurement of the type shown in fig. 1.1 is generally needed to determine the state of polarization of the incident beam. If we were limited to a single measurement, we would *not* in general be able to specify the polarization completely. Such limitations might occur in practice, for example, if the beam consisted of a short pulse of radiation too short for two successive experiments. Such a difficulty might be overcome by splitting the beam into two components, e.g. with a half-silvered mirror, and performing two independent experiments on each component. However, the incident light pulse might be of such low intensity that our detecting apparatus could not measure it accurately if the intensity were further reduced by splitting the beam. In classical physics, these difficulties are only practical ones which can always be surmounted in principle by building better apparatus. It is always possible in classical physics to measure the state of polarization of an incident beam. We shall see that in quantum physics we can be limited to performing only one experiment on a particular incident beam. In such a case, we shall not generally be able to determine its state of polarization completely.

The measuring apparatus shown in fig. 1.1 also modifies the incident beam and determines the state of polarization of each emergent beam. The two beams are always horizontally polarized and vertically polarized respectively, independent of the state of polarization of the incident beam. The incident beam merely determines the relative magnitudes of these two components; their states of polarization are determined by the measuring apparatus; i.e., by the orientation of the Nicol prism. If the Nicol prism is rotated by an angle of 45°, the two beams emerging from the prism are plane polarized at angles of + and −45° with respect to the horizontal, independent of the polarization state of the incident beam. Again, the incident beam determines only the relative magnitudes of these components, the apparatus determines their polarization. This is an interesting example of the interaction between the measuring apparatus and the particular system being measured. The system being measured is changed or transformed by the apparatus in a manner which is characteristic of the apparatus.

In classical physics this interaction and transformation is completely reversible and has no special significance. One could, for example, reconstruct the original incident beam from the two components produced in fig. 1.1 by feeding them through an appropriate optical system into a Nicol prism which combines the two components again as shown in fig. 1.2. If no measurements are made in the space between the two Nicol prisms, the system is simply a complicated black box in which the incident beam enters and exactly the same beam comes out. The intensities of the individual horizontal and vertical components can be measured while they are separated in the space between the two nicols. Thus in classical physics all the information about the incident beam can be measured in an apparatus from which the incident beam emerges unchanged.

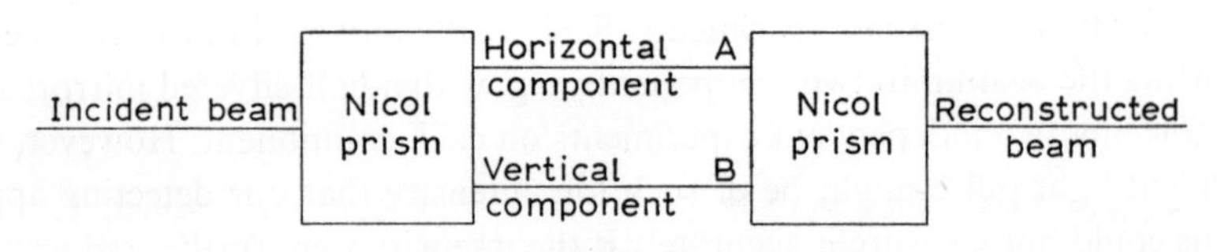

Fig. 1.2

In practice the measurement of the individual components between the nicols in fig. 1.2 requires taking a small amount of energy from each of these components in order to produce a detectable signal in the measuring apparatus. If the incident beam were very weak, this removal of energy might change the relative magnitude and phase of the two components to make the beam emerging from the second nicol quite different from the incident beam. It would therefore not be possible to measure the intensities of the horizontal and vertical components without changing the incident beam. In classical physics such limitations are considered to be due only to the deficiencies of the apparatus. It is always possible in principle to find a better and more sensitive apparatus which can measure the horizontal and vertical components without appreciably affecting the beam. In quantum physics, there are limits to the sensitivities of these measurements. These make it impossible to measure the intensities of both the horizontal and vertical components without perturbing the beam to an extent which prevents reconstructing the incident beam.

There are certain special cases for which much of the above discussion does not apply. In these cases a single measurement suffices to determine the state of polarization of the incident beam completely and the polarization of the incident beam is not changed in passing through the apparatus. For the system shown in fig. 1.1, this occurs when the incident beam is either fully horizontally polarized or fully vertically polarized. The full intensity is observed in one of the two channels and zero intensity in the other. The exact state of polarization of the incident beam is then known from a single measurement and the emerging beam is in exactly the same polarization state as the incident beam. If the nicol were rotated by an angle of 45°, this situation would no longer be true for incident beams which were plane-polarized horizontally or vertically. Instead, plane-polarized beams fully polarized at angles of $\pm 45°$ with respect to the horizontal would now pass through the nicol unchanged, give full intensity in one channel and zero in the other, and allow a complete specification of the polarization of the incident beam to be made on the basis of this single experiment. For any orientation of the nicol, there are always two such states of polarization for the incident beam which satisfy the above conditions. These two states can be conveniently described as the 'eigenstates' of the operation produced by the nicol, as shown above.

1.2 A matrix representation of light polarization

The electric vector for a beam of polarized light propagating in the z-direction can be written in the form

$$\boldsymbol{E} = E_x \mathrm{e}^{\mathrm{i}(\omega t + \phi_x)} \boldsymbol{e}_x + E_y \mathrm{e}^{\mathrm{i}(\omega t + \phi_y)} \boldsymbol{e}_y = \mathrm{e}^{\mathrm{i}\omega t}(E_x \mathrm{e}^{\mathrm{i}\phi_x} \boldsymbol{e}_x + E_y \mathrm{e}^{\mathrm{i}\phi_y} \boldsymbol{e}_y) \quad (1.1)$$

where $\boldsymbol{e}_x$ and $\boldsymbol{e}_y$ are unit vectors in the x and y-directions respectively. A convenient and compact way to represent this is by a vector or column matrix with complex components

$$\xi = \begin{pmatrix} E_x \, \mathrm{e}^{\mathrm{i}\phi_x} \\ E_y \, \mathrm{e}^{\mathrm{i}\phi_y} \end{pmatrix} = \begin{pmatrix} E_1 \\ E_2 \end{pmatrix} \tag{1.2}$$

where E_1 and E_2 are the complex numbers $E_x \mathrm{e}^{\mathrm{i}\phi_x}$ and $E_y \mathrm{e}^{\mathrm{i}\phi_y}$. The x-direction is chosen as horizontal; the y-direction as vertical. In this notation a beam of amplitude E, polarized in the vertical (i.e., y) direction, would be represented by the vector

$$\xi_\mathrm{V} = \begin{pmatrix} 0 \\ E \end{pmatrix}. \tag{1.3}$$

There are two directions of 45° polarization. When the vectors $\boldsymbol{E}_x$ and $\boldsymbol{E}_y$ are equal and *in* phase we call the angle $+45°$; when the vectors $\boldsymbol{E}_x$ and $\boldsymbol{E}_y$ are equal and *opposite* we call the angle $-45°$.

The experiment shown in fig. 1.1 is easily described in terms of these matrices. Let ξ as defined in eq. (1.2) represent the state of the incident beam. Then the two beams emerging at points A and B in fig. 1.1 are represented by the vectors

$$\xi_\mathrm{A} = \begin{pmatrix} E_1 \\ 0 \end{pmatrix} \tag{1.4a}$$

$$\xi_\mathrm{B} = \begin{pmatrix} 0 \\ E_2 \end{pmatrix}. \tag{1.4b}$$

These vectors are related to the vector representing the incident beam by the matrix equations

$$\xi_\mathrm{A} = A\xi \tag{1.5a}$$

$$\xi_\mathrm{B} = B\xi \tag{1.5b}$$

where A and B are the matrices

$$A = \begin{pmatrix} 1 & 0 \\ 0 & 0 \end{pmatrix} \tag{1.6a}$$

$$B = \begin{pmatrix} 0 & 0 \\ 0 & 1 \end{pmatrix}. \tag{1.6b}$$

The intensities of the beams observed at A and B are given by the norms of the vectors ξ_A and ξ_B. From eq. (1.4),

$$I_\mathrm{A} = (\xi_\mathrm{A}, \xi_\mathrm{A}) = |E_1|^2 \tag{1.7a}$$

$$I_\mathrm{B} = (\xi_\mathrm{B}, \xi_\mathrm{B}) = |E_2|^2 \tag{1.7b}$$

as expected. The intensities can also be expressed in terms of the vector ξ denoting the incident beam and the operators A and B characteristic of the apparatus. The operators A and B are both Hermitean and are both projection operators; i.e.,

$$A^\dagger = A = A^2 \tag{1.8a}$$

$$B^\dagger = B = B^2. \tag{1.8b}$$

Thus

$$I_A = (\xi_A, \xi_A) = (A\xi, A\xi) = (\xi, A^2\xi) = (\xi|A|\xi) \tag{1.9a}$$

$$I_B = (\xi_B, \xi_B) = (B\xi, B\xi) = (\xi, B^2\xi) = (\xi|B|\xi). \tag{1.9b}$$

The difference between the intensities observed at A and at B is given by

$$\Delta I = I_A - I_B = (\xi|(A - B)|\xi) = (\xi|M|\xi) \tag{1.10}$$

where the matrix M is defined as

$$M = A - B = \begin{pmatrix} 1 & 0 \\ 0 & -1 \end{pmatrix}. \tag{1.11}$$

The results of these intensity measurements are thus expressed simply in terms of 'diagonal matrix elements' or 'expectation values' of the Hermitean operators A, B, and M with the initial vector ξ. We also see that the intensities as given by eq. (1.7) depend only on the magnitudes of the components E_1 and E_2 and therefore give no information about the relative phase.

In the preceding section we saw that an incident beam which is fully polarized horizontally or vertically is unchanged in passing through the nicol and one intensity measurement gives complete information. This is described very simply in our matrix notation. If the incident beam is fully polarized horizontally, $E_2 = 0$, $\xi_A = \xi$ and $\xi_B = 0$. The incident vector ξ is an eigenvector of the matrices A, B, and M which represent the transformation of the incident beam by the apparatus. When the vector representing the incident beam is an eigenvector, these matrices do not change the vector but merely multiply it by a constant; i.e., they do not change the state of polarization but may change the intensity. Similarly, if the incident beam is fully polarized vertically, $E_1 = 0$, $\xi_A = 0$, $\xi_B = \xi$ and ξ is again an eigenvector of the matrices A, B, and M.

1.3 Matrix transformations

Let us now consider the matrix description of an experiment with the nicol prism of fig. 1.1 rotated by an angle of 45° to make the component polarized at an angle of $+45°$ appear at A and the component polarized at $-45°$ appear at B. Beams of amplitude E, polarized at angles of $+45°$ and $-45°$, respectively, are represented by the vectors

$$\xi_{+45°} = \begin{pmatrix} E/\sqrt{2} \\ E/\sqrt{2} \end{pmatrix} \tag{1.12a}$$

$$\xi_{-45°} = \begin{pmatrix} E/\sqrt{2} \\ -E/\sqrt{2} \end{pmatrix}. \tag{1.12b}$$

If the incident beam is represented by the vector ξ, eq. (1.2), the beams observed at A and B are represented by the vectors

$$\xi_{A'} = \begin{pmatrix} \frac{1}{2}(E_1 + E_2) \\ \frac{1}{2}(E_1 + E_2) \end{pmatrix} \tag{1.13a}$$

$$\xi_{B'} = \begin{pmatrix} \frac{1}{2}(E_1 - E_2) \\ -\frac{1}{2}(E_1 - E_2) \end{pmatrix} \tag{1.13b}$$

where primes are used in order to avoid confusion with the preceding case eq. (1.4). The corresponding intensities are given by the norms of these vectors

$$I_{A'} = (\xi_{A'}, \xi_{A'}) = \tfrac{1}{2}|E_1 + E_2|^2 \tag{1.14a}$$

$$I_{B'} = (\xi_{B'}, \xi_{B'}) = \tfrac{1}{2}|E_1 - E_2|^2. \tag{1.14b}$$

Relations analogous to eqs. (1.5), (1.9) and (1.10) are obtained by defining the analogous matrices

$$A' = \begin{pmatrix} \frac{1}{2} & \frac{1}{2} \\ \frac{1}{2} & \frac{1}{2} \end{pmatrix} \tag{1.15a}$$

$$B' = \begin{pmatrix} \frac{1}{2} & \frac{1}{2} \\ -\frac{1}{2} & \frac{1}{2} \end{pmatrix} \tag{1.15b}$$

$$M' = A' - B' = \begin{pmatrix} 0 & 1 \\ 1 & 0 \end{pmatrix} \tag{1.15c}$$

$$\xi_{A'} = A'\xi \tag{1.16a}$$

$$\xi_{B'} = B'\xi \tag{1.16b}$$

$$I_{A'} = (\xi|A'|\xi) \tag{1.17a}$$

$$I_{B'} = (\xi|B'|\xi) \tag{1.17b}$$

$$I_{A'} - I_{B'} = (\xi|M'|\xi) \tag{1.17c}$$

The matrices A', B' and M' are all Hermitean as before, and A' and B' are both projection operators, although none of the three is diagonal like the corresponding matrices in the preceding simple case. The two vectors (1.12) representing plane-polarized beams at angles of $+45°$ and $-45°$ are the eigenvectors of the matrices A', B', and M'. This is in accord with the physical observations that beams polarized at angles of $+45°$ and $-45°$ pass through the apparatus without being changed and that a single measurement can determine the complete state of polarization of such a beam.

A simpler description of this second experiment is obtained if the base vectors are chosen to represent beams polarized at angles of $+45°$ and $-45°$ rather than horizontally and vertically. In this case, beams of amplitude E polarized at angles of $+45°$ and $-45°$, respectively, are represented by the vectors

$$\xi''_{+45°} = \begin{pmatrix} E \\ 0 \end{pmatrix}. \tag{1.18a}$$

$$\xi''_{-45°} = \begin{pmatrix} 0 \\ E \end{pmatrix}. \tag{1.18b}$$

Let the incident beam be represented by

$$\xi'' = \begin{pmatrix} E''_1 \\ E''_2 \end{pmatrix}. \tag{1.19}$$

Then E''_1 represents the component in the incident beam polarized at $+45°$ and E''_2 represents the component polarized at $-45°$. In this representation the beams observed at A and B are represented by the vectors

$$\xi_{A''} = \begin{pmatrix} E''_1 \\ 0 \end{pmatrix} \tag{1.20a}$$

$$\xi_{B''} = \begin{pmatrix} 0 \\ E''_2 \end{pmatrix} \tag{1.20b}$$

and the corresponding intensities are given by

$$I_{A''} = (\xi_{A''}, \xi_{A''}) = |E''_1|^2 \tag{1.21a}$$

$$I_{B''} = (\xi_{B''}, \xi_{B''}) = |E''_2|^2. \tag{1.21b}$$

The matrices A', B' and M' in this representation now become

$$A'' = \begin{pmatrix} 1 & 0 \\ 0 & 0 \end{pmatrix} \tag{1.22a}$$

$$B'' = \begin{pmatrix} 0 & 0 \\ 0 & 1 \end{pmatrix} \tag{1.22b}$$

$$M'' = \begin{pmatrix} 1 & 0 \\ 0 & -1 \end{pmatrix}. \tag{1.22c}$$

Eqs. (1.18), (1.20), (1.21) and (1.22) look quite different from the corresponding equations (1.12), (1.13), (1.14), and (1.15) which describe exactly the same experiment using a different set of base vectors. However, eqs. (1.19), (1.20), (1.21), and (1.22) look exactly like eqs. (1.2), (1.4), (1.7), (1.6), and (1.11), which describe the experiment with the horizontal–vertical nicol, using a different set of base vectors. That two different experiments have the same formal description is not as strange as it seems, because these two different experiments are not really so different. They differ only in the rotation of the apparatus by 45° relative to our x and y-coordinate axes. If we rotate our coordinate axes along with the apparatus, then the second experiment in the rotated coordinate system looks exactly like the first experiment in the unrotated coordinate system. Our second description of the second experiment using base vectors at angles of $+45°$ and $-45°$ corresponds to rotating the coordinate system along with the apparatus and therefore gives exactly the same formal description as that of the first experiment in the unrotated coordinate system.

The two seemingly different descriptions of the second experiment are related to one another by a rotation of the coordinate axes by 45°. This transformation from one set of base vectors to another should be expressible as a unitary transformation on the vectors and matrices

$$K'' = UK'U^{\dagger} \tag{1.23a}$$

$$\xi'' = U\xi' \tag{1.23b}$$

where K represents any one of the matrices A, B or M, and ξ is an arbitrary vector. The unitary matrix U required to describe the transformation can be easily found. It is just the orthogonal matrix representing a rotation by an angle of 45°

$$U = \begin{pmatrix} \dfrac{1}{\sqrt{2}} & \dfrac{1}{\sqrt{2}} \\ -\dfrac{1}{\sqrt{2}} & \dfrac{1}{\sqrt{2}} \end{pmatrix} = \begin{pmatrix} \cos 45° & \sin 45° \\ -\sin 45° & \cos 45° \end{pmatrix}. \tag{1.24}$$

Substituting the expression (1.24) for the matrix U into eq. (1.23a) shows that the matrix (1.24) does indeed perform the transformation indicated on the matrices A', B', and M'. Substituting eq. (1.24) for U into eq. (1.23b) gives relations between E_1'' and E_2'' and the corresponding quantities E_1 and E_2

$$E_1'' = \frac{E_1 + E_2}{\sqrt{2}} \tag{1.25a}$$

$$E_2'' = \frac{E_1 - E_2}{\sqrt{2}}. \tag{1.25b}$$

Eqs. (1.25) show that the expressions for the intensities observed at A and B in the two descriptions, eqs. (1.14) and (1.21), are equal, although the two sets of equations appear to be quite different. This is characteristic of two descriptions of the same physical experiment in different bases or coordinate systems. The numbers describing the components of a vector or matrix in different coordinate systems (or their representatives in different bases) are, in general, quite different from one another. However, the numbers describing the result of a physical measurement must be the same in any coordinate system.

It is now interesting to examine the unitary matrix (1.24). Since it is unitary, it has eigenvectors and eigenvalues of modulus unity. The eigenvectors of the matrix U are just

$$U' = \begin{pmatrix} \dfrac{1}{\sqrt{2}} \\ \dfrac{i}{\sqrt{2}} \end{pmatrix} \tag{1.26a}$$

$$U'' = \begin{pmatrix} \dfrac{1}{\sqrt{2}} \\ -\dfrac{i}{\sqrt{2}} \end{pmatrix}. \tag{1.26b}$$

The corresponding eigenvalues are $(1+i)/\sqrt{2}=e^{i\pi/4}$ and $(1-i)/\sqrt{2}=e^{-i\pi/4}$ which are indeed of modulus unity. The eigenvectors (1.26) represent those states of polarization which are not changed by the transformation (1.24) which is a rotation by an angle of 45°. Examination of the specific form of the eigenvectors (1.26) shows that these represent the two states of circular polarization. These circularly polarized states each remain circularly polarized in the same sense after a rotation of 45°. The only change produced by the transformation is the introduction of a phase factor of $+45°$ or $-45°$

depending on the sense of polarization. These phase factors $e^{i\pi/4}$ and $e^{-i\pi/4}$ are just the two eigenvalues of modulus unity.

1.4 Unpolarized and partially polarized light

What is unpolarized light? The light from the sun is unpolarized. If sunlight is passed through the apparatus of fig. 1.1, half will come out horizontally polarized and half vertically polarized. If the nicol is rotated by an angle of 45°, half will come out polarized at +45°, half at −45°. If the nicol is rotated by any angle, the beam will always be split into two equal components. So far, the behavior of the beam is like that of a circularly polarized beam, which is also split by a nicol into two equal components, independently of the orientation of the nicol. But if the unpolarized beam is passed through a quarter-wave plate which would change circularly polarized light into plane-polarized light, the unpolarized beam *remains unpolarized*. The unpolarized beam is an equal mixture of horizontally and vertically polarized components, but there is *no definite phase* between the two components.

An unpolarized beam cannot be represented by an expression of the form (1.1) or by the matrix representation (1.2). Such expressions always have a definite phase between the horizontal and vertical components. The existence of unpolarized light already gives an indication of the quantum nature of light. Unpolarized light cannot be described as a single simple classical wave of the form (1.1) or as any linear combination of such waves. Unpolarized light can be described classically as a series of very rapid short bursts or pulses of light, each having a different polarization, with no correlation between the polarization of different pulses. If these pulses and the interval between them are very short compared to the characteristic times of the measuring apparatus, they will be detected as a continuous beam, and any polarization measurement will give an average of the polarizations of the individual pulses.

There are also partially polarized beams. Suppose an unpolarized beam is passed through an apparatus like that of fig. 1.2, and half of the vertical component is absorbed in the space between the two nicols. The reconstructed beam will now have a vertical component which is half the intensity of its horizontal component, but with no definite phase relation between the two components. It can also be considered as a mixture of a horizontally polarized beam and an unpolarized beam. Such a beam is called a *partially polarized* beam.

Let us now examine the matrix description of unpolarized and partially polarized beams. For example, consider a beam of short pulses of which one third are horizontally polarized, one third polarized at $+45°$ and one third polarized at $-45°$. If the total intensity of the beam is E, the three components are represented by vectors

$$\xi^{(1)} = \begin{pmatrix} E/\sqrt{3} \\ 0 \end{pmatrix}; \quad \xi^{(2)} = \begin{pmatrix} E/\sqrt{6} \\ E/\sqrt{6} \end{pmatrix}; \quad \xi^{(3)} = \begin{pmatrix} E/\sqrt{6} \\ -E/\sqrt{6} \end{pmatrix}. \tag{1.27}$$

However, these three components cannot be combined into a single vector because they are *incoherent*. They represent pulses occurring at *different times* which cannot interfere with one another.

The results of any experiment with this beam can be calculated with our matrix formulation by calculating the intensities for each of the three components and then adding the intensities. Thus, the difference between the intensities observed at A and B in the apparatus of fig. 1.1 is given by eq. (1.10) as

$$\begin{aligned} I = I_A - I_B &= \langle \xi^{(1)} | M | \xi^{(1)} \rangle + \langle \xi^{(2)} | M | \xi^{(2)} \rangle + \langle \xi^{(3)} | M | \xi^{(3)} \rangle \\ &= \sum_{\alpha=1}^{\alpha=3} \langle \xi^{(\alpha)} | M | \xi^{(\alpha)} \rangle. \end{aligned} \tag{1.28}$$

This notation becomes very cumbersome if there are many components in the beam. A more convenient description of a partially polarized beam is obtained by use of a matrix instead of a vector.

Let us rewrite eq. (1.28) specifying the matrix indices explicitly

$$I = \sum_{\alpha} \sum_{i,j} \xi_i^{(\alpha)*} M_{ij} \xi_j^{(\alpha)}. \tag{1.29}$$

This expression can be simplified by defining the matrix

$$\varrho_{ji} = \sum_{\alpha} \xi_j^{(\alpha)} \xi_i^{(\alpha)*}. \tag{1.30a}$$

Then

$$I = \sum_{i,j} M_{ij} \varrho_{ji} = \mathrm{Tr}(M\varrho). \tag{1.30b}$$

The matrix ϱ_{ji} is called the 'density matrix'.

Eq. (1.30b) shows that the four elements of the density matrix provide sufficient information about the incident beam for the calculation of the result of the experiment of fig. 1.1. Similarly for any measurement whose result is expressed as the expectation value of a matrix, like eq. (1.10), the corresponding expression for a partially polarized beam is given by the corresponding expression of the form (1.30b); namely the trace of the product of the density matrix and the matrix describing the measurement.

Eq. (1.30b) shows that the density matrix contains all the necessary information about an unpolarized or partially polarized beam to describe all intensity measurements which sum over all the pulses in the beam and do not resolve the individual components. The density matrix is thus a generalization of the vector notation (1.2) which describes a beam that can be an 'incoherent mixture' of several components as well as a 'pure state' of definite polarization.

For an unpolarized beam, which can be considered as half horizontally polarized and half vertically polarized, the density matrix is proportional to the unit matrix,

$$\varrho_{\text{unpol}} = \tfrac{1}{2}E\begin{pmatrix} 1 & 0 \\ 0 & 1 \end{pmatrix}. \tag{1.31}$$

An unpolarized beam can also be considered as half polarized at $+45°$, and half at $-45°$, or as a 50–50 mixture of any two components which are described by two orthogonal vectors. The definition (1.30a) shows that the same expression (1.31) is obtained in all these cases.

The total intensity of a beam is given by the trace of the density matrix,

$$I = \operatorname{Tr} \varrho. \tag{1.32}$$

1.5 Introduction of quanta

We know that electromagnetic radiation cannot be infinitely subdivided into beams of smaller and smaller intensity. The smallest possible unit of radiation is the light quantum which has energy $E=h\nu$, where h is Planck's constant and ν is the frequency of the radiation. These quanta, or photons, are indivisible and any device measuring beam intensity can only detect an integral number of quanta.

What happens when a single photon is introduced as the incident beam into the apparatus of fig. 1.1? This photon cannot be split with one part coming out of the nicol at A and the other part coming out at B. The photon is indivisible; if it comes out at all, it must come out either at A or at B. However, if a classical beam polarized at an angle of 45° is introduced into the apparatus half the intensity comes out at A as horizontally polarized light and half the intensity comes out at B as vertically polarized light. In order to obtain a unified description of light which includes classical beams (many photons) as well as single photons, we must reconcile the classical results with the indivisibility of photons on the quantum level.

The classical experiment uses a beam consisting of a very large number of photons so that the quantum structure is not observed. If this beam is fully polarized at an angle of 45°, all the photons in the beam have a 45° polarization. The classical result tells us that half of the incident photons come out at A as horizontally polarized photons, and the other half come out at B as vertically polarized photons. We can obtain a consistent description both of the single-photon and the many-photon case by saying that a single photon may come out either at A or at B with equal probability. We cannot predict which way a given single photon will go. However, when many photons are incident, on the average half of them come out at A and half come out at B.

This probability or statistical description represents a definite break with concepts of classical physics. If in classical physics we know all possible information about a given system at a given time, we can predict its future behavior exactly by use of the equations of motion. In this example we are given a system where we know everything about it at a given time. There is a single photon polarized at an angle of 45° with respect to the horizontal incident upon a nicol prism. However, we are completely unable to predict whether this photon will come out at A or at B.

One might compare the statistical nature of this description with a random process such as flipping a coin. If the coin is symmetrical and there is no bias in the flipping mechanism, one expects on the average that the number of 'heads' equals the number of 'tails'. On the other hand, one cannot predict the result of a single flip of the coin. If the coin is flipped a number of times, the number of 'heads' should be approximately equal to the number of 'tails' with a difference between them due to the statistical fluctuations which is of the order of the square root of the total number of times the coin is flipped. The percentage fluctuation therefore decreases as the number of flippings increases. If the coin is flipped a million times, the statistical fluctuations are of the order of 0.1 percent. The photon experiment appears to resemble the coin experiment. If many 45° polarized photons are incident upon the Nicol prism, the number observed at A is approximately equal to the number observed at B with statistical fluctuations of the order of the square root of the total number of photons. However, there is a crucial difference between the coin flipping example and the photon experiment. A coin is a macroscopic object which we believe to be governed by the laws of classical physics. If we know the exact position and orientation of the coin, the forces acting upon it at the time it is flipped and the elastic and frictional constants of the coin and of the surface on which it falls, we should in principle be able to predict each time whether it will fall heads or tails. If we are unable to do so and can only give a statistical description, it is only because of our ignorance. We do not know everything that we could in principle know about the coin.

One can ask if the same description applies to the incident photon. It might be something much more complicated than we think, with a complex structure involving other degrees of freedom of which we are as yet unaware. If we knew all about this substructure and knew the values of all these 'hidden variables' we might be able to predict how each individual photon would pass through the nicol and whether it would come out at A or at B. *The quantum theory asserts that this is not the case and that our inability to predict whether an individual photon will come out at A or at B is a fundamental property of nature.* So far there has been no experimental evidence for the existence of these hidden variables and it is generally believed that they do not exist*. On the other hand, the quantum theory which assumes this unpredictability has had great success in describing many phenomena.

The matrix representation developed in treating the classical problem is very useful in discussing the quantum-mechanical case. Let the vector ξ, eq. (1.2), represent the state of polarization of a photon.

$$\xi = \begin{pmatrix} E_1 \\ E_2 \end{pmatrix}. \tag{1.2}$$

Let us normalize the vector ξ for a single photon so that

$$|E_1|^2 + |E_2|^2 = 1. \tag{1.33}$$

The classical results for the intensities obtained at A and B (eqs. (1.7), (1.9), and (1.10)) can now be interpreted as giving the average number of photons observed at A and B and the average value of the difference. If there is only one photon, the average number of photons appearing at any point must be less than or equal to one and this 'average number' is simply the probability that the photon appears at this point.

Thus in the quantum case the diagonal matrix elements, or expectation values, of the Hermitean operators, A, B, and M with the initial vector ξ give

* It is, of course, impossible to prove that hidden variables do not exist. Pauli once remarked, 'hidden variables are like mosquitoes – the more you kill of them, the more there are'. There have been recent works attempting to treat quantum-mechanical problems by using hidden variables. A formulation has been developed using hidden variables which gives all the results of the quantum theory including the result that one cannot predict whether the photon in our example will come out at A or at B. The result is completely determined by the hidden variables, but it is impossible to know all of them because of the interaction of the observer with the measuring apparatus. Since the accepted formulation of the quantum theory obtains these results without the need of hidden variables, we do not consider them further.

the mean values of experimental measurements on a single photon, or the average of a series of measurements on many photons.

1.6 The interaction of the observer with the system

Consider the apparatus of fig. 1.1 as a measuring apparatus to determine the state of an incident photon. Suppose a single photon is observed at A. We know that the incident photon was not fully polarized vertically, because then it would have had to come out at B. However, there is little more that can be said about the state of the incident photon, since any state other than full vertical polarization would give some probability that the photon would be observed at point A. There is no possibility of finding out any more information about this one photon. Once it is observed at point A, it is already fully horizontally polarized, and it is no longer possible to make any measurements to determine what the polarization was before passing through the nicol. Thus the process of measurement has given us *some* information about the state of the photon but has also disturbed this photon in such a way as to destroy all record of its previous state. Suppose we now repeat this experiment many times with many photons all prepared in the same way, so that we know that they are all initially in the same state. We thus measure the probabilities for finding the photon at A and at B. If the incident photon is described by a normalized vector ξ (eq. (1.2)), this experiment repeated on many photons measures the absolute values of the complex numbers E_1 and E_2. The relative phase of these numbers is not measured.

Consider now a single photon incident upon the apparatus shown in fig. 1.2 in which it passes through two nicol prisms, the first of which splits it into horizontal and vertical components, and the second of which brings them together and reconstitutes the incident beam. If no measurements are made in the space between the two nicols, one has classically a complicated black box in which the incident beam enters and exactly the same beam comes out at the other end. Since the classical experiment means an incident beam of many photons *all* having the same polarization and *all* of which emerge from the black box having the same polarization as they had initially, *each individual photon* must come out of the box with the same polarization it had upon entering. Otherwise, there would be an inconsistency between the single photon and the classical result.

One can now ask the question, ‘if a single photon enters the apparatus, polarized at an angle of 45° with respect to the horizontal, and leaves the apparatus also polarized at an angle of 45°, which path did it take in going between the two nicols – did it go through point A or through point B?’ This question cannot be simply answered without immediately encountering contradictions. If the photon went through point A, it must have been horizontally polarized at that time and there was then no photon at point B. How then could the photon regain its 45° polarization in going from A through the second nicol? In the classical problem there is no paradox because the incident beam is a wave which splits into two components after passing through the first nicol and then recombines after passing through the second. In order to explain the observed experimental results with single photons, each individual photon, although it is an indivisible quantum, must still somehow split into two components in passing between the two nicols and these parts must come together again in the second nicol.

The photon thus has both particle and wave properties.

As long as the photon is undisturbed by measuring apparatus which attempts to determine exactly where it is, it propagates as a wave described by Maxwell's equations. It is only when the photon interacts with matter that the particle-like properties manifest themselves in requiring that the total quantum of energy, $E=h\nu$, and of momentum, $p=h/\lambda$, must be absorbed in a single event. The relation between the wave and particle aspects is again provided by the statistical interpretation. The place where the photon will be absorbed cannot be predicted with certainty; rather the probability that the photon is absorbed in any one place is proportional to the intensity of the wave at that point. Once the photon is absorbed, it is gone. The amplitudes indicating the propagation of this photon in other places immediately drop to zero as soon as the photon has been absorbed.

Let us now return to the question of which way the photon went between the two nicols in fig. 1.2. There is no clear answer to this question, somehow it went both ways. That quantum theory cannot tell which way the photon went should not be considered a failure of the theory. Rather, quantum theory says that this question has no meaning. The quantum theory states that a photon is something more complicated than a scaled-down version of a macroscopic particle or of a macroscopic wave which we encounter in our everyday experience. There is really no reason why these sub-microscopic particles which we can never see directly need be scaled-down versions of things which we can see. The photon is thus an object which propagates through space like a wave, but can suddenly appear as a particle in a manner which can only be predicted statistically.

There is an important distinction between the two questions (1) 'Did the photon go through A or through B in fig. 1.2?' and (2) 'Which way will the incident photon come out of the nicol in fig. 1.1 – at A or at B?' Both cannot be answered by the quantum theory. Question (1) is meaningless and based on the erroneous assumption that a photon is a particle like a scaled-down version of macroscopic particles. This question does not concern any quantity which can be measured and is therefore something like the medieval question of how many angels can dance on the point of a pin. Question (2) is meaningful and refers directly to physically measurable quantities; e.g., signals observed in photomultipliers at A and B. That this question cannot be answered is an important property of nature. *It represents a definite break with classical physics which requires that all experimentally observable quantities should be predictable if information completely specifying the state of the system has been given previously.*

Let us now return to the experiment of fig. 1.2 and check whether it is indeed impossible to measure whether the photon has gone by point A or point B. Suppose a very sensitive apparatus is placed at A so that the horizontally polarized photon passing through point A may scatter an electron into a photomultiplier. Because this requires a transfer of energy and momentum from the photon to the electron, the photon's energy and momentum and therefore its frequency and wavelength are changed after the collision with the electron. This change produces a change in the phase of the horizontally polarized wave as it passes through the second nicol. Thus even if both the horizontal and vertical components propagate as classical waves, the two can no longer recombine in the nicol to reproduce the incident beam.

In a classical experiment one can make the perturbation of the horizontal component by the electron scattering negligibly small so that the incident beam is still reproduced. However, there are limits to the smallness of this perturbation because of the quantum nature of the electron: i.e., because there is also wave–particle duality in the electron. The initial position and momentum of the electron are uncertain in a manner described by quantum theory. This uncertainty produces an uncertainty in the position along the photon path where the momentum is transferred and in the amount of momentum transferred. Both of these uncertainties contribute to an uncertainty in the phase of the horizontal component as it passes through the nicol. No matter how one arranges the apparatus it turns out that this uncertainty in phase is so large that all coherence between the horizontal component and the vertical component is destroyed. Thus even if the incident beam is considered as a classical wave which splits into two components it has the phase of the horizontal component altered by its interaction with the electron and the resultant beam emerging from the second nicol is no longer a beam with the same polarization (e.g., 45°) as the incident beam. It is an incoherent superposition of horizontally and vertically polarized components; i.e. an unpolarized beam. In our wave–particle description, we can simply say that the incident photon entered the nicol and propagated like a wave through the two channels A and B. However, as soon as a scattered electron was observed at A we knew that the photon had passed through A and was horizontally polarized. It therefore passed through the second nicol without any vertically polarized companion and emerged as a horizontally polarized photon.

1.7 States which are not changed by measurement

When a single photon enters the apparatus of fig. 1.1 we are in general unable to predict whether it will come out at A or at B. On the other hand, if it does come out at A then it is fully horizontally polarized and has been irreversibly altered by passing through the nicol. A series of such measurements on a number of photons identically prepared generally does not give enough information to allow a determination of the initial state of the photon.

In the classical discussion we noted special cases in which a single measurement sufficed to determine the state of polarization of the incident beam completely, and where the state of polarization of the incident beam was not changed in passing through the apparatus. These special cases have the same properties for photons. In the experiment of fig. 1.1 a photon which enters the nicol fully horizontally polarized, will certainly be observed at point A and its polarization will be unchanged; i.e., it will remain horizontal. An analogous situation obtains for an incident vertically polarized photon which can be predicted with certainty to appear at point B with an unchanged vertical polarization. In an experiment with a number of photons, all prepared in the same way, and horizontally polarized, one would observe that all come out of the nicol at A and none at B. One would then conclude on the basis of this single measurement that these photons were all fully horizontally polarized. Again the situation is analogous for incident vertically polarized photons.

As in the classical case, this no longer occurs if the nicol is rotated by an angle of 45°. Then incident horizontally or vertically polarized photons emerge from the nicol, sometimes at A and sometimes at B, with their polarizations changed to +45° or −45°, respectively. On the other hand, the behavior of photons polarized at +45° can be predicted with certainty. They pass through the nicol without having their state of polarization changed. If a large number of identically prepared 45° polarized photons are measured, they appear either all at A or all at B, depending on whether the initial polarization is +45° or −45°. A complete specification of the state of the incident photons is possible on the basis of this one experiment. For any orientation of the nicol, there are similarly always two orthogonal states of polarization for incident photons which pass through the nicol unchanged and whose observation at A or B can be predicted with certainty. The same is true for combinations of Nicol prisms with quarter-wave plates before and after the prism such that a classical incident beam is split into components at A and at B which are circularly or elliptically polarized. For these cases, there are again two orthogonal photon states, of circular or elliptical polarization respectively, which pass through the apparatus unchanged and for which their observation at A or B can be predicted with certainty.

The above discussion can be stated concisely in the matrix representation. For all orientations of the nicol, or of combinations of nicols and quarter-wave plates, the vectors representing the beam or the photon at A and B can be expressed in terms of some linear operator or matrix operating on the vector denoting the incident beam. The states represented by the eigenvectors of these matrices are those states whose polarization remains unchanged in passing through the apparatus. Since these matrices are all 2×2, they have two linearly independent eigenvectors. There are therefore always just two states for the incident photon which have the special properties discussed above.

For each orientation of the nicol, or arrangement of nicol and quarter-wave plates, there are two different states of polarization for which the result of the measurement can be predicted with certainty. There is *no single state* of polarization for the photon such that results can be predicted with certainty for *each* of *two different types* of experiments.

Consider for example, the experiment with the nicol which separates the incident beam into horizontal and vertical components, and an experiment with quarter-wave plates and a nicol which separates the incident beam into left-circularly and right-circularly polarized components. If the incident photon is plane-polarized, horizontally or vertically, the result of the first experiment can be predicted with certainty, but the result of the second cannot. If the incident photon is right-circularly or left-circularly polarized, the result of the second experiment can be predicted with certainty, but that of the first cannot. *It is impossible to prepare a photon in such a state of polarization that the results of both of these experiments could be predicted with certainty.* This is not surprising since there is no paradox in the statement that a photon cannot be fully plane-polarized and fully circularly polarized at the same time. That these two types of measurements are incompatible is not startling, since full circular polarization and full plane polarization are known to be incompatible even for a classical beam of polarized light. We shall find, however, that this concept of incompatibility extends to other properties which are not incompatible in classical physics. For example, it is impossible to prepare a particle in a state where a measurement of its position and a measurement of its momentum can both be predicted with certainty.

Remember, however, that the inability to predict both polarization measurements with certainty *is a purely quantum effect*, arising from the indivisibility of the photon. Classically one could predict with certainty that a circularly polarized beam incident upon the apparatus of fig. 1.1 would emerge with half of its intensity at A and half at B.

1.8 Measurements with single photons

When a classical beam is passed through the apparatus of fig. 1.1 the fraction of the intensity appearing at A or at B can have any value from 0 to 1, depending upon the incident beam. When a single photon is passed through the apparatus, the only possible values are 0 and 1, and the continuous spectrum of values in between are not allowed. These allowed values for a single photon measurement are very simply expressed in the matrix representation, they are *just the eigenvalues of the matrix* which describe the particular measurement, e.g., the matrices A, B and M, eqs. (1.6) and (1.11).

The use of matrices to describe measurements allows the following properties of measurements involving single photons to be expressed very concisely: Each measurement is described by a Hermitean matrix. The allowed values for such a measurement with a single photon are just the eigenvalues of the matrix. After the measurement the photon is in a state which is an eigenvector of the matrix corresponding to the measured eigenvalue. For a beam of photons in a state described by a vector ξ, the *average* result of a measurement is given by the expectation value of the matrix with the vector ξ. If ξ is an eigenvector of the matrix, the expectation value is equal to the eigenvalue, the result of a measurement is the same for all photons in the beam, and the outgoing photon is in the same state as the incoming photon. If ξ is not an eigenvector of the matrix, the result of a measurement is *not* the same for all photons in the beam, and cannot be predicted with certainty; sometimes it is one eigenvalue, sometimes it is the other.

We have seen that it is usually impossible to prepare a photon in such a state that the results of two experiments can be predicted with certainty. This is easily seen in the matrix representation. Each of these measurements is represented by a matrix, and the result of an experiment can be predicted with certainty if the vector representing the incident photon is an eigenvector of the matrix. If two experiments are to be predicted with certainty the vector representing the photon must be a *simultaneous eigenvector of both matrices*. If the two matrices do not have any simultaneous eigenvectors, they represent *incompatible* measurements.

If two matrices commute, they can be simultaneously diagonalized and can have simultaneous eigenvectors. Thus the compatibility of two measurements is related to the commutator of the associated matrices. Two matrices which commute describe compatible measurements. Two matrices which do not commute describe measurements which are generally incompatible.

1.9 Matrices and measurements in quantum theory

The qualitative features of the matrix representation of polarized photons in quantum theory apply in general to the description of more complicated systems having more degrees of freedom. These are as follows.

The state of a dynamical system (like the polarization state of a photon) is described by a complex vector in a vector space. The vector can have many components, in some cases infinitely many rather than only two.

The dynamical variables which are measured are described by Hermitean matrices. The eigenvalues of these matrices give the allowed values for a single measurement of the corresponding variables (e.g., horizontal or vertical photon polarization, electron momentum, etc). If the vector describing the system (e.g. the photon) is an eigenvector of the matrix, the result of the corresponding measurement can always be predicted with certainty and is just the eigenvalue of the matrix corresponding to this eigenvector. If the vector describing the system is not an eigenvector of the matrix, the result of a single measurement of this quantity cannot be predicted with certainty. There will be a statistical distribution. The average value of this dynamical variable over many measurements is given by the expectation value of the matrix taken with the vector describing the system.

In the heuristic derivation of results for single photons, we have used the principle that the average result of any measurement for a large number of photons must be equal to the classical result. This is a special case of a general principle of quantum theory, called the 'correspondence principle'. In any quantum-mechanical problem, there is always a limit, such as the case of a very large number of photons, where classical physics should hold. The correct quantum-mechanical description must reduce to the classical description in this limit.

PROBLEMS

A beam of vertically polarized light is passed *successively* through
(a) a Nicol prism transmitting only vertically polarized light;
(b) a Nicol prism transmitting only light polarized at an angle of $+45°$;
(c) a Nicol prism transmitting only horizontally polarized light;
(d) a medium in a magnetic field which rotates the polarization plane by $+45°$ by means of the Faraday effect;
(e) a quarter-wave plate which shifts the relative phase of the horizontally and vertically plarized components by $\frac{1}{2}\pi$;
(f) a Nicol prism transmitting only light polarized at an angle of $-45°$.

Questions:

1. If the incident wave has amplitude E, what is the amplitude and polarization of the light after stage a, b, c, d, e, and f?

2. If a single photon of vertically polarized light is introduced into the system, what will be observed after a, b, c, d, e, and f?

3. We wish to measure the polarization of the light using a Nicol prism in the detecting apparatus shown in fig. 1.1. The nicol is oriented in the beam so that its crystal axes are horizontal and vertical. What readings will this apparatus give if placed in the above system after stages a, b, c, d, e, and f respectively? Do these readings give sufficient information to determine the polarization of the light at each stage?

4. Repeat question 3, but with the axes of the Nicol prism oriented at an angle of 45° with respect to the horizontal and vertical.

5. Let the incident beam of amplitude E, polarized in the vertical direction be represented by the vector

$$\xi_0 = \begin{pmatrix} 0 \\ E \end{pmatrix}.$$

Write the vectors ξ_a, ξ_b, ξ_c, ξ_d, ξ_e, and ξ_f describing the light after passing through stages a, b, c, d, e, and f.

6. At each stage a, b, c, d, e, f, the light is transformed from a state represented by one vector to another state represented by a different vector. Each transformation can therefore be represented by a matrix. For example, the transformation a, transmitting only vertically polarized light, is represented by the matrix

$$A = \begin{pmatrix} 0 & 0 \\ 0 & 1 \end{pmatrix}.$$

Write down the matrices B, C, D, E, and F, corresponding to the transformations b, c, d, e, and f.

Starting with the initial vector $\binom{0}{E}$, operate successively with the matrices A, B, C, D, E, and F, and show that the vectors obtained in each stage are just those given as the answer to problem 5.

7. Successive operations can be presented by matrix products. The product BA represents operation first with A and then with B. Form the products BA, DC, and FE. Starting with the initial vector $\binom{0}{E}$ operate successively with these three product matrices and show that the same results are obtained as above. Note that BA is *not* equal to AB which represents first operating with B and then with A. Form the matrix product $FEDCBA$ and show that this product matrix operating on the initial vector $\binom{0}{E}$ gives the final result ξ_{f} directly.

8. Instead of choosing the x and y-directions as coordinate axes, choose axes at an angle of 45°, such that the vector $\binom{0}{E}$ represents a beam polarized at an angle of $-45°$. The vertically polarized incident beam is represented by the vector

$$\xi_0 = \begin{pmatrix} -E/\sqrt{2} \\ E/\sqrt{2} \end{pmatrix}.$$

Repeat exercises 5 and 6 using this choice of axes. Use primes to denote the vectors and matrices for this case (e.g., ξ'_{a}, ξ'_{b}, A', B').

9. What are the eigenvalues and eigenvectors of the matrices A, B, C, D, E, and F? Which of these matrices are orthogonal? What are the eigenvalues of the matrices A', B', C', D', E', and F'?

10. The primed vectors, ξ'_{a}, ξ'_{b}, etc., of exercise 9 are related to the original vectors, ξ_{a}, ξ_{b}, etc., by a 45° rotation of the axes. This can be written in matrix notation

$$\xi'_k = R\xi_k,$$

where $k = 0$, a, b, c, d, e, or f, and R is an orthogonal matrix. Write this matrix R and also write down its inverse R^{-1}. What is the corresponding relation between the matrix A' and the matrix A using the transformation matrix R? Show that this same relation also holds for the matrices B, D, and E.

11. Consider the transformation

$$\xi''_k = U\xi_k$$

where the matrix U is

$$\begin{pmatrix} 1/\sqrt{2} & i/\sqrt{2} \\ i/\sqrt{2} & 1/\sqrt{2} \end{pmatrix}.$$

Write the seven vectors $\xi''_0, \xi''_a, \ldots, \xi''_f$ defined by this transformation and the corresponding matrices $A'', \ldots, F''$. Is the matrix U orthogonal? What is the physical meaning of this transformation? What states of polarization do the base vectors $\binom{1}{0}$ and $\binom{0}{1}$ represent after this transformation?

12. Calculate the norms of the vectors ξ_A, ξ_B, ξ_C, ξ_D, ξ_E, and ξ_F and show that these agree with the results of exercise 1. Calculate the norms also of the vectors ξ'_A, ξ'_C, and ξ'_E of exercise 8 and the vectors ξ''_A, ξ''_C, and ξ''_E of exercise 11. Compare these results with the results for the corresponding unprimed vectors.

13. Calculate the expression

$$\langle M \rangle = \frac{(\xi|M|\xi)}{(\xi|\xi)}$$

for the general vector ξ of exercise 5 and the matrix M defined by eq. (1.11). Calculate $\langle M \rangle$ for the vectors ξ_A, ξ_B, ξ_C, ξ_D, ξ_E, and ξ_F. What is the relation of these results to the results of the corresponding measurements of exercise 3? Calculate $\langle M \rangle$ for the vectors ξ'_A, ξ'_B, ξ'_C, ξ'_D, ξ'_E, and ξ'_F and discuss the relations of these values to those obtained for the results of the measurements in exercise 4.

14. Calculate $\langle M \rangle$ for the vectors ξ''_A, ξ''_B, ξ''_C, ξ''_D, ξ''_E, and ξ''_F. If these results are to be compared with the results of measurements as in exercise 13, what kind of measurements would these be?

15. Calculate the matrix $M'' = UM\tilde{U}^*$, where U is defined in exercise 11. Repeat exercise 14 using the matrix M'' instead of the matrix M and compare the results with the previous two exercises.

16. Classify the matrices A, B, C, D, E, and F defined in exercise 6 as orthogonal, Hermitean, unitary, or projection operators. Discuss the physical meaning of this classification. Classify also the matrices R, U, and M defined in exercises 10, 11, and 13.

17. A beam of polarized light is passed through a medium in a magnetic field which rotates the plane of polarization continuously through an angle of 2π. The beam can be represented by a two-dimensional complex state vector ξ, as defined by eq. (1.2). At any point z in the medium the state of polarization at that point can be represented by a vector $\xi(z)$ which is a function of the variable z. If the length of the medium is L, the plane of polarization of the beam in passing through a distance $\mathrm{d}z$ in the medium is rotated by an angle $2\pi L^{-1}\mathrm{d}z$. Using this infinitesimal rotation, write an expression for the derivative $\mathrm{d}\xi(z)/\mathrm{d}z$. Use the basic vectors as defined in exercise 5. Show that this expression for $\mathrm{d}\xi/\mathrm{d}z$ is a differential equation which can be solved to give an expression $\xi(z) = \mathrm{e}^{\mathrm{i}Kz}\xi(0)$. What kind of a matrix is K? What kind of a matrix is $\mathrm{e}^{-\mathrm{i}Kz}$? Evaluate this matrix.

18. Assuming that $\xi(0)$ is a vertically polarized beam, calculate the expression

$$\langle M \rangle = \frac{(\xi|M|\xi)}{(\xi|\xi)}$$

where M is defined by eq. (1.11). The quantity $\langle M \rangle$ will now be a function of the variable z. What is the relation between the value of $\langle M \rangle$ and a measurement on the beam at the point z?

19. Calculate the derivative $(\mathrm{d}/\mathrm{d}z)\langle M \rangle$, the commutator $[K, M]$ and the expression

$$\langle [K, M] \rangle = \frac{(\xi|[K, M]|\xi)}{(\xi|\xi)}.$$

Show that the relation between these quantities can also be obtained by substituting

$$\xi(z) = \mathrm{e}^{\mathrm{i}Kz}\xi(0)$$

into

$$\langle M \rangle = \frac{\langle \xi|M|\xi \rangle}{(\xi|\xi)}$$

and differentiating with respect to z.

20. Calculate the matrix

$$M(z) = \mathrm{e}^{\mathrm{i}Kz} M \mathrm{e}^{-\mathrm{i}Kz}.$$

Calculate the value of $\langle M \rangle$ using the matrix $M(z)$ and the vector $\xi(0)$. Compare this result with that obtained in exercise 18. Calculate the derivative $(\mathrm{d}/\mathrm{d}z)M(z)$. This can be considered as an equation of motion for the operator $M(z)$. Note that the result of the physical measurement given by the calculation of $\langle M \rangle$ can be obtained either by using a matrix M which does not change with z and a state vector ξ which does change with z, or by using a matrix $M(z)$ which changes with z and a fixed state vector $\xi(0)$. This corresponds to the Schrödinger and Heisenberg pictures respectively in quantum mechanics.

21. Find the eigenvalues and the eigenvectors of the matrix K. What are the physical states represented by the eigenvectors? Find a unitary transformation to a representation in which the eigenvectors of K are the base vectors. Transform the state vector $\xi(z)$ to this representation and notice the simpler character of the dependence upon z. Calculate the transformed matrices M, K, and $M(z)$ in the new representation and note that the matrix M takes on a more complicated form. Calculate $\langle M \rangle$ in the new representation and show that it gives the same result as obtained in exercise 18.

22. Consider a beam of polarized photons incident upon the measuring apparatus of fig. 1.1. If the incident beam is represented by the state vector $\xi = \binom{E_x}{E_y}$ then the difference between the intensities measured in the horizontally and vertically polarized channels is given by the expression $\langle \xi | M | \xi \rangle$ where the matrix M is defined by eq. (1.11). If the incident beam consists of a single photon, under what conditions can the result of the measurement described above be predicted with certainty? Write the state vector for the photon for each case, normalized so that $|E_x|^2 + |E_y|^2 = 1$. What is the relation of these state vectors to the matrix M?

23. Consider a measurement in which the nicol has been rotated by 45° from the position of fig. 1.1; i.e., so that the apparatus now measures the difference between the intensities of the components polarized at +45° and −45°. Find a matrix K such that the expression $\langle \xi | K | \xi \rangle$ describes the result of this experiment. Describe the conditions under which a measurement on a single photon with this apparatus can be predicted with certainty and write down the corresponding state vectors. Can you find a single incident photon state for which the results of both this measurement and the one described in exercise 22 can be predicted with certainty? Discuss the relation of your answer to the preceding question to the value of the commutator $[K, M]$.

24. Repeat exercise 23 for the case where an appropriately oriented quarter-wave plate is placed in front of the nicol so that the apparatus now measures the difference between the intensities of the right-handed and left-handed circularly polarized components of the incident beam. Denote the matrix describing this measurement by L.

25. A beam of polarized photons is passed through the apparatus shown in fig. 1.2.
(a) What will be observed coming out of the second nicol if a single circularly polarized photon is introduced into the first nicol and no measurement or disturbance is made between the two nicols?
(b) What was the path of the photon between the two nicols in part (a)?
(c) What would be observed in part (a) if an opaque object were inserted between the two nicols in such a way as to block the horizontally polarized beam but not the vertical one?
(d) What would be observed if a measuring apparatus were inserted in the horizontally polarized beam in such a way that it gave a signal (e.g. a scattered electron) whenever a photon passed through the horizontal beam, but allowed the photon to continue on its path through the second nicol?

26. Repeat problem 3 for the partially polarized incident beam described by eq. (1.27). Write the density matrices ϱ_0, ϱ_a, ϱ_b, ϱ_c, ϱ_d and ϱ_f describing the beam at each stage.

27. Starting with the density matrix ϱ_0, perform the transformations corresponding to the vector transformations in problems 6 and 7.

MOMENTUM TRANSFER TO BOUND SYSTEMS AND THE MÖSSBAUER EFFECT

This monograph in chs. 2, 3 and 4 introduces the student to interesting problems of 'form factor physics' at the frontier of the present research as well as to many basic concepts and mathematical techniques. It begins from a level requiring only very elementary formalism; namely the description of a free particle and a harmonic oscillator and the transformations between different bases.

The form factor concept arises in all investigations of the structure of complex microscopic systems, e.g. X-ray crystallography, light scattering by atoms, electron scattering from nucleons and nuclei and nuclear beta decay. All these processes are described by a transition amplitude containing a form factor which depends only on the properties of the bound system (e.g. the crystal) and the kinematics of the transition process (momentum and energy transfer or wavelengths and frequencies) and it is independent of the details of the elementary radiation, scattering or absorption process. This separation of the form factor from the elementary process makes possible a derivation of the form factor at a very elementary level and a unified description of the applications in all areas of physics.

Form factor physics provides an excellent introduction to wave–particle duality, complementarity and the uncertainty principle because the emitted, absorbed and scattered radiation can be described as either a classical wave or a beam of classical particles. The result for the form factor is obtainable in the two complementary descriptions, either in terms of wavelengths and frequencies of waves or of momenta and energies of particles. In the wave description the form factor describes the spatial size and structure of the system, determined because the waves coming from different parts of the system have different phases. In the particle description the form factor describes the structure in momentum space, determined because the kinematics of momentum and energy transfer depend upon the momentum distribution in the system. The Mössbauer effect is a particularly simple example. Many of its basic features are already present in a very simplified model in which a nucleus emitting a gamma ray is bound in an external harmonic oscillator potential rather than in a crystal and the radiation process is not described in detail but simply as a sudden change in the momentum of the emitting nucleus.

Chapter 2 begins with an elementary description of the kinematics of momentum conservation and nuclear resonance absorption for a free nucleus, described first classically and then quantum-mechanically using plane waves. These results are then applied to the harmonic oscillator model and the Debye–Waller form factor is derived and interpreted in various ways showing the uncertainty principle, complementarity and wave–particle duality. Elementary examples of operator manipulations and unitary transformations are introduced to calculate moments of the energy spectrum. The thermal shift is first presented as the result of a change in mass between the initial excited state and final ground state of the nucleus emitting the gamma ray. It is then shown to be equivalent to the second order Doppler shift for a moving source of electromagnetic radiation and also to a demonstration of the celebrated twin paradox in which a nucleus oscillating back and forth in a potential well is 'younger' than another nucleus which is at rest for the same period of time. All this treatment is understandable for a student who understands the basic formalism of quantum mechanics and the harmonic oscillator.

Chapter 3 begins with the generalization of the treatment of ch. 2 to the case of the Mössbauer effect in a solid. The effect in a harmonic crystal is calculated in detail by using the normal-mode expansion to reduce the problem to one of a large number of harmonic oscillators, each treated by the methods of ch. 2. The Mössbauer effect for a system in thermal equilibrium provides an interesting example for a discussion of the difference between statistical mixtures and pure states and for extension of the operator manipulations of ch. 2 to obtain expressions involving exponentials of Heisenberg operators and space–time correlation functions. The result is shown to correspond to the classical expression obtained for frequency modulated radiation from a moving oscillator. The exact form of a gamma ray spectrum from a harmonic crystal in thermal equilibrium is then calculated using the solution of the Heisenberg equations of motion for a harmonic oscillator and techniques for handling exponentials of non-commuting operators. Relaxation effects provide a good illustration of the equivalent complementary descriptions in the energy and time domains.

Chapters 2 and 3 never consider the detailed radiation process and describe the change in the state of the radiating nucleus only in terms of conservation of energy and momentum. Chapter 4 treats the radiation processes involving momentum transfer to bound systems, with the aid of the second-quantized formalism described in ch. 5 and the phenomenological Hamiltonians and the 'golden rule' of time-dependent perturbation theory developed in ch. 7. Readers unfamiliar with these topics should read chs. 5 and 7 before ch. 4.

A phenomenological Hamiltonian for radiation from a free nucleus is defined using second-quantized operators for creation and annihilation of photons, of the nucleus in the ground state and of the nucleus in the excited state. Conservation laws and invariance principles are used to restrict the form of the Hamiltonian. The Hamiltonian for the case of the bound nucleus is constructed by adding a potential. The golden rule formula is used to derive the general result that the probability for a transition from a bound nucleus factorizes into the product of the corresponding transition probability for a free nucleus and a form factor which depends only on the momentum transfer and the properties of the bound system and is independent of the details of the radiation process. The same methods are then applied to the scattering of photons by bound atoms, nuclear beta decay and electron scattering by complex nuclei. In each case the transition probability is shown to have the same factorized form, and expressions for the form factor are obtained.

CHAPTER 2

THE MÖSSBAUER EFFECT

2.1 Momentum conservation and nuclear resonance absorption

The Mössbauer effect is an ideal subject for students of Introductory Quantum Mechanics. There is much to learn from a study of the Mössbauer effect. First of all note that a graduate student these days can win the Nobel Prize in physics for work done as a Ph.D. thesis. However, Mössbauer did not work in the glamorous field of theoretical elementary particle physics thinking that he would get the great idea to solve all the problems of the universe. He began working on a difficult and tedious experimental problem which had the very modest aim of measuring the lifetime of a particular excited energy level in the nucleus ^{191}Ir. In the course of his experimental work, he discovered some peculiar phenomena which he did not understand. None of the professors whom he consulted really understood it either, and he went ahead, both experimentally and theoretically with the work that earned him the Nobel Prize.

If any of you have ambitions to win the Nobel Prize, think carefully in choosing your Ph.D. thesis subject, particularly if you have the choice between using elegant mathematical techniques like fancy group theory to find a new theory for elementary particles and an experimental problem of measuring the lifetimes of a few nuclear levels on the Tandem Accelerator. I will not promise you the Nobel Prize if you measure these nuclear lifetimes, but I can guarantee almost with certainty, that anything that you do for a Ph.D. thesis in particle physics will not get you the Nobel Prize.

The basic physics underlying the Mössbauer effect can be perfectly clear to you after a few lectures. It is all elementary quantum mechanics and should have been understood years before the effect was discovered. However the basic principles were not appreciated because it involved the application of elementary quantum mechanics both to nuclear and to solid state physics. Nuclear physicists did not know solid state physics and vice versa.

The Mössbauer effect can be looked at from different points of view to illustrate many of the fundamental principles of quantum mechanics. It is also a special case of a process in which momentum is transferred to a bound system. Such processes are important in a wide variety of problems in all fields of physics. These include the scattering of light by atoms, the scattering of X-rays by crystals, the scattering of very high energy electrons by nuclei and nuclear beta decay.

Consider a nucleus of mass M in an excited state having an energy E_0 above the ground state, as in fig. 2.1. This nucleus can decay to the ground state by emitting a gamma ray. One might expect from conservation of energy that the energy of the gamma ray should be E_0, the energy lost by the nucleus in making the transition. In such a case the gamma ray might be absorbed by another nucleus of the same type in its ground state. The energy E_0 of the gamma ray would be just enough to raise the nucleus to its excited state. The absorption of such a gamma ray having exactly the right energy to cause a transition between two nuclear levels is called resonance absorption.

Fig. 2.1 Fig. 2.2

The process described in the preceding paragraph generally does not occur because the energy of the emitted gamma ray is not exactly equal to E_0. In our calculation of energy conservation, we have neglected momentum conservation and the kinetic energy of the motion of the nucleus. If the nucleus is initially at rest before the emission of the gamma ray, it must recoil after the emission of the gamma ray in order to conserve momentum as shown in fig. 2.2. If the kinetic energy of the recoiling nucleus is R, the energy E_γ of the emitted gamma ray is given by conservation of energy as

$$E_\gamma = E_0 - R. \tag{2.1}$$

If the wave vector of the gamma ray is $\boldsymbol{k}$, its momentum is $\hbar \boldsymbol{k}$ and

$$E_\gamma = h\nu = \hbar kc \tag{2.2}$$

where ν is the frequency of the gamma ray.

By momentum conservation the momentum of the recoil nucleus is $-\hbar \boldsymbol{k}$.

Thus, with the aid of eqs. (2.1) and (2.2) we obtain

$$R = \frac{(\hbar k)^2}{2M} = \frac{E_\gamma^2}{2Mc^2} = \frac{(E_0 - R)^2}{2Mc^2}. \tag{2.3}$$

It is generally not necessary to solve the quadratic equation (2.3) to determine R, since R is so small in comparison with E_0 and E_γ that it can be neglected on the right-hand side of eq. (2.3). Thus

$$R = \frac{E_\gamma^2}{2Mc^2} \approx \frac{E_0^2}{2Mc^2}. \tag{2.4}$$

For a typical case, the 14 keV transition in ^{57}Fe, $E_0 = 14 \times 10^3$ eV and Mc^2 is the rest energy of the ^{57}Fe nucleus, approximately 57×10^9 eV. (It is useful to remember that the rest energy of the nucleon is approximately 1 GeV= $= 10^9$ eV.) Eq. (2.4) then gives $R \approx 2 \times 10^{-3}$ eV and $R/E_0 \approx 1.4 \times 10^{-7}$ is indeed small. Thus the approximate equation (2.4) is clearly valid.

However, R is not small enough to be neglected in considering the process of resonance absorption. Resonance absorption can only occur when the energy of the gamma ray is equal to that required for the transition with a precision of the order of the line width. The natural line widths of nuclear gamma rays are of the order 10^{-5} eV or even smaller, very much smaller than R.

A similar analysis holds for momentum conservation in the absorption process. If the absorbing nucleus is initially at rest, it must have a kinetic energy R after absorbing the gamma ray. Thus the gamma ray energy required for resonance absorption is not E_0 but $E_0 + R$, while the energy emitted is $E_0 - R$.

Thus in this case the gamma ray emitted by the nucleus has too low an energy to produce absorption in another similar nucleus, and resonance absorption simply cannot occur.

2.2 The quantum-mechanical description for a free nucleus

Let us now describe the change in energy and momentum of the nucleus in the language of quantum mechanics. A full description must include the internal structure of the nucleus which gives rise to its energy levels, as well as the quantum theory of the radiation process. Both of these aspects are beyond the scope of our treatment at this stage. However, they are not really necessary for an understanding of the basic features of the Mössbauer effect. We shall consider only the motion of the nucleus as a whole (i.e., the motion of a center of mass).

Let the ket vector $|i\rangle$ represent the initial state of the nuclear motion before the emission of the gamma ray. Let the ket vector $|f\rangle$ represent the final state of nuclear motion after the emission of the gamma ray. We assume that the relation between these two states is determined by the laws of conservation of energy and momentum and do not consider the detailed process of the emission of the gamma ray. We assume that the nucleus is initially moving as a free particle and is therefore described as a plane wave with wave vector $\boldsymbol{k}'$

$$|i\rangle = |\boldsymbol{k}'\rangle. \tag{2.5}$$

After the emission of the gamma ray of wave vector $\boldsymbol{k}$ and momentum $\hbar\boldsymbol{k}$ the momentum of the nucleus must change from $\hbar\boldsymbol{k}'$ to $\hbar(\boldsymbol{k}'-\boldsymbol{k})$ in order to conserve momentum. Thus the final state of the nuclear motion is a plane wave with wave vector $\boldsymbol{k}'-\boldsymbol{k}$.

$$|f\rangle = |\boldsymbol{k}' - \boldsymbol{k}\rangle. \tag{2.6}$$

The Hamiltonian describing the motion of the nucleus is just the kinetic energy

$$H = \frac{\boldsymbol{p}^2}{2M} \tag{2.7}$$

where $\boldsymbol{p}$ is the momentum operator. Both the initial state $|i\rangle$ and the final state $|f\rangle$ are eigenvectors of this Hamiltonian

$$H|i\rangle = E_i|i\rangle \tag{2.8a}$$

$$H|f\rangle = E_f|f\rangle. \tag{2.8b}$$

The eigenvalues are just the kinetic energies of the initial and final states

$$E_i = \frac{\hbar^2 k'^2}{2M} \tag{2.9a}$$

$$E_f = \frac{\hbar^2(\boldsymbol{k}' - \boldsymbol{k})^2}{2M} = E_i - \frac{\hbar^2\boldsymbol{k}'\cdot\boldsymbol{k}}{M} + R \tag{2.9b}$$

where we have used eq. (2.2). By conservation of energy,

$$E_\gamma = E_0 + E_i - E_f = E_0 - R + \frac{\hbar^2\boldsymbol{k}'\cdot\boldsymbol{k}}{M}. \tag{2.10a}$$

This result is a generalization of eq. (2.1) to the case where the nucleus initially has momentum $\hbar\boldsymbol{k}'$ rather than being at rest ($\hbar\boldsymbol{k}' = 0$). Although we have used a quantum-mechanical derivation, eq. (2.10) follows only from energy and momentum conservation, and is obtainable from a classical description as well.

A similar analysis for absorption shows that the gamma ray energy required for resonance absorption by a nucleus with momentum $\hbar \boldsymbol{k}'$ is

$$E_\gamma(\text{abs}) = E_0 + R - \frac{\hbar^2 \boldsymbol{k}' \cdot \boldsymbol{k}}{M}. \tag{2.10b}$$

Eq. (2.10a) shows that the energy of the emitted gamma ray depends upon the initial momentum of the nucleus. Since $\hbar \boldsymbol{k}'/M$ is just the initial velocity $\boldsymbol{v}_\text{i}$ of the nucleus,

$$E_f = E_0 + R - \hbar \boldsymbol{k} \cdot \boldsymbol{v}_\text{i} \tag{2.11a}$$

$$E_\gamma = E_0 - R + \hbar \boldsymbol{k} \cdot \boldsymbol{v}_\text{i} = E_0 - R + E_\gamma \frac{\boldsymbol{v}_\text{i} \cdot \hat{\boldsymbol{k}}}{c} \tag{2.11b}$$

where $\hat{\boldsymbol{k}}$ is a unit vector in the direction of $\boldsymbol{k}$. Thus the dependence of the energy of the emitted gamma ray (and therefore of its frequency) upon the initial velocity of the nucleus has the same form as a classical Doppler shift for a wave emitted by a source moving with velocity $\boldsymbol{v}_\text{i}$. It might seem surprising to see this result from a treatment where the gamma ray was considered to be a particle carrying energy and momentum, and where the relation (2.11) was derived from the kinematics of energy and momentum conservation. This is an example of wave–particle duality ana the relation between the wave and particle pictures of the photon.

To compare the result in eqs. (2.10) and (2.11) for the energy of the gamma ray with experimental measurements we note that experiments are always performed with a large number of nuclei, and a large number of gamma rays are measured. The initial velocity $\boldsymbol{v}_\text{i}$ is not the same for every nucleus. If the nuclei are in a classical gas at thermal equilibrium at some temperature, they have a Maxwell distribution of velocities. The gamma rays emitted would then not have a single energy but would be a continuous spectrum related by eq. (2.11) to the Maxwell distribution for $\boldsymbol{v}_\text{i}$. The mean value and mean square value of E would be given by

$$\langle E_\gamma \rangle = E_0 - R + \left\langle E_\gamma \frac{\boldsymbol{v}_\text{i} \cdot \hat{\boldsymbol{k}}}{c} \right\rangle \tag{2.12a}$$

$$\langle E_\gamma^2 \rangle = \left\langle \left[E_0 - R + E_\gamma \frac{\boldsymbol{v}_\text{i} \cdot \hat{\boldsymbol{k}}}{c} \right]^2 \right\rangle. \tag{2.12b}$$

If the initial velocity distribution is symmetric then $\langle v_{\mathrm{i}} \rangle = 0$ and* $\langle E_\gamma v_{\mathrm{i}} \rangle \approx 0$;

$$\langle E_\gamma \rangle = E_0 - R \tag{2.13a}$$

$$\langle E_\gamma^2 \rangle = (E_0 - R)^2 + \langle E_\gamma^2 v_{\mathrm{i}}^2 \rangle / c^2 \approx \langle E_\gamma \rangle^2 \left(1 + \frac{\langle v_{\mathrm{i}}^2 \rangle}{c^2} \right) \tag{2.13b}$$

and

$$\langle E_\gamma^2 \rangle - \langle E_\gamma \rangle^2 = \langle E_\gamma \rangle^2 \langle v_{\mathrm{i}}^2 \rangle / c^2. \tag{2.13c}$$

Thus the gamma ray spectrum is centered about the value $E_0 - R$, the energy which the gamma ray would have if the nucleus were initially at rest. The spectrum has a width proportional to the mean square velocity in the initial distribution. This is again characteristic of the Doppler broadening of a spectral line due to the thermal motion of the emitting atoms.

Fig. 2.3 shows a typical gamma ray spectrum.

Eq. (2.10b) shows that the absorption spectrum for this case is the same as fig. 2.3 but centered at $E_0 + R$ instead of $E_0 - R$. It is thus a mirror image of fig. 2.3 about the line $E_\gamma = E_0$.

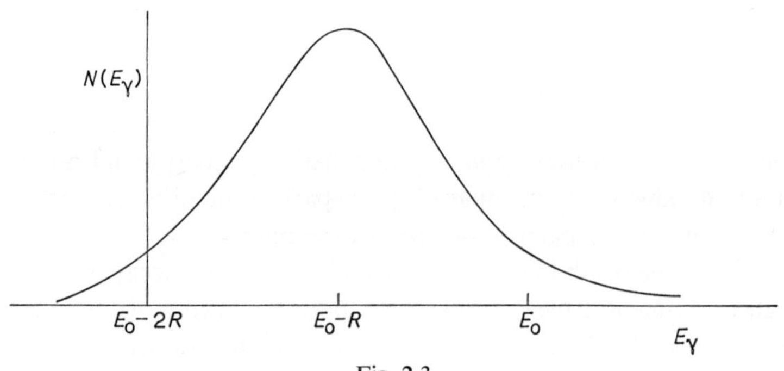

Fig. 2.3

Although the mean and mean square values of E would seem to be sufficient for determining the general shape of a spectrum of the type shown in fig. 2.3 this is not the case for the Mössbauer effect. Other finer details of the spectrum turn out to be very significant. One way to describe these finer details is by use of the higher moments of the spectrum.

$$\langle E_\gamma^m \rangle = \left\langle \left[E_0 - R + E_\gamma \frac{\boldsymbol{v}_{\mathrm{i}} \cdot \hat{\boldsymbol{k}}}{c} \right]^m \right\rangle \tag{2.14}$$

* Since $E_\gamma - \langle E_\gamma \rangle$ and v_{i}/c are both small, $E_\gamma(v_{\mathrm{i}}/c) \approx \langle E_\gamma \rangle (v_{\mathrm{i}}/c)$.

Eq. (2.14) gives the moments of the gamma ray spectrum in terms of the moments of the spectrum of the initial velocity distribution.

Fig. 2.3 shows that the gamma ray spectrum spreads out far enough to include the energy E_0 if the spread of the initial velocities is sufficiently large. A similar spread occurs in the absorption spectrum. There is therefore a very small part of the emission spectrum which overlaps with the resonance absorption spectrum. If the source is in thermal equilibrium, the width of the distribution depends upon the temperature. Increasing the temperature increases the width of the distribution and increases the absorption, Conversely, lowering the temperature decreases the absorption. Early experiments in nuclear resonance absorption used this effect to detect resonance absorption by its temperature dependence. Mössbauer's original experiment was just of this type. However, he found that lowering the temperature increased the absorption rather than decreasing it*.

There have also been experiments in which the probability of resonance absorption was increased by mechanical motion of the source in order to give the gamma rays a Doppler shift in the proper direction. In this case $\langle v_\mathrm{i} \rangle \neq 0$ and the centroid of the spectrum can be shifted from $E_0 - R$ toward E_0 as shown by eq. (2.12a). For the case of ^{57}Fe mentioned above, a Doppler shift of the order of R required velocities of the order of $cR/E_0 \approx 4 \times 10^3$ cm/sec. Such velocities have been obtained by the use of rotating disks and centrifuges.

In all these cases the portion of the spectrum having the proper energy is extremely small and the probability for resonance absorption is thus also extremely small.

2.3 Emission from a bound nucleus. The Mössbauer effect

Mössbauer discovered experimentally that the spectrum of emitted gamma rays can be very different from that shown in fig. 2.3 if the nucleus is bound in a crystal. Under suitable circumstances, there can be an appreciable probability that the gamma ray is emitted with the full energy E_0 and can be absorbed with no energy loss due to the kinetic energy of nuclear recoil. Thus resonance absorption can be observed. To understand the Mössbauer effect, consider a model where the nucleus is not free, but is moving in one dimension in a harmonic oscillator potential. This is not exactly the same as the motion of an atom in a crystal where the forces acting upon it are due to the inter-

* The reason for this peculiar behavior is just the Mössbauer effect, discussed in the following section.

actions with neighboring atoms. However, the simple model of an atom moving in an external harmonic potential already contains the basic features of the Mössbauer effect.

The Hamiltonian describing the motion of the nucleus in a harmonic potential is

$$H = \frac{p^2}{2M} + \tfrac{1}{2}M\omega^2 x^2 \tag{2.15}$$

where $\omega/2\pi$ is the oscillator frequency. The eigenvectors of this Hamiltonian are denoted by the quantum number n and the eigenvalues are $(n+\frac{1}{2})\hbar\omega$

$$H|n'\rangle = (n' + \tfrac{1}{2})\hbar\omega|n'\rangle = E_{n'}|n'\rangle. \tag{2.16}$$

We assume that before the emission of the gamma ray the atom is initially in some eigenstate n' of the harmonic oscillator

$$|i\rangle = |n'\rangle. \tag{2.17}$$

Since this state (2.17) is not an eigenstate of the momentum operator p we cannot immediately write down the form of the final state by applying the laws of momentum conservation. However, the set of plane-wave eigenvectors of the momentum operator constitute a complete orthonormal set. We can therefore expand our initial state vector $|i\rangle$ in plane waves. In the Schrödinger coordinate representation this would be the Fourier transform of the oscillator wave function. In the Dirac notation, we have

$$|i\rangle = \sum_{k'} |k'\rangle\langle k'|i\rangle. \tag{2.18}$$

We now assume that we can use momentum conservation upon each of the individual plane-wave components $|k'\rangle$ separately to obtain the form of the final-state vector after the emission of the gamma ray*. Each plane-wave component k' has its wave vector shifted by an amount k to become $|k'-k\rangle$. The state vector for the final state is thus

$$|f\rangle = \sum_{k'} |k' - k\rangle\langle k'|i\rangle. \tag{2.19}$$

The summation over k' is not easily carried out directly in eq. (2.19). It can be simplified by noting that the operator e^{-ikx} operating on a plane-wave state changes the wave number by an amount $-k$,

$$e^{-ikx}|k'\rangle = |k' - k\rangle. \tag{2.20}$$

* This is a dynamical assumption which implies that the momentum transfer takes place in a single impulse, and not as a series of small impulses. We need not concern ourselves with the validity of the assumption at this stage.

We can therefore rewrite eq. (2.19) using eq. (2.20)

$$|f\rangle = \sum_{k'} e^{-ikx}|k'\rangle\langle k'|i\rangle. \tag{2.21}$$

The summation over k' can now be done directly by closure to obtain

$$|f\rangle = e^{-ikx}|i\rangle. \tag{2.22}$$

With this simple expression for the final state of the nucleus, we can calculate the energy spectrum of the gamma ray by use of conservation of energy. In contrast to the case of the free nucleus, the final state (2.22) is not an eigenstate of the Hamiltonian (2.15) and therefore does not have a well defined energy. This means that we cannot predict the energy of the gamma ray in advance, but can only give a probability description. We can predict the form of the gamma ray spectrum emitted from a large number of nuclei which are initially all in the state $|i\rangle$.

Let us expand the final state vector $|f\rangle$ in the complete set $|n'\rangle$ of orthonormal eigenfunctions of the Hamiltonian H

$$|f\rangle = \sum_{n''} |n''\rangle\langle n''|f\rangle \tag{2.23a}$$

$$= \sum_{n''} |n''\rangle\langle n''|e^{-ikx}|i\rangle \tag{2.23b}$$

$$= \sum_{n''} |n''\rangle\langle n''|e^{-ikx}|n'\rangle \tag{2.23c}$$

where we have substituted eqs. (2.22) and (2.17). The probability that the atom is found in the state $|n''\rangle$ with energy $E_{n''}=(n''+\frac{1}{2})\hbar\omega$ in the final state is given by the square of the appropriate coefficient in the expansion (2.23). Let $p_{n'\to n''}$ represent the probability that the atom will be in the state $|n''\rangle$ after emitting the gamma ray, if it is initially in the state $|n'\rangle$.

$$p_{n'\to n''} = |\langle n''|e^{-ikx}|n'\rangle|^2. \tag{2.24}$$

The energy of the gamma ray is obtained by conservation of energy. If the atom is found in the state $|n''\rangle$ after the emission,

$$E_\gamma = E_0 - \hbar\omega(n'' - n'). \tag{2.25}$$

There is one obvious difference between the gamma ray spectrum for this case and that of fig. 2.3. The present gamma ray spectrum is discrete rather than continuous because of the discrete energy levels of the harmonic oscillator. The energy of the gamma ray can differ from the energy E_0 only by an integral number of oscillator quanta with an uncertainty given by the line width which is very small. The gamma ray spectrum therefore consists of a set of sharp spikes with the intensity of each spike being given by the appropriate probability coefficient $P_{n' \to n''}$.

If the binding is very weak, one should expect the gamma ray spectrum to approach that of fig. 2.3 for a free atom. For weak binding the oscillator frequency is low and one might expect a spectrum consisting of a large number of closely spaced spikes having an envelope roughly like the curve of fig. 2.3. Such a spectrum is shown in fig. 2.4.

The more interesting case is that of strong binding. Let us assume, for example, that $\hbar\omega = 2R$. In this case the spectrum consists of spikes at E_0, $E_0 \pm 2R$, and other spikes further out as shown in fig. 2.5. However, there can be no gamma ray emitted with an energy between E_0 and $E_0 - 2R$, which is the region where the main bulk of the spectrum is found in the case of no binding or weak binding. For this case on might expect an appreciable probability for the gamma ray to come out with the exact energy E_0 required for resonance absorption. This turns out in fact to be the case.

From eq. (2.25) we see that the condition for resonance absorption $E_\gamma = E_0$ requires $n'' = n'$; i.e., the gamma ray comes out with the full energy if the energy of the nuclear motion is the same after the transition as before.

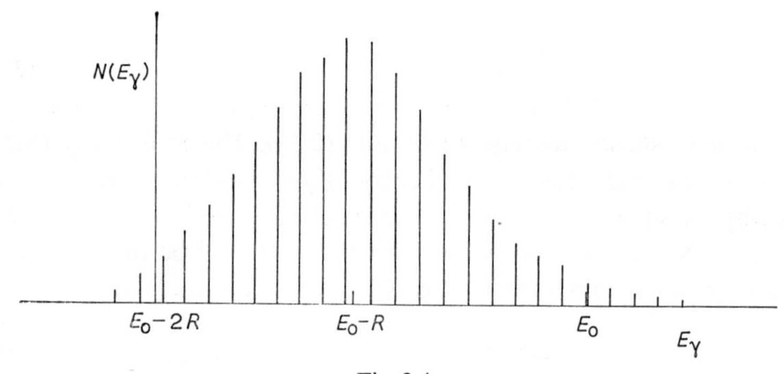

Fig. 2.4

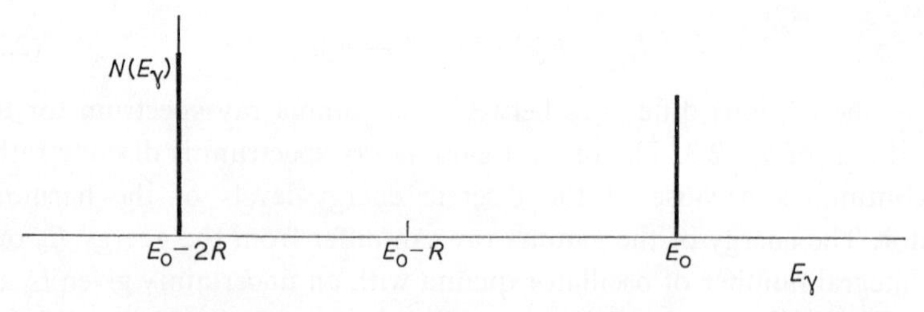

Fig. 2.5

The probability that the gamma ray comes out with the exact energy E_0 is given by setting $n''=n'$ in eq. (2.24).

Phenomena of this kind were well known for many years before the discovery of the Mössbauer effect, although the connection was not appreciated. In ordinary X-ray diffraction by a crystal, the scattered X-ray comes out in a different direction from the incident X-ray. Since the X-ray carries momentum there must be a momentum transfer to the crystal. The crystal consists of many moving atoms and is roughly like a large number of harmonic oscillators. If X-ray diffraction is to be observed, the scattered X-rays emerging from the crystal must all have the same wavelength. They must therefore have the same energy as the incident X-ray. This means that the set of harmonic oscillators of the crystal must remain in the same state after scattering the X-ray as they were before scattering the X-ray, although they absorb the momentum required to change the direction of the X-ray. The probability that this occurs is given by expressions which look very much like our eq. (2.24) where states of the entire crystal appear, rather than those of just a single harmonic oscillator. The relation of this result to momentum transfer was not appreciated because the relation analogous to (2.24) for X-ray diffraction had been derived from a classical wave picture for the X-ray, rather than the particle picture which we have been using in order to consider momentum conservation. It is a consequence of the wave–particle duality of the photon that the same result can be obtained in two such apparently different ways. We shall see this in more detail below when we evaluate the expression (2.24) for the specific case of our single oscillator.

2.4 Explicit calculation of the gamma-ray spectrum

The expression (2.24) can be evaluated in a straightforward manner either by matrix methods or by simple integration. All we need are matrix elements of the operator e^{-ikx} between harmonic oscillator eigenstates. Let us use the x-representation, and let $\psi_{n'}(x)$ represent the normalized wave function for the n' state of the harmonic oscillator. Then eq. (2.14) assumes the form

$$P_{n'\to n''} = |\int \psi^*_{n''}(x)\mathrm{e}^{-ikx}\psi_{n'}(x)\,\mathrm{d}x|^2. \tag{2.26}$$

The probability that the gamma ray has the exact energy E_0 is given by setting $n''=n'$. Since

$$\varrho_{n'}(x) = \psi^*_{n'}(x)\psi_{n'}(x) \tag{2.27}$$

is just the probability density for finding the atom at the point x when the atom is in the state n' of the oscillator we can write

$$P_{n' \to n'} = |\varrho_{n'}(x)\mathrm{e}^{-ikx}\,\mathrm{d}x|^2. \tag{2.28}$$

This is just the square of the kth Fourier component of the Fourier transform of the probability density function $\varrho_{n'}(x)$. Note that this result is very general and does not depend upon the properties of the harmonic oscillator eigenfunctions. In our derivation thus far we have used the quantum numbers n' merely as labels to designate the eigenvectors of the Hamiltonian and have not yet explicitly used the fact that the Hamiltonian is a harmonic oscillator.

The result (2.28) is well known in X-ray crystallography. The intensity of the X-ray beam scattered from a crystal gives a measure of the magnitude of the Fourier transform of the electron density. (It is the *electron* density since in this case the X-ray interacts with electrons rather than with nuclei.) The value of k can be varied experimentally by changing the energy of the incident X-ray beam and changing the angle of scattering, thus allowing the magnitude of the Fourier transform (but not the phase) to be obtained over a range of values of k.

The result (2.28) can also be obtained by a classical wave treatment. The wave function $\psi_{n'}(x)$ describing the motion of the atom is now considered to be a standing wave rather than a moving particle. This standing wave is considered as an extended source emitting gamma radiation which is also considered to be waves. The amplitude of the radiation from the source is obtained by adding up the contributions from small regions in the source. If the size of the source is appreciable in comparison with the wavelength of the gamma radiation, contributions from different regions of the source are not in phase. There is a phase difference depending upon the path difference. The contribution from the region around the point x in the source is proportional to the source density function $\varrho_{n'}(x)$ at the point x and has a phase e^{-ikx}. The sum of the contributions from all regions is therefore an integral with exactly the form (2.28).

We have derived the expression (2.28) from assumptions which are much more general than the particular case of the Mössbauer effect. All that we have assumed is that there is a momentum transfer $\hbar k$ to a particle which is in some kind of bound state and that the probability of a transition from the initial bound state to some final bound state is given by expanding both the initial and final bound states in plane waves and using momentum conservation for the transition. Such a momentum transfer to a bound particle occurs in many fields of physics and can be treated in exactly the same manner. The expression (2.28) is called a form factor or structure factor, because it depends on the form or the structure of the bound state wave function. Very often the main purpose of an experiment is the measurement of this form factor in order to study the properties of the bound system. One example of such a case is high energy electron scattering which has been used to determine the electric charge distribution in nuclei and in the proton neutron.

If the momentum transfer k is small we can evaluate the expression (2.28) by expanding the exponential in powers of k. We assume that the density distribution $\varrho(x)$ is symmetric about the origin and that therefore the terms from odd powers of kx in the expansion vanish. The first non-trivial term is the quadratic term. To this order we obtain:

$$P_{n' \to n'} \approx | \int \varrho_{n'}(x)[1 - \tfrac{1}{2}k^2x^2]\,\mathrm{d}x|^2 \approx 1 - k^2\langle x^2 \rangle_{n'} \tag{2.29}$$

where

$$\langle x^2 \rangle_{n'} = \int \varrho_{n'}(x)x^2\,\mathrm{d}x. \tag{2.30}$$

This is just the mean square deviation of the position of the atom from its equilibrium value or the mean square radius of the wave function. The form factor at small momentum transfers thus depends on the mean square radius of the system. In the particular case of electron scattering it has been used to measure the mean square radius of the nuclear charge distribution.

For higher values of k, the value of the expression (2.28) depends upon other details of the shape of the density function $\varrho_{n'}(x)$ than the mean square radius. If the atom is moving initially in the ground state of a harmonic oscillator potential, $\psi_0(x)$ has the form of a gaussian and therefore:

$$\varrho_0(x) = N\mathrm{e}^{-\alpha x^2} \tag{2.31a}$$

where N and α are constants depending upon the constants M and ω appearing in the oscillator Hamiltonian. In the case of the Mössbauer effect it turns out in practice that the density function $\varrho(x)$ is nearly always gaussian to a good approximation. Thus the expression (2.31a) leads to results which are generally valid even in cases when the system is not initially in the ground state of a harmonic oscillator.

The expression (2.28) is easily evaluated with the density function (2.31a). We can obtain the result immediately without calculation by remembering that the Fourier transform of a gaussian function is also a gaussian and that the result must be equal to the approximate result (2.29) to second order in k. Thus:

$$P_{0 \to 0} = \mathrm{e}^{-k^2\langle x^2 \rangle_0}. \tag{2.31b}$$

The expression (2.31b) has been known for a long time in X-ray diffraction and is usually called the Debye–Waller factor. It was obtained for the X-ray case by a classical wave argument similar to the one given above.

Eq. (2.31b) gives the probability that the gamma ray is emitted with the energy E_0 required for resonance absorption if the atom is initially moving in the lowest state of the harmonic oscillator potential. However, as discussed above, this result has a much more general validity and holds whenever the probability density $\varrho(x)$ is gaussian. It can be assumed to hold in all but exceptional cases for the probability that the gamma ray has the energy E_0.

2.5 Various expressions for the Debye–Waller factor and their interpretation

The result (2.31b) can be rewritten in other ways using the well-known property of the harmonic oscillator that the average kinetic energy is equal to the average potential energy and both are equal to half of the total energy. Since the latter in the lowest state is $\frac{1}{2}\hbar\omega$ we have

$$\tfrac{1}{2}M\omega^2\langle x^2\rangle = \frac{\langle p^2\rangle}{2M} = \tfrac{1}{4}\hbar\omega \tag{2.32a}$$

$$\langle x^2\rangle = \frac{\hbar}{2M\omega} = \frac{\hbar^2}{2M(\hbar\omega)} \tag{2.32b}$$

$$\langle x^2\rangle\langle p^2\rangle = \tfrac{1}{4}\hbar^2. \tag{2.32c}$$

Eq. (2.32c) expresses the well-known property of the harmonic oscillator ground state; namely, that the inequality of the Heisenberg uncertainty relation between position and momentum becomes an equality for this state. Substituting eqs. (2.32b) and (2.32c) into eq. (2.31b) we obtain two alternative expressions for the Debye–Waller factor:

$$P_{0\to 0} = \mathrm{e}^{-\hbar^2k^2/2M\hbar\omega} = \mathrm{e}^{-R/\hbar\omega} \tag{2.33a}$$

$$P_{0\to 0} = \mathrm{e}^{-\hbar^2k^2/4\langle p^2\rangle}. \tag{2.33b}$$

The three expressions (2.31b), (2.33a) and (2.33b) for the same quantity have different intuitive physical interpretations. These are related in a manner typical of the complementarity of quantum mechanics.

Let us first examine eq. (2.31b). This is simply interpreted in the wave picture. The intensity of gamma radiation is large only if $k^2\langle x^2\rangle = (2\pi)^2\langle x^2\rangle/\lambda^2$ is small; that is, if the extension of the source emitting the gamma rays is small compared to the wavelength λ of the gamma rays so that the radiation contributions from different regions of the source are all in phase.

Eq. (2.33a) is easily interpreted in terms of the spectrum shown in fig. 2.5. It says that the intensity of the gamma ray having energy $E = E_0$ is small unless R is small compared to $\hbar\omega$. In other words, the spacing between the energy levels of the oscillator must be of the same order of magnitude as the free recoil energy R or even larger. This leads to the situation shown in fig. 2.5 where the intensity at energy E_0 must be large because the total intensity has to be divided between a few isolated spikes which are far apart compared to the free recoil energy R. When the kinetic energy of the recoil R is not enough to supply energy for a jump from one oscillator orbit to the next, there is a large probability that the atom stays in the same orbit and does not jump.

Eq. (2.33b) shows that the intensity of the gamma ray with energy E_0 is small unless the momentum $\hbar k$ of the gamma ray is small compared to the fluctuations in the momentum of the atom in its initial orbit; i.e., $(\hbar k)^2 < 4\langle p^2 \rangle$. The atom has a good chance of absorbing the momentum of the gamma ray without changing its quantum state if the momentum of the gamma ray is of the same order as the fluctuations in the momentum of the atom due to its zero-point oscillations in the oscillator potential.

Although these three interpretations seem to be quite different, they are related by the typically quantum-mechanical concepts of the uncertainty principle, complementarity and wave–particle duality. The result (2.31b) has been interpreted on a wave picture using the typical wave concepts of wavelength, phase, and interference. The result (2.33b) has been interpreted in a typical particle picture using momentum for the gamma ray and for the emitting atom, and considering the fluctuations in the momentum of the latter. These two descriptions are complementary and the relation between them can be seen directly with the aid of the uncertainty principle (which was in effect actually used in the form of eq. (2.32c) to obtain eq. (2.33b) from eq. (2.31b). If the wave picture says that the radiating atom must be confined to a region which is small compared to the wavelength of the gamma ray, the uncertainty principle tells us on the particle picture that the uncertainty in the momentum of the atom must be large compared to the momentum of the gamma ray. These two statements are thus complementary ways of saying the same thing.

The result (2.33a) is expressed in terms of energies rather than momenta or wavelength. Its relation to the result (2.33b) expressed in terms of momenta is straightforward since the free recoil energy is proportional to the square of the gamma ray momentum and the harmonic oscillator frequency is proportional to the kinetic energy of the atom, and therefore to the square of its momentum.

2.6 Sum rules and moments of the energy spectrum

With the aid of the formalism of quantum mechanics using operators and matrices, we can find expressions for the moments of the gamma ray spectrum from a bound atom and compare them with those given by eqs. (2.21) and (2.14) for the case of gamma ray emission from a free atom. We first verify that the expression (2.24) is consistent with the definition of a probability by checking whether the sum of all the probabilities over all final states n'' is equal to unity.

$$\sum_{n''} P_{n'\to n''} = \sum_{n''} \langle n'|\mathrm{e}^{+ikx}|n''\rangle\langle n''|\mathrm{e}^{-ikx}|n'\rangle$$
$$= \langle n'|\mathrm{e}^{ikx}\cdot\mathrm{e}^{-ikx}|n'\rangle = 1 \tag{2.34}$$

where the sum over n'' is easily done by closure.

The moments of the gamma ray energy spectrum can be obtained by first calculating the moments of the energy spectrum E_f of the final state $|f\rangle$ of the motion of the atom in the oscillator potential. Since the energy of the gamma ray and the energy of the final state are related by conservation of energy

$$E_\gamma = E_0 + E_f - E_{n'} \tag{2.35}$$

and E_0 and $E_{n'}$ are constants, the moments of E and E_f are simply related.

The moments of the energy spectrum of the final state $|f\rangle$ are simply the expectation values of the powers of the Hamiltonian (2.15) in the state $|f\rangle$. Using eq. (2.22) we have

$$\langle E_f^m\rangle = \langle f|H^m|f\rangle = \langle i|\mathrm{e}^{ikx}H^m\mathrm{e}^{-ikx}|i\rangle \tag{2.36a}$$

or, using eq. (2.35), since E_0 and $E_{n'}$ are constants,

$$\langle E_\gamma^m\rangle = \langle f|(E_0 - H + E_{n'})^m|f\rangle = \langle i|\mathrm{e}^{ikx}(E_0 - H + E_{n'})^m\mathrm{e}^{-ikx}|i\rangle. \tag{2.36b}$$

Before evaluating eq. (2.36) we note that the operator e^{-ikx} is obviously unitary, since operator kx is Hermitean. The relation (2.22) between the initial and final state vectors can be considered as a unitary transformation. The expectation value of any operator in the final state $|f\rangle$ can be calculated by taking the expectation value in the initial state $|i\rangle$ and appropriately transforming the operator. From this point of view, eq. (2.36) expresses exactly this unitary transformation. The mean value of E_f^m in the final state can be calculated by either taking the expectation value of H^m in the final state or of the transformed operator $\mathrm{e}^{ikx}H^m\mathrm{e}^{-ikx}$ in the initial state. The evaluation of eq. (2.36) can be facilitated by noting some simple properties of this unitary transformation.

(1) Since e^{ikx} is a function only of x, it commutes with any function of x. Any function of x is therefore left unchanged by the transformation

$$e^{ikx} f(x) e^{-ikx} = f(x). \tag{2.37}$$

(2) The transform of the operator p is easily calculated.

$$\begin{aligned} e^{ikx} p e^{-ikx} &= e^{ikx}\{e^{-ikx} p + [p, e^{-ikx}]\} \\ &= e^{ikx}\left\{e^{-ikx} p + \frac{\hbar}{i} \frac{d}{dx}(e^{-ikx})\right\} \\ &= e^{ikx}\{e^{-ikx} p - \hbar k e^{-ikx}\} = p - \hbar k. \end{aligned} \tag{2.38}$$

The transform of any power of p is therefore

$$e^{ikx} p^m e^{-ikx} = (e^{ikx} p e^{-ikx})^m = (p - \hbar k)^m. \tag{2.39}$$

The transform of the Hamiltonian (2.15) is

$$e^{ikx} H e^{-ikx} = \frac{(p - \hbar k)^2}{2M} + \frac{M\omega^2}{2} x^2 = H - \frac{\hbar k p}{M} + R. \tag{2.40}$$

The transform of any power of the Hamiltonian is thus

$$e^{ikx} H^m e^{-ikx} = (e^{ikx} H e^{-ikx})^m = \left(H - \frac{\hbar k p}{M} + R\right)^m \tag{2.41a}$$

and since E_0 and $E_{n'}$ are constants

$$e^{ikx}(E_0 - H + E_{n'})^m e^{-ikx} = \left(E_0 - H + E_{n'} + \frac{\hbar k p}{M} - R\right)^m. \tag{2.41b}$$

Inserting the expressions (2.41) into eqs. (2.36) we have

$$\langle E_f^m \rangle = \langle i| \left(H - \frac{\hbar k p}{M} + R\right)^m |i\rangle \tag{2.42a}$$

$$\langle E_\gamma^m \rangle = \langle i| \left(E_0 - H + E_{n'} + \frac{\hbar k p}{M} - R\right)^m |i\rangle. \tag{2.42b}$$

The initial state $|i\rangle$ is an eigenfunction of the Hamiltonian H with an eigenvalue which we can call $E_i = E_{n'}$. The first and second moments, $m=1$ and $m=2$, are thus obtained immediately from eqs. (2.42). The operator H operating on the state $|i\rangle$ either from the left or from the right gives the eigenvalue E_i.

$$\langle E_f \rangle = E_i + R - \frac{\hbar k}{M} \langle i|p|i\rangle \tag{2.43a}$$

$$\langle E_f^2 \rangle = \langle i| \left(E_i + R - \frac{\hbar k p}{M} \right)^2 |i\rangle \tag{2.43b}$$

$$\langle E_\gamma \rangle = E_0 - R + \frac{\hbar k}{M} \langle i|p|i\rangle \tag{2.43c}$$

$$\langle E_\gamma^2 \rangle = \langle i| \left(E_0 - R + \frac{\hbar k p}{M} \right)^2 |i\rangle . \tag{2.43d}$$

For $m > 2$ we can write

$$\langle E_f^m \rangle = \langle i| \left(E_i - \frac{\hbar k p}{M} + R \right) \left(H - \frac{\hbar k p}{M} + R \right)^{m-2} \times \\ \times \left(E_i - \frac{\hbar k p}{M} + R \right) |i\rangle \tag{2.44a}$$

$$\langle E_\gamma^m \rangle = \langle i| \left(E_0 - R + \frac{\hbar k p}{M} \right) \left(E_0 - H + E_i + \frac{\hbar k p}{M} - R \right)^{m-2} \times \\ \times \left(E_0 - R + \frac{\hbar k p}{M} \right) |i\rangle . \tag{2.44b}$$

Further simplification is not straightforward since the remaining operators H no longer operate directly on the initial state $|i\rangle$. The operator H does not commute with the operator p and can be brought over to operate directly on the state $|i\rangle$ only at the price of introducing additional terms involving commutators. One might write this as

$$\langle E_\gamma^m \rangle = \langle i| \left(E_0 - R + \frac{\hbar k p}{M} \right)^m |i\rangle + \text{terms involving commutators.} \tag{2.45}$$

Let us now compare these results, eqs. (2.42), (2.43), (2.44), and (2.45) with corresponding expressions obtained for the case where the nucleus emitting the gamma ray was not bound, namely, eqs. (2.12) and (2.14). The operator p/M acting on the initial-state wave function $|i\rangle$ is just the initial-velocity operator v_i. Thus the only difference between the results for the bound atom and the results for the free atom are the terms involving commutators in eq. (2.45) or the non-commutativity of the operator H with the operator p

in eq. (2.44). These differences do not affect the first and second moments which are the same for the bound and unbound cases. They affect only the shape of the spectrum as indicated by the higher moments. Furthermore, since the terms involving commutators are proportional to $\hbar$, they should vanish in the classical limit $\hbar \to 0$; or when the atom is initially moving in a state which has a high quantum number so that the motion is approximately describable in classical terms, i.e., at high temperatures.

It is instructive to re-examine the spectrum shown in fig. 2.3 to see in a qualitative way how it must be modified at low temperatures in order to take the above considerations into account. We first note that the spectrum in fig. 2.3 extends beyond the value of E_0. The region above E_0 corresponds to the case where the gamma ray gains energy at the expense of the kinetic energy of the motion of the atom. This is clearly impossible if the atom is initially at zero temperature; i.e., it is moving in the state of lowest possible energy. In a purely classical picture there is no contradiction because the energy of the particle is initially zero at zero temperature. Its momentum and velocity are therefore certainly zero and there is no Doppler broadening of the spectrum which would then be a delta function at $E_0 - R$ with no region extending beyond E_0. The same would be true for the quantum-mechanical case if the atom were moving freely and its motion were described by a plane wave. In such a case, the lowest state would be the plane wave of zero wave vector, $k=0$, and infinite wavelength which also corresponds to zero velocity and zero kinetic energy. However, if the atom is bound in a potential such as a harmonic oscillator, the kinetic energy in its lowest state is no longer zero. There is a finite kinetic energy and a finite velocity distribution due to the zero-point oscillations in the lowest state.

This zero-point motion is a characteristic feature of bound quantum-mechanical systems. Although an atom moving with this zero-point motion and emitting a gamma ray constitutes a moving gamma ray source, it cannot produce a Doppler shift in the usual sense. Since it is already in its lowest state of oscillation *it cannot lose energy while emitting the gamma ray*. It can only gain energy. This is in contrast to any classical motion where it is always possible for the atom to emit a gamma ray with an energy greater than E_0, using either the wave or the particle picture for the photon. On both pictures, if the atom is moving with high enough velocity in the direction of emission of the photon, the photon energy will be increased on the wave picture by the Doppler effect; on the particle picture by the kinematics of energy and momentum conservation. It is clear that if the atom is moving with quantum-mechanical zero-point motion it is impossible to state that the photon is emitted *at the time when the atom is moving in the direction of emission of the photon*. Such a precise description of the zero-point motion of the atom is not allowed by the uncertainty principle, or saying it in another way, by the wave-like properties *of the atom*, which is a standing wave in the harmonic potential. The uncertainty principle also does not allow the further loss of energy by the motion of the atom. Its kinetic and potential energy determine the fluctuations in position and momentum and these are already the minimum allowed by the uncertainty principle.

However, although one would expect that Doppler shifts and Doppler broadening for an atom undergoing zero-point motion would be quite different from one moving classically, eqs. (2.43) show that the first and second moments of the gamma ray spectrum are just those for an atom moving classically with this velocity distribution. This may have led some theorists to miss discovering the Mössbauer effect. If one simply calculated these two moments and stopped at this point, one might conclude that indeed the spectrum emitted from a bound source moving with quantum-mechanical zero-point motion is also like that of fig. 2.3. It has the same centroid and it has the same width. However, even at this point, without calculating the higher moments one might see that the spectrum shape must be quite different because a spectrum emitted by a bound atom in its lowest state cannot extend beyond E_0. If one simply removes the area in fig. 2.3 to the right of E_0, one is clearly changing the first and second moments of the spectrum. The centroid has been moved to the left and the second moment has been made smaller. To restore these moments to their original values without adding any area to the spectrum to the right of E_0 requires a considerable modification of the spectrum. As we have seen, this occurs in practice by having the Mössbauer effect, that is, by having an appreciable portion of the spectrum concentrated as a spike at the energy E_0.

Note that the difference between the free and the bound case is described by the commutators discussed in relation to eqs. (2.44) and (2.45). The latter were derived without requiring the potential to have any specific form other than to be a function of x alone and not of p. These results are true for any potential, provided that there are no velocity-dependent forces, and the initial state $|i\rangle$ is an eigenvector of the Hamiltonian. If we let H be the free particle Hamiltonian, (2.7), we find that H commutes with p, all the terms involving commutators in eq. (2.45) are zero, and eq. (2.45) reduces to eq. (2.14).

2.7 Review of the original Mössbauer experiments

We can now understand the results first obtained by Mössbauer and see why they were so incomprehensible at the time. The shape of the gamma ray spectrum in his case was intermediate between those shown in figs. 2.4 and 2.5, and consisted of two parts: (1) a thermal distribution centered around $E_0 - R$, (2) a sharp spike at E_0. At that time the existence of the spike was not appreciated, and the experiments were interpreted in terms of the thermal distribution. At room temperature, the Debye–Waller factor is very small in this case, and the spike is indeed negligibly small. Thus considering only the thermal distribution gives an exceedingly good approximation.

As the temperature is lowered, the thermal fluctuations in the nuclear motion decrease. Decreasing the momentum fluctuations narrows the thermal distribution around $E_0 - R$ and reduces the portion of the spectrum in the neighborhood of E_0. Decreasing the fluctuations in position decreases the mean square radius $\langle x^2 \rangle$ and therefore increases the Debye–Waller factor. The intensity of the spike at E_0 thus increases and becomes appreciable at liquid nitrogen temperature.

In a resonance absorption experiment, both the emission and absorption spectrum must be considered. The absorption spectrum is a mirror image about the point E_0 of the emission spectrum from a source at the same temperature. Resonance absorption arises from two different contributions which behave differently as a function of temperature. The tails of the thermal distributions from the emission and absorption spectra have a small overlap near the energy E_0. This overlap *decreases* as the temperature of *either* the source *or* the absorber is lowered. The contribution from the spikes at E_0 in the emission and absorption spectra *increase* when the temperature is lowered. However, because of the exponential dependence of the Debye–Waller factor, an appreciable contribution is obtained only when the temperatures of *both* the source *and* absorber are sufficiently low to give appreciable Debye–Waller factors.

Mössbauer originally planned to determine the natural line width of this particular transition in ^{191}Ir by studying the temperature dependence of the resonance absorption. If this is assumed to be due entirely to the overlap of the thermal distributions, the only unknown parameter in the theoretical description is the natural line width. It can therefore be determined by measurements at several temperatures. Mössbauer found that lowering the temperature of *either* the source *or* the absorber reduced the resonance absorption, as expected from the reduction of the overlap of the thermal distributions. From these measurements he already had enough information to determine the line width. However, when he reduced the temperature of the source and the absorber simultaneously, he found that the resonance absorption *increased*, rather than decreasing even further. This result was completely incomprehensible on the picture of overlapping thermal distributions, but is easily understood when the contribution from the spike is taken into account. The existence of this spike had been pointed out many years earlier in a paper by Lamb on neutron absorption in crystals, but somehow nobody had paid any attention to it.

Once it was realized that the low temperature absorption could come from the overlap of the two narrow spikes in the emission and absorption spectra, Mössbauer saw that this hypothesis could easily be tested by giving the source or absorber a Doppler shift. If either spectrum could be shifted by an amount larger than the natural line width, the two spikes would no longer overlap and the increased absorption should disappear. The velocity required for the Doppler shift would be much smaller than that previously used to overcome the recoil energy loss R, since it need only produce a shift of the order of the natural line width. For the particular case of ^{191}Ir, this turned out to be of the order of a few cm/sec, a velocity easily obtained from a phonograph turn table.

Since Mössbauer's original discovery, the effect has been developed extensively and is now used for many applications.

2.8 The thermal shift

In the preceding treatment the nucleus has been treated as a point particle having mass M and moving under the influence of external forces in a harmonic oscillator potential. This oversimplified picture of the nucleus is justified to a very good approximation in most treatments of the Mössbauer effect. However with improved precision and sensitivity in Mössbauer experiments, effects are observed which cannot be described without a more accurate and detailed treatment of the properties of the nucleus. One such effect is the thermal shift, the change in the energy of the emitted gamma ray with temperature. This effect was discovered experimentally in experiments designed to measure the gravitational red shift and was predicted theoretically by a Cambridge undergraduate student, B. D. Josephson.

The thermal shift can be described with only a small modification of the previous treatment. The nucleus is still considered to be a point particle moving under the influence of external forces before and after emitting the gamma ray. However because of the relation $E = mc^2$ between mass and energy, the mass of the nucleus changes by an amount equal to E_γ/c^2 when the gamma ray is emitted. Although this change in mass is extremely small the precision of the Mössbauer experiments is sufficiently great to detect its effects.

Consider a nucleus moving in an external one-dimensional harmonic potential as discussed in section 2.3. The Hamiltonian is given by eq. (2.15)

$$H = \frac{p^2}{2M} + \tfrac{1}{2}M\omega^2 x^2. \tag{2.15}$$

Let M be the mass of the atom when the nucleus is in its ground state after emission of the gamma ray. The Hamiltonian (2.15) therefore describes the motion of the atom in the potential after the emission of the gamma ray. Before the emission of the gamma ray the mass of the atom is different and is given by

$$M' = M + \delta M \tag{2.46a}$$

where

$$\delta M = E_\gamma / c^2. \tag{2.46b}$$

This change in mass does not affect the forces binding the atom in the harmonic potential. The Hamiltonian describing the motion of the atom before the emission of the gamma ray is therefore obtained from eq. (2.15) by simply changing the mass term in the kinetic energy.

$$H' = \frac{p^2}{2M'} + \frac{M\omega^2}{2} x^2 \tag{2.47a}$$

$$= \frac{p^2}{2M'} + \frac{M'}{2}\left(\frac{M}{M'}\omega^2\right) = \frac{p}{2M'} + \frac{M'\omega'^2}{2} x^2 \tag{2.47b}$$

where

$$\omega' = \sqrt{\frac{M}{M'}}\,\omega. \tag{2.47c}$$

Since $\delta M/M$ is small we expand eq. (2.47c) to obtain

$$\omega' \approx \omega\left(1 - \frac{\delta M}{2M}\right) = \omega\left(1 - \frac{E_\gamma}{2Mc^2}\right). \tag{2.47d}$$

Let the atom be initially in the eigenstate $|n'\rangle$ of the modified Hamiltonian H', eq. (2.47a) with mass M'. If after emitting the gamma ray the atom is in the corresponding state of excitation $|n'\rangle$ of the original Hamiltonian (2.15) with mass M, the gamma ray would have the energy E_0 if the change in nuclear mass were neglected. The mass change produces a slight shift of the gamma ray energy. This can be calculated as follows: The energies of the initial and final states of motion in the oscillator potential are given by

$$E_i = (n' + \tfrac{1}{2})\hbar\omega' \tag{2.48a}$$

$$E_f = (n' + \tfrac{1}{2})\hbar\omega. \tag{2.48b}$$

The energy of the gamma ray is given by

$$E_\gamma = E_0 + E_i - E_f = E_0 + (n' + \tfrac{1}{2})\hbar(\omega' - \omega). \tag{2.49a}$$

Substituting eq. (2.47d) we obtain

$$E_\gamma \approx E_0 - \tfrac{1}{2}(n' + \tfrac{1}{2})\frac{\delta M}{M}\hbar\omega. \tag{2.49b}$$

Substituting eq. (2.46b) then gives

$$E_\gamma \approx E_0\left\{1 - \tfrac{1}{2}(n' + \tfrac{1}{2})\frac{Mc^2}{\hbar\omega}\right\}. \tag{2.49c}$$

The shift in the gamma ray energy thus depends upon the state of excitation $|n'\rangle$ of the motion of the atom in the oscillator potential.

The result (2.49) can also be obtained by the use of perturbation theory. The perturbation calculation is no easier than the exact one for an atom in an external harmonic oscillator potential. However it is of interest for more complicated cases, like that of a crystal, where the exact solution is much more difficult. We therefore perform a perturbation calculation for the external potential in order to have a result which can be extended to other cases.

Let us calculate the gamma ray energy to first order in δM in perturbation theory. We write the Hamiltonian H' as an unperturbed part plus a perturbation

$$H' = H + \left(\frac{1}{2M'} - \frac{1}{2M}\right)p^2. \tag{2.47e}$$

We are interested in calculating the difference in energy between the state $|n'\rangle$ of the Hamiltonian H, eq. (2.15), and the energy of the corresponding state of the Hamiltonian H', eq. (2.47). To first order in δM this is given by the first-order correction to the energy in the treatment of the Hamiltonian (2.47e) by perturbation theory, where the Hamiltonian H is taken as the unperturbed Hamiltonian. The result is

$$\varepsilon_1 = \left(\frac{1}{2M'} - \frac{1}{2M}\right)\langle n'|p^2|n'\rangle, \tag{2.50}$$

where the state $|n'\rangle$ is an eigenstate of the unperturbed Hamiltonian H.

To first order in M this can be rewritten

$$\varepsilon_1 \approx -\frac{1}{2M}\frac{\delta M}{M}\langle n'|p^2|n'\rangle$$

$$= -\frac{\delta M}{M}\langle n'|E_{\text{kin}}|n'\rangle = -\tfrac{1}{2}E_\gamma\langle n'|\frac{v^2}{c^2}|n'\rangle. \tag{2.51a}$$

Thus

$$E_\gamma = E_0\left(1 - \tfrac{1}{2}\langle n'|\frac{\langle v^2\rangle}{c^2}|n'\rangle\right) \tag{2.51b}$$

where E_{kin} is the kinetic energy, $E_{\text{kin}}=p^2/2M$ and v is the velocity $v=p/M$. The result (2.51) is clearly the same as the result (2.49b) obtained from the exact solution by neglecting the terms of higher than first order in δM. This is to be expected since perturbation theory should give a result as a power series in the perturbation and the first-order perturbation result should be exact to first order in δM.

The expression (2.51) has the form of a Doppler shift and is indeed exactly the contribution to the Doppler shift which is of second order in v/c. A simple derivation of this shift in classical relativistic mechanics is given by calculating the time dilatation effect. Consider a short time interval $\mathrm{d}t$ in the laboratory system. Then in the rest frame of the atom, moving with velocity v, the time for the same interval $\mathrm{d}t_0$ is related to $\mathrm{d}t$ by the relativistic time dilatation:

$$\mathrm{d}t = \frac{\mathrm{d}t_0}{(1 - v^2/c^2)^{\frac{1}{2}}} \approx \mathrm{d}t_0\left[1 + \frac{v^2}{2c^2}\right]. \tag{2.52}$$

Integrating this expression over a time interval equal to τ_0 in the rest system, we find the time interval in the laboratory system is

$$\tau = \int_0^{\tau_0}\mathrm{d}t_0\left[1 - \frac{v^2}{2c^2}\right] = \tau_0\left[1 + \frac{\langle v^2\rangle}{2c^2}\right] \tag{2.53}$$

where $\langle v^2\rangle$ is the average over the time interval. The ratio of frequencies in the two systems is the inverse of the ratio of time intervals. This result is the same as eq. (2.51b).

The observation of the thermal shift has been interpreted as experimental verification of the celebrated twin paradox in special relativity. The atom moving back and forth in a potential is like the twin going to the distant star and back. It radiates at a lower frequency than another atom (its twin) which is at rest in the laboratory, by just the factor (2.53). Since the frequency of a given nuclear transition defines a clock, the moving clock is actually running more slowly than the stationary one.

The shift of the gamma ray energy (2.51a) is proportional to the mean kinetic energy of the atom emitting the gamma ray and therefore depends upon the temperature of the system. For this reason it is sometimes called the thermal shift. Experimental detection of the thermal shift requires a source of gamma radiation and also a resonant absorber. The resonant energy in the absorber is also shifted because of the change in mass of the nucleus absorbing the gamma ray as it makes the transition to the excited state. Calculation of the shift in the resonance absorption energy is identical with the calculation above for the shift in the emission energy and gives exactly the same result, eqs. (2.49) and (2.51). Thus if both the source and the absorber are in the same state of excitation (i.e., they have the same mean kinetic energy) the shifts due to the mass change are the same in both source and absorber and resonant absorption occurs as if both shifts were completely absent. If the states of excitation of the source and absorber are different; i.e. if the two are at different temperatures, the shift in the source is not the same as the shift in the absorber and an effect is observed. Resonance absorption then cannot occur unless the difference between the two shifts is compensated by introducing an additional shift such as a Doppler shift.

In practical experiments, the temperatures of the source and the absorber are varied and the absorber is also moved to give a Doppler shift. The resonance absorption is then studied as a function of the source and the absorber temperatures and also as a function of the absorber velocity. The experimental results are consistent with the theoretical prediction (2.51). In a single monatomic harmonic crystal the kinetic energy of any given atom is half the total energy per atom in the crystal. Thus the variation with temperature of the thermal shift is related to the specific heat C_L of the lattice

$$\frac{dE_\gamma}{dT} = -C_L/\mu' c^2 \tag{2.54}$$

where μ' is the gram atomic weight of the lattice substance. The measurement of the thermal shift therefore provides a method for measuring the mean kinetic energy of an atom.

PROBLEMS

1. Estimate the feasibility of a Mössbauer experiment under the following conditions: The nuclei under consideration have a mass $A=100$. The binding forces in the crystals being considered are such that they are roughly equivalent to a harmonic oscillator potential with an energy $\hbar\omega$ equal to the thermal energy kT at room temperature. The experiment is performed at a temperature low compared to room temperature. The experiment will be feasible if the probability that the gamma ray is emitted with the energy E_0 required for resonance absorption is at least equal to 1%.
(a) For what range of gamma ray energies is the experiment feasible?
(b) Another experiment requires that the probability that the gamma ray is emitted with the energy E_0 be at least equal to 30%. For what range of gamma ray energies is such an experiment feasible?
(c) How are the answers to parts (a) and (b) affected if the mass of the nucleus is $A=50$? If it is $A=200$?

2. (a) Assuming that the atom is initially in the ground state of the harmonic oscillator potential at time $t=0$, calculate explicitly the probability that it will be in any state $|n'\rangle$ after emitting the gamma ray. Use the creation and annilation operators $a^\dagger$ and a to simplify the calculation.
(b) Using the results of part (a), verify that the total probability summed over all states $|n'\rangle$ is equal to unity and verify the sum rules for the first and second moments of the gamma ray spectrum (2.43c) and (2.43d).

3. As a simple model of the Mössbauer effect in an anisotropic crystal, consider the Mössbauer effect from an atom bound in an anisotropic three-dimensional harmonic oscillator potential.

$$H = \frac{p_x^2}{2M} + \frac{p_y^2}{2M} + \frac{p_z^2}{2M} + \tfrac{1}{2}M\omega_x^2 x^2 + \tfrac{1}{2}M\omega_y^2 y^2 + \tfrac{1}{2}M\omega_z^2 z^2.$$

The frequencies of oscillation in the three directions ω_x, ω_y, and ω_z are different from one other.
(a) Write down the equations analogous to eq. (2.24), (2.31b) and (2.33a) for the case of the three-dimensional anisotropic potentials.

(b) Using the results of problem 2, and assuming that $\omega_x < \omega_y < \omega_z < 1.2\omega_x$ give expressions for the intensities and energies of the first five lines of the gamma ray spectrum having an energy less than E_0. Assume that the atom is initially in the lowest state of the oscillator potential.
(c) Give intensity and energy of the first line in the gamma ray spectrum having an energy greater than E_0.

CHAPTER 3

THE MÖSSBAUER EFFECT IN A SOLID

3.1 General formulation

When a Mössbauer experiment is performed in the laboratory the nucleus emitting the gamma ray is not moving in an external harmonic oscillator potential, but in a crystal under the influence of the forces due to all the other atoms. The simplified treatment in sections 2.3–2.6 can easily be revised to treat the more realistic case of motion in a crystal. The oscillator Hamiltonian (2.15) must now be replaced by the Hamiltonian for the entire crystal. After this modification, many of the results of sections 2.3–2.6 apply directly to the case of a crystal. We need only interpret the state vectors as representing states of the entire crystal rather than states of a single atom moving in a harmonic oscillator potential.

All results follow from the basic equation (2.22) which relates the initial state of the system and the final state after the emission of the gamma ray. To apply this equation to the case of a crystal, we let $|i\rangle$ and $|f\rangle$ represent the initial and final states of the whole crystal and define more precisely the coordinate x to specify that it is the coordinate of the particular nucleus which emits the gamma ray and not of some other nucleus in the crystal. Let $\boldsymbol{x}_i$ denote the coordinate of any atom i in the crystal. Then $i = 1, 2, \ldots, N$, where N is the number of atoms in the crystal. Let $\boldsymbol{x}_{\mathrm{L}}$ represent the coordinate of the particular nucleus which emits the gamma ray. The coordinates $\boldsymbol{x}_i$ and $\boldsymbol{x}_{\mathrm{L}}$ are vectors in three-dimensional space. The particular component of the vector $\boldsymbol{x}_{\mathrm{L}}$ appearing in the relation (2.22) should be the component in the direction of emission of the gamma ray. Thus the basic equation for the case of the crystal is

$$|f\rangle = \mathrm{e}^{-i\boldsymbol{k}\cdot\boldsymbol{x}_{\mathrm{L}}}|i\rangle. \tag{3.1}$$

The derivation of eq. (3.1) for the case of the crystal follows exactly the derivation of eq. (2.22) for the case of the harmonic oscillator potential. We first extend the concept of expanding the initial state vector $|i\rangle$ in plane waves. The state vector $|\boldsymbol{k}'\rangle$ completely specifies the state of a one-particle system as being a plane wave with wave vector $\boldsymbol{k}'$. We now need a complete specification for states of a crystal with N atoms and $3N$ degrees of freedom. A complete set of commuting observables describing this system consists of $3N$ operators. A complete specification of a state requires $3N$ eigenvalues. For our plane wave expansion we wish $\boldsymbol{p}_L$, the momentum of the nucleus emitting the gamma ray, to be among our complete set of commuting observables. We do not make a specific choice of the other variables as they affect only the motion of the other atoms in the crystal and the details of their motion are not relevant for our purposes. Let $|\boldsymbol{k}'_L, \alpha'\rangle$ define a complete set of states in which the motion of the atom $\boldsymbol{x}_L$ is described by a plane wave of momentum $\hbar\boldsymbol{k}_L$ and the rest of the atoms in the crystal are in a state denoted by α', where α' represents the eigenvalues of some set of $3N-3$ other operators chosen to give a complete specification of the crystal.

We now write the expansion of the initial state in this complete set of states

$$|i\rangle = \sum_{\boldsymbol{k}_L'} \sum_{\alpha'} |\boldsymbol{k}'_L, \alpha'\rangle\langle\boldsymbol{k}'_L, \alpha'|i\rangle. \tag{3.2}$$

Since $\boldsymbol{k}'_L$ is a vector in three-dimensional space it represents 3 quantum numbers. Summations on the index $\boldsymbol{k}'_L$ are written as simple sums, but really represent integrations over a three-dimensional $\boldsymbol{k}$-space. The summation over the index α' means summing over all sets of values of the complete set of quantum numbers represented by α'.

Let us write the Schrödinger wave function for a particular state $|\boldsymbol{k}'_L, \alpha'\rangle$ to illustrate some features of a many-body wave function.

$$\psi_{\boldsymbol{k}'_L,\alpha'} = e^{i\boldsymbol{k}'_L\cdot\boldsymbol{x}_L}\,\phi_{\alpha'}(\boldsymbol{x}_1, \boldsymbol{x}_2, \ldots, \boldsymbol{x}_{i\neq L}, \ldots, \boldsymbol{x}_N) \tag{3.3}$$

where the function $\phi_{\alpha'}$ depends on the coordinates $\boldsymbol{x}_i$ of all the N atoms in the crystal except the atom $\boldsymbol{x}_L$. Our complete set of states $|\boldsymbol{k}'_L, \alpha'\rangle$ has been chosen to separate the motion of the atom $\boldsymbol{x}_L$ from the motion of the other atoms. The Schrödinger wave function is written as a product of two factors, one depending *only* on the coordinate $\boldsymbol{x}_L$ and the other depending on all the other coordinates *except* $\boldsymbol{x}_L$.

The eigenfunctions of the Hamiltonian are certainly *not* of this form (3.3). The motion of the atom $\boldsymbol{x}_L$ is strongly coupled to the motion of all the other atoms, while the product form (3.3) describes a state in which the motion of the atom $\boldsymbol{x}_L$ is independent of the motion of the other atoms. However, since the set of separated states (3.3) forms a complete set of states, any arbitrary initial state can be expressed as a linear combination 3.2) of such states.

We now apply momentum conservation separately to each plane-wave component in the expansion (3.2) by analogy with the case of the single atom in the harmonic oscillator potential. A plane-wave component $|\boldsymbol{k}'_{\mathrm{L}}, \alpha'\rangle$ has its wave vector shifted by the amount $\boldsymbol{k}$ after emitting the gamma ray to become $\boldsymbol{k}'_{\mathrm{L}} - \boldsymbol{k}$, while the motion of the other atoms is left unchanged and they remain in the state α'. The state vector for the final state analogous to (2.19) is thus

$$|f\rangle = \sum_{\boldsymbol{k}'_{\mathrm{L}}} \sum_{\alpha'} |(\boldsymbol{k}'_{\mathrm{L}} - \boldsymbol{k}), \alpha'\rangle\langle \boldsymbol{k}'_{\mathrm{L}}, \alpha'|i\rangle. \tag{3.4}$$

The simplification analogous to eq. (2.20) gives

$$\mathrm{e}^{-\mathrm{i}\boldsymbol{k}\cdot\boldsymbol{x}_{\mathrm{L}}} |\boldsymbol{k}'_{\mathrm{L}}, \alpha'\rangle = |(\boldsymbol{k}'_{\mathrm{L}} - \boldsymbol{k}), \alpha'\rangle. \tag{3.5}$$

The values of the quantum numbers α' are not changed by the operation with $\mathrm{e}^{-\mathrm{i}\boldsymbol{k}\cdot\boldsymbol{x}_{\mathrm{L}}}$ as indicated by the separated product form of the wave function (3.3). Rewriting eq. (3.4) using eq. (3.5) we obtain the analog of eq. (2.21)

$$|f\rangle = \sum_{\boldsymbol{k}'_{\mathrm{L}}} \sum_{\alpha'} \mathrm{e}^{-\mathrm{i}\boldsymbol{k}\cdot\boldsymbol{x}_{\mathrm{L}}} |\boldsymbol{k}'_{\mathrm{L}}, \alpha'\rangle\langle \boldsymbol{k}'_{\mathrm{L}}, \alpha'|i\rangle. \tag{3.6}$$

The summation over the indices $\boldsymbol{k}'_{\mathrm{L}}$ and α' is now performed directly by closure to obtain eq. (3.1). We have not assumed that the initial state $|i\rangle$ was any particular state such as an eigenfunction of a particular Hamiltonian. The result (3.1) thus holds for an arbitrary initial state.

We now determine the gamma ray energy spectrum by use of energy conservation. To calculate the energy transferred to the crystal or the energy of the emitted gamma ray, we need the energies of the initial and the final states. Since these energies are well defined only for eigenfunctions of the Hamiltonian, we choose the initial state to be an eigenfunction of the Hamiltonian of the whole crystal. Let $|n'\rangle$ denote an eigenvector of the Hamiltonian of the whole crystal with energy $E_{n'}$. The letter n' denotes the complete set of quantum numbers required to specify an eigenvector of the Hamiltonian; i.e., the eigenvalues of a complete set of commuting operators which commute with the Hamiltonian of the whole crystal.* With this new definition of the state vector $|n'\rangle$ we can now expand the final state $|f\rangle$ in the set $|n'\rangle$ of eigenvectors of the Hamiltonian in order to obtain the form of the gamma ray energy spectrum.

We can follow the treatment for the single particle in the oscillator potential, eqs. (2.23) and (2.24), to obtain the relation for the expansion of the final state $|f\rangle$ and for the probability that the crystal is in the state

* An example of such a set will be considered in section 3.2.

$|n''\rangle$ after the emission of the gamma ray, if it is initially in the state $|n'\rangle$:

$$|f\rangle = \sum_{n''} |n''\rangle\langle n''|f\rangle \tag{3.7a}$$

$$= \sum_{n''} |n''\rangle\langle n''|\mathrm{e}^{-\mathrm{i}\boldsymbol{k}\cdot\boldsymbol{x}_\mathrm{L}}|i\rangle \tag{3.7b}$$

$$= \sum_{n''} |n''\rangle\langle n''|\mathrm{e}^{-\mathrm{i}\boldsymbol{k}\cdot\boldsymbol{x}_\mathrm{L}}|n'\rangle \tag{3.7c}$$

$$P_{n'\to n''} = |\langle n''|\mathrm{e}^{-\mathrm{i}\boldsymbol{k}\cdot\boldsymbol{x}_\mathrm{L}}|n'\rangle|^2. \tag{3.8}$$

The energy of the gamma ray corresponding to the transition of the crystal from the state $|n'\rangle$ to the state $|n''\rangle$ is obtained from conservation of energy with the energy eigenvalues for the whole crystal. Thus by analogy to eq. (2.25)

$$E_\gamma = E_0 - (E_{n''} - E_{n'}). \tag{3.9}$$

Since the value of E_γ depends on the eigenvalues $E_{n''}$ and $E_{n'}$ of the Hamiltonian for the entire crystal, we cannot find a more detailed expression for E_γ without first assuming some model for the crystal and calculating its energy eigenvalues.

The probability that the gamma ray has exactly the energy E_0 is again given by setting $n''=n'$. This can here also be expressed in terms of the Fourier transform of the probability density function $\varrho_{n'}(\boldsymbol{x}_\mathrm{L})$. However, the probability density function for the atom $\boldsymbol{x}_\mathrm{L}$ must be expressed in terms of the wave functions for the whole crystal. Consider the Schrödinger wave function for the state n' of the whole crystal, $\psi_{n'}(\boldsymbol{x}_1, \boldsymbol{x}_2, \ldots, \boldsymbol{x}_N)$. The square of the wave function represents the probability of finding each atom in the crystal in an elementary volume of a $3N$-dimensional space around the point $(\boldsymbol{x}_1, \boldsymbol{x}_2, \ldots, \boldsymbol{x}_N)$. The probability density for finding the atom emitting the gamma ray at the point $\boldsymbol{x}_\mathrm{L}$ is found by averaging the $3N$-dimensional probability density over the positions of all the other particles.

$$\varrho_{n'}(\boldsymbol{x}_\mathrm{L}) = \int \psi^*_{n'}(\boldsymbol{x}_1, \ldots, \boldsymbol{x}_N)\psi(\boldsymbol{x}_1, \ldots, \boldsymbol{x}_N)\mathrm{d}\boldsymbol{x}_1, \ldots, \mathrm{d}\boldsymbol{x}_i, \ldots \mathrm{d}\boldsymbol{x}_N \quad (i \neq \mathrm{L}). \tag{3.10}$$

With this definition of the probability density we obtain the result analogous to eq. (2.28) that $P_{n'\to n'}$ is the Fourier transform of the density,

$$P_{n'\to n'} = |\textstyle\int \varrho_n(\boldsymbol{x}_\mathrm{L})\mathrm{e}^{-\mathrm{i}\boldsymbol{k}\cdot\boldsymbol{x}_\mathrm{L}}\, \mathrm{d}\boldsymbol{x}_\mathrm{L}|^2. \tag{3.11}$$

The expansion for small k and the particular case of the gaussian density function are exactly analogous to eqs. (2.30) and (2.31).

$$P_{n'\to n'} \approx 1 - \langle \boldsymbol{k}\cdot\boldsymbol{x}_\mathrm{L}\rangle^2 \quad \text{for small } k \tag{3.12a}$$

$$P_{n'\to n'} = \mathrm{e}^{-\langle(\boldsymbol{k}\cdot\boldsymbol{x}_\mathrm{L})^2\rangle} \quad \text{for gaussian density} \tag{3.12b}$$

where

$$\langle(\boldsymbol{k}\cdot\boldsymbol{x}_{\mathrm{L}})^2\rangle = \int \varrho_{n'}(\boldsymbol{x}_{\mathrm{L}})(\boldsymbol{k}\cdot\boldsymbol{x}_{\mathrm{L}})^2\,\mathrm{d}\boldsymbol{x}_{\mathrm{L}}. \tag{3.12c}$$

All the results of section 2.6 for sum rules and moments of the energy spectrum apply equally well to the case of gamma ray emission from an atom bound in a crystal. Eqs. (2.34)–(2.45) can be applied directly by letting all of the state vectors represent states of the entire crystal rather than of a single atom in an external potential, letting H be the crystal Hamiltonian and replacing the coordinate x by the vector coordinate $\boldsymbol{x}_{\mathrm{L}}$, and kx by $\boldsymbol{k}\cdot\boldsymbol{x}_{\mathrm{L}}$. The only property of the Hamiltonian used to obtain the results for the moments of the energy spectrum is eq. (2.40). This can be rewritten for the case of the crystal as

$$\mathrm{e}^{\mathrm{i}\boldsymbol{k}\cdot\boldsymbol{x}_{\mathrm{L}}} H \mathrm{e}^{-\mathrm{i}\boldsymbol{k}\cdot\boldsymbol{x}_{\mathrm{L}}} = H - \frac{\hbar\boldsymbol{k}\cdot\boldsymbol{p}}{M} + R. \tag{3.13}$$

Eq. (3.13) is easily seen to be valid for any Hamiltonian having the form

$$H = \frac{\boldsymbol{p}_{\mathrm{L}}^2}{2M} + \text{terms which commute with } \boldsymbol{x}_{\mathrm{L}}. \tag{3.14}$$

Eq. (3.14) states that the kinetic energy of the atom $\boldsymbol{x}_{\mathrm{L}}$ is the only term in the Hamiltonian which does not commute with the coordinate $\boldsymbol{x}_{\mathrm{L}}$. In practical crystals the Hamiltonian has the form (3.14), since the other terms are the kinetic energies of the other particles which commute with $\boldsymbol{x}_{\mathrm{L}}$, and the potentials of the interatomic forces. If the latter depend only on the positions of the atoms and not on their velocities, they also commute with $\boldsymbol{x}_{\mathrm{L}}$. In practical crystals velocity-dependent forces are negligible. Thus eq. (3.13) is valid for crystals and this together with eq. (3.1) is enough to allow all the results (2.42) to (2.45) for the moments of the energy spectrum to be obtained.

3.2 The Mössbauer effect in a harmonic crystal

To obtain more detailed results for the Mössbauer effect in a crystal we must consider the many-particle dynamics of the crystal itself. We are thus faced with a many-body problem in quantum mechanics. The many-body problem in both quantum and classical mechanics is complicated because it requires the simultaneous treatment of a large number of degrees of freedom, all coupled to one another. One method of attacking the many-body problem is to separate the degrees of freedom; i.e., to look for a suitable transformation of variables so that the different degrees of freedom separate at least approximately after the transformation. One example is the transformation to normal coordinates in a system undergoing small vibrations about equilibrium. If such a transformation separating the variables can be found, it is useful in both classical and quantum mechanics.

One common approximation in treating crystal dynamics is the harmonic approximation in which the forces between the atoms in the crystal are assumed to be harmonic. The potential between two atoms is expanded in powers of the distance between the atoms and all terms of higher order than the quadratic term are neglected. The crystal dynamics then reduces to the well-known case of a system undergoing small vibrations and can be treated by the usual method of normal modes. After the transformation to the normal modes, the Hamiltonian reduces to a set of independent harmonic oscillators.

Consider the Mössbauer effect in a crystal whose dynamics are described in the harmonic approximation. Let all of the atoms in the crystal have the same mass M. The transformation to normal coordinates is a simple orthogonal transformation, and the Hamiltonian expressed in terms of the normal mode variables is simply the sum of a set of harmonic oscillators.

$$H = \sum_{s=1}^{3N} \frac{\pi_s^2}{2M} + \frac{M\omega_s^2 \xi_s^2}{2}, \tag{3.15}$$

where ξ_s, π_s and ω_s are the coordinates, momenta and angular frequencies of the normal modes. The eigenstates of this Hamiltonian can be represented by ket vectors having the form $|\{n_s'\}\rangle$ where $\{n_s'\}$ represents a set of $3N$ numbers describing the state of excitation of each normal mode.

To calculate the Mössbauer effect, we must evaluate expressions like eq. (3.8). The coordinate $\boldsymbol{x}_\mathrm{L}$ is easily expressed in terms of the normal coordinate variables ξ_s. Only the component of the vector $\boldsymbol{x}_\mathrm{L}$ in the direction of the wave vector $\boldsymbol{k}$ of the gamma ray enters in the calculation. Let us therefore write $x_{\mathrm{L}k}$ for this component of x_L so that

$$\boldsymbol{k} \cdot \boldsymbol{x}_\mathrm{L} = k x_{\mathrm{L}k}. \tag{3.16}$$

We now express $x_{\mathrm{L}k}$ as a linear combination of the normal-mode variables

$$x_{\mathrm{L}k} = \sum_{s=1}^{3N} a_{\mathrm{L}s} \xi_s \tag{3.17}$$

where the coefficients $a_{\mathrm{L}s}$ satisfy the orthonormality relation

$$\sum_{s=1}^{3N} a_{\mathrm{L}s}^2 = 1. \tag{3.18}$$

Substituting the expansion (3.17) into eq. (3.8) we obtain

$$P_{\{n'_s\}\to\{n''_s\}} = |\langle\{n''_s\}|\exp(-ik \sum_{s=1}^{3N} a_{Ls}\xi_s)|\{n'_s\}\rangle|^2 \tag{3.19a}$$

$$= |\langle\{n''_s\}| \prod_{s=1}^{3N} e^{-ika_{Ls}\xi_s}|\{n'_s\}\rangle|^2. \tag{3.19b}$$

The matrix element of the operator product (3.19b) can be evaluated by considering each normal mode individually. Since these modes are all independent, the state $|\{n'_s\}\rangle$ can be considered as the direct product of $3N$ individual single normal-mode states. The Schrödinger wave function for this state expressed in terms of the normal coordinates ξ_s is simply a product of harmonic oscillator wave functions.

$$|\{n'_s\}\rangle = \prod_{s=1}^{3N} \psi_{n'_s}(\xi_s) = \prod_{s=1}^{3N} |n'_s\rangle \tag{3.20}$$

where $|n'_s\rangle$ represents a state of a single harmonic oscillator having frequency ω_s. The expression (3.19b) can therefore be expressed as a product of matrix elements each acting on a single normal-mode state.

$$P_{\{n'_s\}\to\{n''_s\}} = \prod_{s=1}^{3N} \left|\langle n''_s| e^{-ika_{Ls}\xi_s}|n'_s\rangle\right|. \tag{3.21}$$

Elegant methods for evaluating such expressions involving exponentials of harmonic oscillator operators are given in section 3.5. At this point we present a very simple approximate method. In a macroscopic crystal N is of the order of 10^{23} and is very large. If all of the constants a_{Ls} are approximately of the same order of magnitude, each one must be of order $(3N)^{-\frac{1}{2}}$ since there are $3N$ of them and the sum of their squares is unity by eq. (3.18). We therefore expand the exponentials in eq. (3.21) to third order in a_{Ls}; neglecting a_{Ls}^4 which is of order $(3N)^{-2}$.

$$P_{\{n'_s\}\to\{n''_s\}} = \prod_{s=1}^{3N} \left|\langle n''_s|1 - ika_{Ls}\xi_s - \tfrac{1}{2}k^2a_{Ls}^2\xi_s^2 + \tfrac{1}{6}ik^3a_{Ls}^3\xi^3|n'_s\rangle\right|^2$$
$$+ O(3N)^{-1}. \tag{3.22}$$

The error is of order $(3N)^{-1}$ since there are $3N$ terms neglected, each of order $(3N)^{-2}$.

Consider the case where the initial and final states are the same, $\{n'_s\}=\{n''_s\}$, and the gamma ray is emitted with the full energy E_0. For this case the matrix elements appearing in eq. (3.22) are diagonal matrix elements; i.e., expectation values in harmonic oscillator eigenstates. The expectation values of odd powers of ξ_s vanish and we obtain

$$P_{\{n's\}\to\{n's\}} = \prod_{s=1}^{3N} \left|1 - \tfrac{1}{2}k^2 a_{Ls}^2 \langle n_s'|\xi_s^2|n_s'\rangle\right|^2 + O(3N)^{-1}. \tag{3.23}$$

This result (3.23) is expressed as the product of a large number of factors, each of which is very close to unity. Such a product is conveniently expressed to a very good approximation by an exponential.

$$P_{\{n's\}\to\{n's\}} \approx \left|\exp(-\tfrac{1}{2}k^2 \sum_{s=1}^{3N} a_{Ls}^2 \langle n_s'|\xi_s^2|n_s'\rangle)\right|^2$$
$$\approx \exp(-k^2 \sum_{s=1}^{3N} a_{Ls}^2 \langle n_s'|\xi_s^2|n_s'\rangle). \tag{3.24}$$

The validity of this approximation can be tested by writing the first few terms of the expansion of (3.23) in the small quantities a_{Ls}^2 which are of order $(3N)^{-1}$

$$P_{\{n's\}\to\{n's\}} = \Big|1 - \tfrac{1}{2}k^2 \sum_{s=1}^{3N} a_{Ls}^2 \langle n_s'|\xi_s^2|n_s'\rangle$$
$$+ \tfrac{1}{8}k^4 \sum_{\substack{t=1\\ t\neq s}}^{3N} \sum_{s=1}^{3N} a_{Lt}^2 a_{Ls}^2 \langle n_t'|\xi_t^2|n_t'\rangle \langle n_s'|\xi_s^2|n_s'\rangle + O(3N)^{-1}\Big|^2. \tag{3.25}$$

The first two terms of the expansion (3.24) are identical to the first two terms of the expansion of the exponential (3.25). The third term of the expansion (3.24) differs from the corresponding term in the expansion of the exponential (3.25) only by the exclusion in the double sum of (3.25) of the term with $t=s$, which would appear in the expansion of the exponential (3.25). The error introduced by the approximation of 'exponentiating' the product (3.24) is thus of order $(3N)^{-1}$, since each term in the double sum is of order $(3N)^{-2}$, and there are $3N$ terms with $t=s$.

The approximation (3.24) is very good since $(3N)^{-1}$ is very small. This exponential technique works very well whenever there are a large number of factors in a product and each of them is very close to unity. We also note by inspection of eq. (3.24) that the approximation is particularly good in the cases of physical interest for the Mössbauer effect, namely, when the result (3.24) is sufficiently large so that there is a measureable probability of the Mössbauer effect. Let us say that the probability is at least 0.1 %. For this case, the exponent in (3.24) is certainly less than 10 and therefore the individual terms in the sum must be very small.

3.3 The Debye–Waller factor and Mössbauer intensity for a harmonic crystal

From eq. (3.24) we can derive the expression (3.12b) for the Debye–Waller factor on a much more general basis than was used in the preceding simple derivation for a single particle in a harmonic oscillator ground state. From eq. (3.17) we can express the mean square fluctuation in x_{Lk} in terms of the mean square fluctuations of the normal coordinates.

$$\langle\{n'_s\}|x^2_{Lk}|\{n'_s\}\rangle = \sum_{s=1}^{3N}\sum_{t=1}^{3N}\langle n'_s|a_{Ls}a_{Lt}\xi_s\xi_t|n'_s\rangle$$
$$= \sum_{s=1}^{3N} a^2_{Ls}\langle n'_s|\xi^2_s|n'_s\rangle. \tag{3.26}$$

Substituting eq. (3.26) into eq. (3.24) and using eq. (3.16) we obtain

$$P_{\{n's\}\to\{n's\}} \approx \exp(-k^2\langle\{n'_s\}|x^2_{Lk}|\{n'_s\}\rangle$$
$$\approx \exp(-\langle\{n'_s\}|(\boldsymbol{k}\cdot\boldsymbol{x}_L)^2|\{n'_s\}\rangle). \tag{3.27}$$

This result (3.27) is the same as the expression (3.12b). This derivation shows that it is valid in any state $\{n'_s\}$ of a crystal lattice, if the following two approximations are valid: (1) The harmonic approximation which leads to the existence of independent normal modes. (2) The assumption that the displacement $\boldsymbol{x}_L$ of the atom emitting the gamma ray is the sum of many small contributions from a large number of independent normal modes.

The previous derivation of this result eq. (3.12b) assumed a gaussian probability density function. The present derivation also implies a gaussian density. Eq. (3.11) shows that the expression (3.27) is also the Fourier transform of the probability density distribution $\varrho_{n'}(\boldsymbol{x}_L)$. Since (3.27) is a gaussian function of $\boldsymbol{k}$, its Fourier transform is also gaussian. That this distribution should be gaussian is reasonable since the contributions of the individual normal modes all add at random to form the displacement $\boldsymbol{x}_L$. It is not surprising that this 'random walk' with an extremely large number of steps should give a normal or gaussian distribution, even though the probability distributions for each of the individual normal coordinates may not be gaussian.

Let us evaluate explicitly the matrix elements appearing in the expression (3.24). From the properties of the harmonic oscillator

$$k^2\langle n'_s|\xi^2_s|n'_s\rangle = k^2\,\frac{\hbar(n'_s+\frac{1}{2})}{M\omega_s} = \frac{R}{\hbar\omega_s}(2n'_s+1) \tag{3.28}$$

where R is the free recoil energy. Substituting eq. (3.28) into eq. (3.24) we obtain

$$P_{\{n'_s\}\to\{n'_s\}} \approx \exp\left\{-R \sum_{s=1}^{3N} \frac{(a_{Ls})^2}{\hbar\omega_s}(2n'_s + 1)\right\}. \tag{3.29}$$

This result (3.29) is a natural generalization of (2.33a) for the ground state of a single harmonic oscillator. For the ground state of the crystal ($n'_s=0$ for all s) eq. (3.29) differs from eq. (2.33a) simply by replacing the factor $1/\omega$ by an appropriate mean of $1/\omega_s$ over all the normal modes of the crystal. The weighting factors $(a_{Ls})^2$ express the importance of each normal mode in the motion of the atom $\boldsymbol{x}_L$. The effect of the excitation for the case where the crystal is not in its ground state is to introduce the factors $(2n'+1)$ in each term in the sum.

The result (3.29) gives the probability that the gamma ray has the exact energy E_0 if the crystal is initially in a particular excited state. In a practical experiment the crystal is not in any particular known excited state but is in some statistical mixture, such as that of thermal equilibrium with a heat reservoir at a temperature T. The result (3.29) should therefore be averaged over all possible initial states with a suitable statistical factor, e.g. the Boltzmann factor giving the probability of finding the system in that state at the temperature T. A detailed analysis of the Mössbauer effect in systems in thermal equilibrium at a given temperature is given in section 3.5. Here we assume that eq. (3.39) can be applied to the case of a crystal in thermal equilibrium by simply taking a thermal average of the expressions (3.23) over the Boltzmann distribution. This has the effect of replacing $\langle n'_s|\xi_s^2|n'_s\rangle$ by its thermal average at temperature T in eq. (3.23) and in all the subsequent derivation, including eq. (3.27):

$$\langle n'_s|\xi_s^2|n'_s\rangle \to \langle\xi_s^2\rangle_T = \frac{\hbar}{M\omega_s}(\langle n'_s\rangle_T + \tfrac{1}{2}) = \frac{\hbar}{2M\omega_s}\coth\left(\frac{\hbar\omega_s}{2kT}\right) \tag{3.30a}$$

where we have used the well-known expression for the mean excitation of a harmonic oscillator in thermal equilibrium,

$$\langle n'_s\rangle_T = \frac{1}{e^{\hbar\omega_s/kT} - 1}. \tag{3.30b}$$

Inserting (3.30a) into eq. (3.27) we obtain the probability that the gamma ray is emitted with the full energy E_0 when the crystal is in thermal equilibrium at temperature T

$$P(E_0, T) = \exp - \langle(\boldsymbol{k}\cdot\boldsymbol{x}_L)^2\rangle_T \tag{3.31}$$

$$P(E_0, T) = \exp\left\{-R \sum_{s=1}^{3N} \frac{(a_{Ls})^2}{\hbar\omega_s} \coth \frac{\hbar\omega_s}{2kT}\right\}. \tag{3.32}$$

The application of this result (3.32) to a particular crystal requires the knowledge of the frequency of the normal modes and of the expansion coefficients a_{Ls}. Before assuming any specific model we can note some general features of the result (3.32). At zero temperature and at temperatures low compared to the quantum energy of most of the relevant normal modes, the hyperbolic cotangent is approximately equal to unity. The exponent appearing in (3.32) is then the ratio of the free recoil energy R to a characteristic frequency of the set of normal modes, namely, a harmonic mean weighted by the coefficients $(a_{Ls})^2$. The quantity $P(E_0, T)$ decreases with increasing temperature. Since the function is exponential, it decreases very rapidly once it becomes appreciably less than unity.

The Debye model for a crystal gives a reasonably good description of many bulk properties of crystal lattices. In this model, the expression (3.32) is evaluated by replacing the sum over the normal modes by an integral over a continuum. The result is expressed in terms of a single parameter, the Debye temperature θ. The evaluation of the expression (3.44) for the Debye model at zero temperature gives the result

$$P(E_0, T)_{\text{Debye}} = \exp\left(-\frac{3R}{2k\theta}\right). \tag{3.33}$$

This shows that the Mössbauer intensity at zero temperature is large when the free recoil energy is smaller than $k\theta$ and goes to zero rapidly when R becomes greater than $k\theta$. Since $k\theta$ is a characteristic lattice vibration energy for the crystal, this is in qualitative agreement with the previous result for a harmonic oscillator.

3.4 Statistical mixtures vs. pure states

We now consider in detail the Mössbauer effect from an initial state which is a statistical mixture rather than an eigenstate of the crystal. For simplicity in notation we consider again the case of a one-dimensional harmonic scillator. The generalization to a crystal is straightforward.

Let $p_{n'}^{(i)}$ be the probability in the initial state of finding the atom in any state $|n'\rangle$ in the oscillator potential. The probability that the atom will be in the state $|n''\rangle$ after emitting the gamma ray if it is initially in the state $|n'\rangle$ is given by eq. (2.24)

$$P_{n' \to n''} = |\langle n''|\mathrm{e}^{-ikx}|n'\rangle|^2. \tag{2.24}$$

The probability that it will be in the state $|n''\rangle$ after the transition is then

$$p_{n''}^{(f)} = \sum_{n'} P_{n' \to n''} p_{n'}^{(i)} = \sum_{n'} \left|\langle n''| e^{-ikx} |n'\rangle\right|^2 p_{n'}^{(i)}. \tag{3.34}$$

Eq. (3.34) is obtained by combining probabilities in the classical manner because the initial distribution is a classical statistical mixture. The reason we can only specify a probability distribution instead of specifying the initial state exactly is only our incomplete knowledge of the system. We could in principle measure the initial state exactly without disturbing the system. The reason we do not know the exact state is a practical experimental one and not one which is related to the fundamental laws of quantum mechanics.

A completely different situation occurs when the system is initially in some single pure quantum state which is not an eigenstate of the oscillator Hamiltonian. We can then expand this initial state $|a\rangle$ in the eigenfunctions $|n'\rangle$ of the harmonic oscillator

$$|a\rangle = \sum_{n'} |n'\rangle\langle n'|a\rangle. \tag{3.35}$$

The probability that the system in the state $|a\rangle$ is in the state $|n'\rangle$ of the oscillator is

$$p_{n'}^{(a)} = \left|\langle n'|a\rangle\right|^2. \tag{3.36}$$

This expression (3.36) depends only on the absolute magnitudes of the coefficients in the expansion (3.35). Thus, if we know that the system is initially in the state $|a\rangle$, eq. (3.35), we have more information about the state than just the probability distribution (3.36). We also know the phases of the expansion coefficients $\langle n'|a\rangle$. The initial state $|a\rangle$ is a coherent superposition of the different oscillator eigenstates $|n'\rangle$ and interference effects can be observed between the different components in the expansion. The statistical distribution created by eq. (3.34) is an incoherent mixture of different harmonic oscillator eigenstates $|n'\rangle$. It is not a pure quantum state and interference effects cannot be observed between the different components.

Let us now calculate the probability that the system will be in the state $|n''\rangle$ after emitting the gamma ray if it is initially in the pure quantum state $|a\rangle$ before emitting the gamma ray. This is given using eq. (2.22) or (2.24), by

$$P_{a \to n''} = \left|\langle n''| e^{-ikx} |a\rangle\right|^2 \tag{3.37a}$$

$$= \left|\sum_{n'} \langle n''| e^{-ikx} |n'\rangle\langle n'|a\rangle\right|^2 \tag{3.37b}$$

$$= \sum_{n'n'''} \langle n''| e^{-ikx} |n'\rangle\langle n'''| e^{ikx} |n''\rangle\langle n'|a\rangle\langle a|n'''\rangle. \tag{3.37c}$$

Eq. (3.37) for the initial pure quantum state is similar to eq. (3.34) for the incoherent statistical mixture but there is one important difference. In eq. (3.37b) the sum over the state $|n'\rangle$ is performed and then the whole sum is squared. If the individual terms in the summation are each squared before performing the sum then one would obtain exactly the result (3.34) with $p_{n'}^{(i)}$ given by eq. (3.36). The difference between eq. (3.37) and eq. (3.34) is exhibited clearly in eq. (3.37c) where the square of the sum is written out explicitly as a double sum. The result (3.34) is obtained from those terms in the double sum (3.37c) having the index $n'''=n'$. The difference between the coherent superposition and the incoherent mixture arises from the 'cross terms' in which $n'\neq n'''$. These clearly depend upon the phases of the expansion coefficients $\langle n'|a\rangle$. Let us write

$$\langle n'|a\rangle = \left|\langle n'|a\rangle\right| \mathrm{e}^{\mathrm{i}\phi_{n'}}.$$

We then have

$$P_{a\to n''} = \sum_{n'n'''} \langle n''|\mathrm{e}^{-\mathrm{i}kx}|n'\rangle\langle n'''|\mathrm{e}^{\mathrm{i}kx}|n''\rangle \left|\langle n'|a\rangle\langle a|n'''\rangle\right| \mathrm{e}^{\mathrm{i}(\phi_{n'}-\phi_{n'''})}. \quad (3.37\mathrm{d})$$

The result (3.34) for an incoherent mixture can be obtained from eq. (3.37d) by arguing that the phases $\phi_{n'}$ are not known, or that we are averaging measurements made over a large number of pure states, each having different values of the phases $\phi_{n'}$. If we average eq. (3.37d) over all values of $\phi_{n'}$ and $\phi_{n'''}$ all terms with $n'\neq n'''$ average out to zero and eq. (3.34) is obtained.

3.5 The Mössbauer effect for a system in thermal equilibrium

The result (3.34) is not of practical interest since the probability distribution for the final state is not measured in any experiment. The gamma ray energy depends upon the *difference in energy* between the final state and the initial state. The probability that the gamma ray is emitted with the energy E_0 required for resonant absorption is the probability that the final state is the same as the initial state. This can be obtained by setting n' equal to n'' in eq. (2.24) and averaging over the distribution of n' in the initial state.

$$P(E_\gamma = E_0) = \sum_{n'} \left|\langle n'|\mathrm{e}^{-\mathrm{i}kx}|n'\rangle\right|^2 p_{n'}^{(i)}. \quad (3.38)$$

Let us now attempt to calculate the form of the gamma ray spectrum. The gamma ray energy E_γ can take on the values $E_0+n\hbar\omega$, where n is any integer from minus infinity to plus infinity. Let $P(n\hbar\omega)$ be the probability that the gamma ray is emitted with the energy $E_\gamma=E_0+n\hbar\omega$. Unfortunately $P(n\hbar\omega)$

is not easily expressed or evaluated since it involves averaging eq. (2.24) over all pairs of states $|n'\rangle$ and $|n''\rangle$ having a constant energy difference $E_\gamma - E_0$.

The necessity for summing over selected pairs of initial and final states satisfying certain energy conditions can be avoided by first evaluating expressions of the form obtained in section 2.6 for moments of the energy spectrum. These results depend only on expectation values of certain operators in the initial state. The corresponding results for a statistical distribution of initial states can therefore be obtained simply by averaging the expectation values over the appropriate initial state. Thus from eq. (2.36b) we have

$$\langle E_\gamma^m \rangle = \sum_{n'} \langle n' | e^{ikx} (E_0 - H + E_{n'})^m e^{-ikx} | n' \rangle p_{n'}^{(i)}. \tag{3.39}$$

For a system in thermal equilibrium at temperature T, $p_{n'}^{(i)}$ is given by the Boltzmann factor

$$p_{n'}^{(i)} = N e^{-\beta E_{n'}} \tag{3.40}$$

where $\beta = 1/kT$ and N is a normalization factor. We can therefore write

$$\langle E_\gamma^m \rangle_T = N \sum \langle n' | e^{ikx} (E_0 - H + E_{n'})^m e^{-ikx} e^{-\beta H} | n' \rangle \tag{3.41}$$

where we have used the fact that the Hamiltonian H acting on the state n' is equal to the eigenvalue $E_{n'}$.

Eq. (3.39) gives the mean value of any power of E_γ over the energy spectrum. However, the expression (3.39) is not convenient for evaluation because the index n' occurs in the operator as well as in the label of the state vector. This makes the summation difficult. A more convenient function of E_γ can be found by noting that eq. (3.39) enables us to write expressions for the mean value of any function.

$$\langle f(E_\gamma) \rangle = N \sum_{n'} \langle n' | e^{ikx} f(E_0 - H + E_{n'}) e^{-ikx} e^{-\beta H} | n' \rangle \tag{3.42}$$

and in particular

$$\langle e^{-iE_\gamma t/\hbar} \rangle = N \sum_{n'} \langle n' | e^{ikx} e^{-i(E_0 - H + E_{n'})t/\hbar} e^{-ikx} e^{-\beta H} | n' \rangle \tag{3.43}$$

where t is a parameter having the dimensions of time. We shall see later that this time variable has a simple physical interpretation. However, for the present we simply note that time is canonically conjugate to energy, and that a Fourier transform of an energy spectrum is always expressed in terms of a variable with the dimensions of time. Eq. (3.43) thus gives the Fourier transform of the energy spectrum. Eq. (3.43) can be simplified because the expression $E_0 - H + E_{n'}$ appears in an exponent and the exponential can now be broken up into three separate factors. Since E_0 and $E_{n'}$ are numbers and not operators they can be moved to convenient points in the expression without worrying about commutators. Let us bring the factor $e^{-iE_0 t}$ over to the left-hand side of the equation and the factor $e^{iE_{n'}t}$ to the right. Then we can replace $E_{n'}$ by the operator H as in the case of the Boltzmann factor to obtain

$$\langle e^{-i(E_\gamma - E_0)t/\hbar} \rangle = N \sum_{n'} \langle n' | e^{ikx} e^{iHt/\hbar} e^{-ikx} e^{-(it/\hbar + \beta)H} | n' \rangle. \tag{3.44}$$

The index n' no longer appears explicitly in the operator. Since the sum of all of the diagonal elements of an operator is simply the trace of the operator, eq. (3.44) can also be written

$$\langle e^{-i(E_\gamma - E_0)t/\hbar} \rangle = N \, \mathrm{Tr}\{ e^{ikx} e^{iHt/\hbar} e^{-ikx} e^{-(it/\hbar + \beta)H} \}. \tag{3.45}$$

We have derived eq. (3.45) without assuming explicitly that the atom is moving in a harmonic oscillator potential. The result (3.45) is valid in general provided that H is the Hamiltonian operator describing the whole system in which the atom is moving. In particular, eq. (3.45) also holds for the Mössbauer effect in a crystal.

Eq. (3.45) has the characteristic form for the mean value of an operator in a statistical mixture, using the density operator formalism. The density operator for the initial state of the system is just $\varrho = N e^{-\beta H}$. Thus eq. (3.45) can also be written

$$\langle e^{-i(E_\gamma - E_0)t/\hbar} \rangle = \mathrm{Tr}\{ \varrho e^{ikx} e^{iHt/\hbar} e^{-ikx} e^{-iHt/\hbar} \}. \tag{3.46}$$

This can also be written

$$= \mathrm{Tr}\{ \varrho e^{ikx_{\mathrm{H}}(0)} e^{-ikx_{\mathrm{H}}(t)} \} \tag{3.47}$$

where $x_{\mathrm{H}}(t)$ is the operator x at time t in the Heisenberg representation.

This expression (3.47) is evaluated exactly in section 3.7 for the case of a harmonic oscillator or a harmonic lattice. We first consider an interesting physical interpretation of (3.47) as describing radiation from a moving source, whose position is given by $x_{\mathrm{H}}(t)$. If the source is stationary $x_{\mathrm{H}}(t)$ is independent of t, and eq. (3.47) shows that the Fourier transform of the energy spectrum referred to E_0 is also independent of t. The energy spectrum thus consists of a single frequency of radiation, with frequency $E_0/\hbar$ in agreement with the classical picture. If $x_{\mathrm{H}}(t)$ is a function of t, the result is a 'frequency modulation' of the radiated wave.

3.6 Radiation from a classical moving oscillator

The very close analog between the exact quantum-mechanical result (3.47) and this classical picture can be seen by examining in detail the classical electromagnetic radiation from a point source moving in a complicated manner while radiating. For simplicity we consider motion in one dimension. Let $x(t)$ be the position of this source at time t, ω_0 the frequency of the emitted radiation and k the wave vector at a large distance from the source where the radiation is observed. The velocity of the source is assumed to be small compared to c. Because of the motion of the source, the path length between the source and the detector changes as a function of time and introduces a time variation in the phase of the observed electromagnetic wave. The electric field at the detector at time t is then given by

$$E(k, t) = E_0 \mathrm{e}^{\mathrm{i}[(\omega_0 t - kx(t)]} \tag{3.48}$$

where E_0 is a constant. Because x is a function of time, the electromagnetic wave observed at the detector is not a simple periodic wave of frequency ω_0. It has a complicated frequency spectrum $g(k, \omega)$ given by the Fourier transform of eq. (3.49)

$$g(k, \omega) = N \int_{-\infty}^{\infty} E_0 \mathrm{e}^{\mathrm{i}[(\omega_0 - \omega)t - kx(t)]} \mathrm{d}t \tag{3.49}$$

where N is a normalization factor. The absolute square of eq. (3.49) gives the intensity of the Fourier component of frequency ω. This can be written as a double integral over two time variables

$$\begin{aligned} I(k, \omega) &= g^*(k, \omega) g(k, \omega) \\ &= N^2 E_0^2 \int_{-\infty}^{\infty} \mathrm{d}t' \int_{-\infty}^{\infty} \mathrm{d}t'' \mathrm{e}^{\mathrm{i}(\omega_0 - \omega)(t' - t'')} \mathrm{e}^{-\mathrm{i}k[x(t') - x(t'')]}. \end{aligned} \tag{3.50}$$

Let us set $t = t' - t''$, then

$$I(k, \omega) = N^2 E_0^2 \int_{-\infty}^{\infty} \mathrm{d}t \, \mathrm{e}^{\mathrm{i}(\omega_0 - \omega)t} \int_{-\infty}^{\infty} \mathrm{e}^{-\mathrm{i}k[x(t'+t) - x(t')]} \mathrm{d}t' \tag{3.51a}$$

$$= N^2 E_0^2 \int_{-\infty}^{\infty} \mathrm{d}t \, \mathrm{e}^{-\mathrm{i}(\omega - \omega_0)t} \int_{-\infty}^{\infty} \mathrm{d}t' \mathrm{e}^{\mathrm{i}kx(t')} \mathrm{e}^{-\mathrm{i}kx(t'+t)}. \tag{3.51b}$$

This result (3.51a) has the form of a Fourier transform of a space-time self-correlation function. The argument of the exponential in the second integral is the change $\delta x(t, t')$ in the position x in a time interval of duration t beginning at the time t'. The integral over t' gives a time average of $e^{-ik\delta x(t,t')}$ over the path of the source. This average is the Fourier transform in space of a probability distribution function $P(\delta x, t)$ defined as the probability that the source has moved a distance δx in a time interval t. Since the first integral is just a Fourier transform with respect to the time interval t, the expression (3.51a) is proportional to the Fourier transform in space and time of this space-time correlation function.

The form (3.51b) is seen to be the classical equivalent of the quantum-mechanical result (3.47). The classical time average over the path is equivalent to the quantum-mechanical average obtained by taking the trace of the product with the density operator. Since the motion is independent of the choice of the time origin, the quantum-mechanical average can be taken at $t'=0$. This can be seen explicitly by replacing $x_H(0)$ and $x_H(t)$ in eq. (3.47) by $x_H(t')$ and $x_H(t+t')$. The value of the expression is independent of t'. Eq. (3.51b) gives the frequency spectrum of the radiated intensity, while eq. (3.47) gives the Fourier transform of the spectrum. Eq. (3.51b) is just the Fourier transform of the classical average of the same quantity whose quantum average is given by eq. (3.47).

This treatment is easily generalized to the case of a system of particles in three dimensions all radiating coherently and all moving. This is relevant to coherent scattering of radiation by a crystal, where all the atoms together are considered as sources. If $\boldsymbol{x}_i$ represents the coordinate of the ith radiating particle, eqs. (3.48) and (3.49) are easily generalized by adding an index and performing the appropriate sum.

$$E(\boldsymbol{k}, t) = E_0 \sum_i e^{i[(\omega_0 t - \boldsymbol{k}\cdot\boldsymbol{x}_i(t)]} \tag{3.52a}$$

$$I(\boldsymbol{k}, \omega) = N^2 E_0^2 \int_{-\infty}^{\infty} dt\, e^{i(\omega_0-\omega)t} \int_{-\infty}^{\infty} \sum_{i,j} e^{-i\boldsymbol{k}\cdot[\boldsymbol{x}_i(t'+t)-\boldsymbol{x}_j(t')]} dt'. \tag{3.52b}$$

This result is expressed in terms of a pair correlation function since the argument of the exponent in the second integral is the difference between the coordinates of two different particles at times $t'+t$ and t', rather than being the coordinates of the same particle at different times.

Thus we see that the radiation from a classical bound system is determined by the two-body space-time correlation function. Conversely, a study of the radiation as a function of the wave number k and frequency ω can give information about this space-time pair correlation function. This is also true in quantum mechanics.

3.7 The Mössbauer effect for a harmonic crystal in thermal equilibrium

Let us now evaluate the expression (3.47) explicitly for the case of a nucleus moving in a one-dimensional harmonic oscillator potential. For this case the Heisenberg equations of motion can be solved exactly to give the operators at time t in terms of the operators at $t=0$

$$x_{\mathrm{H}}(t) = x_{\mathrm{H}}(0) \cos \omega t + \frac{1}{M\omega} p_{\mathrm{H}}(0) \sin \omega t. \tag{3.53}$$

The commutator of $x_{\mathrm{H}}(0)$ and $x_{\mathrm{H}}(t)$ is then

$$[x_{\mathrm{H}}(0), x_{\mathrm{H}}(t)] = \frac{i\hbar}{M\omega} \sin \omega t. \tag{3.54}$$

Since this commutator is a number (not an operator) which commutes with both $x_{\mathrm{H}}(0)$ and $x_{\mathrm{H}}(t)$, we can reduce the product of exponentials appearing in eq. (3.47) to a single exponential

$$\mathrm{e}^{ikx_{\mathrm{H}}(0)}\mathrm{e}^{-ikx_{\mathrm{H}}(t)} = \mathrm{e}^{ik[x_{\mathrm{H}}(0)-x_{\mathrm{H}}(t)]} \exp\left(\frac{ik^2\hbar \sin \omega t}{2m\omega}\right), \tag{3.55}$$

where we have used the identity

$$\mathrm{e}^{A+B} = \mathrm{e}^{A}\mathrm{e}^{B}\mathrm{e}^{-\frac{1}{2}[A,B]} \tag{3.56}$$

which holds when

$$[A, [A, B]] = [B, [A, B]] = 0.$$

Thus

$$\mathrm{e}^{ikx_{\mathrm{H}}(0)}\mathrm{e}^{-ikx_{\mathrm{H}}(t)} = \exp ik\left(x_{\mathrm{H}}(0)(1 - \cos \omega t) - \frac{1}{M\omega} p_{\mathrm{H}}(0) \sin \omega t\right) \times \exp\left(\frac{ik^2\hbar \sin \omega t}{2M\omega}\right). \tag{3.57}$$

Substituting eq. (3.57) into (3.47) we obtain

$$\langle \mathrm{e}^{-\mathrm{i}(E_\gamma - E_0)t/\hbar}\rangle = N \operatorname{Tr}\left[\exp ik\left(x_{\mathrm{H}}(0)(1 - \cos \omega t) - \frac{1}{M\omega} p_{\mathrm{H}}(0) \sin \omega t\right) \times \mathrm{e}^{-\beta H}\right] \exp\left(\frac{ik^2\hbar \sin \omega t}{2M\omega}\right). \tag{3.58}$$

The expression (3.58) is the mean value for a harmonic oscillator in thermodynamic equilibrium of an exponential function of a linear combination of

x and p. The general expression for such a mean value is

$$N\ \mathrm{Tr}[\mathrm{e}^{\mathrm{i}\xi(ux-vp)}\mathrm{e}^{-\beta H}]$$
$$= \exp\left[-\tfrac{1}{2}\xi^2\left(\frac{\hbar}{2M\omega}\right)(u^2 + M^2\omega^2v^2)\coth(\tfrac{1}{2}\hbar\omega\beta)\right]. \quad (3.59)$$

Substituting this result (3.59) into eq. (3.58) we obtain

$$\langle \mathrm{e}^{-\mathrm{i}(E_\gamma - E_0)t/\hbar}\rangle = \exp\left[-\tfrac{1}{2}k^2\left(\frac{\hbar}{M\omega}\right)[(1-\cos\omega t)\coth(\tfrac{1}{2}\hbar\omega\beta) - \mathrm{i}\sin\omega t]\right]$$

$$= \exp\left[\frac{R}{2\hbar\omega\sinh(\frac{1}{2}\hbar\omega\beta)}[\mathrm{e}^{\mathrm{i}\omega t+\frac{1}{2}\hbar\omega\beta} + \mathrm{e}^{-\mathrm{i}\omega t-\frac{1}{2}\hbar\omega\beta} - 2\cosh(\tfrac{1}{2}\hbar\omega\beta)]\right]. \quad (3.60)$$

We can now return to the problem of determining the explicit form $P(n\hbar\omega)$ of the gamma ray spectrum. The Fourier transform (3.60) can also be expressed in terms of $P(n\hbar\omega)$:

$$\langle \mathrm{e}^{-\mathrm{i}(E_\gamma - E_0)t/\hbar}\rangle = \sum_{n=-\infty}^{n=+\infty} P(n\hbar\omega)\mathrm{e}^{-\mathrm{i}n\omega t}. \quad (3.61)$$

Comparing eq. (3.61) and eq. (3.60) suggests that we try to reduce the complicated expression (3.60) to a simple Fourier series of the type (3.61). The coefficients of this series then gives directly the probabilities $P(n\hbar\omega)$. The appearance of exponentials in an exponent suggests an expansion in Bessel functions. We find that eq. (3.60) resembles the generating function for Bessel functions.

$$\mathrm{e}^{\frac{1}{2}Z(\zeta-\zeta^{-1})} = \sum_{n=-\infty}^{n=+\infty} J_n(Z)\zeta^n. \quad (3.62)$$

The first part of the exponent in eq. (3.60) has the form of eq. (3.62) if we set

$$Z = \frac{\mathrm{i}R}{\hbar\omega\sinh(\frac{1}{2}\hbar\omega\beta)} \quad (3.63\mathrm{a})$$

$$\zeta = -\mathrm{i}\mathrm{e}^{-(\mathrm{i}\omega t+\frac{1}{2}\hbar\omega\beta)}. \quad (3.63\mathrm{b})$$

We also note from eq. (3.30a) that

$$\frac{R}{\hbar\omega}\coth\tfrac{1}{2}\hbar\omega\beta = \frac{\hbar k^2}{2M\omega}\coth\tfrac{1}{2}\hbar\omega\beta = \langle k^2x^2\rangle. \quad (3.64)$$

Thus

$$\langle e^{-i(E_\gamma - E_0)t/\hbar} \rangle = e^{-\langle k^2 x^2 \rangle} \sum_{n=-\infty}^{n=+\infty} J_n \left[\frac{iR}{\hbar\omega \sinh(\frac{1}{2}\hbar\omega\beta)} \right] (-i)^n e^{-\frac{1}{2}n\hbar\omega\beta - in\omega t} \tag{3.65a}$$

$$= e^{-\langle k^2 x^2 \rangle} \sum_{n=-\infty}^{n=+\infty} I_n \left[\frac{R}{\hbar\omega \sinh(\frac{1}{2}\hbar\omega\beta)} \right] e^{-\frac{1}{2}n\hbar\omega\beta} e^{-in\omega t}. \tag{3.65b}$$

Eq. (3.65b) now has the same form as eq. (3.61). Equating the coefficients in the two series we obtain an expression for the probability $P(n\hbar\omega)$ that the gamma ray is emitted with an energy $E_\gamma = E_0 + n\hbar\omega$.

$$P(n\hbar\omega) = e^{-\langle k^2 x^2 \rangle} I_n \left(\frac{R}{\hbar\omega \sinh(\frac{1}{2}\hbar\omega\beta)} \right) e^{-\frac{1}{2}n\hbar\omega\beta} \tag{3.66a}$$

and in particular

$$P(0) = e^{-\langle k^2 x^2 \rangle} I_0 \left(\frac{R}{\hbar\omega \sinh(\frac{1}{2}\hbar\omega\beta)} \right). \tag{3.66b}$$

Let us examine eq. (3.66) in the low temperature limit $T \to 0$, $\beta \to \infty$. For all positive values of n, eq. (3.66) vanishes in this limit. This is to be expected at zero temperature, since the atom is moving in the lowest state of the oscillator potential and can only gain energy during the transition and cannot lose energy. Thus the gamma ray can only come out with the energy E_0 or with a lower energy but not with a higher energy than E_0. Thus eq. (3.66) gives a non-vanishing result at zero temperature only for negative values of n. Eq. (3.66) can be simplified in the low temperature limit by expanding the Bessel function I_n in a series and keeping only the first term.

$$P(n\hbar\omega) \xrightarrow[\beta \to \infty]{} e^{-\langle k^2 x^2 \rangle} \frac{1}{n!} \left(\frac{R}{\hbar\omega} \right)^n \qquad n \leqq 0. \tag{3.67}$$

In the zero temperature limit, the atom must be in the lowest state of the oscillator potential and is thus in a *pure* quantum state. The result (3.67) therefore must agree with the calculation of the energy spectrum in problem 2a of ch. 2.

This result is easily generalized to the case of a harmonic crystal, because of the separation of the normal modes. For this case, we can substitute into eq. (3.46) the expansion (3.17) for x in normal mode variables, and express H and ϱ as

$$H = \sum_s H_s \tag{3.68a}$$

$$\varrho = \prod_s \varrho_s \tag{3.68b}$$

where H_s and ϱ_s are the Hamiltonian and Boltzmann factor for the individual normal mode s. The Hamiltonian for the harmonic crystal is a simple sum and the Boltzmann factor is a simple product. Thus eq. (3.46) for a crystal separates into a product of factors each for a single normal mode

$$\langle \mathrm{e}^{-\mathrm{i}(E_\gamma - E_0)t/\hbar} \rangle = \prod_s \mathrm{Tr}(\varrho_s \mathrm{e}^{\mathrm{i}ka_{\mathrm{L}s}\xi_s} \mathrm{e}^{\mathrm{i}H_s t/\hbar} \mathrm{e}^{-\mathrm{i}ka_{\mathrm{L}s}\xi_s} \mathrm{e}^{-\mathrm{i}H_s t/\hbar}). \tag{3.69}$$

Each factor has exactly the same form as (3.46) for a single oscillator. Thus we can immediately apply the result (3.65b) to the crystal by replacing x by ξ_s, ω by ω_s, k by $ka_{\mathrm{L}s}$ and R by

$$R_s = \frac{\hbar^2 a_{\mathrm{L}s}^2 k^2}{2M} \tag{3.70}$$

$$\langle \mathrm{e}^{-\mathrm{i}(E_\gamma - E_0)t/\hbar} \rangle = \prod_s \exp(-\langle k^2 a_{\mathrm{L}s}^2 \xi_s^2 \rangle)$$

$$\times \sum_{n=-\infty}^{n=+\infty} I_n \left[\frac{R_s}{\hbar\omega_s \sinh(\frac{1}{2}\hbar\omega_s\beta)} \right] \mathrm{e}^{-\frac{1}{2}n\hbar\omega_s\beta} \mathrm{e}^{-\mathrm{i}n\omega_s t} \tag{3.71a}$$

$$= \mathrm{e}^{-\langle(\boldsymbol{k}\cdot\boldsymbol{x}_{\mathrm{L}})^2\rangle} \sum_{\mathrm{all}\,\{n_s\}} \left[\prod_s I_{n_s} \left\{ \frac{R_s}{\hbar\omega_s \sinh(\frac{1}{2}\hbar\omega_s\beta)} \right\} \times \mathrm{e}^{-\frac{1}{2}n_s\hbar\omega_s\beta} \right] \mathrm{e}^{-\mathrm{i}\sum_s n_s\omega_s t}. \tag{3.71b}$$

The gamma energy spectrum for the case of a crystal is much more complicated than that for the single oscillator. The variable n_s represents the change in energy of the sth normal mode, and each set of $3N$ values $\{n_s\}$ defines an energy transfer to the crystal of $\sum_s \hbar n_s \omega_s$. Eq. (3.71b) shows each such term explicitly. For the case where the crystal remains in the same state after the emission of the gamma ray, $n_s = 0$ for all s, and we obtain, as a generalization of eq. (3.66b)

$$P(0) = \mathrm{e}^{-\langle(\boldsymbol{k}\cdot\boldsymbol{x}_{\mathrm{L}})^2\rangle} \prod_s I_0 \left\{ \frac{R_s}{\hbar\omega_s \sinh(\frac{1}{2}\hbar\omega_s\beta)} \right\}. \tag{3.72}$$

Since R_s is of order $a_{\mathrm{L}s}^2$ which is of order $(3N)^{-1}$ the Bessel functions can be expanded. The first term gives exactly the result (3.31) of our previous approximate treatment

$$P(E_0, T) = \exp -\langle(\boldsymbol{k}\cdot\boldsymbol{x}_{\mathrm{L}})^2\rangle_T. \tag{3.31}$$

The corrections are again of order $(3N)^{-1}$ since there are $3N$ terms, each quadratic in R_s and therefore of order $(3N)^{-2}$.

Other aspects of the spectrum can be considered either by numerical evaluation of eq. (3.71b) or by various approximations. One possible approximation is to separate the terms according to the 'number of excitation phonons'. The term with all $n_s=0$ considered above is the 'no-phonon' term. The one-phonon term has $n_s=\pm 1$ for one value of s, all other $n_s=0$. The two-phonon term has $n_s=\pm 1$ for two values of s, etc.

We note that in a peculiar crystal, where theɪe is a single normal mode with $a_{Ls}\approx 1$, there may be a correction to the conventional result (3.31) due to the Bessel function. This might occur when the source is an impurity in a crystal, and the dominant mode is a 'localized mode' in which the impurity atom oscillates in a potential well due to the rest of the crystal, with only very weak coupling to the motion of the other atoms. However, cases where the Bessel function gives appreciable corrections to the Debye–Waller factor have not yet been found. The following section shows that relaxation effects suppress the Bessel function. Thus it probably will not be observed in practical cases.

3.8 The thermal shift and relaxation effects

The expression (2.51) for the thermal shift (second-order Doppler shift) is easily generalized to the case of a crystal. The perturbation theory expression can be applied directly to give the shift in terms of the mean kinetic energy or the mean square velocity or momentum of the atom emitting the gamma ray.

For the case of a system in thermal equilibrium, we encounter the following difficulty. The system is not in a definite quantum state but is a statistical mixture. For each component in the mixture, the thermal shift has a different value, since $\langle E_{\text{kin}}\rangle$, $\langle p^2\rangle$ and $\langle v^2\rangle$ are different for different states. Does this mean that the Mössbauer line will be split into components with different thermal shifts? Or will there be a single line with a thermal shift given by averaging the theoretical value over the Boltzmann distribution.

To resolve this question let us consider the case of a nucleus moving in a harmonic oscillator potential in thermal equilibrium. If it is in a well-defined quantum state at the time of emission of the gamma ray, it will have a well-defined thermal shift, given by eq. (2.51). Consider a large number of such nuclei, all in well-defined quantum states at the time of emission of the gamma ray, but with a Boltzmann distribution for the probability of finding a nucleus in a given state. Then each gamma ray has a well-defined thermal shift, and there is a spectrum of thermal shifts, corresponding to different initial states $|n'\rangle$, with the intensity of each line proportional to the Boltzmann factor for the corresponding initial state.

However, one can ask about the mechanism of thermal equilibrium. The oscillator must somehow be coupled to a heat reservoir at temperature T, and be allowed to exchange energy to come to thermal equilibrium. If the oscillator is exchanging energy, it cannot remain free even in the same eigenstate. If it is actually jumping from one state to another during the emission of the gamma ray, how does this affect the thermal shift?

To get a better understanding of this problem, consider the case of a nucleus moving in a harmonic oscillator potential and weakly coupled to a heat reservoir represented by a very large system with many degrees of freedom. We first assume no interaction between the oscillator and the reservoir. Let $|n'\alpha'\rangle$ define a complete set of stationary states for the entire system, where α' denotes a complete set of quantum numbers required to specify the eigenfunctions of the Hamiltonian of the reservoir. The eigenvalues of the total Hamiltonian are very highly degenerate, since the large system has many degrees of freedom and the states $|n', \alpha'\rangle$ and $|n'', \alpha''\rangle$ are degenerate for all reservoir states α' and α'' whose energies satisfy the relation $E_{\alpha''} - E_{\alpha'} = (n' - n'')\hbar\omega$.

Let us now 'turn on' a very weak interaction between the oscillator and the reservoir, sufficiently weak to be treated by first-order degenerate perturbation theory. The first-order eigenfunctions of the coupled system are linear combinations of states which are all degenerate in the unperturbed system.

$$|\xi\rangle = \sum_{n'\alpha'} C^{(\xi)}_{n'\alpha'} |n'\alpha'\rangle \tag{3.73}$$

where the coefficients $C^{(\xi)}_{n'\alpha'}$ are obtained by diagonalizing the interaction in the subspace of degenerate unperturbed states.

Consider the emission of a Mössbauer gamma ray by a nucleus in this state $|\xi\rangle$. The probability that the system remains in the same state after emission of the gamma ray is given by substituting (3.73) into eq. (3.8)

$$P_{\xi\to\xi} = |\langle\xi|e^{-ikx}|\xi\rangle|^2 = \Big|\sum_{n'n''} C^{(\xi)*}_{n'\alpha'} C^{(\xi)}_{n''\alpha''} \langle n'\alpha'|e^{-ikx}|n''\alpha''\rangle\Big|^2. \tag{3.74}$$

Since the operator e^{-ikx} acts only on the nucleus in the oscillator potential and not on the reservoir, the matrix element vanishes unless $\alpha' = \alpha''$. However, if $\alpha' = \alpha''$, then n' and n'' must also be equal since all the states in the expansion (3.73) have the same unperturbed energy.

Thus

$$\langle n'\alpha'|e^{-ikx}|n''\alpha''\rangle = \langle n'|e^{-ikx}|n''\rangle\,\delta_{\alpha'\alpha''} = \langle n'|e^{-ikx}|n'\rangle\,\delta_{n'n''}\delta_{\alpha'\alpha''} \tag{3.75}$$

and

$$P_{\xi\to\xi} = \left| \sum_{n'} \sum_{\alpha'} \left| C^{(\xi)}_{n'\alpha'} \right|^2 \langle n' | \mathrm{e}^{-ikx} | n' \rangle \right|^2 . \tag{3.76}$$

This can be written

$$P_{\xi\to\xi} = \left| \sum_{n'} \langle n' | \mathrm{e}^{-ikx} | n' \rangle p^{(\xi)}_{n'} \right|^2 \tag{3.77a}$$

where

$$p^{(\xi)}_{n'} = \sum_{\alpha'} |C^{(\xi)}_{n'\alpha'}|^2 \tag{3.77b}$$

is the probability that the oscillator is in the eigenstate $|n'\rangle$ when the whole system is in the state $|\xi\rangle$. This result (3.77) resembles the expression (3.38) for the case of a statistical mixture; however, the two expressions are not identical. In eq. (3.77) the matrix element is averaged over the probability distribution *before* squaring, rather than averaging the squared matrix element as in (3.38).

Before evaluating the expression (3.77) let us note some properties of a statistical mixture of states of the whole system in thermal equilibrium at a temperature T. The mean energy per degree of freedom is kT, with quantum corrections at low temperatures. The total energy of the system is of order NkT, where N is the number of degrees of freedom. The spread in the energy of the whole system over the ensemble is of order $N^{\frac{1}{2}}kT$. Thus the spread in the energy per degree of freedom is $N^{-\frac{1}{2}}kT$, and is negligible. The ensemble of states of the large system in thermal equilibrium thus consists of a large number of states all with very nearly the same total energy.

The particular state $|\xi\rangle$ considered above is a linear combination of a large number of degenerate eigenstates of the unperturbed system. Because of the large number of states α' appearing in the summation (3.73) the value of the probability $p^{(\xi)}_{n'}$ is determined by statistical considerations and is the same for all states $|\xi\rangle$ having the same energy. Thus the probability $p^{(\xi)}_{n'}$ is just a Boltzmann factor corresponding to the temperature which gives a total energy for the system equal to the energy of the state $|\xi\rangle$. The expression (3.77) thus holds for an ensemble in thermal equilibrium, since the values of $p^{(\xi)}_{n'}$ are the same for all states in the ensemble.

The expression (3.77) is the square of a thermal average of a harmonic oscillator matrix element having the form (3.59). This can be evaluated immediately and gives the usual result $\exp(-k^2\langle x^2\rangle)$ without the additional Bessel function factor appearing in eq. (3.66b).

The thermal shift for the state $|\xi\rangle$ is given by evaluating eq. (2.51) in this state. It therefore depends upon the mean kinetic energy of the Boltzmann distribution, and gives a single unsplit line for the case of a system in thermal equilibrium.

However, in obtaining this result, we have not really answered the question posed at the beginning of this section regarding the mechanism of thermal equilibrium. Rather, we have avoided it by cheating. The expression (3.74) does not necessarily give the intensity of the Mössbauer line. The final state need not be exactly the same as the initial state; it is sufficient that the two states have the same energy. In previous discussions, the presence of accidentally degenerate states played no role, since they would generally have a very different structure from the initial state and therefore have a very small matrix element for the operator e^{-ikx}. In the present case this is not true because the interaction of the oscillator with the reservoir mixes a given oscillator state into a large number of states of the whole system.

The probability $P(E_0, \xi)$ that the gamma ray is emitted with the energy E_0 when the system is in the state $|\xi\rangle$ is given by

$$P(E_0, \xi) = \sum_{\xi'} \left|\langle \xi' | e^{-ikx} | \xi \rangle\right|^2 \tag{3.78a}$$

where the summation is over the subspace, which we denote by ξ of all states $|\xi'\rangle$ which are degenerate with $|\xi\rangle$. The average of the expression (3.78a) over all states ξ in the subspace is thus a trace in this subspace,

$$\langle P(E_0, \xi)\rangle_\xi = (N_\xi)^{-1} \operatorname{Tr}_\xi\{(e^{ikx})_\xi (e^{-ikx})_\xi\}, \tag{3.78b}$$

where N_ξ is the number of states in this subspace, $(e^{+ikx})_\xi$ is the matrix of the operator in the subspace, and Tr_ξ is the trace in the subspace.

Let us now assume that the interaction between the oscillator and the reservoir is so small that the states which are degenerate in the absence of interaction are still very nearly degenerate and are included in the subspace ξ. Since the trace is invariant under unitary transformations in the subspace, we choose the simpler $|n'\alpha'\rangle$ basis to evaluate the trace, rather than the $|\xi\rangle$ basis. This gives

$$\begin{aligned} \langle P(E_0, \xi)\rangle_\xi &= (N_\xi)^{-1} \sum_{n'\alpha' n''\alpha''} \langle n'\alpha' | e^{ikx} | n''\alpha''\rangle \langle n''\alpha'' | e^{-ikx} | n'\alpha'\rangle \\ &= (N_\xi)^{-1} \sum_{n'\alpha'} \left|\langle n'\alpha' | e^{-ikx} | n'\alpha'\rangle\right|^2 \end{aligned} \tag{3.79}$$

where we have used eq. (3.75).

This result leads to exactly the result (3.38) for a statistical mixture. The matrix element is squared *before* taking the thermal average, and contains the additional Bessel function factor found in eq. (3.66b). The Bessel function is always greater than one, and the difference is due to the contribution of the other final states $|\xi'\rangle \neq |\xi\rangle$ in eq. (3.78a).

The question whether eq. (3.74) or (3.78) gives the correct result depends upon whether all the states $|n'\alpha'\rangle$ which are degenerate in the absence of the coupling between the oscillator and the reservoir can still be considered degenerate in the presence of the interaction; i.e. whether the splitting of these levels by the interaction is small or large compared to the natural line width of the gamma ray. If the splitting is small, the result (3.78) is valid and the Bessel function should be included in eq. (3.66b). If the splitting is large, eq. (3.74) applies and the Bessel function should be dropped.

The question of the relative sizes of the splitting and line width can also be stated in terms of characteristic *times*, since time is canonically conjugate to energy. Let us suppose that the system is in a definite state $|n'\alpha'\rangle$ at the time $t=0$, before the gamma ray is emitted. This state can be expanded in the eigenfunctions of the interaction $|\xi'\rangle$,

$$|n'\alpha'\rangle = \sum_{\xi'} C_{n'\alpha'}^{(\xi')*} |\xi'\rangle. \tag{3.80a}$$

At a later time t, the system is in the state

$$\mathrm{e}^{-\mathrm{i}Ht/\hbar} |n'\alpha'\rangle = \sum_{\xi'} \mathrm{e}^{-\mathrm{i}E't/\hbar} C_{n'\alpha'}^{(\xi')*} |\xi'\rangle \tag{3.80b}$$

where E' is the eigenvalue of H for the state $|\xi'\rangle$.

As time passes, the different components in the wave packet (3.80b) get out of phase with one another, and the system is no longer in the state $|n'\alpha'\rangle$. To get an idea of the 'lifetime' of the initial state, let us take the overlap of the states (3.80b) and the initial state (3.80a) as a function of time.

$$\langle n'\alpha' | \mathrm{e}^{-\mathrm{i}Ht/\hbar} | n'\alpha' \rangle = \mathrm{e}^{-\mathrm{i}\bar{E}t/\hbar} \sum_{\xi'} |C_{n'\alpha'}^{(\xi')}|^2 \mathrm{e}^{-\mathrm{i}(E'-\bar{E})t/\hbar} \tag{3.81a}$$

where $\bar{E}$ is the mean energy of the packet,

$$\bar{E} = \sum_{\xi'} |C_{n'\alpha'}^{(\xi')}|^2 E'. \tag{3.81b}$$

Expanding the exponential for small t, we have

$$\langle n'\alpha' | \mathrm{e}^{-\mathrm{i}Ht/\hbar} | n'\alpha' \rangle \approx \mathrm{e}^{-\mathrm{i}\bar{E}t/\hbar} [1 - |C_{n'\alpha'}^{(\xi')}|^2 (E' - \bar{E})^2 t^2/\hbar^2] \tag{3.82a}$$

$$\approx \mathrm{e}^{-\mathrm{i}\bar{E}t/\hbar} [1 - (\Delta E)^2 t^2/\hbar^2] \tag{3.82b}$$

where $(\Delta E)^2$ is the mean square width,

$$(\Delta E)^2 \equiv \sum_{\xi'} |C_{n'\alpha'}^{(\xi')}|^2 (E' - \bar{E})^2. \tag{3.82c}$$

The quantity ΔE is an energy which characterizes the splitting of the degenerate levels by the interaction. Eq. (3.82) shows that if the system is in the state $|n'\alpha'\rangle$ at time $t=0$, it has a very high probability of still being in this state for times t short compared to $\hbar/\Delta E$, but that for times long compared to $\hbar/\Delta E$, it has a very low probability of being in this state. The time $\hbar/\Delta E$ thus defines a 'relaxation time' for the coupling of the oscillator to the heat reservoir.

We have seen that the usual expression for a statistical mixture (3.38) applies when the splitting ΔE is small compared to the natural line width Γ of the nuclear gamma ray. This means that the relaxation time $\hbar/\Delta E$ is long compared to the lifetime $\hbar/\Gamma$ of the excited nuclear state. The other expression (3.74) applies when the relaxation time is short compared to the nuclear lifetime. This has a simple physical interpretation. When the relaxation time is long compared to the nuclear lifetime, the nucleus remains in the same state $|n'\rangle$ of the oscillator potential during the gamma ray emission process. If the relaxation time is short, the nucleus jumps from one oscillator state to another during the emission of the gamma ray. In the first case, the result (3.38) for a statistical mixture applies, because the emission always takes place from a well-defined state of the oscillator. The averaging accounts for different transitions taking place from different initial oscillator states. In the second case, the result (3.77) applies because the nucleus is *not* in a well-defined oscillator state during the emission of the gamma ray. The averaging over oscillator states must be done in computing the matrix element for a *single* gamma ray transition.

This example illustrates the importance of the characteristic relaxation time of the mechanism which brings about thermal equilibrium. The emission of the Mössbauer gamma ray from a system in thermal equilibrium depends upon whether the relaxation time is short or long compared with the lifetime of the nuclear state. If the relaxation time is short, then the thermal average is taken in computing the matrix element before squaring. If the relaxation time is long, the matrix element is squared before averaging.

The same picture applies to the thermal shift. If the relaxation time is short, then the thermal shift gives a single line, and the magnitude of the shift is given by the average kinetic energy in the Boltzmann distribution. If the relaxation time is long, there are many different values of the thermal shift, and the line will appear either split or broadened. This picture is characteristic of many relaxation phenomena which involve transitions between split levels. With long relaxation times, split spectral lines are observed, because each transition is well defined, and the spectrum reproduces the level splittings. With short relaxation times, only a single line is observed, with the natural line width, and a position which is determined by the centroids of the split levels.

In a physical crystal, the relaxation mechanism is the interaction between the different normal modes which is neglected in the harmonic approximation. The relaxation time is the 'phonon lifetime', the time that a given lattice vibration exists before being scattered or dissipated into other modes. This lifetime is short compared to characteristic nuclear lifetimes. Thus in practical cases the conventional expression for the Debye–Waller factor (3.31) applies without the Bessel function and the thermal shift gives a well-defined single line with the natural width.

CHAPTER 4

THE MÖSSBAUER EFFECT AND MOMENTUM TRANSFER TO BOUND SYSTEMS

4.1 The Mössbauer radiation process for a free nucleus

The preceding treatment of the Mössbauer effect considered only the consequences of momentum transfer to the nucleus. Let us now also consider the description of the emission of the gamma ray by the nucleus. Although the full treatment requires radiation theory and quantization of the electromagnetic field, we shall use a simpler approach and construct a phenomenological Hamiltonian which describes all the essential features of the process,* using the second quantized notation and conservation laws.

Let $a^{\dagger}_{\boldsymbol{k}e}$ be a creation operator for the nucleus in its excited state with center-of-mass momentum $\hbar\boldsymbol{k}$ and let $a^{\dagger}_{\boldsymbol{k}g}$ be the creation operator for the nucleus in its ground state with momentum $\hbar\boldsymbol{k}$. The nucleus has many other states besides the ground state and the particular excited state from which the Mössbauer gamma ray is emitted. Creation operators can be defined for the complete set of states. However, only the particular excited state from which the gamma ray is emitted is of interest. Thus we ignore the existence of the other excited states and treat the internal degrees of freedom of the nucleus as if there were only two states, the ground state and one particular excited state.

Let $b^{\dagger}_{\boldsymbol{k}s}$ represent the creation operator for a photon of wave vector $\boldsymbol{k}$ and polarization s. The two polarization states are not important for the present calculation and are included just for completeness. The index s thus takes on two values which can be horizontal and vertical, left and right circularly polarized or any other two orthogonal states. In the following treatment, physical effects are independent of polarization and the index s always appears in a summation for which any convenient basis can be used.

We first consider a system of free photons and nuclei without the possibility

* This producedure is commonly used in problems of solid state and particle physics where it is impossible or unpractical to obtain a Hamiltonian from first principles. See, for example, ch. 7.

of photon emission and absorption by the nuclei. The Hamiltonian describing this non-interacting system is just

$$H = \sum_{ks} (\hbar kc) b^{\dagger}_{ks} b_{ks} + \sum_{k} E_{ke} a^{\dagger}_{ke} a_{ke} + \sum_{k} E_{kg} a^{\dagger}_{kg} a_{kg} \tag{4.1}$$

where $\hbar kc$ is the energy of a photon of wave vector $\boldsymbol{k}$ and E_{ke} and E_{kg} are the energies in the excited or ground state, respectively, of nuclei with total momentum $\hbar\boldsymbol{k}$. Since the energies E_{ke} and E_{kg} appear as parameters in this Hamiltonian, the entire treatment can be carried through without specification of the particular form of these energies. In evaluation of results, one can insert either the exact relativistic values including the effect of the relativistic mass difference between the ground and excited states or any suitable approximation.

We now wish to include in the Hamiltonian an interaction describing the transition of the nucleus from the excited state to the ground state with the emission of a photon. If we are to conserve momentum, this interaction must consist of terms having the form

$$a^{\dagger}_{(\boldsymbol{k}-\boldsymbol{q})g} a_{\boldsymbol{k}e} b^{\dagger}_{\boldsymbol{q}s}$$

in which the nucleus in the excited state with momentum $\hbar\boldsymbol{k}$ emits a photon of momentum $\hbar\boldsymbol{q}$ and therefore makes the transition to the ground state and a total momentum $\hbar(\boldsymbol{k}-\boldsymbol{q})$. The most general interaction consists of a sum of terms of this type and those where a photon of momentun $-\hbar\boldsymbol{q}$ is absorbed.

$$H_{\text{int}} = \sum_{\boldsymbol{kq}s} C_{\boldsymbol{kq}s} a^{\dagger}_{(\boldsymbol{k}-\boldsymbol{q})g} a_{\boldsymbol{k}e} (b^{\dagger}_{\boldsymbol{q}s} + b_{\boldsymbol{q}s}) + \text{h.c.} \tag{4.2}$$

where $C_{\boldsymbol{kq}s}$ is a coefficient whose dependence on the indices is yet unspecified. The requirement that the Hamiltonian be Hermitean requires the addition of Hermitean conjugate terms describing the transition from the ground state to the excited state with the absorption of a photon.

Now consider the case where initially one nucleus is present in the excited state with momentum $\hbar\boldsymbol{k}$ and no photons are present

$$|i\rangle = a^{\dagger}_{\boldsymbol{k}e}|0\rangle. \tag{4.3}$$

We now treat the total Hamiltonian (4.1) and (4.2) by time-dependent perturbation theory using the initial state (4.3). The only relevant final states have the form

$$|f_s\rangle = b^{\dagger}_{\boldsymbol{q}s} a^{\dagger}_{(\boldsymbol{k}-\boldsymbol{q})g}|0\rangle \tag{4.4}$$

where the value of $\boldsymbol{q}$ is determined by energy conservation and results automatically from time-dependent perturbation treatment

$$\hbar cq = E_{ke} - E_{(\boldsymbol{k}-\boldsymbol{q})g}. \tag{4.5}$$

The effect of recoil on the photon energy appears on the right-hand side of eq. (4.5) as the difference between the energy of a nucleus with momentum $\hbar\boldsymbol{k}$ and a nucleus with momentum $\hbar(\boldsymbol{k}-\boldsymbol{q})$. Eq. (4.5) restricts the photon energy to a unique value for a given angle between $\boldsymbol{k}$ and $\boldsymbol{q}$. All angles are possible. The transition probability per unit time is given by the usual 'golden rule' formula of time-dependent perturbation theory

$$W^{\mathrm{F}}_{i\to f_s} = \frac{2\pi}{\hbar}\,|\langle f_s|H_{\mathrm{int}}|i\rangle|^2 \varrho_f = \frac{2\pi}{\hbar}\,|C_{\boldsymbol{kq}s}|^2 \varrho_f \tag{4.6}$$

where ϱ_f is the density of final states. We have added the superscript F for 'free' to distinguish this case from the Mössbauer case of emission from a bound nucleus. The total transition probability is obtained from eq. (4.6) by summing over both polarization states s and photon momenta $\boldsymbol{q}$ consistent with energy conservation (4.5). The sum over $\boldsymbol{q}$ is thus an integration over the angles between $\boldsymbol{k}$ and $\boldsymbol{q}$.

It is interesting to extend this treatment to the case where photons are present in the initial state. Consider the case where there are n photons of just the right wave vector $\boldsymbol{q}$ and polarization s present in the initial state in addition to the excited nucleus

$$|i_n\rangle = (n!)^{-\frac{1}{2}}(b^\dagger_{\boldsymbol{q}s})^n a^\dagger_{\boldsymbol{k}\mathrm{e}}|0\rangle. \tag{4.7}$$

The relevant final state is then

$$|f_{ns}\rangle = [(n+1)!]^{-\frac{1}{2}}(b^\dagger_{\boldsymbol{q}s})^{(n+1)} a^\dagger_{(\boldsymbol{k}-\boldsymbol{q})\mathrm{g}}|0\rangle. \tag{4.8}$$

If these states (4.7) and (4.8) are inserted in the golden rule (4.6), the matrix element is multiplied by the additional factor

$$|\langle 0|(b_{\boldsymbol{q}s})^{n+1}(b^\dagger_{\boldsymbol{q}s})^{n+1}|0\rangle[(n+1)!n!]^{-\frac{1}{2}}|^2 = n+1 \tag{4.9}$$

The transition probability is increased by a factor $n+1$ when photons are originally present in the proper state. This 'stimulated emission' of photons is a direct consequence of the Bose statistics or boson commutation rules of photons and does not require detailed radiation theory. Stimulated emission is so far of very little interest in connection with the Mössbauer effect and is presented here only as an interesting digression. It is of course of great interest in other applications, and is the basic physical principle underlying the operation of masers and lasers.

The interaction Hamiltonian (4.2) was chosen to be the most general one satisfying conservation of momentum. The coefficient $C_{\boldsymbol{kq}s}$ can depend in an arbitrary way on the indices. The transition probability (4.6) depends upon these coefficients and its dependence on the momenta $\boldsymbol{k}$ and $\boldsymbol{q}$ and the polarization s is still unspecified except for the restrictions imposed by energy conservation (4.5).

In constructing phenomenological interactions of the type (4.2), one should use as many symmetries and invariance principles as possible to restrict the values of the parameters appearing in the Hamiltonian. We have used momentum conservation, we now consider relativistic invariance. We are concerned primarily with nuclei whose velocities are very small compared to the velocity of light. Thus we can use Galilean invariance rather than Lorentz invariance and require that the interaction (4.2) be invariant under a transformation to a coordinate system moving with a small velocity v, and neglect terms of order v/c. Let us examine the effect of such a transformation on the interaction (4.2). The photon momentum $\boldsymbol{q}$ and polarization s are not changed to this approximation since the Doppler shift is of order v/c. On the other hand, the initial momentum $\hbar\boldsymbol{k}$ of the nucleus is changed appreciably since the initial velocity of the nucleus may well be smaller than the velocity v of the coordinate transformation while still leaving $v \ll c$. Thus if the coefficient $C_{\boldsymbol{kq}s}$ depends upon the initial momentum $\hbar\boldsymbol{k}$, the Hamiltonian is not invariant under Galilean transformations. Galilean invariance therefore requires that $C_{\boldsymbol{kq}s}$ be independent of $\boldsymbol{k}$. More generally one can say that Lorentz invariance requires that $C_{\boldsymbol{kq}s}$ be independent of $\boldsymbol{k}$ with corrections of the order of v/c.

The physical meaning of this requirement is simply that in a nonrelativistic description, the probability of decay of a nucleus (4.6) must be independent of the motion of the observer relative to the nucleus; i.e. of the momentum $\hbar\boldsymbol{k}$ of the nucleus in any particular coordinate system. Relativistic corrections can be incorporated by requiring the dependence of $C_{\boldsymbol{kq}s}$ on $\boldsymbol{k}$ to give the correct relativistic dependence of the decay probability or lifetime of the state on the motion of the nucleus relative to the observer; i.e., it must give the effect of the relativistic time dilatation.

We now consider the variation of $C_{\boldsymbol{kq}s}$ on the variables $\boldsymbol{q}$ and s. The range of values of $\boldsymbol{q}$ of interest for the Mössbauer effect is restricted by the requirement of energy conservation (4.5) to what is commonly called the 'energy shell'. This shell is very nearly a sphere in momentum space because the allowed photon energy varies with angle only as a result of very small recoil effects. The magnitude of $\boldsymbol{q}$ varies by less than 10^{-6} over the range allowed. Phenomenological interactions of the type (4.2) are assumed to be reasonable and smooth functions unless there are good physical reasons for believing otherwise. One would not expect the value of $C_{\boldsymbol{kq}s}$ to jump appreciably as a result of a change in the magnitude of $\boldsymbol{q}$ by one part per million. We therefore neglect the variation of C_{kqs} on the magnitude of $\boldsymbol{q}$ in the region relevant to the Mössbauer effect.

If the nucleus is initially at rest and unpolarized, no preferred direction in space is defined. Rotational invariance therefore requires equal probability for photon emission in all directions and also equal probabilities for both polarizations. For an initially polarized nucleus the requirement of rotational and space inversion invariance are simply expressed in terms of angular momentum and parity conservation and place restrictions on the angular momentum and parity (i.e., the multipolarity) of the photon. If only one multipole is relevant, the angular distribution and polarization of the photon is uniquely determined. If more than one multipole is possible, the angular distribution and polarization of the photon are determined by the 'mixing ratio' of the multipoles, a quantity determined by the nuclear structure. These are standard problems in the theory of angular correlations and are not considered further here.

The variation of the coefficient C_{kqs} on its indices is thus determined completely by conservation laws and invariance principles. Neglecting relativistic corrections and angular correlation effects arising in the case of an initially polarized nucleus, we can neglect completely the variation of C_{kqs} on its indices in the golden rule formula (4.6) and can set

$$C_{kqs} \approx C. \tag{4.10a}$$

For the more general case of a polarized nucleus, we drop only the dependence on $\boldsymbol{k}$.

$$C_{kqs} \approx C_{qs}. \tag{4.10b}$$

Note that the dependence on $\boldsymbol{q}$ can be dropped only in the golden rule formula and not in the interaction (4.2), since the interaction also describes virtual processes which do not conserve energy. However eq. (4.10b) can be used in the interaction, as the independence of $\boldsymbol{k}$ does not assume energy conservation.

4.2 Radiation from a bound nucleus

We now consider the case where the nucleus emitting the gamma ray is not free but is bound in some external potential. The treatment for a free nucleus can be modified by analogy with the simpler treatment of section 2.3. The binding in the external potential is described by adding to our phenomenological Hamiltonian a potential expressed as the most general single-particle operator in the second-quantized notation

$$V = \sum_{kk'} V_{kk'} a_{k'}^{\dagger} a_{k}, \tag{4.11}$$

With the nucleus bound in an external potential, the plane-wave basis is not convenient. We should instead transform to the basis of single-particle states which are the eigenstates of the motion of the nucleus in the external potential (i.e., harmonic oscillator functions if the external potential is a harmonic oscillator potential). Let $a_{ng}^{\dagger}$ and $a_{ne}^{\dagger}$ represent creation operators for a nucleus in its internal ground or excited state respectively and in the nth state of excitation in the external potential.

The Hamiltonian for the system can now be written

$$H = \sum_{ks} (\hbar kc) b_{ks}^{\dagger} b_{ks} + \sum_{n} E_{ne} a_{ne}^{\dagger} a_{ne} + \sum_{n} E_{ng} a_{ng}^{\dagger} a_{ng} + H_{\text{int}} \tag{4.12}$$

where E_{ne} and E_{ng} are the energies respectively in the excited or ground state of nuclei in the nth state of excitation in the external potential and H_{int} is the interaction (4.2). We must now express H_{int} in the new basis. We first consider the transformation to the new basis of the operator product

$$a_{(\boldsymbol{k}-\boldsymbol{q})\mathrm{g}}^{\dagger} a_{\boldsymbol{k}\mathrm{e}} = \sum_{n'n''} \langle n''\mathrm{g}|(\boldsymbol{k}-\boldsymbol{q})\mathrm{g}\rangle\langle \boldsymbol{k}\mathrm{e}|n'\mathrm{e}\rangle a_{n''\mathrm{g}}^{\dagger} a_{n'\mathrm{e}} \tag{4.13}$$

where the transformation coefficients $\langle n''g|(\boldsymbol{k}-\boldsymbol{q})g\rangle$ and $\langle ke|n'e\rangle$ are just the transformation coefficients from single-particle plane-wave states to single-particle states bound in the external potential (4.11). We now suppress the indices g and e in these transformation coefficients since they refer to the motion of the nucleus as a whole in the external potential and are negligibly affected by the state of internal excitation of the nucleus. We therefore neglect the mass difference responsible for the thermal shift. The mass difference can be treated by perturbation theory as in the simpler treatment in section 2.8. By analogy with eq. (2.20),

$$\begin{aligned}\langle n''|\boldsymbol{k}-\boldsymbol{q}\rangle &= \int \psi_{n''}^{*}(\boldsymbol{x}) \mathrm{e}^{\mathrm{i}(\boldsymbol{k}-\boldsymbol{q})\cdot\boldsymbol{x}} \mathrm{d}x \\ &= \int \psi_{n''}^{*}(\boldsymbol{x}) \mathrm{e}^{-\mathrm{i}\boldsymbol{q}\cdot\boldsymbol{x}} \mathrm{e}^{\mathrm{i}\boldsymbol{k}\cdot\boldsymbol{x}} \mathrm{d}x = \langle n''|\mathrm{e}^{-\mathrm{i}\boldsymbol{q}\cdot\boldsymbol{x}}|\boldsymbol{k}\rangle \end{aligned} \tag{4.14}$$

where $\boldsymbol{x}$ is the center-of-mass coordinate of the whole nucleus and $\psi_{n''}(\boldsymbol{x})$ is the Schrödinger wave function for the state $|n''\rangle$. The operator $\mathrm{e}^{-\mathrm{i}\boldsymbol{q}\cdot\boldsymbol{x}}$ changes a plane wave of wave vector $\boldsymbol{k}$ into a plane wave of wave vector $\boldsymbol{k}-\boldsymbol{q}$. A simple expression for the sum of eq. (4.13) over the index $\boldsymbol{k}$ is obtained by using (4.14) and closure

$$\sum_{k} a^{\dagger}_{(k-q)\mathrm{g}} a_{k\mathrm{e}} = \sum_{n'n''k} \langle n''|\mathrm{e}^{-\mathrm{i}q\cdot x}|k\rangle\langle k|n'\rangle a^{\dagger}_{n''\mathrm{g}} a_{n'\mathrm{e}}$$
$$= \sum_{n'n''} \langle n''|\mathrm{e}^{-\mathrm{i}q\cdot x}|n'\rangle a^{\dagger}_{n''\mathrm{g}} a_{n'\mathrm{e}}. \tag{4.15}$$

Substituting eq. (4.15) into eq. (4.2), we obtain a simple expression for the interaction in the new basis,

$$H_{\mathrm{int}} = \sum_{n'n''qs} C_{qs}\langle n''|\mathrm{e}^{-\mathrm{i}q\cdot x}|n'\rangle a^{\dagger}_{n''\mathrm{g}} a_{n'\mathrm{e}} (b^{\dagger}_{qs} + b_{qs}) + \mathrm{h.c.} \tag{4.16}$$

where we have used eq. (4.10b).

The 'golden rule' formula of time-dependent perturbation theory gives the transition probability per unit time for the emission of the photon from a nucleus initially in the state $|n'\rangle$ and finally in the state $|n''\rangle$.

$$W_{n'\to n''} = \frac{2\pi}{\hbar} |C_{qs}|^2 |\langle n''|\mathrm{e}^{-\mathrm{i}q\cdot x}|n'\rangle|^2 \varrho_f. \tag{4.17a}$$

Substituting the result (4.6) for a free nucleus

$$W_{n'\to n''} = W^{\mathrm{F}}_{i\to f_s} |\langle n''|\mathrm{e}^{\mathrm{i}q\cdot x}|n'\rangle|^2. \tag{4.17b}$$

To compare this result with the simple treatment of section 2.3, we calculate the probability $P_{n'\to n''}$ that the nucleus is in the state $|n''\rangle$ after emitting the gamma ray if it is in the state $|n'\rangle$ before. This is obtained by normalizing the result (4.17). Since the energy differences between the different final states n'' are very small with respect to the energy of the gamma ray, we neglect the effect of the change in the gamma ray energy on the density of final states ϱ_f, thus ϱ_f is independent of n'', and eq. (4.17) can be summed over all final states n''. The total transition probability per unit time from the initial state n' is

$$\sum_{n''} W_{n'\to n''} = \frac{2\pi}{\hbar} |C_{qs}|^2 \varrho_f. \tag{4.18}$$

Thus,

$$P_{n'\to n''} = \frac{W_{n'\to n''}}{\sum_{n''} W_{n'\to n''}} = |\langle n''|\mathrm{e}^{-\mathrm{i}q\cdot x}|n'\rangle|^2. \tag{4.19}$$

This result is in exact agreement with the corresponding result (2.24) of the simple treatment which did not consider the radiation process. Since all the results for the gamma ray energy spectrum obtained in sections 2.3–2.8 follow from this eq. (4.19) and the conservation of energy (2.25), the same results follow from the present treatment which includes a description of the radiation process. The only difference is that energy conservation as described in the simple treatment does not hold exactly because of the finite width of the gamma ray line. The Hamiltonian (4.12) does not assume energy conservation, and a proper solution of the Schrödinger equation defined by this Hamiltonian includes all the effects of the finite line width.

The extension of the simple treatment in ch. 3 to the Mössbauer effect in a solid and systems of thermal equilibrium are essentially independent of the radiation process and can be incorporated in the present treatment in essentially the same way. Once the interaction is expressed in the convenient form (4.16) in which the states $|n'\rangle$ and $|n''\rangle$ are any single-particle basis, it can be transformed to any other basis, such as the normal modes of a crystal. The results are the same as in ch. 3.

The relation (4.17b) between the transition probabilities from free and bound nuclei illustrates a very important and general feature of momentum transfer to a member of a bound system. The result (4.17b) for bound nuclei is the product of the result for free nuclei and a factor $|\langle n''|\mathrm{e}^{-\mathrm{i}\boldsymbol{q}\cdot\boldsymbol{x}}|n'\rangle$ which depends only on the properties of the bound states $|n''\rangle$ and $|n''\rangle$ and the momentum transfer $\hbar\boldsymbol{q}$, and is independent of the radiation process. This factor is called a form factor or structure factor and appears in many processes in many areas of physics. In the remaining sections of this chapter, we shall show how the same approach can be used in three widely different areas of physics: (1) scattering of photons by bound atoms, (2) nuclear beta decay and (3) high-energy electron scattering. In all these the golden rule formula can be written in a form similar to (4.17b).

4.3 Scattering of photons by bound atoms

The interaction (4.16) can describe the scattering as well as the emission of photons by atoms bound in an external potential or in a crystal. Consider an initial state containing a photon in a plane-wave state with wave vector $\boldsymbol{q}_i$ and the nucleus in its ground state moving in a crystal lattice whose state is denoted by the quantum number n_i.

$$|i\rangle = |\boldsymbol{q}_i, g, n_i\rangle. \tag{4.20}$$

A scattering process is a transition from this state (4.20) to a final state with a photon in another plane wave state $\boldsymbol{q}_f$, with the nucleus again in its ground state, but with the crystal in another state $|n_f\rangle$ which may or may not be the same as the initial state.

$$|f\rangle = |\boldsymbol{q}_f, g, n_f\rangle. \tag{4.21}$$

The transition is described by second-order time-dependent perturbation theory, since the interaction must act twice, once to annihilate the incident photon and once to create the scattered photon. In the intermediate state the nucleus is in an excited state $|e\rangle$. The interaction (4.16) can produce two types of intermediate states which contribute to the transition: those with no photons present, and those with both the incident and scattered photons present. The scattered photon can be emitted either before or after the absorption of the incident photon. In the intermediate state the crystal lattice can be in any arbitrary state $|n'\rangle$ and the nucleus can also be in any arbitrary excited state $|e\rangle$. We therefore generalize the interaction (4.16) by summing over all excited states $|e\rangle$ of the nucleus and replacing the coefficient C_{qs} with a coefficient C_{qse} depending also on the particular excited state $|e\rangle$. Since the only values of $\boldsymbol{q}$ which enter in the calculation are $\boldsymbol{q}_i$ and $\boldsymbol{q}_f$, we write C_{ie} and C_{fe} for $C_{q_i s_i e}$ and $C_{q_f s_f e}$ in the subsequent calculation. The transition probability is given by the golden rule formula for a second order process

$$W_{i\to f} = \frac{2\pi}{\hbar}\left|\sum_e \sum_{n'}\left\{C_{fe}C_{ie}^* \frac{\langle n_f|\mathrm{e}^{-\mathrm{i}\boldsymbol{q}_f\cdot\boldsymbol{x}}|n'\rangle\langle n'|\mathrm{e}^{\mathrm{i}\boldsymbol{q}_i\cdot\boldsymbol{x}}|n_i\rangle}{E_{n_i g} + \hbar q_i c - E_{n'e}} + C_{fe}^* C_{ie} \frac{\langle n_f|\mathrm{e}^{\mathrm{i}\boldsymbol{q}_i\cdot\boldsymbol{x}}|n'\rangle\langle n'|\mathrm{e}^{-\mathrm{i}\boldsymbol{q}_f\cdot\boldsymbol{x}}|n_i\rangle}{E_{n_i g} - \hbar q_f c - E_{n'e}}\right\}\right|^2 \varrho_f. \tag{4.22}$$

This is result is general and not limited to Mössbauer scattering. In fact, it applies strictly only to non-resonant scattering because the energy denominator in the first term vanishes in the case of resonance.

For non-resonant scattering, eq. (4.22) can be simplified because the energy denominators are of the order of nuclear energies and are very large compared to the lattice energies. The dependence of the energy denominator upon the particular state $|n'\rangle$ of the crystal lattice in the intermediate state can therefore be neglected and the sums over the intermediate lattice state carried out by closure.

$$W_{i\to f} = W^{\mathrm{F}}\left|\langle n_f|\mathrm{e}^{\mathrm{i}(\boldsymbol{q}_i-\boldsymbol{q}_f)\cdot\boldsymbol{x}}|n_i\rangle\right|^2, \tag{4.23}$$

where

$$W^{\mathrm{F}} = \frac{2\pi}{\hbar}\left|\sum_e\left\{\frac{C_{fe}C_{ie}^*}{E_g + \hbar q_i c - E_e} + \frac{C_{fe}^* C_{ie}}{E_g - E_e - \hbar q_f c}\right\}\right|^2 \varrho_f \tag{4.24}$$

is the transition probability for scattering from a free nucleus. This result is directly analogous to the result (4.17) obtained for the emission process. The transition probability is a product of the transition probability for the corresponding process with a free nucleus, and a factor depending only on the momentum transfer and the initial and final states of the lattice. This factor is just the Debye–Waller factor and was, in fact, first introduced for scattering processes rather than emission in considering the scattering of X-rays by atoms in a crystal.

The results (4.23) and (4.24) apply equally well to non-resonant scattering of X-rays by *atoms* bound in a crystal as well as by nuclei. Let the indices g and e refer to the ground and excited states of an atom rather than of a nucleus. Although the excitation energies of atomic levels are much lower than those of nuclear levels, they are still several orders of magnitude greater than the energies of lattice vibrations and the closure approximation is still valid for eq. (4.22). One further modification is necessary for the application of the result (4.23) to describe X-ray diffraction by crystals. Since the photon can be scattered by any one of the atoms in the crystal, the transition matrix element consists of a sum of terms of the type (4.23), one for each atom. Thus if $\boldsymbol{x}_k$ represents the coordinate of the kth atom, we have

$$W_{i\to f} = W^{\mathrm{F}} \Big| \sum_k \langle n_f |^{\mathrm{i}(\boldsymbol{q}_i - \boldsymbol{q}_f)\cdot \boldsymbol{x}_k} | n_i \rangle \Big|^2. \tag{4.25}$$

The sum is clearly coherent and leads to the usual constructive interference at the Bragg angles if the scattering is *elastic*; i.e. $|n_f\rangle = |n_i\rangle$, and there is no energy transfer to the crystal. For the inelastic processes, $|n_f\rangle \neq |n_i\rangle$, the scattered X-rays have an essentially continuous energy spectrum because of the energy transfer to the lattice. Because of the corresponding continuous spectrum of wavelengths, this scattering is incoherent.

Note the two equivalent interpretations of (4.25) related by wave–particle duality. Our approach has been to consider the photon as a particle with momentum and energy transfer. The phase factors appearing in (4.25) result from momentum transfer, and the incoherent scattering results from inelastic energy transfer. The conventional interpretation of Bragg scattering uses a wave picture for the photon. The phase factors in (4.25) result from path differences in waves scattered from different atoms. The incoherent scattering results from fluctuations in the positions of the scattering atoms which lead to randomness in the phases of the scattered waves.

For the case of resonance scattering, such as occurs in the Mössbauer effect, the treatment must be modified to take into account radiation damping effects. These modify the singular term by adding an imaginary part to the denominator. We shall not consider this treatment in detail here as we are only interested in the difference between the resonance scattering by a free nucleus and one bound in a crystal. For the resonant case, the entire contribution can be considered to come from the particular term in (4.22) for which the energy denominator vanishes. This is a term involving one particular excited state $|e\rangle$ of the nucleus and one particular lattice intermediate state $|n'\rangle=|n_i\rangle$. (Although it is possible in principle to have resonance scattering from a particular lattice state $|n'\rangle$ which is not the same as the initial state, the matrix element is so small for any *one* state except the initial state that this can be neglected.) Since $|n'\rangle=|n_i\rangle$ the lattice energies do not affect the energy denominator nor the necessary modifications due to damping. The result thus again factors into the transition probability for a free nucleus multiplied by a factor expressing the effects of the binding:

$$W_{i\to f,n_f} = W^{\mathrm{F}}\left|\langle n_f|\mathrm{e}^{-\mathrm{i}\boldsymbol{q}_f\cdot\boldsymbol{x}}|n_i\rangle\langle n_i|\mathrm{e}^{\mathrm{i}\boldsymbol{q}_i\cdot\boldsymbol{x}}|n_i\rangle\right|^2. \tag{4.26}$$

This result contains *two* momentum transfer factors like the Debye–Waller factor, one for the incident momentum $\boldsymbol{q}_i$ and one for the final momentum $\boldsymbol{q}_f$. In non–resonant X-ray scattering there is only a single factor depending only on the momentum transfer. This difference results from the requirement that a particular state of the lattice be the intermediate state in the resonance scattering and that the momentum of the incident photon be absorbed by the lattice. In non-resonant scattering, there is no requirement on the intermediate state and the momentum transfer to the lattice need not be elastic in the intermediate state.

The case $|n_f\rangle=|n_i\rangle$ in eq. (4.26) gives the probability of a 'double Mössbauer effect' in which there is no energy loss due to recoil, neither in the absorption of the photon nor in the re-emission. For this case the emitted photon comes out with the same energy as the incident photon. Such elastically scattered radiation can be used in Bragg scattering experiments and can be used as a source for a further Mössbauer experiment. In most Mössbauer experiments, the energy of the scattered photon is not measured and the transition probability is effectively summed over all possible final states. The sum over all states $|n_f\rangle$ is easily done in eq. (4.26) by closure to give

$$\sum_f W_{i\to f} = W^{\mathrm{F}}\left|\langle n_i|\mathrm{e}^{\mathrm{i}\boldsymbol{q}_i\cdot\boldsymbol{x}_i}|n_i\rangle\right|^2. \tag{4.27}$$

Here we have the usual momentum transfer factor for the Mössbauer effect, a single factor in which the total momentum of the gamma ray appears. In this case the momentum transfer factor refers to the absorption process alone since the emission process is summed over all lattice states and includes inelastic as well as elastic components.

4.4 Nuclear beta decay

We now apply the same approach to nuclear beta decay. Consider first the decay of a neutron into a proton, electron and neutrino. An interaction describing this process has the form

$$H_{\text{int}} = \sum_{kqs} C_{kqs} a^{\dagger}_{\text{p},k-q} a^{\dagger}_{\text{e},q-s} a^{\dagger}_{\nu s} a_{\text{n}k} + \text{h.c.} \tag{4.28}$$

where we have introduced creation and annihilation operators for protons, neutrons, electrons and neutrinos (p, n, e and ν) in plane-wave states and disregard the spin degrees of freedom of these particles. We have assumed momentum conservation in writing the interaction (4.28) and thus have an interaction coupling C_{kqs} depending upon three variables, the momentum $\hbar\boldsymbol{k}$ of the neutron in the initial state, the momentum transfer $\boldsymbol{q}$ and the neutrino momentum $\boldsymbol{s}$. The dependence of C_{kqs} on the initial momentum $\boldsymbol{k}$ is again determined by Lorentz invariance, since this dependence describes the relation between the decay processes seen by observers moving with different velocities relative to the decaying neutron. For non-relativistic neutron velocities, Lorentz invariance reduces to Galilean invariance, and again requires that C_{kqs} be independent of $\boldsymbol{k}$, neglecting corrections of order v/c. If beta decay is described by first-order perturbation theory, the transition probability is given by the golden rule and is proportional to the square of the matrix element of the interaction (4.28) between the initial and final states of the unperturbed Hamiltonian.

Let us now consider the beta decay of a complex nucleus rather than that of a free neutron. The interaction (4.28) remains the same. We need only to modify the unperturbed Hamiltonian to include the nuclear binding forces. We neglect spin, the Coulomb interaction between the emitted electron and the positively charged nucleus, and relativistic corrections which are of order v/c, where v is the velocity of the neutron in the nucleus. The unperturbed Hamiltonian thus includes all the dynamics of nuclear structure. Its eigenfunctions describe systems of complex nuclei and free non-interacting electrons and neutrinos. Consider the transition between an initial state $|n_i\rangle$ of a single complex nucleus with no electrons or neutrinos present and a final state $|n_f, \boldsymbol{qs}\rangle$, including a complex nucleus in a state $|n_f\rangle$ along with a neutrino of momentum $\hbar\boldsymbol{s}$ and an electron of momentum $\hbar(\boldsymbol{q}-\boldsymbol{s})$. The transition probability per unit time is given by the golden rule formula

$$W_{n_i \to n_f, q, s} = \frac{2\pi}{\hbar} |C_{qs}|^2 \, |\langle n_f| \sum_{k} a^{\dagger}_{\text{p},k-q} a_{\text{n}k} |n_i\rangle|^2 \, \varrho_f . \tag{4.29}$$

We again find that the probability for a transition separates into two factors. One, $|C_{qs}|^2$ depends only on the beta decay interaction and is independent of states of the complex nucleus. The second factor depends only on the states of the complex nucleus and the momentum transfer $\boldsymbol{q}$. Let us compare this result (4.29) with the corresponding result (4.17) for the Mössbauer effect. The interaction coupling C_{qs} here depends upon the indices $\boldsymbol{q}$ and $\boldsymbol{s}$, because two particles, an electron and a neutrino, are emitted in beta decay rather than a single photon. The momenta of the emitted electron and neutrino are not determined completely by energy and momentum conservation; they can share the energy and momentum between them in different ways. However the allowed values of $\boldsymbol{q}$ and $\boldsymbol{s}$ are restricted by the requirement of energy conservation which has not yet been considered in eq. (4.29).

The nuclear transition matrix element in (4.29) consists of a momentum transfer factor similar to that of the Mössbauer effect. However, a complex nucleus contains many neutrons, any one of which may make the transition to a proton, whereas in the Mössbauer effect one particular nucleus in the crystal makes the transition from the excited state to the ground state. This is shown explicitly by rewriting the result (4.29) using the Schrödinger representation for the transition operator

$$W_{n_i \to n_f, \boldsymbol{q}, \boldsymbol{s}} = \frac{2\pi}{\hbar} |C_{qs}|^2 \, |\langle n_f| \sum_j e^{-i\boldsymbol{q}\cdot\boldsymbol{x}_j} \tau_{+j} |n_i\rangle|^2 \, \varrho_f \tag{4.30}$$

where τ_{+j} is the isospin operator which changes the jth particle from a neutron to a proton.

In nuclear beta decay, the energies of the emitted electron and neutrino are of the same order of magnitude as the spacing between the energy levels of the complex nucleus. The density of final states ϱ_f thus varies considerably with n_f, in contrast to the case of the Mössbauer effect. It is therefore not possible to obtain simple expressions analogous to (4.19) for the relative probabilities of different transitions. However, other simplifying factors exist in conventional beta decay. The electrons and neutrinos emitted in beta decay have energies of the order of 1 MeV. Their wavelengths are much larger than the nuclear radius. The quantity $\boldsymbol{q}\cdot\boldsymbol{x}_j$ is therefore very much smaller than unity for values of $\boldsymbol{x}_j$ within the nuclear radius. The probability of finding a nucleon outside the nuclear radius decreases to zero very sharply at the nuclear surface. Thus the exponential factor in eq. (4.30) can be expanded in powers of the argument. The transition probability is given to a good approximation by the first term in the expansion having a non-vanishing matrix element between the two nuclear states.

The strongest transitions are those for which the contribution of the first term in the expansion does not vanish. These terms are called allowed transitions. The transition probability for allowed transitions is obtained by replacing the exponential in eq. (4.30) by unity.

$$W_{n_i \to n_f, q, s}(\text{allowed}) = \frac{2\pi}{\hbar} |C_{qs}|^2 \, |\langle n_f | T_+ | n_i \rangle|^2 \, \varrho_f \tag{4.31}$$

where

$$T_+ = \sum_j \tau_{+j} \tag{4.32}$$

is the plus component of the total isospin of the whole nucleus. Since T_+ commutes with both the operators of the total isospin and those of the total ordinary angular momentum, the final nuclear state $|n_f\rangle$ must have the same isospin and the same angular momentum as the initial state $|n_i\rangle$ in an allowed transition described by eq. (4.31). These transitions are called Fermi transitions and the restrictions on isospin and angular momentum are usually expressed as the following selection rules:

$$\Delta J = 0; \quad \Delta T = 0 \quad \text{in allowed Fermi transitions,} \tag{4.33}$$

where J is the total angular momentum and T is the total isospin of the nucleus.

If spin is included (still nonrelativistically) in this treatment, two types of nuclear matrix elements should appear in eq. (4.29); one in which the spin of the created proton is oriented in the same direction as that of the annihilated neutron and one with a 'spin flip' in which the spin of the proton is reversed from the direction of the neutron spin. The matrix element considered above in (4.29) and (4.30) corresponds to no spin change. A matrix element describing spin flip is constructed by adding a spin-dependent factor which turns out to be a Pauli spin operator. Transitions governed by the additional spin factor are called Gamow–Teller transitions. The nuclear matrix element for Gamow–Teller transitions is

$$\langle n_f | GT | n_i \rangle = \langle n_f | \sum_j \mathrm{e}^{-i\boldsymbol{q}\cdot\boldsymbol{x}_j} \boldsymbol{\sigma}_j \tau_{+j} | n_i \rangle \tag{4.34}$$

where $\boldsymbol{\sigma}_j$ is a Pauli spin operator.

The factor $\mathrm{e}^{i\boldsymbol{q}\cdot\boldsymbol{x}_j}$ can be expanded in the same manner as for Fermi transitions. The first term in the expansion describes allowed Gamow–Teller transitions. The operator $\sum_i \boldsymbol{\sigma}_j \tau_{+j}$ is no longer the total isospin of the system but transforms like a vector in both ordinary angular momentum space and isospin space. We thus obtain the following selection rules for allowed

Gamow–Teller transitions:

$$\begin{aligned} &\Delta J = 0 \quad \text{or} \quad 1, \qquad \Delta T = 0 \quad \text{or} \quad 1 \quad \text{in allowed GT transitions} \\ &J = 0 \rightarrow J = 0 \quad \text{excluded} \end{aligned} \tag{4.35}$$

In the Fermi theory of beta decay, the coupling coefficient C_{qs} is assumed to be independent of the momentum transfer $\boldsymbol{q}$. This follows from the assumption that the interaction (4.28) is a 'local' interaction; i.e. that the proton, electron and neutrino appear at the same point $\boldsymbol{x}$ in configuration space where the neutron disappears. The expressions for the interaction in configuration space and momentum space are related by simple Fourier transforms. It is easily shown that the requirement of a local interaction in configuration space results in the independence of the coefficient C_{kqs} on the momenta $\boldsymbol{q}$ and $\boldsymbol{s}$.

Since the dependence of the nuclear matrix element on the momentum transfer $\boldsymbol{q}$ drops out for allowed transitions, the only momentum-dependent factor remaining in the expression (4.34) is the density ϱ_f of the final states. The energy spectrum of the electrons and neutrinos emitted in beta decay therefore depend only on the statistical factor ϱ_f. The mass of the nucleus is very large. Thus the kinetic energy of the nuclear recoil can be neglected in energy conservation calculations. The density ϱ_f thus refers to the density of states for the electron–neutrino system subject to the constraint from energy conservation on the total energy of the electron and neutrino. The shape of the beta ray spectrum in allowed transitions is thus determined completely by the total energy available for the electron and neutrino and is the same for all allowed transitions having the same energy. This result has been confirmed by experimental studies of beta decay.

The interaction (4.28) is easily generalized to include all types of 'four-fermion weak interactions', such as the decay of a proton into a neutron plus positron and neutrino, the capture of an electron by a proton to produce a neutron and neutrino, and similar reactions involving μ-mesons instead of electrons. For positron decay and electron capture, the results are similar to those for negative electron emission and lead to the same selection rules (4.33) and (4.35).

For μ-meson capture by a complex nucleus, an expression similar to (4.30) is obtained, with corrections for the effects of spin and relativity. The emitted neutrino has a very high momentum since it carries away the main part of the rest energy of the muon, about 100 MeV. Expansion of the exponential factor in the matrix element (4.30) is not useful since the exponent is large. The nuclear matrix elements normally classified as allowed and forbidden are of the same order of magnitude. The form of the nuclear matrix element resembles very much the transition matrix element for the Mössbauer effect. If spins are neglected and the nuclear states are described by the harmonic oscillator shell model, the transition matrix element looks exactly like that of the Mössbauer effect. The transition corresponds to a proton in a particular harmonic oscillator orbit, changing into a neutron and either remaining in the same orbit or going into another orbit. If it remains in the same orbit, the matrix element is just the Debye–Waller factor (2.28).

In the case of the Mössbauer effect, the results depend upon the binding of the nucleus in the lattice only in the form of matrix elements of the momentum transfer factor $e^{i\boldsymbol{q}\cdot\boldsymbol{x}}$ between various states; i.e. upon the Fourier transform of the transition probability density corresponding to the particular transition. This is also expressed in terms of the space-time correlation function for the motion of the nucleus in the lattice. Conversely we see that Mössbauer measurements as a tool to investigate the lattice dynamics cannot give any more information than the set of all transition probability densities or the space-time correlation function. In the case of μ-capture, where any proton in the nucleus can turn into a neutron, additional information about nuclear structure is in principle obtainable, in addition to the space-time correlation function for a given neutron. To see this, let us first write the square of the transition matrix element.

$$\begin{aligned}|M_{n_{i,\mu}\to n_{f,\nu}}|^2 &= |\langle n_f|\sum_j e^{-i\boldsymbol{q}\cdot\boldsymbol{x}_j}\tau_{-j}|n_i\rangle|^2 \\ &= \langle n_i|\sum_k e^{i\boldsymbol{q}\cdot\boldsymbol{x}_k}\tau_{+k}|n_f\rangle\langle n_f|\sum_j e^{-i\boldsymbol{q}\cdot\boldsymbol{x}_j}\tau_{-j}|n_i\rangle. \qquad (4.36)\end{aligned}$$

We now obtain a sum rule by summing the result (4.36) over all possible nuclear final states n_f and evaluating the sum of the right-hand side by closure to obtain

$$\sum_{n_f} |M_{n_{i,\mu}\to n_{f,\nu}}|^2 = \sum_{jk} \langle n_i|e^{-i\boldsymbol{q}\cdot(\boldsymbol{x}_j-\boldsymbol{x}_k)}\tau_{+k}\tau_{-j}|n_i\rangle. \qquad (4.37)$$

This corresponds to the sum (2.34) in the Mössbauer effect which is always equal to unity because there is only one nucleus in the crystal which makes the transition; i.e. $j=k$. Here we have the expectation value of an operator which tells us something about the structure of the nuclear ground state. In the absence of the isospin operators, it would be a Fourier transform of the 'pair correlation function' defined as the probability distribution for finding two particles in the nucleus at a given distance from one another. The isospin factors introduce an additional charge exchange between a proton and a neutron. For other processes, such as high-energy electron scattering, where the momentum is transferred to the nucleons in the nucleus without any charge exchange, the corresponding expression gives the Fourier transform of the pair correlation function directly. These experiments have been proposed as a way to measure the pair correlation function.

4.5 Electron scattering by complex nuclei

We now consider electron scattering by a nucleus. The interaction responsible for the scattering is just the Coulomb interaction between the electron and the individual protons in the nucleus. This interaction is a two-body operator which is easily expressed in the second-quantized notation:

$$H_{\mathrm{C}} = \sum_{k'k''q} V_{k'k''q} a^{\dagger}_{k'+q} c^{\dagger}_{k''-q} c_{k''} a_{k'} \tag{4.38}$$

where $a^{\dagger}_{k}$ and $c^{\dagger}_{k}$ are proton and electron creation operators respectively. The coefficient $V_{k'k''q}$ is just the Fourier transformation of the Coulomb interaction in configuration space.

$$V_{k'k''q} = \int \frac{\mathrm{d}\boldsymbol{r}_{\mathrm{p}}}{\Omega} \int \frac{\mathrm{d}\boldsymbol{r}_{\mathrm{e}}}{\Omega} \,\mathrm{e}^{-\mathrm{i}(\boldsymbol{k}'+\boldsymbol{q})\cdot \boldsymbol{r}_{\mathrm{p}}} \mathrm{e}^{-\mathrm{i}(\boldsymbol{k}''-\boldsymbol{q})\cdot \boldsymbol{r}_{\mathrm{e}}} V(|\boldsymbol{r}_{\mathrm{e}} - \boldsymbol{r}_{\mathrm{p}}|)\, \mathrm{e}^{\mathrm{i}\boldsymbol{k}''\cdot \boldsymbol{r}_{\mathrm{e}}} \mathrm{e}^{\mathrm{i}\boldsymbol{k}'\cdot \boldsymbol{r}_{\mathrm{p}}} \tag{4.39}$$

where Ω is the volume of the box in which the wave functions are normalized. The factors depending on $\boldsymbol{k}'$ and $\boldsymbol{k}''$ drop out of the integrand, which is seen to depend only on the momentum transfer $\boldsymbol{q}$ and the relative coordinate

$$\boldsymbol{r} = \boldsymbol{r}_{\mathrm{e}} - \boldsymbol{r}_{\mathrm{p}}. \tag{4.40a}$$

Thus, we can write

$$V_{k'k''q} = V_q = \int \frac{\mathrm{d}\boldsymbol{r}}{\Omega} \,\mathrm{e}^{\mathrm{i}\boldsymbol{q}\cdot \boldsymbol{r}} V(r). \tag{4.40b}$$

By the symmetry of the integrand $V_{k'k''q}$ depends only on the magnitude of $\boldsymbol{q}$.

Let us examine the scattering of an electron by a complex nucleus in first-order time-dependent perturbation theory, analogous to our treatment of beta decay. We consider a transition between an initial state $|n_i, k''\rangle$ containing a single complex nucleus in the state $|n_i\rangle$ and an electron with momentum $\hbar \boldsymbol{k}''$, and a final state $|n_f, \boldsymbol{k}''-\boldsymbol{q}\rangle$ containing a complex nucleus in a state $|n_f\rangle$ and an electron with momentum $|\hbar(\boldsymbol{k}''-\boldsymbol{q})\rangle$. The transition probability per unit time is given by the golden rule formula:

$$W_{(n_i, k'')\to(n_f, k''-q)} = \left\{\frac{2\pi}{\hbar} |V_q|^2 \varrho_f\right\} |\langle n_f| \sum_{k'} a^{\dagger}_{k'+q} a_{k'} |n_i\rangle|^2. \tag{4.41a}$$

By analogy with eq. (4.17) this can be separated into two factors.

$$W_{(n_i,\boldsymbol{k}'')\to(n_f,\boldsymbol{k}''-\boldsymbol{q})} = W^{\mathrm{F}}\left|\langle n_f|\sum_{\boldsymbol{k}'} a^{\dagger}_{\boldsymbol{k}'+\boldsymbol{q}}a_{\boldsymbol{k}'}|n_i\rangle\right|^2 \tag{4.41b}$$

where

$$W^{\mathrm{F}} = \frac{2\pi}{\hbar}|V_q|^2 \varrho_f. \tag{4.41c}$$

The first factor W^{F} in eq. (4.41b) depends only on the electron–proton interaction and on the density of final states of the electron and is determined by the scattering of electrons on free protons. The second factor depends only on the states of the complex nucleus and the momentum transfer. This second factor is called the square of the electric form factor of the nucleus. The form factor can be written in the Schrödinger representation by analogy with eq. (4.30)

$$\langle n_f|\sum_{\boldsymbol{k}'} a^{\dagger}_{\boldsymbol{k}'+\boldsymbol{q}}a_{\boldsymbol{k}'}|n_i\rangle = \langle n_f|\sum_{j=1}^{Z} \mathrm{e}^{i\boldsymbol{q}\cdot\boldsymbol{x}_j}|n_i\rangle \tag{4.42}$$

where Z is the number of protons in the nucleus, and the summation is over all protons.

For the case of elastic scattering, $n_f = n_i$, the form factor can be expressed as:

$$\langle n_i|\sum_{j=1} \mathrm{e}^{i\boldsymbol{q}\cdot\boldsymbol{x}_j}|n_i\rangle = \int \varrho(x)\mathrm{e}^{i\boldsymbol{q}\cdot\boldsymbol{x}}\mathrm{d}\boldsymbol{x} \tag{4.43a}$$

where

$$\varrho(x) = \sum_{j=1}^{Z} \mathrm{d}\boldsymbol{x}_1 \ldots \mathrm{d}\boldsymbol{x}_A \psi^*_{n_i}\delta(\boldsymbol{x} - \boldsymbol{x}_j)\psi_{n_i} \tag{4.43b}$$

is the electric charge density.

The electric form factor is seen to be the Fourier transform of the electric charge density in the nucleus. Electron scattering experiments thus measure the charge distribution in a nucleus. Just as in the case of the Mössbauer effect, at small momentum transfers one measures the mean square of the charge distribution. At higher momentum transfers, one obtains more information about the shape of the charge distribution.

In this simple treatment we have disregarded the spin of the electron and proton and their magnetic moments and magnetic interactions. When these are properly taken into account one finds that there are two contributions to the scattering process which can be described as electric and magnetic. The electric form factor depends upon the charge distribution of the nucleus, as we have seen. The magnetic form factor depends on the magnetic moment density distribution of the nucleus.

In this case, in contrast to beta decay, we know the exact form of the interaction (4.38) from first principles, because we know the form of the Coulomb interaction and we know the electric charges of the electron and the proton. We are therefore in a position to evaluate the expression (4.41) completely and obtain the cross section for Coulomb scattering of an electron by a proton to first order in perturbation theory.

We first must relate eq. (4.41) for the transition probability per unit time to the scattering cross section. Since the initial-state wave function is normalized in a box of volume Ω, the probability per unit time that an electron will pass through a unit area normal to the direction of motion is v_i/Ω, where v_i is the initial velocity of the electron. We assume at this stage that the nucleus is infinitely heavy so that its recoil velocity can be neglected. The scattering cross section is the ratio of the transition probability per unit time to the probability per unit per unit time that an electron passes through a unit area normal to its motion. Thus:

$$\sigma(n_i, \boldsymbol{k}'' \to n_f, \boldsymbol{k}'' - \boldsymbol{q}) = \frac{2\pi\Omega}{\hbar v_i} |V_q|^2 \varrho_f |\langle n_f| \sum_{\boldsymbol{k}'} a^\dagger_{\boldsymbol{k}'+\boldsymbol{q}} a_{\boldsymbol{k}'} |n_i\rangle|^2. \tag{4.44}$$

Let us consider the cross section for the scattering of an electron into an infinitesimal element of solid angle $\sin\theta \, d\theta \, d\phi$ in the direction specified by polar angles (θ, ϕ). The density of final states is:

$$\varrho_f \, dE_f = \frac{\Omega}{(2\pi)^3} k_f^2 \, dk_f \sin\theta \, d\theta \, d\phi \tag{4.45a}$$

$$\varrho_f = \frac{\Omega}{\hbar^2 c^2 (2\pi)^3} k_f E_f \sin\theta \, d\theta \, d\phi \tag{4.45b}$$

where $\hbar k_f$ and E_f are the momentum and energy of the electron final state, related by relativistic energy and momentum conservation relations

$$k_f = |\boldsymbol{k}'' - \boldsymbol{q}| \tag{4.46a}$$

$$\hbar c k_f d(\hbar c k_f) = E_f dE_f. \tag{4.46b}$$

The velocity of the electron in the final state v_f is given by

$$\frac{v_f}{c} = \frac{\hbar k_f c}{E_f}. \tag{4.46c}$$

The differential cross section for scattering into a unit solid angle about the polar angles θ, ϕ is obtained by substituting the relations (4.45) and (4.46) into eq. (4.44).

$$\sigma_{n_i \to n_f}(\theta, \phi) = \frac{1}{(2\pi\hbar^2)^2} \left(\frac{v_f}{v_i}\right) \left(\frac{E_f}{c}\right)^2 |\Omega V_q|^2 \, |\langle n_f| \sum_{\boldsymbol{k}'} a^\dagger_{\boldsymbol{k}'+\boldsymbol{q}} a_{\boldsymbol{k}'} |n_i\rangle|^2. \tag{4.47}$$

This expression can be separated into three factors, each having a different physical interpretation. The first is a kinematic factor depending on the velocities and energies of the initial and final states, and independent of all other details of the scattering process. The second factor depends upon the electron–proton interaction. From eq. (4.39) we see that the volume of the box cancels out of this expression as it should, and that it depends only on the momentum transfer $\boldsymbol{q}$. This result is general and does not depend upon the specific form of the Coulomb potential. This can be seen by substituting a general potential $U(r)$ which depends only upon the distance between the proton and the electron in place of the Coulomb potential $1/r$. The third factor is the form factor for the nucleus which depends only on the momentum transfer $\boldsymbol{q}$ and on the properties of the bound state. It is independent of the other kinematical variables or details of the elementary scattering process. The result (4.47) is nonrelativistic only because we have used the nonrelativistic matrix element (4.40). All other factors are relativistic; thus the relation (4.47) can easily be made relativistic by inserting an expression for $V_{\boldsymbol{k}'\boldsymbol{k}''\boldsymbol{q}}$ calculated from a relativistic description.

The principal approximation which has gone into the expression (4.47) has been the assumption that first-order perturbation theory gives an adequate answer; i.e. the neglect of higher-order terms. This approximation in the calculation of a scattering cross section is commonly called the Born approximation. A characteristic of the Born approximation is the dependence exhibited in eq. (4.47) on the kinematic factors. The quantity v_i/E_f depends only on the momentum transfer $\boldsymbol{q}$, regardless of the form of the interaction or the bound-state form factor. Thus the validity of the Born approximation can be tested by measurement of this quantity for different values of incident energy and scattering angle adjusted to keep the same momentum transfer. The result should always be the same if the Born approximation is valid. When the recoil motion of the nucleus and relativistic effects are taken into account, it is still possible to separate the Born approximation expression for the scattering cross section into the product of a known kinematic factor and an unknown function of the momentum transfer $\boldsymbol{q}$, and therefore still possible to test the validity of the Born approximation by separating out the known kinematic factor.

Let us know apply the formula (4.47) to the case of nonrelativistic elastic Coulomb scattering. Then $v_i = v_f$ and $E_f = mc^2$, where m is the electron mass. We evaluate the integral (4.40b) by first converting to polar coordinates with the polar axis in the direction of the $\boldsymbol{q}$; the integration over the angular variables is easily carried out to give:

$$V_{k'k''q} = -e \int_{\Omega} \frac{e^{iqr\cos\theta}}{r} r^2 \sin\theta \, d\theta \, d\phi = \frac{-4\pi e^2}{q} \int_0^R \sin qr \, dr. \tag{4.48}$$

The radial integral seems to be very sensitive to the value of the upper limit R, which is associated with the size of the box. This difficulty is clearly unphysical and is avoided by the following trick. We multiply the integrand by a screening factor $e^{-\kappa r}$ and let the size of the box go to infinity. We then take the limit as $\kappa \to 0$. We thus obtain:

$$V_{k'k''q} = -\frac{4\pi e^2}{q} \lim_{\kappa \to 0} \int_0^\infty e^{-\kappa r} \sin qr \, dr = \lim_{\kappa \to 0} \left(-\frac{4\pi e^2}{q^2 + \kappa^2} \right) \tag{4.49a}$$

$$= -\frac{4\pi e^2}{q^2}. \tag{4.49b}$$

The trick has a simple physical interpretation. The difficulty depends on the existence of the Coulomb interaction at very large distances. In all practical cases the potential will not go out to very large distances because the Coulomb field of a nucleus is always screened by its atomic electrons. Even if it is an ion in a crystal lattice or in an ionic beam, the field is screened at large distances by the walls of the apparatus. Thus the expression (4.49a) with a small finite value for κ is probably closer to reality than the unscreened expression (4.48). The form of the result (4.49a) indicates that the effect of this screening on the scattering process will be small when $\kappa \ll \boldsymbol{q}$; i.e. if the screening factor changes very slowly with respect to the period of oscillation of $\sin \boldsymbol{q}r$. For this case the expression (4.49b) is a satisfactory approximation. The result (4.49a) with a finite value of κ can be used to describe the scattering by a screened atom, where the value of κ is obtained from some model of the atom. We assume that the screening radius is large and use the expression (4.49b).

Substituting into (4.47), we obtain for the elastic scattering cross section

$$\sigma_{\text{el}}(\theta, \phi) = \left(\frac{2me^2}{q^2} \right)^2 \left| \langle n_i | \sum_{\boldsymbol{k}'} a^\dagger_{\boldsymbol{k}'+\boldsymbol{q}} a_{\boldsymbol{k}'} | n_i \rangle \right|^2. \tag{4.50a}$$

This is expressed in terms of the electron velocity and scattering angle by substituting

$$\hbar q = 2mv \sin \tfrac{1}{2}\theta. \tag{4.50b}$$

Thus

$$\sigma_{\text{el}}(\theta, \phi) = \left(\frac{e^2}{2mv^2} \right)^2 (\operatorname{cosec}^4 \tfrac{1}{2}\theta) \left| \langle n_i | \sum_{\boldsymbol{k}'} a^\dagger_{\boldsymbol{k}'+\boldsymbol{q}} a_{\boldsymbol{k}'} | n_i \rangle \right|^2. \tag{4.50c}$$

This is just the Rutherford formula for Coulomb scattering multiplied by the square of the form factor for momentum transfer $\boldsymbol{q}$ to the protons in the nucleus. For the case of a 'point nucleus' where the charge density is a delta function the form factor is just equal to the constant Z, independent of $\boldsymbol{q}$, and the result is the Rutherford formula for the scattering of an electron by a Coulomb potential with charge Ze.

MANY-PARTICLE QUANTUM MECHANICS FOR PEDESTRIANS

This monograph presents the quantum mechanics of many-particle systems both as the basis of modern atomic, nuclear and solid state physics and as a laboratory for the study of basic concepts used in relativisistic quantum mechanics and field theory. The monograph is divided into two parts. Chapters 5 and 6 present the second-quantized notation at a very early level and develop its applications in an elementary manner. Chapters 9, 10 and 11 consider more sophisticated applications and methods.

Second quantization is introduced without any relation to field theory as a notation for describing states of identical and indistinguishable particles without giving them unphysical labels. The student thus begins his study of many-particle systems with a formalism in which the *indistinguishability is built in from the beginning* and does not need to be put in afterwards by symmetry conditions on wave functions. The consequences of this indistinguishability are then developed at an elementary level to obtain boson and fermion commutation and anticommutation relations, the restrictions on symmetries of two-particle states, the enhancement of boson matrix elements with many particles in the same state, and the presence of 'exchange terms' in matrix elements of two-body operators.

The special role of the vacuum is discussed with a view to future applications. The mathematical technique of evaluating matrix elements by reduction to vacuum expectation values is presented. The concept of the 'relevant vacuum' is introduced first as a device to eliminate irrelevant 'particles on the moon' and then as a useful approximate description for closed shell atomic and nuclear configurations. This picture of a 'formal vacuum' really describing systems of many degrees of freedom which can be polarized and excited is a useful guide to the student's intuition in future problems.

The power of the second-quantized notation in treating systems of different numbers of particles on the same footing and in treating composite systems is illustrated in ch. 6. The simple question 'is the deuteron a boson or a fermion', is answered by explicitly constructing the appropriate creation and annihilation operators and evaluating commutators. The result that deuterons are 'almost but not quite bosons' is interpreted in two directions. The corrections to boson behavior are seen to be negligible at the low densities present in all physical problems involving physical deuterons, and the general rule that systems containing odd numbers of fermions behave like fermions and those containing even numbers behave like bosons is seen to be justified as long as the overlaps of wave functions of different composite systems can be neglected. However, the case of high density overlapping bound fermion pairs irrelevant for deuterons is shown to be interesting for the case of pairing correlations in superconductors and nuclei. Many of the qualitative features of the BCS description are shown to be a natural result of the intermediate behavior of these pairs, being neither bosons nor fermions but something in between. A simplified model exactly soluble by means of the quasispin techniques commonly used in nuclear physics shows much of the physics of seniority in nuclei and of superconductivity in metals.

The more advanced treatment begins in ch. 9 with a discussion of the kind of information desired in a many-body problem since the enormous amount of information contained in a complete many-body wave function cannot be handled and is not really desirable. The non-interacting Fermi gas is treated and the concepts of particles and holes introduced. The Thomas–Fermi and Hartree–Fock approximations are introduced in an unorthodox manner using the second-quantized notation, followed by a demonstration of the equivalence of this approach to a self-consistent field picture. Two classes of applications to physical systems are discussed: 1. Atomic and nuclear systems which have a small number of particles, spherical symmetry, single-particle orbitals which are angular momentum eigenfunctions and shell structure. 2. Solids and nuclear matter which contain very large numbers of particles, translational symmetry and orbitals which are eigenfunctions of the linear momentum. Two examples, a finite nucleus and the electron gas are carried through the treatment to display the similarities and the differences in the applications of techniques which are formally similar in small finite systems and infinite systems.

The effects of correlations omitted in the Hartree–Fock approximation are discussed and some of the methods for dealing with them are presented in chs. 10 and 11.

Chapter 10 presents the basic method of the BCS theory of superconductivity for the treatment of pairing correlations. The strong-coupling limit is shown to be equivalent to the simplified model discussed in ch. 6. The BCS wave function and the Bogoliubov–Valatin transformation are presented as generalizations of the simple treatment and simple applications to superconductors and nuclei are discussed.

Chapter 11 considers the particle–hole excitations commonly treated by the random-phase approximation. The concept of elementary excitation is developed in which the desired information is not the properties of the exact ground or excited states of the system but rather the properties of excitations produced by simple operators corresponding to simple physical excitation mechanisms such as photon absorption. Giant resonances are discussed and collective and single-particle excitations.

The random-phase approximation is developed in the formalism of linearizing the equations of motion and applied to the two cases of a finite nucleus and the electron gas. The monograph ends with a brief introduction to more sophisticated methods; the time-dependent Hartree–Fock approximation, sum rules, time-dependent formulations, spectral functions and Green's functions.

Many of the concepts developed in this treatment are not only useful in many-body problems but provide a basis for the intuition in dealing with relativistic quantum mechanics and field theory. The use of the second-quantized notation considerably simplifies the treatment of time-dependent perturbation theory in ch.7 by providing a formalism which can treat processes in which the number of particles changes. The concepts of the relevant vacuum, particles and holes and the time-dependent formulations of elementary excitations are used in chs. 12 and 13.

CHAPTER 5

IDENTICAL PARTICLES AND SECOND QUANTIZATION

5.1 Introduction

In quantum mechanics the state of a two particle system can be described by a Schrödinger wave function $\psi(x_1, x_2)$ which depends on the coordinates of the two particles. The square of the wave function gives the probability that particle number one is found at the point x_1 and that simultaneously particle number two is found at the point x_2. Consider now two identical and indistinguishable particles like the two electrons in the helium atom. There are no labels on the two electrons. There is no way to specify which electron is number one and which is number two. The two electron state is completely described by saying that there is one electron at point a and another electron at point b. The notation $\psi(x_1, x_2)$ is redundant; both $\psi(a, b)$ and $\psi(b, a)$ give the probability amplitude for finding one electron at point a and one at point b. Thus $\psi(a, b)$ and $\psi(b, a)$ are not independent. They can differ only by a phase factor, since their magnitudes describe the same physical quantity.

$$|\psi(a, b)| = |\psi(b, a)|. \tag{5.1}$$

This redundance is generally removed by requiring $\psi(x_1, x_2)$ to be either symmetric or antisymmetric under the exchange of x_1 and x_2.

For systems of more than two identical particles the notation $\psi(x_1, x_2, \ldots, x_n)$ becomes even more cumbersome and redundant. The absolute value of the function should not change under any permutation of the variables, since such a permutation leads to another expression for the same physical probability amplitude:

$$|\psi(x_1, x_2, \ldots, x_n)| = |\psi(x_2, x_1, \ldots, x_n)| \text{ etc.} \tag{5.2}$$

The redundance is conventionally removed by requiring the function to have some kind of permutation symmetry, such as being symmetric or antisymmetric in the variables. A better description of systems of identical particles would state directly the probability amplitude for finding one particle at point a and another particle at point b, rather than first introducing unphysical labels 1 and 2 for the particles and then removing the unphysical redundance by requiring the function not to change its absolute magnitude under permutations.

The Schrödinger many-particle wave functions are also unsuitable for describing processes in which the number of particles can change; e.g. the emission and absorption of photons or mesons and the creation and annihilation of electron–positron pairs. To describe such processes states having different numbers of particles must be in the same Hilbert space and be connected by operators which change the number of particles. These processes are usually treated in the framework of quantum field theory.

We shall now develop the formalism usually called second quantization or the occupation number representation, which avoids these difficulties of the Schrödinger representation. The concept of operators which create and annihilate particles can be introduced directly at an elementary level simply as a convenient notation to describe systems of identical particles, without going into field theory. This formalism has no redundance in describing systems of identical particles and can treat processes which change the number of particles.

5.2 The vacuum and one-particle states

We begin by writing a state vector representing the vacuum. This is denoted as

$$|0\rangle$$

and represents the state with no particles.

We now consider all possible states for a single particle. Using Schrödinger wave functions we can write:

$$|\boldsymbol{k}'\rangle = \mathrm{e}^{\mathrm{i}\boldsymbol{k}'\cdot\boldsymbol{x}}. \tag{5.3}$$

This defines a complete set of plane-wave states. In the second-quantized notation we write these states as:

$$|\boldsymbol{k}'\rangle \equiv a^{\dagger}_{\boldsymbol{k}'}|0\rangle. \tag{5.4}$$

The 'creation' operator $a^{\dagger}_{\boldsymbol{k}'}$ 'creates' a particle in the plane-wave state of wave vector $\boldsymbol{k}'$. A state of one particle is represented by a creation operator acting on the vacuum. The set of one particle states defined by all possible values of the wave vector $\boldsymbol{k}'$ defines a complete set. Thus, eq. (5.4) defines a complete set of one particle states.

The bra vector corresponding to the state (5.4) is denoted by

$$\langle \boldsymbol{k}'| \rightarrow \langle 0|a_{\boldsymbol{k}'} \tag{5.5}$$

where the operator $a_{\boldsymbol{k}'}$ is the Hermitean conjugate of the operator $a_{\boldsymbol{k}'}^{\dagger}$. If the states defined by the relation (5.4) constitute an orthonormal set,

$$\langle 0|a_{\boldsymbol{k}'}a_{\boldsymbol{k}''}^{\dagger}|0\rangle = \delta_{\boldsymbol{k}'\boldsymbol{k}''}. \tag{5.6}$$

The relations (5.4), (5.5) and (5.6) define a new notation for one-particle states which is exactly equivalent to the Schrödinger notation (5.3) and does not make calculations particularly simpler. Only in systems of two or more particles does the second-quantized notation become useful and more convenient.

It is not necessary to choose the plane-wave basis for the definition of second-quantized creation operators. Any other basis which we might prefer can always be expressed as a linear combination of plane-wave states. Similarly, creation operators corresponding to this other basis can also be defined as linear combinations of the operators $a_{\boldsymbol{k}'}^{\dagger}$. For example, let us define a new set of base vectors $|n'\rangle$ by the relation:

$$|n'\rangle = \sum_{\boldsymbol{k}'} |\boldsymbol{k}'\rangle\langle \boldsymbol{k}'|n'\rangle. \tag{5.7}$$

We can define in the same way the corresponding set of creation operators in terms of the operators $a_{\boldsymbol{k}'}^{\dagger}$:

$$a_{n'}^{\dagger} = \sum_{\boldsymbol{k}'} a_{\boldsymbol{k}'}^{\dagger}\langle \boldsymbol{k}'|n'\rangle \tag{5.8}$$

where the coefficients in the expansion (5.8) are the same as the ones in the expansion (5.7). The operator $a_{n'}^{\dagger}$ thus creates a particle in the state $|n'\rangle$.

One particular set of base vectors often used is:

$$|\boldsymbol{x}'\rangle = (2\pi)^{-\frac{3}{2}} \int_{-\infty}^{\infty} e^{-i\boldsymbol{k}'\cdot\boldsymbol{x}'}|\boldsymbol{k}'\rangle d\boldsymbol{k}'. \tag{5.9}$$

The corresponding creation operators are

$$a^{\dagger}(\boldsymbol{x}') = (2\pi)^{-\frac{3}{2}} \int_{-\infty}^{\infty} e^{-i\boldsymbol{k}'\cdot\boldsymbol{x}'} a_{\boldsymbol{k}'}^{\dagger} d\boldsymbol{k}'. \tag{5.10}$$

$a^{\dagger}(\boldsymbol{x}')$ thus represents an operator creating a particle at the point $\boldsymbol{x}'$ in space. There is no particular reason why $\boldsymbol{x}'$ must be written as an argument in parentheses while $\boldsymbol{k}'$ is written as a subscript; however, this is the conventional notation.

5.3 Two-particle and many-particle states. Operator commutation relations

Let us now consider all possible states for a two-particle system. Using Schrödinger wave functions we can write a complete set of plane-wave states for two free particles:

$$|\boldsymbol{k}', \boldsymbol{k}''\rangle = e^{i\boldsymbol{k}'\cdot\boldsymbol{x}_1} e^{i\boldsymbol{k}''\cdot\boldsymbol{x}_2}. \tag{5.11}$$

These describe states where particle number 1 has wave number $\boldsymbol{k}'$ and particle number 2 has wave number $\boldsymbol{k}''$. For a given values of $\boldsymbol{k}'$ and $\boldsymbol{k}''$ there are two states, one in which particle number 1 has wave number $\boldsymbol{k}'$ and particle number 2 has wave number $\boldsymbol{k}''$, and vice versa. We now postulate that the two particles are indistinguishable and that there is only *one* state where one particle has wave number $\boldsymbol{k}'$ and the other has wave number $\boldsymbol{k}''$. In the Schrödinger representation we must decide how to handle the two different wave functions of the form (5.11) which represent the same physical state. This problem is avoided in the second-quantized notation.

We define the second-quantized notation for the two-particle states by the relation:

$$|\boldsymbol{k}', \boldsymbol{k}''\rangle \to a^{\dagger}_{\boldsymbol{k}'} a^{\dagger}_{\boldsymbol{k}''}|0\rangle. \tag{5.12a}$$

This expression (5.12a) has two creation operators acting on the vacuum state and thus creates one particle with wave vector $\boldsymbol{k}'$ and one with wave vector $\boldsymbol{k}''$. The labels 1 and 2 for the particles are not introduced and need not confuse us. However, the same two-particle state can be written in two ways. For given values of $\boldsymbol{k}'$ and $\boldsymbol{k}''$ we have the state (5.12a) and also the state:

$$|\boldsymbol{k}'', \boldsymbol{k}'\rangle \to a^{\dagger}_{\boldsymbol{k}''} a^{\dagger}_{\boldsymbol{k}'}|0\rangle. \tag{5.12b}$$

The state vectors (5.12a) and (5.12b) represent the same physical state and will be equal if the two creation operators $a^{\dagger}_{\boldsymbol{k}'}$ and $a^{\dagger}_{\boldsymbol{k}''}$ commute. However the two expressions (5.12a) and (5.12b) need not be equal to describe the same physical state; they can differ by a factor. They will differ by a factor F if the operators satisfy the following commutation relation:

$$a^{\dagger}_{\boldsymbol{k}''} a^{\dagger}_{\boldsymbol{k}'} = F a^{\dagger}_{\boldsymbol{k}'} a^{\dagger}_{\boldsymbol{k}''}. \tag{5.13a}$$

This relation can also be written:

$$a^{\dagger}_{\boldsymbol{k}'} a^{\dagger}_{\boldsymbol{k}''} = a^{\dagger}_{\boldsymbol{k}''} a^{\dagger}_{\boldsymbol{k}'}/F. \tag{5.13b}$$

These commutation relations should hold in any basis and be invariant under the transformation (5.8) from one basis to another. This requires that the factor be the same for all states. Comparing eqs. (5.13a) and (5.13b) we see that only the factors ± 1 are allowed. Thus there are two possible commutation relations for the creation operators, namely,

$$a^\dagger_{k''}a^\dagger_{k'} = a^\dagger_{k'}a^\dagger_{k''} \tag{5.14a}$$

$$a^\dagger_{k''}a^\dagger_{k'} = -a^\dagger_{k'}a^\dagger_{k''}. \tag{5.14b}$$

Two creation operators can either *commute* (5.14a) or *anticommute* (5.14b). Particles for which the creation operators *commute* are called *bosons*; particles for which the creation operators *anticommute* are called *fermions.*

These simple rules lead immediately to some basic physical properties of systems of bosons and fermions. We first examine the case where $\boldsymbol{k}''=\boldsymbol{k}'$ in eqs. (5.14). For bosons, eq. (5.14a) becomes a trivial identity. However for fermions eq. (5.14b) becomes:

$$a^\dagger_{k'}a^\dagger_{k'} = -a^\dagger_{k'}a^\dagger_{k'} = 0. \tag{5.15}$$

Thus two fermions cannot occupy the same quantum state.

A further interesting property is found if we express a two-particle state in terms of center-of-mass and relative coordinates. Let $2\boldsymbol{K}$ be the wave vector of the center-of-mass motion and $2\boldsymbol{q}$ the wave vector of the relative motion. The commutation rules (5.14) then give:

$$a^\dagger_{K+q}a^\dagger_{K-q} = \pm a^\dagger_{K-q}a^\dagger_{K+q} \tag{5.16}$$

where the $+$ refers to bosons and the $-$ to fermions. We now construct a two-particle state which is a plane wave for the center-of-mass motion and a wave packet for the relative motion.

$$\sum_q g(\boldsymbol{q})a^\dagger_{K+q}a^\dagger_{K-q}|0\rangle = \pm \sum_q g(\boldsymbol{q})a^\dagger_{K-q}a^\dagger_{K+q}|0\rangle \tag{5.17a}$$

$$= \pm \sum_q g(-\boldsymbol{q})a^\dagger_{K+q}a^\dagger_{K-q}|0\rangle \tag{5.17b}$$

where $g(\boldsymbol{q})$ is any function of $\boldsymbol{q}$. Eq. (5.17a) follows from eq. (5.16). Eq. (5.17b) is obtained by changing the sign of the dummy variable $\boldsymbol{q}$ in the summation (5.17a). The function $g(\boldsymbol{q})$ must therefore be an *even function* of $\boldsymbol{q}$ for bosons and an *odd function* for fermions. (We neglect here the possible presence of other degrees of freedom required to describe the particles, such as spins. For particles with spin eqs. (5.16) are valid for the case where both particles are in the same spin state.)

Since the parity of the function $g(\boldsymbol{q})$ is directly related to the relative orbital angular momentum, it follows that the relative orbital angular momentum must be even for bosons and odd for fermions. Note that reversing the sign of the relative momentum is the same as interchanging the particles and that the function $g(\boldsymbol{q})$ is just the Schrödinger wave function for the relative motion in momentum space. Thus we can also say that boson wave functions must be symmetric with respect to interchange of the particles while fermion wave functions must be antisymmetric.

These symmetry properties are introduced as additional postulates in the Schrödinger representation. In dealing with Schrödinger wave functions describing several particles it is always necessary to check them to see that they satisfy the symmetry postulate. When the second-quantized notation is used, the symmetry properties result automatically from the commutation relations (5.14). Any state vector written in the second-quantized notation automatically satisfies the symmetry requirements.

The bra vector corresponding to the ket vector (5.12a) is defined by the Hermitean conjugate of eq. (5.12a).

$$\langle \boldsymbol{k}', \boldsymbol{k}''| \rightarrow \langle 0|a_{\boldsymbol{k}''}a_{\boldsymbol{k}'}. \tag{5.18}$$

The norm of this state is the inner product of eqs. (5.18) and (5.12a).

$$\langle 0|a_{\boldsymbol{k}''}a_{\boldsymbol{k}'}a^{\dagger}_{\boldsymbol{k}'}a^{\dagger}_{\boldsymbol{k}''}|0\rangle \equiv N_{\boldsymbol{k}'\boldsymbol{k}''} \tag{5.19a}$$

where the value of the normalization constant $N_{\boldsymbol{k}'\boldsymbol{k}''}$ can be chosen for convenience as shown below. A more general relation using orthogonality is:

$$\langle 0|a_{\boldsymbol{k}'''}a_{\boldsymbol{k}'}a^{\dagger}_{\boldsymbol{k}'}a^{\dagger}_{\boldsymbol{k}''}|0\rangle = N_{\boldsymbol{k}'\boldsymbol{k}''}\delta_{\boldsymbol{k}'''\boldsymbol{k}''}. \tag{5.19b}$$

The expectation value in eq. (5.19b) is the scalar product of two two-particle states. However we can also consider the first operator operating to the left on the vacuum bra and the remaining three operating to the right on the vacuum ket. With this interpretation the expectation value is the scalar product of two one-particle states, and the equation is an orthonormality relation. The state defined by the three operators acting to the right on the vacuum is thus seen to be

$$a_{\boldsymbol{k}'}a^{\dagger}_{\boldsymbol{k}'}a^{\dagger}_{\boldsymbol{k}''}|0\rangle = N_{\boldsymbol{k}'\boldsymbol{k}''}a^{\dagger}_{\boldsymbol{k}''}|0\rangle. \tag{5.20}$$

From eqs. (5.18), (5.19) and (5.20) we see that the operators $a_{\boldsymbol{k}'}$ can be considered either as creation operators acting to the left on a bra vector or as annihilation operators acting to the right on a ket vector. Acting to the left they *increase* by one the number of particles in the system; acting to the right, they *decrease* the number by one. An annihilation operator acting on the vacuum state must give zero since it cannot reduce the number of particles, while acting on a one-particle state it can only give the vacuum state or zero. Thus,

$$a_{k'}|0\rangle = 0 \tag{5.21a}$$

$$a_{k'}a_{k''}^{\dagger}|0\rangle = \delta_{k'k''}|0\rangle \tag{5.21b}$$

where the coefficient on the right-hand side of eq. (5.21b) is obtained from the orthonormality relation (5.6) for one-particle states.

We now investigate the commutation relations for the annihilation operators. The relations for two annihilation operators are obtained directly by taking the Hermitean conjugates of the corresponding relations (5.14) for the creation operators. Two annihilation operators are seen to commute for bosons and to anticommute for fermions; i.e. they behave in the same manner as the corresponding creation operators.

We now consider the commutation relations for a creation operator and an annihilation operator. From eq. (5.21) we obtain:

$$(a_{k'}a_{k''}^{\dagger} + a_{k''}^{\dagger}a_{k'})|0\rangle = (a_{k'}a_{k''}^{\dagger} - a_{k''}^{\dagger}a_{k'})|0\rangle = \delta_{k'k''}|0\rangle \tag{5.22}$$

for both bosons and fermions. This equation (5.22) gives the values of the commutators and anticommutators acting on the vacuum state but is clearly insufficient to establish the commutation relations. We next evaluate the boson commutator or the fermion anticommutator acting on a one-particle state. Using eqs. (4.14) (4.20) and (4.21) we obtain:

$$(a_{k'}a_{k''}^{\dagger} \mp a_{k''}^{\dagger}a_{k'})a_{k'}^{\dagger}|0\rangle = (\pm a_{k'}a_{k'}^{\dagger}a_{k''}^{\dagger} \mp a_{k''}^{\dagger})|0\rangle = \pm(N_{k'k''}-1)a_{k''}^{\dagger}|0\rangle \tag{5.23}$$

with the upper sign for bosons and lower sign for fermions. It is desirable to choose the normalization constant $N_{k'k''}$ to make the boson commutator or fermion anticommutator a c-number whose value is independent of the state upon which it acts. Comparing eqs. (5.22) and (5.23) we see that this can be achieved for the vacuum and one-particle states by setting

$$N_{k'k''} = 1 \pm \delta_{k'k''}. \tag{5.24}$$

Thus $N_{k'k''}=1$ except for the case $k'=k''$, where $N_{k'k''}=2$ for bosons and vanishes for fermions. The fermion result is expected, since two fermions cannot occupy the same state. The factor of two for the boson case has interesting physical implications which are seen in detail below. Qualitatively, one can say that two bosons 'like to occupy the same quantum state', while two fermions are unable to do so.

The complete set of commutation or anticommutation rules for bosons and fermions is then:

$$a_{k'}^{\dagger}a_{k''}^{\dagger} \mp a_{k''}^{\dagger}a_{k'}^{\dagger} = 0 = a_{k'}a_{k''} \mp a_{k''}a_{k'} \tag{5.25a}$$

$$a_{k'}a_{k''}^{\dagger} \mp a_{k''}^{\dagger}a_{k'} = \delta_{k'k''} \tag{5.25b}$$

where we use the + sign for fermions, − sign for bosons. These commutation rules (5.25) are invariant under a transformation such as eq. (5.8) to a new set of base vectors.

Eq. (5.24) gives the interesting result for bosons:

$$\langle 0|a_{\boldsymbol{k}'}a_{\boldsymbol{k}'}a^{\dagger}_{\boldsymbol{k}'}a^{\dagger}_{\boldsymbol{k}'}|0\rangle = 2\langle 0|a_{\boldsymbol{k}'}a^{\dagger}_{\boldsymbol{k}'}|0\rangle = 2 \tag{5.26a}$$

$$\langle 0|(a_{\boldsymbol{k}'})^n(a^{\dagger}_{\boldsymbol{k}'})^n|0\rangle = n!. \tag{5.26b}$$

These results look very much like a harmonic oscillator. A normalized state vector describing a state with n bosons in the same quantum state $\boldsymbol{k}'$ must have a normalization factor $(n!)^{-\frac{1}{2}}$.

$$(n!)^{-\frac{1}{2}}(a^{\dagger}_{\boldsymbol{k}'})^n|0\rangle. \tag{5.27}$$

In general, the normalized state vector for a state in which there are n' bosons in the state $|\boldsymbol{k}'\rangle$, n'' bosons in the state $|\boldsymbol{k}''\rangle$, etc., has the form

$$(n'!n''! \ldots)^{-\frac{1}{2}}(a^{\dagger}_{\boldsymbol{k}'})^{n'}(a^{\dagger}_{\boldsymbol{k}''})^{n''} \ldots |0\rangle. \tag{5.28}$$

For fermions the problem of the normalization factor does not exist since there can never be more than a single particle in the same quantum state.

States of more than two particles are written by applying the appropriate number of creation operators to the vacuum state. Once the commutation relations (5.25) for these operators have been established there are no further ambiguities.

5.4 Linear single-particle operators in the second-quantized notation

We now express operators representing observables in the second-quantized notation. Consider first states of only a single particle. Any linear operator in a one-particle system is completely specified by its matrix elements in some particular complete set of states. The operation of a given operator A on an arbitrary state vector $|\xi\rangle$ is given by the relation:

$$A|\xi\rangle = \sum_{\boldsymbol{k}'\boldsymbol{k}''} |\boldsymbol{k}''\rangle\langle\boldsymbol{k}''|A|\boldsymbol{k}'\rangle\langle\boldsymbol{k}'|\xi\rangle. \tag{5.29}$$

In the second-quantized notation we write:

$$|\xi\rangle \rightarrow \sum_{\boldsymbol{k}'} \langle\boldsymbol{k}'|\xi\rangle a^{\dagger}_{\boldsymbol{k}'}|0\rangle \tag{5.30a}$$

and from eq. (5.29):

$$A|\xi\rangle \rightarrow \sum_{\boldsymbol{k}'\boldsymbol{k}''} \langle\boldsymbol{k}''|A|\boldsymbol{k}'\rangle\langle\boldsymbol{k}'|\xi\rangle a^{\dagger}_{\boldsymbol{k}''}|0\rangle. \tag{5.30b}$$

This can be rewritten, using eq. (5.21b)

$$A|\xi\rangle \to \sum_{k''k'''} \langle k''|A|k'''\rangle (a^\dagger_{k''} a_{k'''}) \sum_{k'} \langle k'|\xi\rangle a^\dagger_{k'}|0\rangle \tag{5.30c}$$

The operator A is thus represented in second quantization as:

$$A|\xi\rangle \to \sum_{k'k''} \langle k''|A|k'\rangle a^\dagger_{k''} a_{k'}. \tag{5.31}$$

Eq. (5.30) states that the operator A is a sum of terms which annihilate a particle in one state $\boldsymbol{k}'$ and create it in another $\boldsymbol{k}''$. The magnitude of the term is just the matrix element of the operator between these two states in the Schrödinger representation. In other words, the matrix element of the operator A between the states $|\boldsymbol{k}'\rangle$ and $|\boldsymbol{k}''\rangle$ is just equal to the component of the state $A|\boldsymbol{k}'\rangle$ which is 'in the direction' of $|\boldsymbol{k}''\rangle$.

One example of a single-particle operator is the momentum:

$$\boldsymbol{p} = \hbar \sum_{k'} \boldsymbol{k}' a^\dagger_{k'} a_{k'}. \tag{5.32}$$

Eq. (5.32) is a single sum because the momentum operator is diagonal in this representation. Another example is:

$$\cos \boldsymbol{k}\cdot\boldsymbol{x} = \tfrac{1}{2}(\mathrm{e}^{\mathrm{i}\boldsymbol{k}\cdot\boldsymbol{x}} + \mathrm{e}^{-\mathrm{i}\boldsymbol{k}\cdot\boldsymbol{x}}) = \tfrac{1}{2} \sum_{k'} a^\dagger_{k'+k} a_{k'} + a^\dagger_{k'-k} a_{k'}. \tag{5.33}$$

This operator is the sum of two terms. The first increases the momentum of the particle by $\boldsymbol{k}$; the other decreaes the momentum of the particle by $\boldsymbol{k}$. The matrix elements are equal for all values of $\boldsymbol{k}'$.

We next examine the operation on two-particle states of the operator A defined by eq. (5.31). Consider the two-particle state:

$$|\boldsymbol{k}''', \boldsymbol{k}^{\mathrm{iv}}\rangle \to a^\dagger_{k'''} a^\dagger_{k^{\mathrm{iv}}}|0\rangle. \tag{5.34}$$

We can write:

$$A|\boldsymbol{k}''', \boldsymbol{k}^{\mathrm{iv}}\rangle \to \sum_{k'k''} \langle k''|A|k'\rangle a^\dagger_{k''} a_{k'} a^\dagger_{k'''} a^\dagger_{k^{\mathrm{iv}}}|0\rangle. \tag{5.35}$$

The annihilation operator $a_{k'}$ gives zero when operating directly on the vacuum state $|0\rangle$. Thus, all terms in the sum (5.35) vanish except those with $\boldsymbol{k}'=\boldsymbol{k}'''$ or $\boldsymbol{k}'=\boldsymbol{k}^{\mathrm{iv}}$. These terms are reduced by the commutation rules (5.25) to give

$$A|\boldsymbol{k}'''\boldsymbol{k}^{\mathrm{iv}}\rangle = \sum_{k''} \{\langle k''|A|k'''\rangle a^\dagger_{k''} a^\dagger_{k^{\mathrm{iv}}} \pm \langle k''|A|k^{\mathrm{iv}}\rangle a^\dagger_{k'''} a^\dagger_{k''}|0\}\rangle \tag{5.36a}$$

with the + sign for bosons, − sign for fermions. Both boson and fermion cases can be written in the same way by interchanging the order of the operators in the second term of eq. (5.36a)

$$A|\boldsymbol{k}'''\boldsymbol{k}^{\text{iv}}\rangle = \sum_{\boldsymbol{k}''} \{\langle \boldsymbol{k}''|A|\boldsymbol{k}'''\rangle a^{\dagger}_{\boldsymbol{k}''} a^{\dagger}_{\boldsymbol{k}^{\text{iv}}} + \langle \boldsymbol{k}''|A|\boldsymbol{k}^{\text{iv}}\rangle a^{\dagger}_{\boldsymbol{k}''} a^{\dagger}_{\boldsymbol{k}'''}\} |0\rangle. \quad (5.36\text{b})$$

The two terms in eqs. (5.36) have a very simple interpretation. The operator A, being a single-particle operator, changes the state of only one particle. In a two-particle system it can change the state of either of the two particles. In the first term of eq. (5.36) the operator A changes the state of the particle which was originally in state $\boldsymbol{k}'''$. In the second term the operator A changes the state of the particle which was initially in the state $\boldsymbol{k}^{\text{iv}}$. In the Schrödinger representation the wave functions are of the form (5.11), the two particles are labeled 1 and 2, and the operator A, eq. (5.30), would be represented as the sum of an operator acting only on the first particle and one acting only on the second.

$$A = A(x_1, p_1) + A(x_2, p_2). \quad (5.37)$$

The first term is a function of the dynamical variables of the first particle and the second is exactly the same function for the second particle.

The operation of the operator A, eq. (5.31) on an N-particle state is a natural generalization of eq. (5.36) and is given by N terms, each corresponding to a change in the state of one of the N particles. In the Schrödinger representation the generalization of eq. (5.37) to an N-particle system is a symmetric sum of N single-particle operators.

$$A = \sum_{i=1}^{N} A(x_i, p_i). \quad (5.38)$$

Operators of the type (5.31) are therefore called single-particle operators. Their action on any state changes the state of each particle singly in the manner indicated by the single-particle operator.

The terms in eq. (5.36) with $\boldsymbol{k}''=\boldsymbol{k}'''$ and $\boldsymbol{k}''=\boldsymbol{k}^{\text{iv}}$ are of particular interest. For fermions, the term with $\boldsymbol{k}''=\boldsymbol{k}'''$ in the second sum of the right-hand side of eq. (5.36) has two fermions in the same state and vanishes. Similarly, the term in the first sum with $\boldsymbol{k}''=\boldsymbol{k}^{\text{iv}}$ vanishes. Thus, although the operator A acts on each single particle individually, the many-particle character of the system manifests itself when the operator A tries to put one particle into a state which is already occupied by another particle. In a fermion system, this is impossible and the term vanishes.

Peculiar things also occur in boson systems when the operator A attempts to put a particle into a state already occupied by other particles. For bosons this transition is not forbidden; on the contrary, it is even encouraged. This can be seen from the normalization relations (5.26). The norm of the state:

$$a^{\dagger}_{k'''}a^{\dagger}_{k''}|0\rangle \tag{5.39}$$

is unity if $\boldsymbol{k}''\neq\boldsymbol{k}'''$. However, if $\boldsymbol{k}''=\boldsymbol{k}'''$ the norm of the state is 2. Consider the matrix elements of the operator (5.31) between the state $|\boldsymbol{k}'''\boldsymbol{k}^{\text{iv}}\rangle$ and the states $|\boldsymbol{k}''\boldsymbol{k}^{\text{iv}}\rangle$ and $|\boldsymbol{k}''\boldsymbol{k}'''\rangle$. These are given by the scalar products of the latter states with the state (5.36). If none of the individual $\boldsymbol{k}$ states are the same,

$$\langle\boldsymbol{k}''\boldsymbol{k}^{\text{iv}}|A|\boldsymbol{k}'''\boldsymbol{k}^{\text{iv}}\rangle = \langle\boldsymbol{k}''|A|\boldsymbol{k}'''\rangle \tag{5.40a}$$

$$\langle\boldsymbol{k}''\boldsymbol{k}'''|A|\boldsymbol{k}'''\boldsymbol{k}^{\text{iv}}\rangle = \langle\boldsymbol{k}''|A|\boldsymbol{k}^{\text{iv}}\rangle. \tag{5.40b}$$

The matrix elements are equal to the matrix elements of A between corresponding single particle states. On the other hand for a final state with two bosons in the same state, the normalization relation (5.26a) gives

$$\langle\boldsymbol{k}^{\text{iv}}\boldsymbol{k}^{\text{iv}}|A|\boldsymbol{k}'''\boldsymbol{k}^{\text{iv}}\rangle = \sqrt{2}\langle\boldsymbol{k}^{\text{iv}}|A|\boldsymbol{k}'''\rangle. \tag{5.41}$$

An additional factor of $\sqrt{2}$ occurs when the two bosons are in the same state. The same effect of course appears in the Schrödinger representation where one labels the particles and uses symmetrized functions for bosons and antisymmetrized functions for fermions. The symmetry properties of the initial wave function then lead to the result that two fermions cannot be in the same state, whereas two bosons have 'a greater probability of being in the same state'.

The enhancement of the probability that two bosons are in the same state becomes greater as the number of particles is increased. The norm of the state $(n!)^{-\frac{1}{2}}(a^{\dagger}_{k'''})^{n}a^{\dagger}_{k''}|0\rangle$ is unity if $\boldsymbol{k}''\neq\boldsymbol{k}'''$, but is $n+1$ if $\boldsymbol{k}''=\boldsymbol{k}'''$. By the same argument as given in eqs. (4.40) and (4.41), the result (4.40) is generalized to the case where n particles are already in the state $\boldsymbol{k}''$. A factor of $(n+1)^{\frac{1}{2}}$ enhances the matrix element when the particle whose state is changed by the operator A goes into a state already occupied by n particles.

It is instructive at this point, to evaluate explicitly the matrix element of a single-particle operator A between a pair of two-particle states. For simplicity, we choose the matrix element $\langle\boldsymbol{k}^{\text{v}}\boldsymbol{k}'''|A|\boldsymbol{k}'''\boldsymbol{k}^{\text{iv}}\rangle$. We have kept one of the plane-wave states, $\boldsymbol{k}'''$; to be the same in both the initial and final state, since we have seen from eq. (5.36) that the operator A can change the state of only one particle at a time. We also require that $\boldsymbol{k}^{\text{iv}}\neq\boldsymbol{k}'''$ and $\boldsymbol{k}^{\text{iv}}\neq\boldsymbol{k}^{\text{v}}$. Using the expression (5.31) for the operator A in the second-quantized notation we obtain:

$$\langle \boldsymbol{k}^{\mathrm{v}}\boldsymbol{k}'''|A|\boldsymbol{k}'''\boldsymbol{k}^{\mathrm{iv}}\rangle = \langle 0|a_{\boldsymbol{k}^{\mathrm{v}}}a_{\boldsymbol{k}'''} \sum_{\boldsymbol{k}'\boldsymbol{k}''} \langle \boldsymbol{k}''|A|\boldsymbol{k}'\rangle a^{\dagger}_{\boldsymbol{k}''}a_{\boldsymbol{k}'}a^{\dagger}_{\boldsymbol{k}'''}a^{\dagger}_{\boldsymbol{k}^{\mathrm{iv}}}|0\rangle \tag{5.42a}$$

$$= \sum_{\boldsymbol{k}'\boldsymbol{k}''} \langle \boldsymbol{k}''|A|\boldsymbol{k}'\rangle\langle 0|a_{\boldsymbol{k}^{\mathrm{v}}}a_{\boldsymbol{k}'''}a^{\dagger}_{\boldsymbol{k}''}a_{\boldsymbol{k}'}a^{\dagger}_{\boldsymbol{k}'''}a^{\dagger}_{\boldsymbol{k}^{\mathrm{iv}}}|0\rangle. \tag{5.42b}$$

The expression for the matrix element of an operator between two two-particle states is thus expressed as the expectation value of a more complicated operator in the vacuum state. This is characteristic of the second-quantized formulation where the vacuum state can always be made to appear on both ends of any matrix element. In field-theoretical calculations of physical processes, results are always expressed in terms of such 'vacuum expectation values' of products of second-quantized operators. The standard method for evaluating such expectation values, known as Wick's theorem, is now used to evaluate equation (5.42).

The essential feature of the method is to rearrange the order of the operators using the commutation or anticommutation relations (5.25) to move the annihilation operators to the right and the creation operators to the left. The vacuum expectation value of an operator product in this order always vanishes since the last annihilation operator on the right acting on the vacuum ket gives zero as well as the first creation operator on the left acting on the vacuum bra. In this way, all the vacuum expectation values of operators are eliminated and only the products of delta functions from commutators or anticommutators remain.

We begin the evaluation of the vacuum expectation value in the expression (5.42) by bringing the last creation operator to the left. We use the notation indicating + for bosons and − for fermions whenever the two cases require different signs.

$$\langle 0|a_{\boldsymbol{k}^{\mathrm{v}}}a_{\boldsymbol{k}'''}a^{\dagger}_{\boldsymbol{k}''}a_{\boldsymbol{k}'}a^{\dagger}_{\boldsymbol{k}'''}a^{\dagger}_{\boldsymbol{k}^{\mathrm{iv}}}|0\rangle = \pm\langle 0|a_{\boldsymbol{k}^{\mathrm{v}}}a_{\boldsymbol{k}'''}a^{\dagger}_{\boldsymbol{k}''}a_{\boldsymbol{k}'}a^{\dagger}_{\boldsymbol{k}^{\mathrm{iv}}}a^{\dagger}_{\boldsymbol{k}'''}|0\rangle$$

$$= \langle 0|a_{\boldsymbol{k}^{\mathrm{v}}}a_{\boldsymbol{k}'''}a^{\dagger}_{\boldsymbol{k}''}a^{\dagger}_{\boldsymbol{k}^{\mathrm{iv}}}a_{\boldsymbol{k}'}a^{\dagger}_{\boldsymbol{k}'''}|0\rangle \pm \delta_{\boldsymbol{k}'\boldsymbol{k}^{\mathrm{iv}}}\langle 0|a_{\boldsymbol{k}^{\mathrm{v}}}a_{\boldsymbol{k}'''}a^{\dagger}_{\boldsymbol{k}''}a^{\dagger}_{\boldsymbol{k}'''}|0\rangle$$

$$= \pm\langle 0|a^{\dagger}_{\boldsymbol{k}^{\mathrm{iv}}}a_{\boldsymbol{k}^{\mathrm{v}}}a_{\boldsymbol{k}'''}a^{\dagger}_{\boldsymbol{k}''}a_{\boldsymbol{k}'}a^{\dagger}_{\boldsymbol{k}'''}|0\rangle \pm \delta_{\boldsymbol{k}'\boldsymbol{k}^{\mathrm{iv}}}\langle 0|a_{\boldsymbol{k}^{\mathrm{v}}}a_{\boldsymbol{k}'''}a^{\dagger}_{\boldsymbol{k}''}a^{\dagger}_{\boldsymbol{k}'''}|0\rangle. \tag{5.43a}$$

The first term vanishes because of the creation operator acting on the vacuum bra. The remaining term is reduced further by the same procedure and using eq. (5.21b).

$$\pm\,\delta_{\boldsymbol{k}'\boldsymbol{k}^{\mathrm{iv}}}\langle 0|a_{\boldsymbol{k}^{\mathrm{v}}}a_{\boldsymbol{k}'''}a^{\dagger}_{\boldsymbol{k}''}a^{\dagger}_{\boldsymbol{k}'''}|0\rangle =$$

$$= \delta_{\boldsymbol{k}'\boldsymbol{k}^{\mathrm{iv}}}[\langle 0|a_{\boldsymbol{k}^{\mathrm{v}}}a^{\dagger}_{\boldsymbol{k}''}a_{\boldsymbol{k}'''}a^{\dagger}_{\boldsymbol{k}'''}|0\rangle \pm \delta_{\boldsymbol{k}''\boldsymbol{k}'''}\langle 0|a_{\boldsymbol{k}^{\mathrm{v}}}a^{\dagger}_{\boldsymbol{k}'''}|0\rangle]$$

$$= \delta_{\boldsymbol{k}'\boldsymbol{k}^{\mathrm{iv}}}[\langle 0|a_{\boldsymbol{k}^{\mathrm{v}}}a^{\dagger}_{\boldsymbol{k}''}|0\rangle \pm \delta_{\boldsymbol{k}''\boldsymbol{k}'''}\langle 0|a_{\boldsymbol{k}^{\mathrm{v}}}a^{\dagger}_{\boldsymbol{k}'''}|0\rangle] = \delta_{\boldsymbol{k}'\boldsymbol{k}^{\mathrm{iv}}}[\delta_{\boldsymbol{k}^{\mathrm{v}}\boldsymbol{k}''} \pm \delta_{\boldsymbol{k}''\boldsymbol{k}'''}\delta_{\boldsymbol{k}'''\boldsymbol{k}^{\mathrm{v}}}]. \tag{5.43b}$$

Thus

$$\langle 0|a_{k^{\mathrm{v}}}a_{k'''}a^{\dagger}_{k''}a_{k'}a^{\dagger}_{k'''}a^{\dagger}_{k^{\mathrm{iv}}}|0\rangle = \delta_{k'k^{\mathrm{iv}}}\delta_{k^{\mathrm{v}}k''}[1 \pm \delta_{k'''}\delta_{k^{\mathrm{v}}}] \tag{5.44a}$$

and

$$\langle k^{\mathrm{v}}k'''|A|k'''k^{\mathrm{iv}}\rangle = \langle k^{\mathrm{v}}|A|k^{\mathrm{iv}}\rangle\{1 \pm \delta_{k'''k^{\mathrm{v}}}\}. \tag{5.44b}$$

Eq. (5.44a) presents a typical evaluation of a vacuum expectation value. Eq. (5.44b) is just the expected result. The matrix element between two-particle states of a single-particle operator is just equal to the matrix element of the operator between those single-particle states which are different in the two states provided that both particles are not in the same quantum state. When both particles are in the same quantum state, the additional term in equation (5.44) makes the matrix element vanish for the fermion case and introduces the additional factor of 2 for the boson case as we have already seen from eq. (5.41).

5.5 Number operators

A single-particle operator of particular interest is:

$$n_k = a^{\dagger}_k a_k. \tag{5.45}$$

The commutation rules (5.26) for the creation and annihilation operators give the following commutation relations for the operator n_k.

$$[n_k, a^{\dagger}_{k'}] = a^{\dagger}_k \delta_{kk'} \tag{5.46a}$$

$$[n_k, a_{k'}] = -a_k \delta_{kk'} \tag{5.46b}$$

$$[n_k, (a^{\dagger}_k)^m] = m(a^{\dagger}_k)^m. \tag{5.46c}$$

These relations hold for both boson and fermion operators. The operator n_k is of second degree in these operators; thus all the $\pm$ signs in eqs. (5.26) disappear in the calculation of the commutators (5.46). From eq. (5.46c):

$$n_k(a^{\dagger}_k)^m|0\rangle = m(a^{\dagger}_k)^m|0\rangle. \tag{5.47}$$

The state describing m particles in the quantum state $\boldsymbol{k}$ is thus an eigenfunction of the operator n_k with the eigenvalue m. The operator n_k is called the 'number operator' for the state $\boldsymbol{k}$. Its eigenvalue is the number of particles in the state $\boldsymbol{k}$. Any state in which there is a definite number of particles in the state $\boldsymbol{k}$ is an eigenfunction of the operator n_k. By eq. (5.46a) the operator n_k commutes with all creation operators creating particles in states $\boldsymbol{k}'$ orthogonal to $\boldsymbol{k}$. The operator n_k therefore gives the number of particles in the state $\boldsymbol{k}$ irrespective of the existence of particles in other states.

A state need not be an eigenfunction of the operator $n_{\boldsymbol{k}}$. States which are linear combinations of eigenfunctions of $n_{\boldsymbol{k}}$ with different eigenvalues do not have a definite number of particles in the state $\boldsymbol{k}$. Such states might describe a system of particles which are continually making transitions from one state to another as a result of mutual interactions.

Now consider the sum of the number of operators over the complete set of states $\boldsymbol{k}$.

$$N = \sum_{\text{all } \boldsymbol{k}} n_{\boldsymbol{k}}. \tag{5.48}$$

This operator gives the total number of particles in the system. Any state of a system containing a definite number of particles must be described by a state vector which is an eigenfunction of the operator N with an eigenvalue equal to the total number of particles. In those physical problems where particles are not created nor destroyed, such as electrons in an atom or a solid, or nucleons in a nucleus, the exact eigenfunctions for the system must be eigenfunctions of the operator N.

It is sometimes useful to consider state vectors which are not eigenfunctions of the operator N and describe a system which does not have a definite number of particles even in systems where particles are not created nor destroyed. Although a linear combination of states having different numbers of particles has no direct physical interpretation in such systems the use of these wave functions may make calculations easier. A similar situation arises in the use of the grand canonical ensemble in statistical mechanics.

5.6 Two-particle operators in the second-quantized notation

Let us now consider operators in a two-particle system which can change the state of both the particles simultaneously. In the Schrödinger representation such an operator has the matrix elements of the form:

$$\langle \boldsymbol{k}^{\text{iv}} \boldsymbol{k}''' | B | \boldsymbol{k}'' \boldsymbol{k}' \rangle. \tag{5.49a}$$

In the Schrödinger representation where the two particles carry labels one can write another matrix element of the type (5.49) with the wave vectors $\boldsymbol{k}'''$ and $\boldsymbol{k}^{\text{iv}}$ interchanged.

$$\langle \boldsymbol{k}''' \boldsymbol{k}^{\text{iv}} | B | \boldsymbol{k}'' \boldsymbol{k}' \rangle. \tag{5.49b}$$

To illustrate the meaning of these two matrix elements let us consider a specific example. Consider the operator describing the electrostatic Coulomb interaction

$$V = e^2/r_{12} \tag{5.50}$$

between two particles of charge e separated by a distance r_{12}. Let us write the matrix elements of the operator (5.50) using a plane-wave basis for the states of the two particles.

First consider the case where one particle is a proton and the other is a positive pion. The matrix elements for the proton–pion system are:

$$\langle \boldsymbol{k}_\pi^{\text{iv}} \boldsymbol{k}_\text{p}''' | V | \boldsymbol{k}_\text{p}'' \boldsymbol{k}_\pi' \rangle = \int \mathrm{e}^{-i\boldsymbol{k}^{\text{iv}}\cdot \boldsymbol{r}_1} \mathrm{e}^{-\boldsymbol{k}'''\cdot \boldsymbol{r}_2} \frac{e^2}{r_{12}} \mathrm{e}^{i\boldsymbol{k}''\cdot \boldsymbol{r}_2} \mathrm{e}^{i\boldsymbol{k}'\cdot \boldsymbol{r}_1} \mathrm{d}\boldsymbol{r}_1 \mathrm{d}\boldsymbol{r}_2 \tag{5.51a}$$

$$\langle \boldsymbol{k}_\pi''' \boldsymbol{k}_\text{p}^{\text{iv}} | V | \boldsymbol{k}_\text{p}'' \boldsymbol{k}_\pi' \rangle = \int \mathrm{e}^{-i\boldsymbol{k}'''\cdot \boldsymbol{r}_1} \mathrm{e}^{-i\boldsymbol{k}^{\text{iv}}\cdot \boldsymbol{r}_2} \frac{e^2}{r_{12}} \mathrm{e}^{i\boldsymbol{k}''\cdot \boldsymbol{r}_2} \mathrm{e}^{i\boldsymbol{k}'\cdot \boldsymbol{r}_1} \mathrm{d}\boldsymbol{r}_1 \mathrm{d}\boldsymbol{r}_2. \tag{5.51b}$$

The wave vectors carry the additional subscript p or π for proton or pion. The two matrix elements (5.51a) and (5.51b) are physically different; the state on the left-hand side in (5.51a) has the proton in the state with wave vector $\boldsymbol{k}'''$ and the pion in the state with wave vector $\boldsymbol{k}^{\text{iv}}$; whereas the state on the left-hand side of the matrix element (5.51b) has the proton in the state with wave vector $\boldsymbol{k}^{\text{iv}}$ and the pion in the state with wave vector $\boldsymbol{k}'''$. The two states on the left-hand sides of the matrix elements in eqs. (5.51) are related by an interchange of the proton and the pion. Since the proton and pion are different distinguishable particles these two states are physically different and have different properties.

Consider now the case when the two particles are identical, both pions or both protons. Both integrals (5.51a) and (5.51b) can be written for any set of values of the four wave vectors $\boldsymbol{k}'$, $\boldsymbol{k}''$, $\boldsymbol{k}'''$, and $\boldsymbol{k}^{\text{iv}}$. The two integrals have different values. But if the particles are identical both describe matrix elements between the same indistinguishable states. One can now ask, 'how does one combine these two matrix elements (5.51) to take into account the indistinguishability of the particles?'

By analogy with the one-particle operators we write the operator V in the second-quantized notation:

$$V \to \tfrac{1}{2} \sum_{\substack{k'k'' \\ k'''k^{\text{iv}}}} \langle \boldsymbol{k}^{\text{iv}} \boldsymbol{k}''' | V | \boldsymbol{k}'' \boldsymbol{k}' \rangle a^\dagger_{k^{\text{iv}}} a^\dagger_{k'''} a_{k''} a_{k'}. \tag{5.52}$$

Each term in eq. (5.52) describes the annihilation of two particles in states $\boldsymbol{k}'$ and $\boldsymbol{k}''$ and the creation of two particles in states $\boldsymbol{k}'''$ and $\boldsymbol{k}^{\text{iv}}$. This is expected for a two-particle operator; by annihilating two particles and creating two others it effectively changes the states of two particles. However, this operator does not recognize which particle appears in the state $\boldsymbol{k}'''$, the one originally in the state $\boldsymbol{k}'$ or the one originally in the state $\boldsymbol{k}''$. This is consistent with the indistinguishability of the identical particles.

The expression (5.52) contains four terms all describing the annihilation of a pair of particles in the same given pair of states $\boldsymbol{k}'$ and $\boldsymbol{k}''$ and the creation of a pair of particles in the same states $\boldsymbol{k}'''$ and $\boldsymbol{k}^{\text{iv}}$.

$$\langle \boldsymbol{k}^{\text{iv}}\boldsymbol{k}'''|V|\boldsymbol{k}''\boldsymbol{k}'\rangle a^{\dagger}_{\boldsymbol{k}^{\text{iv}}}a^{\dagger}_{\boldsymbol{k}'''}a_{\boldsymbol{k}''}a_{\boldsymbol{k}'} \tag{5.53a}$$

$$\langle \boldsymbol{k}^{\text{iv}}\boldsymbol{k}'''|V|\boldsymbol{k}'\boldsymbol{k}''\rangle a^{\dagger}_{\boldsymbol{k}^{\text{iv}}}a^{\dagger}_{\boldsymbol{k}'''}a_{\boldsymbol{k}'}a_{\boldsymbol{k}''} \tag{5.53b}$$

$$\langle \boldsymbol{k}'''\boldsymbol{k}^{\text{iv}}|V|\boldsymbol{k}''\boldsymbol{k}'\rangle a^{\dagger}_{\boldsymbol{k}'''}a^{\dagger}_{\boldsymbol{k}^{\text{iv}}}a_{\boldsymbol{k}''}a_{\boldsymbol{k}'} \tag{5.53c}$$

$$\langle \boldsymbol{k}'''\boldsymbol{k}^{\text{iv}}|V|\boldsymbol{k}'\boldsymbol{k}''\rangle a^{\dagger}_{\boldsymbol{k}'''}a^{\dagger}_{\boldsymbol{k}^{\text{iv}}}a_{\boldsymbol{k}'}a_{\boldsymbol{k}''}. \tag{5.53d}$$

The terms (5.53a) and (5.53c) involve matrix elements of the potential corresponding respectively to (5.51a) and (5.51b) for the case of distinguishable particles. These two matrix elements are different. Because of the difference in the order of the two creation operators in eq. (5.53a) and (5.53c) the two terms combine in the sum (5.52) as either the sum of the two matrix elements or the difference depending upon whether the particles are bosons or fermions. A similar statement can be made for the other two terms. This can be seen explicitly by writing the contribution of the four terms (5.53) to the sum (5.52) after rearranging the order of the operators to put them all in the same standard form, such as that of eq. (5.53a). For the fermion case, the anticommutation rule requires a change in sign for each permutation. The result obtained is

$$[\{\langle \boldsymbol{k}^{\text{iv}}\boldsymbol{k}'''|V|\boldsymbol{k}''\boldsymbol{k}'\rangle + \langle \boldsymbol{k}'''\boldsymbol{k}^{\text{iv}}|V|\boldsymbol{k}'\boldsymbol{k}''\rangle\} \\ \pm\{\langle \boldsymbol{k}^{\text{iv}}\boldsymbol{k}'''|V|\boldsymbol{k}'\boldsymbol{k}''\rangle + \langle \boldsymbol{k}'''\boldsymbol{k}^{\text{iv}}|V|\boldsymbol{k}''\boldsymbol{k}'\rangle\}]a^{\dagger}_{\boldsymbol{k}^{\text{iv}}}a^{\dagger}_{\boldsymbol{k}'''}a_{\boldsymbol{k}''}a_{\boldsymbol{k}'} \tag{5.54}$$

(+ for bosons and − for fermions). If the interaction is symmetric with respect to permutation of the two particles, as is the case for the Coulomb interaction (5.50), the matrix element does not change if the quantum numbers of the two particles are interchanged in *both* the initial *and* the final states. For this case the first two terms in eq. (5.54) are equal to one another and the third and fourth terms are also equal to one another.

Of particular interest is the case where the two states created are the same as the two states annihilated. Consider, for example, the case where $\boldsymbol{k}'''=\boldsymbol{k}''$ and $\boldsymbol{k}^{\text{iv}}=\boldsymbol{k}'$. For distinguishable particles, such as a proton and pion, eq. (5.51a) describes a diagonal matrix element, whereas (5.51b) describes a non-diagonal matrix element in which the same plane-wave states appear on both sides but the proton and the pion have been interchanged. The state on the right-hand side of the matrix element has the pion in the state $\boldsymbol{k}'$ and the proton in the state $\boldsymbol{k}''$ whereas the state on the left-hand side has the reverse. For indistinguishable particles, both matrix elements are diagonal since interchanging indistinguishable particles, does not change their state. Thus, the diagonal matrix elements of a two-body operator for identical particles is either the sum or the difference of the corresponding two matrix elements for distinguishable particles; the sum if the particles are bosons and the difference if they are fermions. The term corresponding to eq. (5.51a) which is diagonal also for distinguishable particles is often called the 'direct' term, whereas the term corresponding to eq. (5.51b) which involves exchanging the particles is often called the 'exchange' term. The sign of the exchange term is seen to be positive for bosons and negative for fermions as a result of the commutation rules.

In the Schrödinger representation, these exchange terms arise because of the necessity to use symmetrized or antisymmetrized wave functions. In the second-quantized notation the exchange terms arise naturally with their proper signs from the commutation rules. All effects of Bose statistics or of the Pauli exclusion principle are built into the commutation and anticommutation relations (5.26).

5.7 Spin, other degrees of freedom and different kinds of particles

Spin and other internal degrees of freedom of a particle are easily included in the second-quantized description. Consider the case of a particle of spin s. The orientation of the spin is specified by a quantum number which has $2s+1$ values representing the projection of the spin on some axis. One can define a complete set of second-quantized creation operators $a^{\dagger}_{\boldsymbol{k}\sigma}$ creating a particle in plane-wave states of wave vector $\boldsymbol{k}$ and in the spin state σ. The same commutation or anticommutation relations (5.26) hold for this case with the addition of the spin indices σ and an additional delta function in eq. (5.26b), requiring the spin states to be the same as well as the wave vectors.

The extension to other internal degrees of freedom such as isospin is straightforward. Systems of neutrons and protons can be described by nucleon creation and annihilation operators $a^{\dagger}_{\boldsymbol{k}\sigma\tau}$ and $a_{\boldsymbol{k}\sigma\tau}$ for nucleons in states of wave vector $\boldsymbol{k}$, spin σ and charge τ. The index τ takes on two values corresponding to neutrons and protons.

Consider now a system containing different kinds of identical particles, such as hydrogen atoms containing electrons and protons. Both electrons and protons are fermions and can be described by separate sets of second-quantized operators which satisfy the fermion anticommutation rules among themselves. What should be the appropriate relations between proton and electron operators; should they commute or anticommute? Since electrons and protons are completely different the operators might be expected to commute, as in the case of the Schrödinger operators referring to different independent degrees of freedom. On the other hand, since both protons and electrons are fermions their operators might be expected to anticommute with one another as well as among themselves. This is the case for neutrons and protons in the isospin formalism discussed above.

To deal with this problem we must carefully consider the physical difference between the statements: 'two completely different kinds of particles', or 'two different states of the same particle'. In quantum mechanics two different states of the same particle can be combined in linear combinations, where the *relative phase* of the two components is physically measurable in interference experiments. For electrons and protons such a relative phase is not measurable and has no physical meaning. A particle is *either an electron or a proton*. It cannot be a coherent superposition of the two with a physically measurable phase. There is said to be a 'superselection rule' between states having different values of the electric charge. The same is true for neutrons and protons, and illustrates an important physical difference between isospin and ordinary spin.

The specification of commutation or anticommutation rules between different kinds of particles can be determined by examining the behavior under a change of base vectors. If we adopt the convention that the electron operators commute with proton operators although they anticommute among themselves, difficulties arise in changing to a new set of base vectors which are linear combinations of electron and proton states. The commutation relations for the new basis would be very complicated and quite different from eq. (5.26). For electrons and protons this difficulty is purely formal because a state which is a linear combination of an electron and a proton has no physical significance. There is no physical significance in the relative phase of an electron state vector and proton state vector. However, if anticommutation relations are assumed throughout for electrons and protons no formal difficulties are encountered even when considering the non-physical linear combinations of neutrons and protons.

The difference between the spin and isospin is seen as follows. Nucleons can be considered either as two different kinds of particles, neutrons and protons, or as two states of the same particle. We might also consider electrons as two different kinds of particles: (1) those having their spin pointing 'up' and (2) those having their spin pointing 'down'. As long as there are no spin-dependent forces it makes no difference whether we consider all electrons as identical particles or consider the 'spin-up' particles as different from the 'spin-down' particles. In the presence of spin-dependent forces this picture breaks down because of the necessity to consider electrons having their spins oriented in other directions than up or down; i.e. states which are linear combinations of spin-up and spin-down. Formally this results from the existence of physical dynamical variables which do not commute with the z-component of the spin. There are no such physically measurable operators which do not commute with the electric charge or the z-component of the isospin. The operators τ_x and τ_y are purely formal operators, convenient for calculations, but do not correspond to observable dynamical variables.

If two kinds of fermions are *really* different in the sense that a *coherent linear combination* of states of the fermions has no physical significance, one can either choose the convention that their second-quantized operators commute or that they anticommute. As long as one chooses one of these two conventions and uses it consistently, physical results obtained are independent of which convention is chosen. Formal differences appear only as unobservable phases of state vectors.

In systems of different types of particles where some of the particles are bosons, the bosons introduce no ambiguity. Boson operators commute with all other second-quantized operators.

We can now generalize the result (5.17) requiring the spatial symmetry of a two-particle state to be symmetric for bosons and antisymmetric for fermions. For two particles with spin, we can construct a two-particle state analogous to (5.17)

$$\sum_{\boldsymbol{q},\sigma',\sigma''} g(\boldsymbol{q})h(\sigma',\sigma'')a^{\dagger}_{\boldsymbol{K}-\boldsymbol{q},\sigma'}a^{\dagger}_{\boldsymbol{K}+\boldsymbol{q},\sigma''} = \pm \sum_{\boldsymbol{q},\sigma',\sigma''} g(-\boldsymbol{q})h(\sigma'',\sigma')a^{\dagger}_{\boldsymbol{K}-\boldsymbol{q},\sigma'}a^{\dagger}_{\boldsymbol{K}+\boldsymbol{q},\sigma''} \tag{5.55}$$

where $h(\sigma', \sigma'')$ describes the spins of the two particles in the wave packet, the right-hand side of eq. (5.55) is obtained by interchanging the order of the creation operators, changing the sign of $\boldsymbol{q}$ and interchanging σ' and σ''. Thus for bosons, $g(\boldsymbol{q})$ is an even function if h is symmetric in the spin and $g(\boldsymbol{q})$ is odd if h is antisymmetric. For fermions $g(\boldsymbol{q})$ is odd if h is symmetric and $g(\boldsymbol{q})$ is even if h is antisymmetric.

This result is simply expressed in terms of the spin and orbital angular momenta. For bosons $h(\sigma', \sigma'')$ is symmetric for even total spin and antisymmetric for odd spin, while for fermions $h(\sigma', \sigma'')$ is antisymmetric for even spin and symmetric for odd spin. This is easily remembered by noting that for any spin s the state of *maximum* spin $2s$ is always symmetric, and $2s$ is even for bosons (integral spin) and odd for fermions (half-integral spin). Thus for both bosons and fermions, $g(\boldsymbol{q})$ is even for even spin and odd for odd spin. Thus even spin requires even parity and orbital angular momentum, while odd spin implies odd parity and orbital angular momentum for two identical particles, whether they are bosons or fermions.

Similar arguments hold for other degrees of freedom, such as isospin. However note that strange bosons and fermions can both have integral or half-integral isospin.

5.8 The relevant vacuum

The vacuum is defined as the state containing no particles. It therefore satisfies eq. (5.21a) for any state $\boldsymbol{k}'$. However, in most physical applications we are not really interested in a state which satisfies eq. (5.21a) for *every* possible state $\boldsymbol{k}'$. Suppose for example, that we are studying the three electrons in a lithium atom on the earth when there may be other electrons on the moon. We are not interested in the electrons on the moon and know that their effect on the lithium atom on earth should be negligible. Yet in the Schrödinger representation wave functions must be properly antisymmetrized with respect to all electrons. It can be shown that unless there are very peculiar strong long range interactions the electrons on the moon can be disregarded since their motion does not affect the motion of the electrons of the lithium atom on earth. In the second-quantized notation the problem of 'particles on the moon' is handled directly by defining the vacuum state to satisfy eq. (5.21a) only for operators which describe particles moving on the earth. We do not care what equation it satisfies for operators which create particles on the moon. The vacuum state is then not uniquely defined; it can describe any arbitrary state of particles on the moon. However, this specification of the vacuum is sufficient to give unique predictions for all observable quantities on the earth. These are expressible as vacuum expectation values analogous to eq. (5.42) of products of operators which annihilate and create particles on earth. These vacuum expectation values are uniquely defined if the vacuum state is specified to satisfy eq. (5.21a) for all states creating particles on the earth and are independent of the description of particles on the moon.

This concept of the 'relevant vacuum' which satisfies eq. (5.21a) only for operators creating particles in quantum states relevant for the problem under consideration can be extended in many ways. For example, many atomic and nuclear problems consider sets of particles in inert closed shells plus a few external valence particles in an unfilled shell. All the properties of interest are determined by the valence particles. A vacuum state can then be defined which satisfies eq. (5.21a) only for the states of the active valence shell. In contrast to the case of the moon where we do not care about the other states, we define the vacuum here uniquely as having all the states in the closed shells occupied. This allows us to take into account some of the properties of the particles in the closed shell; e.g., the screening of the Coulomb potential due to the nucleus by electrons in a closed atomic shell or the average attractive interaction of the closed shell nucleons in a nucleus with the valence nucleons. This use of the vacuum is discussed in more detail in ch. 9.

The use of the vacuum to describe a closed shell nucleus is qualitatively different from the case of the state with no particles on earth and particles on the moon. Adding particles on the earth is not expected to affect the motion of the particles on the moon. In a closed-shell atom or nucleus the assumption that the addition of a few valence particles does not affect the inert closed-shell core is only an approximation. The addition of the valence particles can polarize the core. In the description where the state with the closed-shell core is called the vacuum, these polarization effects are called vacuum polarization.

The system of a few particles on the earth and many particles on the moon has many degrees of freedom of which only a few are relevant to the physical phenomena under consideration. The irrelevant degrees of freedom are eliminated by including them in the vacuum. A many-electron atom or a nucleus with many nucleons is also a system with many degrees of freedom in which a large number (those describing the closed-shell particles) are approximately irrelevant to the physical phenomena under consideration. The elimination of these degrees of freedom by including them in the vacuum is just an approximation. In higher approximations these additional degrees of freedom must be considered. They appear as polarization and excitation of the 'vacuum'. However, the approximation is still useful if these polarization and excitation effects are described more simply in this formalism than in one treating all the degrees of freedom on an equal footing. These effects are discussed in more detail in chs. 10 and 11. In field theory where the systems considered always have an infinite number of degrees of freedom the vacuum is always used in this context and polarization and excitation of the vacuum are inherent features of the theory and observable experimentally.

PROBLEMS

1. Write an expression for the Hamiltonian of a particle moving in an external electromagnetic field, using the second-quantized notation.

2. Evaluate the matrix element $\langle \boldsymbol{k}'\boldsymbol{k}''|A|\boldsymbol{k}'''\boldsymbol{k}^{\text{iv}}\rangle$ where all values of the wave numbers are possible.

3. Two identical particles of mass m move under influence of a harmonic interparticle potential $\frac{1}{2}\omega^2|\boldsymbol{r}_1 - \boldsymbol{r}_2|^2$.

First, consider motion in one dimension, then repeat for three dimensions.
(a) Write in the second-quantized notation the wave function for the state of total momentum $2\hbar k$ having the lowest energy (1) for bosons, (2) for fermions.
(b) Define creation and annihilation operators for the harmonic oscillator quanta describing the relative motion of two particles. Write a general expression for the complete set of stationary states of the system using these operators operating on the wave functions of problem 3a.

4. Evaluate the following vacuum expectation values of products of fermion operators:

$$\langle 0|a_j a_k^\dagger a_m^\dagger a_j^\dagger|0\rangle, \qquad \langle 0|a_j a_k a_m^\dagger a_j^\dagger|0\rangle.$$

Take into account the possibility that $k=m$, $j=k$, or $j=m$.

5. Give an example of a single-particle operator and an example of a two-particle operator in the Schrödinger representation. Write these operators in the second-quantized notation, using a set of creation and annihilation operators $a_k^\dagger$ and a_k for plane-wave states.

6. Write the number operator for the set of plane-wave states of problem 5, whose eigenvalues are the total number of particles in a state. Which of the following states are eigenfunctions of the number operator? Give the eigenvalues for these cases:

(a) $$a_k^\dagger a_m^\dagger a_n^\dagger|0\rangle$$

(b) $$(a_k^\dagger + a_m^\dagger a_n^\dagger)|0\rangle$$

(c) $$\prod_k (u_k + a_k^\dagger a_{-k}^\dagger)|0\rangle$$

(d) $$(\sum_k v_k a_k^\dagger a_{-k}^\dagger)^n|0\rangle$$

7. Consider the motion of two electrons in a helium atom, treating the electrostatic repulsion between the electrons as a perturbation. One possible excited state for the unperturbed system has one electron in the lowest single particle level, the 1s level, and the second electron in the first excited level of zero angular momentum, the 2s level. Let $a_{1+}^\dagger$, $a_{1-}^\dagger$, $a_{2+}^\dagger$, and $a_{2-}^\dagger$ represent creation operators for electrons in the 1s and 2s levels with spins oriented 'parallel' and 'antiparallel' to the positive z-axis.
(a) Using the second-quantized notation write the wave functions for all possible states of the helium atom (neglecting the electron–electron repulsion) in which one electron is in the 1s state and the other is in the 2s state and all possible spin orientations are considered.
(b) The Coulomb interaction is a two-body operator which is *independent* of the spins of the particles. Write expressions for all the non-vanishing matrix elements of the Coulomb interaction between the states given in (a).
(c) Write down the appropriate linear combinations of the states given in (a) which are approximate eigenfunctions of the Hamiltonian in degenerate-perturbation theory. Indicate which of these states remain degenerate, and which of the states has the lowest energy.

8. Let $a_{pk\sigma}^\dagger$ and $a_{nk\sigma}^\dagger$ represent creation operators for protons and neutrons with wave vector k and spin projection σ on the z-axis.
(a) Which of the following states are eigenfunctions of the total spin? Of the total isospin? Give eigenvalues. Assume $k' \neq k''$, $k' \neq k'''$, $k'' \neq k'''$.

$a_{pk'\uparrow}^\dagger a_{nk''\uparrow}^\dagger|0\rangle$ $\qquad$ $a_{pk'\uparrow}^\dagger a_{pk'\downarrow}^\dagger a_{nk'\uparrow}^\dagger|0\rangle$

$\{a_{pk'\uparrow}^\dagger a_{nk''\uparrow}^\dagger - a_{pk''\uparrow}^\dagger a_{nk'\uparrow}^\dagger\}|0\rangle$ $\qquad$ $a_{pk'\uparrow}^\dagger a_{nk''\uparrow}^\dagger a_{nk'''\uparrow}^\dagger|0\rangle$

$\{a_{pk'\uparrow}^\dagger a_{pk''\downarrow}^\dagger + a_{pk''\uparrow}^\dagger a_{pk'\downarrow}^\dagger\}|0\rangle$ $\qquad$ $a_{pk'\uparrow}^\dagger a_{pk'\downarrow}^\dagger a_{nk''\uparrow}^\dagger a_{nk''\downarrow}^\dagger|0\rangle$

$a_{pk'\uparrow}^\dagger a_{nk'\uparrow}^\dagger|0\rangle$ $\qquad$ $a_{pk'\uparrow}^\dagger a_{nk'\uparrow}^\dagger a_{pk''\uparrow}^\dagger a_{nk''\downarrow}^\dagger|0\rangle$

$a_{pk'\uparrow}^\dagger a_{pk'\downarrow}^\dagger|0\rangle$ $\qquad$ $a_{pk'\uparrow}^\dagger a_{pk'\downarrow}^\dagger a_{pk''\uparrow}^\dagger a_{pk''\downarrow}^\dagger a_{pk'''\uparrow}^\dagger a_{pk'''\downarrow}^\dagger|0\rangle$

$a_{pk'\uparrow}^\dagger a_{pk'\downarrow}^\dagger a_{nk''\uparrow}^\dagger|0\rangle$ $\qquad$ $a_{pk'\uparrow}^\dagger a_{nk'\uparrow}^\dagger a_{pk''\uparrow}^\dagger a_{pk''\downarrow}^\dagger a_{pk'''\uparrow}^\dagger a_{pk'''\downarrow}^\dagger|0\rangle$

(b) Which of the following states are eigenfunctions of the parity. Give eigenvalues.

$a_{pk'\uparrow}^\dagger a_{n,-k'\uparrow}^\dagger|0\rangle$ $\qquad$ $a_{pk'\uparrow}^\dagger a_{p,-k'\uparrow}^\dagger a_{nk''\downarrow}^\dagger a_{n,-k''\downarrow}^\dagger|0\rangle$

$a_{pk'\uparrow}^\dagger a_{p,-k'\uparrow}^\dagger|0\rangle$ $\qquad$ $a_{pk'\uparrow}^\dagger a_{n,-k'\uparrow}^\dagger a_{nk''\uparrow}^\dagger a_{p,-k''\uparrow}^\dagger|0\rangle.$

CHAPTER 6

IDENTICAL COMPOSITE PARTICLES AND BOUND SYSTEMS

6.1 Introduction. Description in second quantization

Many physical problems involve large numbers of identical atoms, molecules or nuclei which are not fundamental particles but are known to be bound systems composed of several fermions in a bound state. The internal structure of these particles is completely unimportant in processes where they always move as indivisible units and do not break up. Identical composite particles are certainly indistinguishable; e.g. one cannot distinguish between one deuteron and another. One can apply the same reasoning as used for fundamental particles and search for a description of a system of deuterons which does not label the individual deuterons explicitly. One might define creation and annihilation operators for deuterons, examine the commutation rules, and arrive at the conclusion that they should be either fermions or bosons.

Experiments show that systems of composite particles are indeed correctly described by the quantum mechanics of identical particles and that these particles turn out to be either bosons or fermions. In those experiments where the composite particles move as indivisible units their internal structure does not affect the dynamics of the system but does determine the statistics of the particle; i.e. whether they are bosons or fermions. Atoms or nuclei containing an even number of protons, neutrons and electrons are bosons; those containing an odd number are fermions.

However, the deuteron is not a fundamental particle but a bound state of two fermions. We are therefore not at liberty to define creation and annihilation operators for deuterons as we please and to postulate commutation rules. All deuteron states are really complicated states of neutrons and protons and can be completely described by using the neutron and proton operators which have already been defined. Since the indistinguishability of the individual neutrons and protons is already taken into account, this includes the indistinguishability of deuterons made up of these neutrons and protons. One should therefore find all of the experimentally observed statistical properties of composite particles by examining a description using the second-quantized notation for the constituent fundamental particles.

Let us examine the description of composite systems in the second-quantized notation. We consider first a bound two-particle system, such as a deuteron, in a state having a total momentum $2\hbar\boldsymbol{K}$. We denote this state by $|\mathrm{D}, 2\boldsymbol{K}\rangle$. The Schrödinger wave function describing this system is expressed in terms of the coordinates $\boldsymbol{r}_1$ and $\boldsymbol{r}_2$ of the two particles, and is most conveniently written as a function of the center-of-mass coordinate $\boldsymbol{R}$ and the relative coordinate $\boldsymbol{r}$

$$\langle \boldsymbol{R}, \boldsymbol{r}|\mathrm{D}, 2\boldsymbol{K}\rangle = \mathrm{e}^{2\mathrm{i}\boldsymbol{K}\cdot\boldsymbol{R}}\phi(\boldsymbol{r}) \tag{6.1}$$

where the function $\phi(\boldsymbol{r})$ describes the relative motion. It is convenient to expand $\phi(\boldsymbol{r})$ in plane waves. We assume that the system is in a large box of finite volume and use periodic boundary conditions. The expansion in plane waves is a sum rather than an integral.

$$\langle \boldsymbol{R}, \boldsymbol{r}|\mathrm{D}, 2\boldsymbol{K}\rangle = \mathrm{e}^{2\mathrm{i}\boldsymbol{K}\cdot\boldsymbol{R}} \sum_{\boldsymbol{q}} g_{\boldsymbol{q}} \mathrm{e}^{\mathrm{i}\boldsymbol{q}\cdot\boldsymbol{r}} \tag{6.2a}$$

$$\langle \boldsymbol{R}, \boldsymbol{r}|\mathrm{D}, 2\boldsymbol{K}\rangle = \sum_{\boldsymbol{q}} g_{\boldsymbol{q}} \mathrm{e}^{\mathrm{i}(\boldsymbol{K}+\boldsymbol{q})\cdot\boldsymbol{r}_1} \mathrm{e}^{\mathrm{i}(\boldsymbol{K}-\boldsymbol{q})\cdot\boldsymbol{r}_2} \tag{6.2b}$$

$$\sum_{\boldsymbol{q}} |g_{\boldsymbol{q}}|^2 = 1 \tag{6.2c}$$

where $g_{\boldsymbol{q}}$ is the Fourier transform of $\phi(\boldsymbol{r})$. From eq. (6.2) we write the second-quantized form for the 'deuteron' state vector

$$|\mathrm{D}, 2\boldsymbol{K}\rangle \to \sum_{\boldsymbol{q}} g_{\boldsymbol{q}} a^{\dagger}_{\boldsymbol{K}+\boldsymbol{q}} a^{\dagger}_{\boldsymbol{K}-\boldsymbol{q}} |0\rangle. \tag{6.3}$$

Eq. (6.3) describes two identical particles in the bound state described by the Schrödinger function $\phi(\boldsymbol{r})$. In the deuteron the two particles are not identical but are a neutron and a proton. Let us use the isospin formalism and define $a^{\dagger}_{m\uparrow}$ and $a^{\dagger}_{m\downarrow}$ as operators creating a proton and a neutron (isospin 'up' and 'down') respectively in the state m. We assume that the proton and neutron operators anticommute with one another, and neglect ordinary spin. Let us define a 'deuteron creation operator'

$$D^{\dagger}_{2\boldsymbol{K}} \equiv \sum_{\boldsymbol{q}} g_{\boldsymbol{q}} a^{\dagger}_{(\boldsymbol{K}+\boldsymbol{q})\uparrow} a^{\dagger}_{(\boldsymbol{K}-\boldsymbol{q})\downarrow}. \tag{6.4a}$$

Thus

$$|\mathrm{D}, 2\boldsymbol{K}\rangle = D^{\dagger}_{2\boldsymbol{K}}|0\rangle. \tag{6.4b}$$

The corresponding annihilation operator is then given by the Hermitean conjugate of (6.4a)

$$D_{2\boldsymbol{K}} = \sum_{\boldsymbol{q}} g_{\boldsymbol{q}} a_{(\boldsymbol{K}-\boldsymbol{q})\downarrow} a_{(\boldsymbol{K}+\boldsymbol{q})\uparrow}. \tag{6.5}$$

We have assumed the coefficients $g_{\boldsymbol{q}}$ to be real to simplify the notation. The treatment is easily generalized to the case of complex g by putting asterisks in the appropriate places.

To check whether deuterons behave like bosons we examine the commutation relations between the deuteron creation and annihilation operators. We first consider the commutator of the individual nucleon creation and annihilation operators with the deuteron creation operator (6.4a). By straightforward application of the anticommutation relations (5.25) we find that

$$\begin{aligned}[a_{\boldsymbol{m}\uparrow}, D^{\dagger}_{2\boldsymbol{K}}] &= \sum_{\boldsymbol{q}} g_{\boldsymbol{q}} \{a_{\boldsymbol{m}\uparrow} a^{\dagger}_{(\boldsymbol{K}+\boldsymbol{q})\uparrow} a^{\dagger}_{(\boldsymbol{K}-\boldsymbol{q})\downarrow} - a^{\dagger}_{(\boldsymbol{K}+\boldsymbol{q})\uparrow} a^{\dagger}_{(\boldsymbol{K}-\boldsymbol{q})\downarrow} a_{\boldsymbol{m}\uparrow}\} \\ &= \sum_{\boldsymbol{q}} g_{\boldsymbol{q}} \{a_{\boldsymbol{m}\uparrow} a^{\dagger}_{(\boldsymbol{K}+\boldsymbol{q})\uparrow} + a^{\dagger}_{(\boldsymbol{K}+\boldsymbol{q})\uparrow} a_{\boldsymbol{m}\uparrow}\} a^{\dagger}_{(\boldsymbol{K}-\boldsymbol{q})\downarrow} = \sum_{\boldsymbol{q}} g_{\boldsymbol{q}} \delta_{\boldsymbol{m},\boldsymbol{K}+\boldsymbol{q}} a^{\dagger}_{(\boldsymbol{K}-\boldsymbol{q})\downarrow} \\ &= g_{\boldsymbol{m}-\boldsymbol{K}} a^{\dagger}_{(2\boldsymbol{K}-\boldsymbol{m})\downarrow} \end{aligned} \tag{6.6a}$$

and similarly

$$[a_{\boldsymbol{m}\downarrow}, D^{\dagger}_{2\boldsymbol{K}}] = -g_{(\boldsymbol{K}-\boldsymbol{m})} a^{\dagger}_{(2\boldsymbol{K}-\boldsymbol{m})\uparrow} \tag{6.6b}$$

$$[a^{\dagger}_{\boldsymbol{m}\uparrow}, D_{2\boldsymbol{K}}] = -g_{(\boldsymbol{m}-\boldsymbol{K})} a_{(2\boldsymbol{K}-\boldsymbol{m})\downarrow} \tag{6.6c}$$

$$[a^{\dagger}_{\boldsymbol{m}\downarrow}, D_{2\boldsymbol{K}}] = g_{(\boldsymbol{K}-\boldsymbol{m})} a_{(2\boldsymbol{K}-\boldsymbol{m})\uparrow} \tag{6.6d}$$

$$[a^{\dagger}_{\boldsymbol{m}\downarrow}, D^{\dagger}_{2\boldsymbol{K}}] = [a_{\boldsymbol{m}\downarrow}, D_{2\boldsymbol{K}}] = [a^{\dagger}_{\boldsymbol{m}\uparrow}, D^{\dagger}_{2\boldsymbol{K}}] = [a_{\boldsymbol{m}\uparrow}, D_{2\boldsymbol{K}}] = 0. \tag{6.6e}$$

The commutation relations (6.6) are not quite those expected for the commutator between 'nucleon' creation and annihilation operators and 'deuteron' creation and annihilation operators if the nucleons and the deuterons are different types of particles, and the deuterons are bosons. If they were indeed different, all the commutators (6.6) should be zero. Instead we find that some are proportional to the nucleon creation and annihilation operators with a constant of proportionality equal to one of the coefficients $g_{\boldsymbol{q}}$ appearing in the Fourier expansion (6.1a). These non-vanishing commutators express the fact that deuterons are composite particles containing nucleons. We should expect them to be negligibly small in all physical applications where the composite structure is not important.

The commutators (6.6) can be seen to be very small. The size of the box in which the system is confined is very large compared to the size of the deuteron; thus a large number of plane-wave states is required to construct the wave function for the deuteron. The individual coefficients $g_{\boldsymbol{q}}$ appearing in the expansion must therefore all be very small because of the normalization condition (6.2c). The matrix elements of fermion creation and annihilation operators are always of order unity because of the Pauli exclusion principle. Thus all of the commutators (6.6) are small and can probably be neglected.

This picture should break down if the box is packed so full of deuterons that an appreciable number of nucleon states is occupied. These nucleon states can then not be occupied by free nucleons because of the Pauli exclusion principle. The motion of a system of free nucleons and deuterons should also be affected because deuterons contain neutrons and protons. To see this consider a state in which N deuterons are present

$$(D^{\dagger}_{2\boldsymbol{K}})^N|0\rangle. \tag{6.7}$$

From eq. (6.6a) the commutator

$$[a_{\boldsymbol{m}\uparrow}, (D^{\dagger}_{2\boldsymbol{K}})^N] = Ng_{\boldsymbol{m}-\boldsymbol{K}}(D^{\dagger}_{2\boldsymbol{K}})^{N-1}a^{\dagger}_{(2\boldsymbol{K}-\boldsymbol{m})\uparrow}. \tag{6.8}$$

This commutator is small only if the product $Ng_{\boldsymbol{m}-\boldsymbol{K}}$ is small. Thus if N is small the commutator (6.8) is still negligible. However as soon as N becomes too large; i.e., when the number of deuterons in the box is comparable to the number of plane-wave states needed to build up the deuteron, the picture breaks down and we can no longer consider nucleons and deuterons as different kinds of particles.

We now consider the commutators of the deuteron creation and annihilation operators among themselves. From the commutators (6.6a) and (6.6b) we obtain

$$[D_{2\boldsymbol{K}}, D_{2\boldsymbol{K}'}] = [D^{\dagger}_{2\boldsymbol{K}}, D^{\dagger}_{2\boldsymbol{K}'}] = 0 \tag{6.9a}$$

$$\begin{aligned}
[D_{2\boldsymbol{K}'}, D^{\dagger}_{2\boldsymbol{K}}] &= \sum_{\boldsymbol{q}} g_{\boldsymbol{q}}\{a_{(\boldsymbol{K}'-\boldsymbol{q})\downarrow}[a_{(\boldsymbol{K}'+\boldsymbol{q})\uparrow}, D^{\dagger}_{2\boldsymbol{K}}] + [a_{(\boldsymbol{K}'-\boldsymbol{q})\downarrow}, D^{\dagger}_{2\boldsymbol{K}}]a_{(\boldsymbol{K}'+\boldsymbol{q})\uparrow}\} \\
&= \sum_{\boldsymbol{q}} g_{\boldsymbol{q}}\{g_{(\boldsymbol{K}'-\boldsymbol{K}+\boldsymbol{q})}a_{(\boldsymbol{K}'-\boldsymbol{q})\downarrow}a^{\dagger}_{(2\boldsymbol{K}-\boldsymbol{K}'-\boldsymbol{q})\downarrow} - \\
&\quad - g_{(\boldsymbol{K}'-\boldsymbol{K}-\boldsymbol{q})}a^{\dagger}_{(2\boldsymbol{K}-\boldsymbol{K}'+\boldsymbol{q})\uparrow}a_{(\boldsymbol{K}'+\boldsymbol{q})\uparrow}\} \\
&= \delta_{\boldsymbol{K}\boldsymbol{K}'} - \Delta_{\boldsymbol{K}\boldsymbol{K}'}
\end{aligned} \tag{6.9b}$$

where $\Delta_{\boldsymbol{K}\boldsymbol{K}'}$ is defined by

$$\begin{aligned}
\Delta_{\boldsymbol{K}\boldsymbol{K}'} = \sum_{\boldsymbol{q}} g_{\boldsymbol{q}}\{&g_{(\boldsymbol{K}'-\boldsymbol{K}+\boldsymbol{q})}a^{\dagger}_{(2\boldsymbol{K}-\boldsymbol{K}'-\boldsymbol{q})\downarrow}a_{(\boldsymbol{K}'-\boldsymbol{q})\downarrow} \\
&+ g_{(\boldsymbol{K}'-\boldsymbol{K}-\boldsymbol{q})}a^{\dagger}_{(2\boldsymbol{K}-\boldsymbol{K}'+\boldsymbol{q})\uparrow}a_{(\boldsymbol{K}'+\boldsymbol{q})\uparrow}\}.
\end{aligned} \tag{6.10}$$

The commutators (6.9) are just those for boson creation and annihilation operators provided that $\Delta_{KK'}$ can be neglected. Since $\Delta_{KK'}$ is proportional to the small quantities g_q, a similar argument should hold for this case as for the case of the commutators (6.6) of nucleon and deuteron operators.

The expression (6.10) has a particularly simple form for $\boldsymbol{K}=\boldsymbol{K}'$ where the creation and annihilation operator products reduce to number operators of the form (5.45)

$$\Delta_{KK} = \sum_q g_q^2 (n_{(K-q)\downarrow} + n_{(K+q)\uparrow}) \tag{6.11a}$$

$$= \sum_q g_{K-q}^2 n_{q\downarrow} + g_{q-K}^2 n_{q\uparrow}. \tag{6.11b}$$

Eq. (6.11) is just the average number of nucleons in a particular state $\boldsymbol{q}$ averaged over all states $\boldsymbol{q}$ with the normalized weighting factors g_{K-q} and g_{q-K}. The order of magnitude of Δ_{KK} is thus given by the ratio of the number of nucleons in the system to the total number of plane-wave states used to make the deuteron wave packet. Again we see that the departure of deuterons from boson commutation rules is small if the number of deuterons present in the system is small compared to the number of states available.

Let us assume that the deuteron is made up of states within a momentum range Δp. The total number of states in this momentum range is just the number of states in a cube in momentum space $(\Delta p)^3 V/h^3$ where V is the volume of the box. On the other hand the uncertainty principle tells us that the quantity $h/\Delta p$ is of the order of Δx, the dimensions of the deuteron. We see therefore that the total number of states available is of the order of the ratio of the volume of the box to the volume of the deuteron. Thus eq. (6.11) tells us that deuterons behave like bosons if the total volume occupied by the deuterons is small compared to the volume of the box; i.e., the average distance between deuterons is large compared to the deuteron size and there is no appreciable overlap between the wave functions of different deuterons.

A similar treatment of composite systems containing three or more fermions yields the result that these behave like bosons or fermions, depending upon whether the number of fermions in the system is even or odd, provided that the overlapping of the different composite systems can be neglected; i.e., the size of the particle is small compared to the average distance between them.

6.2 Properties of overlapping fermion pairs

One can ask what happens in a system of bound pairs of fermions such as deuterons when the density is so high that there is a considerable overlap of the wave functions of different pairs; i.e., when the size of the pair is no longer negligible compared to the average distance between them. Such bound pairs have recently been shown to be important in a number of physical phenomena. Overlapping bound or correlated pairs of nucleons are present in complex nuclei and their properties are important for an understanding of nuclear structure. In metals, overlapping bound or correlated electron pairs of opposite spin play a crucial role in the phenomena of superconductivity.

Some of the basic physical properties of these overlapping pairs can be obtained from the following simplified example. Let us consider a simplified 'deuteron' wave packet (6.2) built up of waves over some finite region of $\boldsymbol{q}$ with equal amplitudes $g_{\boldsymbol{q}}$ for all these waves. This represents the Fourier transform $g_{\boldsymbol{q}}$ of the realistic deuteron wave function $\phi(\boldsymbol{r})$ by a rectangular momentum distribution as shown in fig. 6.1. We use the term deuteron to describe such a bound pair and the results would give an approximate treatment of a very dense deuteron gas. Such dense deuteron gases are not of direct physical interest. However our treatment is general and applies to the other more interesting cases of overlapping fermion pairs mentioned above and discussed in more detail below.

Let N be the number of particles in the system and Ω be the number of plane-wave components or values of $\boldsymbol{q}$. Then

$$N = \sum_{\boldsymbol{q}} n_{\boldsymbol{q}\uparrow} + n_{\boldsymbol{q}\downarrow}, \tag{6.12a}$$

$$g_{\boldsymbol{q}} = \Omega^{-\frac{1}{2}} \tag{6.12b}$$

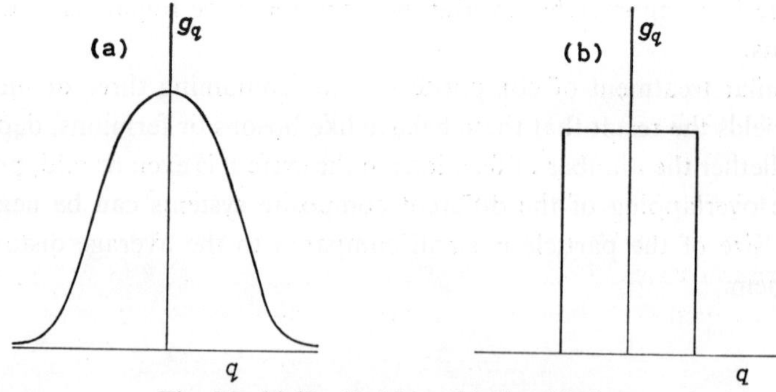

Fig. 6.1. (a) 'Realistic' $g_{\boldsymbol{q}}$, (b) Simplified $g_{\boldsymbol{q}}$.

and eq. (6.11) becomes

$$\Delta_{\boldsymbol{KK}} = N/\Omega. \tag{6.13}$$

The simplicity of the commutation relations of the number operators (6.12a) with all the creation and annihilation operators enables us to immediately write the commutators

$$[\Delta_{\boldsymbol{KK}}, D_{2\boldsymbol{K}}^{\dagger}] = 2D_{2\boldsymbol{K}}^{\dagger}/\Omega \tag{6.14a}$$

$$[\Delta_{\boldsymbol{KK}}, D_{2\boldsymbol{K}}] = -2D_{2\boldsymbol{K}}/\Omega. \tag{6.14b}$$

From eqs. (6.9) and (6.14) we see that for each value of $\boldsymbol{K}$ we can define three operators which satisfy commutation rules like angular momenta

$$S_z = -\tfrac{1}{2}\Omega(1 - \Delta_{\boldsymbol{KK}}) = \tfrac{1}{2}(N - \Omega) \tag{6.15a}$$

$$S_+ = \Omega^{\frac{1}{2}} D_{2\boldsymbol{K}}^{\dagger} = \sum_{\boldsymbol{q}} a_{(\boldsymbol{K}+\boldsymbol{q})\uparrow}^{\dagger} a_{(\boldsymbol{K}-\boldsymbol{q})\downarrow}^{\dagger} \tag{6.15b}$$

$$S_- = \Omega^{\frac{1}{2}} D_{2\boldsymbol{K}} = \sum_{\boldsymbol{q}} a_{(\boldsymbol{K}-\boldsymbol{q})\downarrow} a_{(\boldsymbol{K}+\boldsymbol{q})\uparrow} \tag{6.15c}$$

$$[S_z, S_+] = S_+; \qquad [S_z, S_-] = -S_-; \qquad [S_+, S_-] = 2S_z. \tag{6.16}$$

Operators of this type which satisfy angular momentum commutation rules are sometimes called quasi-spins. By analogy with angular momentum we define the total quasi-spin operator

$$S^2 = \tfrac{1}{2}(S_+S_- + S_-S_+) + S_z^2. \tag{6.17}$$

By analogy with angular momentum we know that the operator S^2 defined by eq. (6.17) commutes with all the quasi-spin operators (6.15) and has eigenvalues of the form $S(S + 1)$ where S is either an integer or half integer.

Let us now consider the state

$$(D_{2\boldsymbol{K}}^{\dagger})^m |0\rangle. \tag{6.18}$$

This state appears to contain m deuterons in the same state with momentum $\hbar\boldsymbol{K}$. In the treatment above, eqs. (6.9)–(6.12) show that the system behaves like one containing m bosons if N/Ω is small. However, when $m = \Omega$, the number of fermions in the state (6.18) is equal to the number of available single-fermion states. There is clearly only one such state possible, namely the one in which each state is occupied by a single fermion.

$$\prod_{\boldsymbol{q}} a_{(\boldsymbol{K}+\boldsymbol{q})\uparrow}^{\dagger} a_{(\boldsymbol{K}-\boldsymbol{q})\downarrow}^{\dagger}. \tag{6.19}$$

It is clear upon inspection that the states defined by eqs. (6.18) and (6.19) are identical for $m=\Omega$. Because the square of any fermion creation operator vanishes, the only terms which contribute to the function (6.18) are those in which no single creation operator appears twice. The total number of creation operators appearing in the product is equal to the number of available states; thus each creation operator appears once. The order of the operators is not important since interchanging the order of two creation operators only changes the state by a phase factor.

Thus for $m \ll \Omega$ the state (6.18) describes a small number of bosons but for $m=\Omega$ it describes a completely filled set of single-fermion states. The interesting region occurs around $m=\frac{1}{2}\Omega$ where the boson approximation is no longer valid and the states are not completely filled. This region can be treated in our simple model because of the simple angular momentum commutation rules of the operators (6.15).

Consider the 'deuteron number' operator

$$N_D = D_{2K}^\dagger D_{2K} = \Omega^{-1} \sum_{qq'} a_{(K+q)\uparrow}^\dagger a_{(K-q)\downarrow}^\dagger a_{(K-q')\downarrow} a_{(K+q')\uparrow}. \tag{6.20}$$

This operator gives the number of deuterons in the state $\boldsymbol{K}$ as long as the boson description of deuterons is valid; i.e., as long as the number of deuterons in the system is small. When this operator is viewed as a function of the fermion creation and annihilation operators, however, it is seen to be a two-body operator having the form (5.52) which annihilates two fermions and creates two fermions in other states. Such an operator could also describe a two-body force between the fermions.

6.3 A simple model for overlapping bound fermion pairs

Consider the operator

$$V = -\varepsilon_D N_D = -\frac{\varepsilon_D}{\Omega} \sum_{qq'} a_{(K+q)\uparrow}^\dagger a_{(K-q)\downarrow}^\dagger a_{(K-q')\downarrow} a_{(K+q')\uparrow} \tag{6.21}$$

where ε_D is a constant. This operator can describe a two-particle interaction having a strength $G=\varepsilon_D/\Omega$. The single-deuteron state with momentum $2\hbar\boldsymbol{K}$ is seen to be eigenvector of the operator V with eigenvalue $-\varepsilon_D$:

$$\begin{aligned} -\varepsilon_D N_D D_{2K}^\dagger|0\rangle &= -\varepsilon_D D_{2K}^\dagger D_{2K} D_{2K}^\dagger|0\rangle = -\varepsilon_D D_K^\dagger[D_{2K}, D_{2K}^\dagger]|0\rangle \\ &= -\varepsilon_D D_{2K}^\dagger|0\rangle. \end{aligned} \tag{6.22}$$

On the other hand a two-fermion state which is orthogonal to the one-deuteron state with momentum $\hbar\boldsymbol{K}$ is an eigenfunction of V with eigenvalue zero. This is shown more precisely below, using the quasi-spin algebra but can be also understood in the approximation that deuterons behave like bosons, independent of the individual fermions, when the number of particles is small.

The operator V, eq. (6.21), thus describes a two-particle interaction which gives rise to binding of the particles into bound pairs or deuterons. If the number of particles in the system is small the eigenfunctions of V are states consisting of a certain number of 'bound pairs' or 'deuterons' and a certain number of unbound particles. The eigenvalue of V is just $-\varepsilon_D$ multiplied by the number of deuterons present in the state. The quantity ε_D thus represents the binding energy of a deuteron. The simple form of the interaction V allows us to find its exact eigenfunctions and eigenvalues and therefore to investigate the properties of systems of 'overlapping bound pairs' when the number of particles is no longer small compared to the number of available states.

In this simplified description we have picked one particular value $2\boldsymbol{K}$ of the deuteron wave vector to define the bound state. A physical deuteron can have any value for its total momentum and a two-particle interaction which describes the binding of physical deuterons must be the sum over all values of $\boldsymbol{K}$ of interactions of the type (6.21). In our model all two-particle states orthogonal to the one-deuteron state with momentum $2\hbar\boldsymbol{K}$ are unbound, and we neglect the binding of deuteron states having momenta $2\hbar\boldsymbol{K}'$ different from the value $2\hbar\boldsymbol{K}$ which we have picked for our interaction (6.21). This does not seriously affect the essential physical features of our argument, which concerns the case where many fermion pairs occupy the *same* state. We choose this state to be the state of momentum $2\hbar\boldsymbol{K}$. Our interaction (6.21) counts the number of pairs in this state.

The interaction V can be expressed in terms of the quasi-spin operators (6.15)

$$V = -\varepsilon_D S_+ S_- / \Omega = -\varepsilon_D \{S^2 - S_z^2 + S_z\}/\Omega. \tag{6.23}$$

Thus all states which are simultaneous eigenfunctions of the operators S^2 and S_z are also eigenfunctions of the interaction V. Since S_z is just a function of the number of particles and the number of states, any state which has a definite number of particles is an eigenfunction of S_z. In particular consider the vacuum state and the states (6.18) containing m deuterons

$$S_z (D_{2\boldsymbol{K}}^\dagger)^m |0\rangle = \tfrac{1}{2}(2m - \Omega)(D_{2\boldsymbol{K}}^\dagger)^m |0\rangle \tag{6.24a}$$

$$S_z |0\rangle = -\tfrac{1}{2}\Omega |0\rangle. \tag{6.24b}$$

The operator S_- consists only of annihilation operators and therefore gives zero when acting on the vacuum state.

$$S_-|0\rangle = 0. \tag{6.25}$$

The vacuum state must be an eigenfunction of the operator S^2 since it is an eigenfunction of S_z and is 'annihilated' by the lowering operator S_-. The vacuum state therefore is an eigenvector of S^2 corresponding to $S=\frac{1}{2}\Omega$. (We use the conventional designation S' for the eigenvalue $S'(S'+1)$ for the operator S^2.) The operator $D^\dagger_{2\mathbf{K}}$ differs from a quasi-spin operator only by a constant factor and must commute with the total quasi-spin S^2.

$$[S^2, D^\dagger_{2\mathbf{K}}] = 0. \tag{6.26a}$$

Thus all the m-deuteron states (6.18) are also eigenfunctions of the operators S^2 with the same eigenvalue:

$$S^2(D^\dagger_{2\mathbf{K}})^m|0\rangle = (D^\dagger_{2\mathbf{K}})^m S^2|0\rangle = \tfrac{1}{2}\Omega(\tfrac{1}{2}\Omega + 1)(D^\dagger_{2\mathbf{K}})^m|0\rangle. \tag{6.26b}$$

The m-deuteron states (6.18) are therefore also eigenfunctions of the interaction V

$$\begin{aligned} V(D^\dagger_{2\mathbf{K}})^m|0\rangle \\ &= -(\varepsilon_\mathrm{D}/\Omega)\left\{\frac{\Omega}{2}\left(\frac{\Omega}{2}+1\right) - \left(\frac{2m-\Omega}{2}\right)^2 + \frac{2m-\Omega}{2}\right\}(D^\dagger_{2\mathbf{K}})^m|0\rangle \\ &= -m\varepsilon_\mathrm{D}\left\{1 - \frac{m-1}{\Omega}\right\}(D^\dagger_{2\mathbf{K}})^m|0\rangle. \end{aligned} \tag{6.27}$$

The vacuum state has eigenvalue zero, and the one-deuteron state has the exact eigenvalue $-\varepsilon_\mathrm{D}$. The m-deuteron state has an eigenvalue approximately $-m\varepsilon_\mathrm{D}$ with corrections which are small when m is small compared to Ω. These corrections have the *opposite* sign from the main term; they *decrease* the binding energy.

The additional correction term can be interpreted as the effect of the Pauli exclusion principle. The deuterons are not really bosons but are made up of individual fermions which cannot occupy the same quantum state. The effect of the Pauli exclusion principle can already be seen in the two-deuteron state. If the operator $(D^\dagger_\mathbf{K})^2$ is written formally there are some terms in which creation operators for the same quantum state appear twice.

If our deuterons consisted of elementary bosons or of particles which were not identical these terms in which two fermions occupy the same state would be allowed and would contribute. However because the deuterons do consist of equivalent fermions these terms in which two fermions occupy the same state must vanish. Formally of course they vanish because the square of any creation operator is zero. The proportion of such terms to the total number of terms is just $1/\Omega$, and it is just these terms which give the correction of $1/\Omega$ in the expression (6.27) for the energy of the two-deuteron state. As more deuterons are added to the system this correction increases and can be interpreted physically as follows. The individual nucleons in a particular 'deuteron' are not able to form the exact wave packet which gives the minimum energy because they are not free to occupy *all* the quantum states in the manner which minimizes the energy. They are constrained by the presence of the fermions in the other deuterons to keep out of those states which are already occupied.

Let us now consider a two-particle state $|2\rangle$ which is orthogonal to the one-deuteron state. Like any two-particle state it is an eigenfunction of S_z with the eigenvalue $\frac{1}{2}(2-\Omega)$

$$S_z|2\rangle = (-\tfrac{1}{2}\Omega + 1)|2\rangle. \tag{6.28}$$

From the orthogonality to the one-deuteron state,

$$\langle 0|D_{2\mathbf{K}}|2\rangle = 0 = \langle 0|S_-|2\rangle/\Omega^{\frac{1}{2}} \tag{6.29a}$$

$$S_-|2\rangle = 0. \tag{6.29b}$$

The state $|2\rangle$ is thus a simultaneous eigenfunction of S^2 and S_z with eigenvalue $S=\frac{1}{2}\Omega-1$. Substituting these eigenvalues into eq. (6.23) shows that this state is an eigenvector of V with eigenvalue zero.

$$V|2\rangle = 0. \tag{6.30}$$

This result is expected since the state contains no deuterons but only two nucleons in an unbound state. Now consider the state $D^\dagger_{2\mathbf{K}}|2\rangle$ obtained by adding a deuteron to the two unbound nucleons in the state $|2\rangle$. By the commutation relation (6.26a) this state is an eigenfunction of S^2 with the same eigenvalue as the state $|2\rangle$. As a four-particle state, it is an eigenfunction of the operator S_z with the eigenvalue $-\frac{1}{2}\Omega+2$. Thus

$$VD^\dagger_{2\mathbf{K}}|2\rangle = -\varepsilon_{\mathrm{D}}(1 - 2\Omega^{-1})D^\dagger_{2\mathbf{K}}|2\rangle. \tag{6.31}$$

The presence of the two unbound particles is seen to reduce the binding energy of the deuteron by a factor $(1-2\Omega^{-1})$. This is double the reduction for the four-particle state where both particles are deuterons. The effect of two unbound particles on the deuteron binding is double that of a bound pair. This is reasonable since two particles which are unbound and therefore uncorrelated occupy two independent states and close two states to both members of the bound pair. If one particle is in a state $(\boldsymbol{K}+\boldsymbol{q}_1)\uparrow$ and the other in state $(\boldsymbol{K}-\boldsymbol{q}_2)\downarrow$, the states $(\boldsymbol{K}-\boldsymbol{q}_1)\downarrow$ and $(\boldsymbol{K}+\boldsymbol{q}_2)\uparrow$ are also closed to the bound pair, since when one particle of the bound pair is in the state $(\boldsymbol{K}+\boldsymbol{q})\uparrow$ the other must be in the state $(\boldsymbol{K}-\boldsymbol{q})\downarrow$. On the other hand if the particles are bound they only close off a single-pair state because they themselves are correlated to be in conjugate states. If they are in states $(\boldsymbol{K}+\boldsymbol{q})\uparrow$ and $(\boldsymbol{K}-\boldsymbol{q})\downarrow$ they only close off these states to the other bound pair.

Let us now generalize our treatment to apply to any number of deuterons and unbound particles. Consider first a state which contains no deuterons and v unbound particles. Such a state satisfies the relation

$$D_{2\boldsymbol{K}}|v\rangle = 0 = S_-|v\rangle. \tag{6.32}$$

The deuteron annihilation operator acting on such a state gives zero since it contains no nucleons bound in deuterons. Such a state is a simultaneous eigenfunction of S and S_z with eigenvalues $S=-S_z=\frac{1}{2}(\Omega-v)$.

From the state (6.32) we construct a state having v free particles and m deuterons by operating on the state m times with the deuteron creation operator.

$$|m, v\rangle = (D^{\dagger}_{2\boldsymbol{K}})^m|v\rangle. \tag{6.33}$$

This state is an eigenfunction of S^2 with the same eigenvalue as the state (6.32). It is an eigenfunction of S_z with the eigenvalue $\frac{1}{2}(v+2m-\Omega)$.

$$S^2|m, v\rangle = \left(\frac{\Omega - v}{2}\right)\left(\frac{\Omega - v}{2} + 1\right)|m, v\rangle \tag{6.34a}$$

$$S_z|m, v\rangle = \left(\frac{2m + v - \Omega}{2}\right)|m, v\rangle. \tag{6.34b}$$

The state $|m, v\rangle$ is therefore an eigenfunction of V with the eigenvalue

$$\begin{aligned} V|m, v\rangle &= -m\varepsilon_{\mathrm{D}}\left\{1 - \frac{(m-1+v)}{\Omega}\right\}|m, v\rangle \\ &= -mG\Omega\left\{1 - \frac{(m-1+v)}{\Omega}\right\}|m, v\rangle. \end{aligned} \tag{6.35}$$

This is just m times the deuteron binding energy multiplied by a factor which expresses the effect of the Pauli principle. This factor is proportional to $m-1+v$; i.e. to the number of other deuterons plus the number of unbound particles.

This expression (6.35) gives a complete solution of the Schrödinger equation defined by the interaction V, (6.21). The states (n, V) define a complete set of eigenfunctions of the interaction and eq. (6.35) gives all the eigenvalues. Our method of solution illustrates the power of algebraic methods which are useful in many applications. We first found a set of operators satisfying simple commutation relations (in this case those of angular momentum). We were then able to use the algebraic results previously obtained in a very different physical context. Although there is no physical angular momentum in the quasi-spin variables (6.15), the fact that they satisfy angular momentum commutation rules is sufficient to allow us to use all the algebraic results previously obtained for physical angular momenta.

6.4 Superconductivity and seniority

The BCS theory describes the electron wave function for a superconducting ground state as one in which there are a large number of overlapping, correlated or bound pairs. A characteristic feature of the superconductor is the 'energy gap' between the ground state and the lowest lying excited states. The energy gap arises from the necessity to break the pairing correlation in producing an excited state. The essential features of this energy gap are present in our simple model. The ground state for the system of $2m$ particles is the state $|m, 0\rangle$ with all particles bound into deuterons. From eq. (6.35) we can calculate the excitation energy of the first excited state $|m-1, 2\rangle$ produced by breaking up one deuteron and producing two unbound particles.

$$E_{m-1,2} - E_{m,0} = \varepsilon_D \left[-(m-1)\left\{1 - \frac{m-2+2}{\Omega}\right\} + m\left\{1 - \frac{m-1}{\Omega}\right\}\right]$$

$$= \varepsilon_D \left\{\left(1 - \frac{m}{\Omega}\right) + \frac{m}{\Omega}\right\} = \varepsilon_D = G\Omega. \tag{6.36}$$

Although this result (6.30) is just equal to the binding energy of a deuteron, the naive conclusion that breaking up a deuteron costs exactly the deuteron binding energy ε_D is not correct. The binding energy per deuteron is not equal to ε_D but is reduced by the effect of the Pauli principle as shown in eq. (6.35). The result (6.36) comes from two contributions.

Breaking up a deuteron into two free particles has two effects: The reduction of the number of bound deuterons by one costs only the average deuteron binding energy, reduced by the Pauli principle; namely $\varepsilon_D(1-m/\Omega)$. There is also an increased effect of the Pauli principle which appears as an increase in the factor $(m-1+v)/\Omega$ appearing in eq. (6.35). This decreases the binding of all of the other deuterons and gives an energy change of $\varepsilon_D m/\Omega$. The sum of these two happens to be equal to ε_D in this simple case. When $m=\frac{1}{2}\Omega$; i.e., half of the available states are filled, the contributions of these two effects are equal. Only half of the energy required to break up a deuteron comes up from the loss of that deuteron. The remaining half of the energy comes from the decrease in the binding of the other deuterons produced by the increased Pauli principle effect when a deuteron is converted into two free particles.

In a more realistic model, the bound deuteron state need not have the particular wave number $2\boldsymbol{K}$ which we have assumed in our model, but could have any value of the center-of-mass momentum. In such a model, we might expect the ground state to be the one in which all the deuterons have zero momentum, $\boldsymbol{K}=0$. We would then expect to find an excited state in which one of the deuterons is in a bound deuteron state with $\boldsymbol{K}\neq 0$. If deuterons behaved like free bosons, we would expect a continuous excitation spectrum with no energy gap, since the excitations are produced by giving individual bosons a small momentum and kinetic energy without breaking up the deuteron. Thus we would not lose the binding energy of the deuteron.

However, for the case of fermion pairs, we still lose the energy from the Pauli principle effect. The binding of any given deuteron is reduced by the presence of other fermions. This reduction is minimized when the other fermions are all present as deuterons *with the same total momentum*. The reduction produced by a fermion pair which does not have the same total momentum as the deuteron is double that of a deuteron with the same total momentum. Thus changing the momentum of one of the deuterons in the ground state increases the Pauli principle effect on *all* the deuterons and requires a finite excitation energy, even if the change in deuteron momentum is small. The Pauli principle effect thus produces an energy gap in the excitation spectrum. This effect is characteristic of overlapping fermion pairs, and would be absent if the fermions behaved like simple bosons.

Consider the case where all the deuterons are in a single state $\boldsymbol{K}\neq 0$. The system then has a finite momentum and carries a finite current. This current resembles the persistent current in a superconductor, which is described in the BCS theory as due to bound pairs of electrons with opposite spins. Our model Hamiltonian (6.21) is actually a simplified version of the BCS model and has many of its qualitative features. The BCS model Hamiltonian differs from our Hamiltonian (6.21) in two ways. The kinetic energy of the individual electrons is also included and the different terms in the interaction are not all equal as in (6.21) but have coefficients depending upon

$\boldsymbol{q}$ and $\boldsymbol{q}'$. The latter is equivalent to using a realistic deuteron bound-state function fig. 6.1a rather than the flat distribution of fig. 6.1b.

A current of bound pairs is qualitatively very different from the current carried by conduction electrons in a normal metal. The electrons carrying a normal current have a statistical distribution of momenta and rapidly reach thermal equilibrium with the lattice ions as a result of the interactions between individual electrons and the lattice vibrations (phonons). A supercurrent of bound pairs all having exactly the same momentum is not an equilibrium state but is metastable because the transition to the equilibrium state involves simultaneously changing the common momentum of all the pairs. The interactions with the lattice vibrations destroy a normal current by scattering individual electrons, one at a time. Such interactions can only scatter electrons in bound pairs by breaking pairs one at a time.

Since breaking a pair requires a finite excitation energy of the order of the energy gap, the number of unbound electrons present is relatively small at low temperatures where the thermal energy is small compared with the energy gap. The exchange of energy between the electrons and the lattice then results from scattering of unbound electrons and breaking and recombination of bound pairs. This leads to a metastable equilibrium state with the number of broken pairs described by a Boltzmann distribution. The common momentum of all the bound pairs cannot be changed by this relaxation mechanism if the average number of broken pairs is relatively small. The probability of a fluctuation with a macroscopic number of broken pairs is negligible. At higher temperatures where the thermal energy becomes sufficiently large to break all the pairs, the coherence of the supercurrent is destroyed and a transition to a normal equilibrium state occurs.

In the nuclear shell model one considers to the lowest approximation that the individual nucleons are each moving in independent orbits in a central field produced by the average of the interactions of all the other particles. One then has a set of single-particle levels obtained by solving the Schrödinger equation in the appropriate self-consistent potential. If the potential is spherically symmetric each energy level is degenerate; a state of angular momentum j has a $(2j+1)$-fold degeneracy. If the central potential used is that of the three-dimensional harmonic oscillator there are additional degeneracies. When there are several particles to be distributed among a number of degenerate single-particle states a large degeneracy results for the many-particle system because of the different ways one can put particles into these degenerate states.

In the next approximation this degeneracy is removed by taking into account the 'residual interaction'; namely the difference between the real nucleon–nucleon force and the average used as a central potential. It is found that in many cases an important part of the residual interaction has a form similar to the operator V of our simplified model defined by eqs. (6.20) and (6.21) where $\boldsymbol{K}=0$ and the quantum number $\boldsymbol{q}$ refers to a component of the angular momentum rather than the linear momentum.

$$V = -G \sum_{\substack{m>0 \\ m'>0}} a^{\dagger}_{njm} a^{\dagger}_{nj,-m} a_{nj,-m'} a_{njm'} = -GS_{+}S_{-} \tag{6.37a}$$

where $a^{\dagger}_{njm}$ creates a nucleon in a shell model state with quantum numbers n, j, m, and one considers only identical nucleons, *either* neutrons *or* protons. In the original seniority scheme only one shell with a definite value of n and j is considered. There are also 'generalized seniorities' which include a summation over several shells.

The quasi-spin raising operator

$$S_{+} = \sum_{m>0} a^{\dagger}_{njm} a^{\dagger}_{nj,-m} \tag{6.37b}$$

is a creation operator for a pair of nucleons in a state of total angular momentum zero.

One finds that correlated or bound nucleon pairs often exist in nuclei analogous to the 'deuteron' in our simplified model. These nucleon pairs however are not physical deuterons but correlated states of two identical nucleons having a total angular momentum of zero. The interaction (6.37a) is called a 'pairing force'. The description of states $|n, v\rangle$ in our simplified model is directly analogous to the description of nuclear states in the so-called 'seniority' classification. A state of n nucleons containing v particles which are not in pairs and $m=\frac{1}{2}(n-v)$ pairs is called a state of seniority v.

The interaction (6.37) has the same simple form as the model interaction (6.20) and (6.21). However, the operator (6.37b) can be interpreted as a creation operator for a pair of nucleons in the angular momentum zero only if a unorthodox phase convention is used which differs from the accepted Condon–Shortley phase convention for coupling angular momenta. With the Condon–Shortley phase convention, we should write

$$S_{+} = \sum_{m>0} (-1)^{j-m} a^{\dagger}_{njm} a^{\dagger}_{nj,-m}. \tag{6.38a}$$

The interaction then has the form

$$V = -GS_{+}S_{-} = -G \sum_{\substack{m>0 \\ m'>0}} (-1)^{m'+m} a^{\dagger}_{njm} a^{\dagger}_{nj,-m} a_{njm'} a_{nj,-m'}. \tag{6.38b}$$

For the 'generalized seniority', which includes several shells, one can define

$$S_+ = \sum_{n,j} \sum_{m>0} C_{nj}(-1)^{j-m} a^\dagger_{njm} a^\dagger_{nj,-m} \tag{6.39a}$$

where the summation is over several shells, and C_{nj} is a phase factor which depends on n and j, but *not* on m. Then

$$V = -G \sum_{n,j,n',j'} \sum_{\substack{m>0 \\ m'>0}} C_{nj} C^*_{n'j'} (-1)^{m'+m} a^\dagger_{njm} a^\dagger_{nj,-m} a_{n'j'm'} a_{n'j',-m'}. \tag{6.39b}$$

KAON DECAY FOR PEDESTRIANS

Chapter 7 presents kaon decay as an introduction to the treatment of time-dependent transition processes in quantum mechanics and to the 'dirty' real world of everyday applications of quantum mechanics to frontier research problems. Up to this point in his education the student has been impressed with the clarity and elegance of classical mechanics and electrodynamics and of the beautiful treatments of elementary quantum mechanics like the solution of the hydrogen atom. Now comes the rude awakening. The student who wishes to embark on a career of original research must leave behind him the elegant world of completely solved problems. At the frontier of open unsolved and half-solved problems there is no immediate hope for deriving a Hamiltonian from first principles, writing down a Schrödinger equation and then solving it either exactly or by a series of justified successive approximations. One must be satisfied with gathering the available bits and pieces of already established knowledge and trying to fit them together to make progress in the search for new knowledge.

Kaon decay is an excellent simple example of the use of quantum mechanics in frontier research. It has many interesting aspects and the absence of spin simplifies the formal treatment. We begin by accepting the existence of kaons and pions as input and the fact that a kaon decays into pions. This input is used without any derivation from first principles to construct a phenomenological Hamiltonian. The previous introduction of the second-quantized notation of ch. 5 makes this possible at an early stage in the course. Constraints imposed by conservation laws and invariance principles are introduced in the Hamiltonian. The resulting well-defined Schrödinger equation is then solved by time-dependent perturbation theory. The perturbation theory is presented first in the old-fashioned formalism of Heitler with its simple physical interpretation of time-dependent amplitudes. The golden rule formula is derived and the role of energy conservation is discussed, in particular the way in which the uncertainty principle appears automatically in the results.

The neutral kaon system is then considered and shows the interesting consequences following from the single input that two kinds of neutral kaons exist which can both decay into two pions. The relations between the neutral kaon system and the simple polarized photon system of ch. 1 is discussed and the ideas of quantum-mechanical coherence and interference are extended to apply to the correlated decays of two neutral kaons at different points in space, thus providing an instructive example of the Einstein–Podolsky–Rosen paradox.

The kaon decay process is then used as an example of more advanced problems in time-dependent quantum mechanics. The K_1–K_2 mass difference is calculated by second-order perturbation theory using the more modern integral formalism which is more compact and convenient for higher-order calculations than the simultaneous differential equations of Heitler which have a clearer physical interpretation. The kaon mass difference, the exponential decay and the Lorentzian resonance curve for the decay pions are all derived. The treatment also shows simple examples of infrared and ultraviolet divergences encountered in field theory and their physical origin. Finally, the role of the discrete symmetries of parity, charge conjugation and time reversal in kaon decay are presented and serve as an elementary introduction to the detailed treatment of the symmetries in ch. 14.

CHAPTER 7

KAON DECAY

The K-meson (kaon) and the π-meson (pion) are 'pseudoscalar' bosons; i.e. they have zero spin and odd intrinsic parity. The mass of the kaon (≈ 500 MeV) is much larger than the mass of two pions (each ≈ 140 MeV) and the kaon is observed experimentally to decay into two pions. Although a detailed dynamical description of the decay process requires the application of relativistic quantum field theory, many of the interesting properties of this decay can be obtained from a 'phenomenological' treatment using the second-quantized formalism but no field theory. In this treatment we construct a model Hamiltonian to give the experimentally observed properties of free kaons and pions and to provide a mechanism for the decay. The decay process can then be treated using time-dependent perturbation theory.

7.1 A simplified model with a single-charge state

Kaons and pions have several possible charge states: positive, negative and neutral. We first neglect the charge states and assume that there is only one kind of kaon and one kind of pion. We can then define the second-quantized creation operators $a^{\dagger}_{\mathrm{K}\boldsymbol{q}}$ and $a^{\dagger}_{\pi\boldsymbol{q}}$ as creation operators for a kaon and a pion respectively in states with momentum $\hbar\boldsymbol{q}$.

The Hamiltonian for a system of non-interacting kaons can be written in the second-quantized notation as

$$H_{\mathrm{K}} = \sum_{\boldsymbol{q}} E_{\mathrm{K}\boldsymbol{q}} a^{\dagger}_{\mathrm{K}\boldsymbol{q}} a_{\mathrm{K}\boldsymbol{q}} \tag{7.1}$$

where $E_{\mathrm{K}\boldsymbol{q}}$ is the energy of a single kaon with momentum $\hbar\boldsymbol{q}$. We assume that the system is in a very large box, with periodic boundary conditions, to give a discrete spectrum for $\boldsymbol{q}$. We can choose $E_{\mathrm{K}\boldsymbol{q}}$ to fit the experimentally observed properties of kaons. The relativistic expression for the total energy of the kaon is

$$E_{\mathrm{K}\boldsymbol{q}} = [(\hbar q)^2 c^2 + M_{\mathrm{K}}^2 c^4]^{\frac{1}{2}} \tag{7.2}$$

where M_K is the kaon rest mass and c is the velocity of light. We must use the relativistic energy, including the rest energy, to discuss processes in which particles are created and destroyed, as the rest energy enters into the relations for conservation of energy. The explicit form (7.2) can be derived from a fundamental relativistic theory. We simply assume it and plug it into the Hamiltonian (7.1).

Similar relations describe a system of non-interacting pions

$$H_\pi = \sum_{q} E_{\pi q} a_{\pi q}^\dagger a_{\pi q} \tag{7.3}$$

$$E_{\pi q} = [(\hbar q)^2 c^2 + M_\pi^2 c^4]^{\frac{1}{2}}. \tag{7.4}$$

We can therefore define an unperturbed Hamiltonian H_0 describing a system of free non-interacting kaons and pions

$$H_0 = H_K + H_\pi. \tag{7.5}$$

The eigenvectors of the unperturbed Hamiltonian H_0 are all possible states containing a particular number of kaons and a particular number of pions in plane-wave states; e.g.

$$|K_{q'} K_{q''} \pi_{q'''}\rangle = a_{Kq'}^\dagger a_{Kq''}^\dagger a_{\pi q'''}^\dagger |0\rangle. \tag{7.6}$$

This state describes two kaons and one pion in plane-wave states of momentum $\hbar \boldsymbol{q}'$, $\hbar \boldsymbol{q}''$ and $\hbar \boldsymbol{q}'''$ respectively. The energy of this state is the sum of the energies of these three particles.

$$H_0|K_{q'} K_{q''} \pi_{q'''}\rangle = (E_{Kq'} + E_{Kq''} + E_{\pi q'''})|K_{q'} K_{q''} \pi_{q'''}\rangle. \tag{7.7}$$

This relation (7.7) can be verified directly using the commutation rules for boson creation and annihilation operators.

We have here a complete description of all possible systems containing kaons and pions in the approximation where we consider them as stable particles and neglect their decays. We have the Hamiltonian and the Schrödinger equation. We can write down explicitly all solutions and eigenvalues of the Schrödinger equation. However, all this is purely formal. There is no new physics in it. The only properties of pions and kaons which we can obtain from this description are those which we have already put in.

Let us now include in this description the decay of a kaon into two pions. We use a phenomenological approach typical of the application of quantum mechanics in current research problems which are much too difficult to be treated exactly from first principles. In nuclear and subnuclear physics we cannot write down a Hamiltonian from first principles as one does for the hydrogen atom, because the basic laws of force are still unknown. In many problems arising in solid state physics, the full description of the many-body system is too complicated to be treated by known methods. In such cases we can construct simplified phenomenological models designed to give certain desired features, e.g. the existence of pions and kaons with their appropriate masses and the decay of a kaon into two pions. The form of the model Hamiltonian is restricted by conservation laws and invariance principles such as conservation of momentum, angular momentum and electric charge, and relativistic invariance. We then solve the Schrödinger equation for the model Hamiltonian and look for other implications of the theory beyond those which the Hamiltonian was explicitly constructed to give. We shall see this explicitly in the example of kaon decay.

To describe the decay of a kaon into two pions we need a wave function which initially at time $t=0$ is a one-kaon state, but which develops a two-pion component at later times. The two-pion component describes the probability that the kaon has decayed. For a kaon with zero momentum in the initial state, such a wave function would have the form

$$|\psi(t)\rangle = \alpha(t) a_{\mathrm{K}0}^{\dagger}|0\rangle + \sum_{\boldsymbol{q}''} \beta_{\boldsymbol{q}''}(t) a_{\pi,-\boldsymbol{q}''}^{\dagger} a_{\pi\boldsymbol{q}''}^{\dagger}|0\rangle \tag{7.8a}$$

where $\alpha(t)$ and $\beta_{\boldsymbol{q}''}(t)$ are time-dependent coefficients satisfying the initial conditions

$$\alpha(0) = 1 \tag{7.8b}$$

$$\beta_{\boldsymbol{q}''}(0) = 0. \tag{7.8c}$$

We now wish to add to our unperturbed Hamiltonian H_0 an interaction term V, which should describe the decay; i.e. the wave function (7.8) should be a solution of the time-dependent Schrödinger equation with the Hamiltonian H_0+V.

$$\begin{aligned} i\hbar \frac{\partial}{\partial t}|\psi\rangle &= i\hbar \frac{\partial \alpha}{\partial t} a_{\mathrm{K}0}^{\dagger}|0\rangle + i\hbar \sum_{\boldsymbol{q}''} \frac{\partial \beta_{\boldsymbol{q}''}}{\partial t} a_{\pi,-\boldsymbol{q}''}^{\dagger} a_{\pi\boldsymbol{q}''}^{\dagger}|0\rangle = (H_0 + V)|\psi\rangle \\ &= E_{\mathrm{K}0}\alpha a_{\mathrm{K}0}^{\dagger}|0\rangle + 2\sum_{\boldsymbol{q}''} E_{\mathrm{K}\boldsymbol{q}''}\beta_{\boldsymbol{q}''} a_{\pi,-\boldsymbol{q}''}^{\dagger} a_{\pi\boldsymbol{q}''}^{\dagger}|0\rangle + V|\psi\rangle. \end{aligned} \tag{7.9}$$

At $t=0$, $\beta_{\boldsymbol{q}''}=0$, and there is no pion component in the wave function. In order to create a pion component at a later time, the operator V must have matrix elements connecting the one-kaon and two-pion states; i.e. it must have terms of the form

$$a_{\pi\boldsymbol{q}'''}^{\dagger} a_{\pi\boldsymbol{q}''}^{\dagger} a_{\mathrm{K}\boldsymbol{q}'} \tag{7.10}$$

which annihilate a kaon and create two pions. However, a term of this kind is not Hermitean, as required for conservation of probability. Thus, if we add a term of the form (7.10) we must also add its Hermitean conjugate:

$$a^{\dagger}_{\mathrm{K}\boldsymbol{q}'}a_{\pi\boldsymbol{q}''}a_{\pi\boldsymbol{q}'''}. \tag{7.11}$$

This term (7.11) describes the annihilation of two pions and the creation of a kaon. Thus conservation of probability in quantum mechanics requires the existence of the inverse process of kaon production by two pions in any description where kaon decay into two pions can take place. This consequence of quantum mechanics is something new which we have not put in from the beginning.

A further restriction on the perturbation V follows from conservation of momentum (or translation invariance of the Hamiltonian). The momentum of the center-of-mass of the two pions must be equal to the kaon momentum. With these restrictions the most general interaction containing terms of the type (7.10) and (7.11) has the form:

$$\begin{aligned} V = \sum_{\boldsymbol{q}'\boldsymbol{q}''} & g(\boldsymbol{q}', \boldsymbol{q}'')a^{\dagger}_{\pi(\boldsymbol{q}'-\boldsymbol{q}'')}a^{\dagger}_{\pi(\boldsymbol{q}'+\boldsymbol{q}'')}a_{\mathrm{K},\,2\boldsymbol{q}'} \\ & + g^{*}(\boldsymbol{q}', \boldsymbol{q}'')a^{\dagger}_{\mathrm{K},\,2\boldsymbol{q}'}a_{\pi(\boldsymbol{q}'+\boldsymbol{q}'')}a_{\pi(\boldsymbol{q}'-\boldsymbol{q}'')} \end{aligned} \tag{7.12}$$

where $g(\boldsymbol{q}', \boldsymbol{q}'')$ is some function of the center-of-mass momentum $2\hbar\boldsymbol{q}'$ and the relative momentum of the two pions $2\hbar q''$, expressing the strength of the interaction or the 'coupling constant', as well as the momentum dependence. The explicit form of the function g is restricted further by the requirements of rotational and relativistic invariance and is determined completely by the specific form of field theory used to describe the process.

We can use rotational invariance to determine the dependence of $g(0, \boldsymbol{q}'')$ on the direction of the vector $\boldsymbol{q}''$, which is the direction of emission of the two pions. For $\boldsymbol{q}'=0$, the kaon is initially at rest and has zero angular momentum. The state is therefore invariant under rotation and $g(0, \boldsymbol{q}'')$ is independent of the direction of $\boldsymbol{q}''$.

Relativity determines the dependence of the functions $g(\boldsymbol{q}', \boldsymbol{q}'')$ on $\boldsymbol{q}'$. A change in $\boldsymbol{q}'$ is just a change of the momentum of the kaon; i.e. a change of the velocity of the kaon relative to the coordinate system. An arbitrary change in $\boldsymbol{q}'$ can therefore be produced by a Lorentz transformation. Thus if the value of the function $g(\boldsymbol{q}', \boldsymbol{q}'')$ is known for anyone particular value of $\boldsymbol{q}'$, e.g. $g(0, \boldsymbol{q}'')$, it can be obtained for all values of $\boldsymbol{q}'$ by means of Lorentz transformations. Thus only the dependence of $g(0, \boldsymbol{q}'')$ on the magnitude of $\boldsymbol{q}''$ remains undetermined.

The complete dependence of the function $g(\boldsymbol{q}', \boldsymbol{q}'')$ can be determined by assuming a specific form of field theory. Consider for example the additional requirement that the interaction be 'local'; i.e. that the two pions are produced at exactly the same point in configuration space where the kaon disappears. This is expressed very simply by writing the interaction in terms of creation operators $a^\dagger(\boldsymbol{x})$ which create a particle at the point $\boldsymbol{x}$ in configuration space. The interaction then has the form

$$V = g \int \mathrm{d}\boldsymbol{x}\, a_\pi^\dagger(\boldsymbol{x}) a_\pi^\dagger(\boldsymbol{x}) a_\mathrm{K}(\boldsymbol{x}) + \text{h.c.} \tag{7.13a}$$

where g is a constant and h.c. denotes Hermitean conjugate. There is no additional dependence on $\boldsymbol{x}$ in the interaction since all points in configuration space are equivalent. The local interaction (7.13a) can be expressed in terms of operators in momentum space by writing each of the three operators as a Fourier transform

$$V = g' \int \mathrm{d}\boldsymbol{x} \sum_{\boldsymbol{q}'\boldsymbol{q}''\boldsymbol{q}'''} \mathrm{e}^{-\mathrm{i}\boldsymbol{q}'\cdot\boldsymbol{x}} \mathrm{e}^{-\mathrm{i}\boldsymbol{q}''\cdot\boldsymbol{x}} \mathrm{e}^{\mathrm{i}\boldsymbol{q}'''\cdot\boldsymbol{x}}\, a_{\pi\boldsymbol{q}'}^\dagger a_{\pi\boldsymbol{q}''}^\dagger a_{\mathrm{K}\boldsymbol{q}'''} + \text{h.c.} \tag{7.13b}$$

where g' is the product of g and the various normalization factors of the Fourier transform. The integration over $\boldsymbol{x}$ is easily carried out and leads to a delta function requiring momentum conservation.

$$V = g'' \sum_{\boldsymbol{q}'\boldsymbol{q}''} a_{\pi\boldsymbol{q}'}^\dagger a_{\pi\boldsymbol{q}''}^\dagger a_{\mathrm{K}(\boldsymbol{q}'+\boldsymbol{q}'')} + \text{h.c.} \tag{7.13c}$$

where g'' is a constant. Comparing the local interaction (7.13) with the general interaction (7.12) shows that the requirement of a local interaction determines the form of the function $g(\boldsymbol{q}', \boldsymbol{q}'')$ to be a constant independent of $\boldsymbol{q}'$ and $\boldsymbol{q}''$.

Although the requirements of momentum conservation have been specifically inserted into the interaction (7.12) the requirements of energy conservation have not. The momenta of the two pions produced are uniquely determined in the decay of a kaon of momentum $2\hbar\boldsymbol{q}'$ by energy and momentum conservation. One might think that V should only include those terms corresponding to the values of the relative pion momentum allowed by energy and momentum conservation. This, however, is not the case. The exact energy of a system is defined uniquely only when the system is in an eigenstate of the Hamiltonian, and the Hamiltonian *includes* the pertrubation V. We shall be considering a state which is not an eigenfunction or stationary state of the total Hamiltonian but is a solution of the *time-dependent* Schrödinger equation which begins as a kaon and ends up as two pions. Since this state is a momentum eigenfunction we *can* use momentum conservation. However, it is not an energy eigenfunction. Furthermore we only know the *unperturbed* energies and *do not know the exact energies* which are eigenvalues of the full Hamiltonian including V. Thus we cannot plug energy conservation in at the beginning. We shall see that it comes out in the calculation.

One might think that we are violating some basic principle of relativity by considering energy and momentum on different footings. This is not true. It is possible to introduce a covariant formulation in which energy and momentum are treated in exactly the same way, but it is not necessary. Although the laws of nature are believed to be Lorentz-invariant, experiments are always performed in a particular Lorentz frame. It is convenient to choose our description to be simple in this frame, although it may be complicated in other frames. When we say that the state (7.8) is certainly a kaon at time $t=0$, we are choosing a particular Lorentz frame. In other frames this initial condition will not be at a fixed time, but at a time depending upon the position in space. Since we violate the symmetry between space and time in our choice of initial conditions, we naturally destroy the symmetry between energy and momentum.

7.2 Time-dependent perturbation theory in the simplified model

We now insert the interaction (7.12) into the time-dependent Schrödinger equation (7.9) and examine the solution of the equation. Only the terms in (7.12) with $\boldsymbol{q}'=0$ give a non-vanishing contribution when acting on the state (7.8). All others terms contain one annihilation operator which commutes with all the creation operators in the wave function (7.8). They can be moved to the right past the creation operators to act directly on the vacuum and give zero. Thus

$$V|\psi\rangle = \sum_{\boldsymbol{q}''} g(0, q'')\alpha a^{\dagger}_{\pi,-\boldsymbol{q}''} a^{\dagger}_{\pi\boldsymbol{q}''}|0\rangle + \sum_{\boldsymbol{q}''} g^*(0, q'')\beta_{\boldsymbol{q}''} a^{\dagger}_{\mathrm{K}0}|0\rangle. \tag{7.14}$$

Substituting eq. (7.14) into eq. (7.9) and equating the coefficients for each state, we obtain:

$$\mathrm{i}\hbar\frac{\partial\alpha}{\partial t} = E_{\mathrm{K}0}\alpha + \sum_{\boldsymbol{q}''} g^*(0, q'')\beta_{\boldsymbol{q}''} \tag{7.15a}$$

$$\mathrm{i}\hbar\frac{\partial\beta_{\boldsymbol{q}''}}{\partial t} = 2E_{\pi q}\beta_{\boldsymbol{q}''} + g(0, q'')\alpha. \tag{7.15b}$$

The formal integration of these equations, using the initial conditions (7.8b) and (7.8c) gives

$$\alpha\mathrm{e}^{\mathrm{i}E_{\mathrm{K}0}t/\hbar} - 1 = -\frac{i}{\hbar}\sum_{\boldsymbol{q}''} g^*(0, q'')\int_0^t \mathrm{d}t'\beta_{\boldsymbol{q}''}(t')\mathrm{e}^{\mathrm{i}E_{\mathrm{K}0}t'/\hbar} \tag{7.15c}$$

$$\beta_{q''} e^{2iE_{\pi q}t/\hbar} = -\frac{i}{\hbar} g(0, q'') \int_0^t dt' \alpha(t') e^{2iE_{\pi q''}t'/\hbar}. \tag{7.15d}$$

We now make the approximation that V is a perturbation and is small. For sufficiently small values of t, the coefficients $\beta_{q''}$ are also small. Eq. (7.15) are easily solved approximately to first order in these small quantities. The right-hand side of eq. (7.15c) is of second order and is neglected. Thus

$$\alpha = e^{-iE_{K0}t/\hbar}. \tag{7.16a}$$

Substituting this result (7.16a) into eq. (7.15d) and integrating, we obtain

$$\beta_{q''} = g(0, q'') \left\{ \frac{e^{-i(E_{K0} - 2E_{\pi q''})t/\hbar} - 1}{E_{K0} - 2E_{\pi q''}} \right\} e^{-2iE_{\pi q''}t/\hbar}. \tag{7.16b}$$

This first-order result is valid only for times which are short compared to the mean lifetime of the decay, as eq. (7.16a) shows that the magnitude of α does not change. This neglects the decrease in the probability of the initial state due to the decay. However this approximation is sufficient for obtaining the probability of finding the system in a two-pion state with momenta $\boldsymbol{q}''$ and $-\boldsymbol{q}''$ after a short time t,

$$|\beta_{q''}|^2 = \frac{4|g(0, q'')|^2}{(E_{K0} - 2E_{\pi q''})^2} \sin^2\{(E_{K0} - 2E_{\pi q''})t/2\hbar\}. \tag{7.17}$$

This result exhibits the characteristic form of the energy conservation law, subject to the uncertainty principle, in time-dependent perturbation theory with a time-*independent* perturbation. The time-dependent wave function contains non-vanishing amplitudes for some states, where the pion momenta do not have the proper value to conserve energy. These, however, do not increase with time but vary periodically and remain small. Only for values for $\boldsymbol{q}''$ which conserve energy; i.e. where $2E_{\pi q''} = E_{K0}$, does the amplitude increase with time. However, for those components which almost conserve energy, the period of the oscillation is very long. These components increase monotonically with time for a quarter of the period. Thus at any time t, there are still some components which do not quite conserve energy which are still increasing; namely those for which

$$(E_{K0} - 2E_{\pi q''})t/2\hbar < \tfrac{1}{2}\pi. \tag{7.18a}$$

The energy of the two-pion state is thus not sharply defined. The width of the energy distribution decreases with time. Its order of magnitude is seen from eq. (7.18a) to be

$$\Delta E_\pi \approx \hbar/t. \tag{7.18b}$$

This is just the spread given by the uncertainty principle.

The total probability that the kaon has decayed into two pions at time t is obtained by summing the expression (7.17) over all values of $\boldsymbol{q}''$. If the system is in a very large box, the sum can be replaced by an integral. Thus

$$\sum_{\boldsymbol{q}''} |\beta_{\boldsymbol{q}''}|^2 = \int |\beta_{\boldsymbol{q}''}|^2 \varrho(E_f)\mathrm{d}E_f \tag{7.19a}$$

where $\varrho(E_f)$ is the density of final states per unit energy E_f and $E_f = 2E_{\pi q''}$. Only a very narrow range of E_f contributes to the integral (7.19a). The quantities $g(0, q'')$ and $\varrho(E_f)$ are assumed to vary smoothly in this range. Thus we can replace them by their values at the energy $E_f = E_{\mathrm{KO}}$ where energy is conserved, denoted by $g(0, q_{\mathrm{K}})$ and $\varrho(E_{\mathrm{KO}})$, and take them outside the integral. Substituting eq. (7.17) into eq. (7.19a), we obtain

$$\sum_{\boldsymbol{q}''} |\beta_{\boldsymbol{q}''}|^2 = 4|g(0, q_{\mathrm{K}})|^2 \varrho(E_{\mathrm{KO}}) \int_{-\infty}^{\infty} \frac{t}{2\hbar} \frac{\sin^2 x}{x^2} \mathrm{d}x$$

$$= \frac{2\pi}{\hbar} |g(0, q_{\mathrm{K}})|^2 \varrho(E_{\mathrm{KO}})t \tag{7.19b}$$

where

$$x = (E_{\mathrm{KO}} - 2E_{\pi q''})t/2\hbar = (E_{\mathrm{KO}} - E_f)t/2\hbar. \tag{7.19c}$$

The total transition probability is seen to be proportional to the time t, as expected at short times when the decay of the initial state is neglected. (A better approximation, which we shall discuss below, leads to the expected exponential decay.) The transition probability per unit time is then

$$W_{\mathrm{K}\to 2\pi} = \frac{2\pi}{\hbar} |g(0, q_{\mathrm{K}})|^2 \varrho(E_{\mathrm{KO}}) = \frac{2\pi}{\hbar} |\langle f|V|i\rangle|^2 \varrho(E_f), \tag{7.20}$$

where $|i\rangle$ and $|f\rangle$ denote the initial and final states. The value of the matrix element of V between these states is just $g(0, q_{\mathrm{K}})$.

The result (7.20) is characteristic of all radiation and decay processes, and has been called the 'golden rule' formula. The transition probability is proportional to the square of a matrix element, which may be momentum-dependent, and a phase space factor which is the energy density of final states. We can evaluate this energy density by noting that the density of states in momentum space is given by

$$\varrho(p)\mathrm{d}p = \frac{4\pi p^2 \mathrm{d}p}{h^3} \Omega \tag{7.21}$$

if the system is enclosed in a box of volume Ω. The density of the states for the two-pion system can then be calculated from eq. (7.21) and substituted into eq. (7.20). One should not be concerned by the appearance of the volume of the box Ω in the result although it is obvious on physical grounds that the rate of decay of a kaon at rest cannot depend upon the size of the box in which the system is enclosed. There is another volume factor hidden in the matrix element which is calculated using plane-wave functions normalized in a box of volume Ω. A factor $1/\Omega$ thus appears in the square of the matrix element in eq. (7.19) and cancels out the volume factor in the phase space.

Let us now return to the assumption that $g(0, q'')$ varies slowly in the narrow range of E_f which contributes to the integral in eq. (7.19a), and can therefore be taken outside the integral. For a local interaction of the form (7.13) $g(0, q'')$ is rigorously independent of q'' and can be taken outside the integral without any approximation. For this case the two pions in the final state have a 'zero range' interaction; i.e. they behave as free pions except when they are exactly at the same point in space.

Suppose instead that the interaction exists over a finite region in space; i.e. that two pions can turn into a kaon when they are separated by a small finite distance which we denote by Δx. Then $g(0, q'')$ will no longer be a constant independent of q'' but will be the Fourier transform of the function specifying the strength of the interaction in configuration space. For very high values of q'', where there are many wavelengths within the interaction region, the contributions to the Fourier transform from different parts of the region cancel one another, and $g(0, q'')$ becomes very small. Thus $g(0, q'')$ can be expected to be reasonably constant for small values of q'', less than $1/\Delta x$, where the wavelength is longer than the range of the interaction, but to drop rapidly to zero for larger values of q''.

Thus $g(0, q'')$ is constant in the range $\Delta q''$ of the integral (7.19a) if $\Delta q'' \ll 1/\Delta x$. Since $\hbar q''$ is the pion momentum, p_π, this condition can be rewritten

$$\Delta x \ll \frac{\hbar}{\Delta p_\pi} = \frac{\hbar v_\pi}{\Delta E_\pi} \approx v_\pi t, \tag{7.22}$$

where we have used the relation between the pion energy E_π, velocity v_π and momentum, and the relation (7.18b) between the time and the energy spread. The condition thus states that sufficient time must have elapsed so that the emitted pions have moved away from one another a distance large compared to the range of the interaction.

7.3 The decay of neutral kaons

In nature four different kinds of kaons and three kinds of pions are found. There are a positively charged kaon, a negatively charged kaon and two neutral kaons while there are one positive, one negative and one neutral pion. The decays of charged kaons into a charged pion and a neutral pion are adequately described by the treatment in the preceding section where the charge states were not considered. It is simply necessary to add appropriate indices to the creation and annihilation operators to define the proper charge of the particle and to require that the Hamiltonian conserve charge. The decay of the neutral kaon introduces additional complicating and interesting features because of the existence of two states having the same charge.

The two neutral kaons are usually called the K^0 and the $\overline{K}^0$. These two particles have the same charge, mass, spin and parity and are distinguishable from one another only by the way in which they are produced in strong interactions. This is described by the use of a quantum number called 'strangeness' which is conserved in strong interactions. The K^0 and the K^+ have strangeness $S=+1$; the $\overline{K}^0$ and K^- have strangeness $S=-1$. The pions all have strangeness $S=0$. However, strangeness is conserved only in strong interactions and not in the weak interaction which gives rise to the decay of the kaon. In fact conservation of strangeness is manifestly violated in the decay of the kaon into two pions which have strangeness $S=0$.

Let us now consider the decay of neutral kaons into two pions. We shall neglect the different charge states of the pion (a neutral kaon can clearly decay either into two neutral pions or into a π^+ and a π^-) but we shall take into account the existence of the two kaon states K^0 and $\overline{K}^0$.

We first consider how to generalize the interaction V, eq. (7.12) to take into account the decay of two kinds of kaons. For simplicity we consider the decay of a kaon at rest and also disregard the momenta of the decay pions. The only relevant pion momenta are those which approximately conserve energy. The momentum dependence and energy conservation follow in exactly the same way here as in the simplified uncharged model. We therefore consider only a perturbation term of the form

$$g a_\pi^\dagger a_\pi^\dagger a_K + g^* a_K^\dagger a_\pi a_\pi \tag{7.23}$$

where we have suppressed the momentum variables. A natural generalization of this interaction to one describing the decay of two kinds of kaons would have the form

$$V = g_1 a_\pi^\dagger a_\pi^\dagger a_{K^0} + g_2 a_\pi^\dagger a_\pi^\dagger a_{\overline{K}^0}^\dagger + g_1^* a_{K^0}^\dagger a_\pi a_\pi + g_2^* a_{\overline{K}^0}^\dagger a_\pi a_\pi \tag{7.24}$$

where we have introduced the notation $a^\dagger_{\mathrm{K}^0}$ and $a^\dagger_{\overline{\mathrm{K}}^0}$ as creation operators for the K^0 and $\overline{\mathrm{K}}^0$ and have introduced two coupling constants g_1 and g_2 to describe the decay.

One might try to treat the interaction (7.24) by time-dependent perturbation theory in exactly the same way as in the simplified model. This would give K^0-decay with coupling constant g_1 and $\overline{\mathrm{K}}^0$-decay with a coupling constant g_2. However, this is not correct because the K^0 and $\overline{\mathrm{K}}^0$ are *degenerate* states and are *mixed* by the perturbation V, which can cause a K^0 to decay into two pions and then cause the two pions to make a $\overline{\mathrm{K}}^0$. It is possible to have a kaon in a state which is neither a K^0 or a $\overline{\mathrm{K}}^0$ but is a linear combination of the two. The simple treatment of the K^0 and $\overline{\mathrm{K}}^0$ independently neglects the degeneracy and the possibility of mixing. A correct treatment requires the use of *degenerate* perturbation theory; i.e., choosing the correct linear combinations of K^0 and $\overline{\mathrm{K}}^0$ which diagonalize the perturbation.

Consider the particular linear combinations

$$a^\dagger_{\mathrm{K}_1} = (|g_1^2| + |g_2^2|)^{-\frac{1}{2}}\{g_1 a^\dagger_{\mathrm{K}^0} + g_2 a^\dagger_{\overline{\mathrm{K}}^0}\} \tag{7.25a}$$

$$a^\dagger_{\mathrm{K}_2} = (|g_1^2| + |g_2^2|)^{-\frac{1}{2}}\{g_2 a^\dagger_{\mathrm{K}^0} - g_1 a^\dagger_{\overline{\mathrm{K}}^0}\}. \tag{7.25b}$$

Let us operate with the operator V on a one-meson state of the particular linear combination K_2 defined by eq. (7.25b).

$$Va^\dagger_{\mathrm{K}_2}|0\rangle = \frac{g_1 g_2}{(|g_1^2| + |g_2^2|)^{\frac{1}{2}}}\, a^\dagger_\pi a^\dagger_\pi|0\rangle - \frac{g_2 g_1}{(|g_1^2| + |g_2^2|)^{\frac{1}{2}}}\, a^\dagger_\pi a^\dagger_\pi|0\rangle = 0. \tag{7.26a}$$

Then

$$Ha^\dagger_{\mathrm{K}_2}|0\rangle = (H_0 + V)a^\dagger_{\mathrm{K}_2}|0\rangle = H_0 a^\dagger_{\mathrm{K}_2}|0\rangle = E_{\mathrm{K}^0} a^\dagger_{\mathrm{K}_2}|0\rangle. \tag{7.26b}$$

The state K_2 is thus an eigenvector of the perturbed as well as of the unperturbed Hamiltonian and therefore *does not decay at all* if the decay mechanism is described by the interaction (7.24).

It is therefore more sensible to write the interaction (7.24) in terms of the linear combinations K_1 and K_2 defined by eq. (7.25) rather than in terms of the states K^0 and $\overline{\mathrm{K}}^0$ which are eigenfunctions of the strangeness.

$$V = g a^\dagger_\pi a^\dagger_\pi a_{\mathrm{K}_1} + g^* a^\dagger_{\mathrm{K}_1} a_\pi a_\pi \tag{7.27a}$$

where

$$g = (|g_1^2| + |g_2^2|)^{\frac{1}{2}}. \tag{7.27b}$$

This interaction describes the decay of a neutral kaon in the state K_1 into two pions with a coupling constant g and implies that the other orthogonal linear combination K_2 does not decay at all.

Experimentally it was first found that indeed only one linear combination of K^0 and $\bar{K}^0$ decayed into two pions. The other orthogonal linear combination did not decay into two pions and eventually decayed into three pions or other multiparticle states with a much longer lifetime. More recently the longer lived kaon which decays into three pions was observed also to decay into two pions with a very small probability. This observation created a great sensation at the time because it also constituted proof of the violation of *CP*-invariance in kaon decays. We neglect this small *CP*-violation in this section and consider its effects in §§ 7.8 and 7.11.

This is a typical example of the use of quantum mechanics in present-day research problems. We constructed a Hamiltonian which would allow the two kinds of neutral kaons to decay into two pions and would also give the observed properties of free kaons and pions; namely the relations between their energy and momentum. We found that quantum mechanics introduced a new feature which was not inserted at the beginning; namely that different linear combinations of the two kaon states had to be considered and that one particular combination could not decay into two pions.

This new feature follows directly from two basic principles of quantum mechanics: (1) the description of quantum-mechanical probabilities as squares of amplitudes, (2) the principle of superposition. If the system is initially in a state which we call a K^0-meson, the probability that it decays into two pions after a time t is expressed by a complex amplitude, $A(t)$, whose square gives the probability that a two-pion state will be found at time t. If the system is initially in a state which we call a $\bar{K}^0$-meson the probability that it decays into two pions after a time t is described by a different complex amplitude $\bar{A}(t)$ whose square is the probability that a two-pion state will be found at time t. A priori there is no relation between the two amplitudes $A(t)$ and $\bar{A}(t)$. The principle of superposition tells us that if the system is initially in a state which is a linear combination of the states K^0 and $\bar{K}^0$ the probability amplitude for two-pion decay is given by the corresponding linear combination of the two amplitudes $A(t)$ and $\bar{A}(t)$. The probability for the decay is given by the square of this linear combination.

$$P\{|xK^0 + y\bar{K}^0\rangle \to |2\pi\rangle\} = |xA + y\bar{A}|^2. \tag{7.28}$$

Thus for all possible values of A and $\bar{A}$ there exists a particular linear combination of K^0 and $\bar{K}^0$ for which the decay amplitude vanishes. Thus we can always find some linear combination of the state K^0 and $\bar{K}^0$ such that the probability of observing two decay pions after a very long time t is zero; i.e. that this particular combination does not decay into two pions.

This case also provides an instructive example of quantum-mechanical interference. In the one-particle Schrödinger wave equation, the possibilities of constructive and destructive interference have a simple pictorial interpretation. Waves arriving at a given point in space by different paths combine with appropriate phases depending upon the path differences. The use of complex numbers to describe these amplitudes appears as a convenient mathematical device to keep track of the phases. In many-body problems or in problems like kaon decay where particles are created and destroyed, this simple wave picture is no longer clear. One cannot think of many particles, each described by a different wave, and these waves interfering with one another. The different ways for getting from one state to another can no longer be represented as different paths for a wave to propagate in three-dimensional space. They are, perhaps, paths in some other space. However complex amplitudes are still used to describe the probabilities for processes going in different 'paths' (e.g., the K^0-'path' and the $\overline{K}^0$-'path' in the above example). These amplitudes are added and can interfere constructively or destructively in a manner analogous to waves. The probability of the process is then given by the square of the resulting amplitude.

7.4 Neutral kaons and polarized photons

The properties of neutral kaons remind us of the passage of polarized photons through polaroids oriented at various angles. If a beam of polarized light is passed through polaroids oriented horizontally or vertically it is convenient to consider the incident photon as some linear combination of horizontal polarization and vertical polarization. One component of this then passes through the polaroid and the other does not. Suppose that after passing through such a polaroid the beam of polarized photons encounters a polaroid oriented at an angle of 45° with respect to the vertical. It is now convenient to consider the photon as a linear combination of components polarized at $\pm 45°$. One of *these* components now passes through the polaroid and the other does not. Consider a beam of polarized photons passing through three polaroids which are successively oriented horizontally, at an angle of 45°, and vertically. The vertically polarized component in the beam is first removed by the horizontal polaroid, but is 'regenerated' by passage through the 45° polaroid. A vertically polarized component thus remains which passes through the third vertically polarized polaroid.

An analogous situation exists with the two possible orthogonal sets of linear combinations of neutral kaon states: K^0 and $\overline{K}^0$ which are the convenient states for the strong interactions and K_1 and K_2 which are the convenient states for the weak interactions. A beam of K^0-mesons rapidly becomes a beam of K_2-mesons because the K_1-component decays into two pions in a short time, while the K_2-component does not decay into two pions and the lifetime for the three-pion decay mode is sufficiently long to be neglected. If this beam of K_2-mesons is now passed through a slab of matter, the K^0 and $\overline{K}^0$ linear combinations are relevant for describing the strong interactions with nucleons. The $\overline{K}^0$ is more strongly absorbed leaving an emergent beam of K^0-mesons. The K_1-component in this beam soon decays again leaving a K_2-beam.

In this way we find that the $(K^0, \overline{K}^0)$ basis is convenient for describing the passage of a kaon beam through matter where it is absorbed by strong interactions while the (K_1, K_2) basis is convenient for describing the passage of the beam through free space where it decays by weak interactions. This is directly analogous to the case of polarized photons, where the horizontal–vertical basis is convenient for describing the passage through horizontal and vertical polaroids while the $\pm 45°$ basis is convenient for describing the passage through $\pm 45°$ polaroids. A beam of pure K_2-mesons has a K_1-component introduced without adding anything to the beam, simply by absorbing the $\overline{K}^0$ component in matter. Similarly a purely horizontally polarized light beam has a vertically polarized component introduced by absorbing a 45° component in a polaroid.

These relations can be expressed formally by defining an operator S for strangeness which is conserved in strong interactions and an operator CP which is conserved in weak interactions (neglecting CP-violation). The operation of these operators on the kaon states is defined by the relations

$$S a^\dagger_{K^0}|0\rangle = +a^\dagger_{K^0}|0\rangle \tag{7.29a}$$

$$S a^\dagger_{\overline{K}^0}|0\rangle = -a^\dagger_{\overline{K}^0}|0\rangle \tag{7.29b}$$

$$CP a^\dagger_{K^0}|0\rangle = -a^\dagger_{\overline{K}^0}|0\rangle \tag{7.29c}$$

$$CP a^\dagger_{\overline{K}^0}|0\rangle = -a^\dagger_{K^0}|0\rangle. \tag{7.29d}$$

The interaction (7.24) is shown below to conserve CP if $g_1 = -g_2$. For this case, eqs. (7.25) become

$$a^\dagger_{K_1} = \frac{1}{\sqrt{2}}\{a^\dagger_{K^0} - a^\dagger_{\overline{K}^0}\} \tag{7.30a}$$

$$a^\dagger_{K_2} = \frac{1}{\sqrt{2}}\{a^\dagger_{K^0} + a^\dagger_{\overline{K}^0}\} \tag{7.30b}$$

and

$$S a^\dagger_{K_1}|0\rangle = a^\dagger_{K_2}|0\rangle \tag{7.31a}$$

$$S a^\dagger_{K_2}|0\rangle = a^\dagger_{K_1}|0\rangle \tag{7.31b}$$

$$CP a^\dagger_{K_1}|0\rangle = +a^\dagger_{K_1}|0\rangle \tag{7.31c}$$

$$CP a^\dagger_{K_2}|0\rangle = -a^\dagger_{K_2}|0\rangle. \tag{7.31d}$$

The states K^0 and $\overline{K}^0$ are the eigenvectors of S and the states K_1 and K_2 are the eigenvectors of CP. The two-pion state produced in kaon decay is an eigenvector of CP with an eigenvalue of $+1$. Thus we see that if CP is conserved in the weak interaction the K_1 decays into two pions and the K_2 does not.

7.5 Systems of two neutral kaons

A pair of kaons can be produced either in an elementary-particle reaction such as proton–antiproton annihilation, or in the decay of an unstable meson resonance such as the ϕ-meson. The system of two neutral kaons has interesting properties because the mesons are produced as K^0 and $\overline{K}^0$ in strong interactions but they decay as K_1 or K_2 in weak interactions. This feature allows conclusions to be drawn from the experimental results about the properties of the initial state which decayed into the two kaons.

Consider the system of a K^0 and $\overline{K}^0$ produced, for example, by the decay of a strongly interacting particle of zero strangeness. If the center-of-mass momentum of the pair is zero we can describe this state as

$$|\psi\rangle = \sum_{\boldsymbol{q}} f(\boldsymbol{q}) a^{\dagger}_{K^0\boldsymbol{q}} a^{\dagger}_{\overline{K}^0, -\boldsymbol{q}} |0\rangle \tag{7.32}$$

where $f(\boldsymbol{q})$ is some function of the relative momentum $2\boldsymbol{q}$. If this system is produced by the decay of a single particle with a well-defined mass the magnitude of $\boldsymbol{q}$ is determined by energy conservation (with a spread depending upon the natural width of the decaying state). The angular dependence of $\boldsymbol{q}$ depends upon the angular momentum of the initial state and is in general some kind of spherical harmonic. In particular, if the state has even parity, $f(\boldsymbol{q})$ is an even function of $\boldsymbol{q}$, whereas if the state has odd parity, $f(\boldsymbol{q})$ is an odd function of $\boldsymbol{q}$. Since kaons have no spin, $f(\boldsymbol{q})$ is even if the angular momentum J of the state is even; and $f(\boldsymbol{q})$ is odd if J is odd.

$$\text{even } J, \text{ even parity} \Rightarrow f(\boldsymbol{q}) = f(-\boldsymbol{q})$$

$$\text{odd } J, \text{ odd parity} \Rightarrow f(\boldsymbol{q}) = -f(-\boldsymbol{q}).$$

The state (7.32) decays into pions which are observed experimentally. Let us express this state in terms of the states K_1 and K_2, which are more convenient for description of the decay, by substituting eq. (7.30) into eq. (7.32).

$$|\psi\rangle = \sum_{\boldsymbol{q}} \tfrac{1}{2} f(\boldsymbol{q}) (a^{\dagger}_{K_1\boldsymbol{q}} + a^{\dagger}_{K_2\boldsymbol{q}})(a^{\dagger}_{K_2, -\boldsymbol{q}} - a^{\dagger}_{K_1, -\boldsymbol{q}}) |0\rangle \tag{7.33a}$$

$$|\psi\rangle = \sum_{q} \tfrac{1}{2} f(q)\{[a^{\dagger}_{K_1 q} a^{\dagger}_{K_2,-q} - a^{\dagger}_{K_2 q} a^{\dagger}_{K_1,-q}] - [a^{\dagger}_{K_1 q} a^{\dagger}_{K_1,-q} - a^{\dagger}_{K_2 q} a^{\dagger}_{K_2,-q}]\}\,|0\rangle. \tag{7.33b}$$

Since q is a dummy index in the sum we can replace q by $-q$ everywhere.

$$|\psi\rangle = \sum_{q} \tfrac{1}{2} f(-q)\{[a^{\dagger}_{K_1,-q} a^{\dagger}_{K_2 q} - a^{\dagger}_{K_2,-q} a^{\dagger}_{K_1 q}] - [a^{\dagger}_{K_1,-q} a^{\dagger}_{K_1 q} - a^{\dagger}_{K_2,-q} a^{\dagger}_{K_2 q}]\}|0\rangle. \tag{7.33c}$$

Comparing eqs. (7.33b) and (7.33c) we note that the first two terms in the square brackets in both equations are equal because boson creation operators commute, but have the opposite sign. The last two terms in the brackets in eqs. (7.33b) and (7.33c) are equal. Thus if $f(q)$ is an odd function of q the last two terms in eq. (7.33c) must vanish whereas if $f(q)$ is even the first two terms must vanish. It therefore follows that if the state $|\psi\rangle$ has even parity and even angular momentum we have

$$|\psi_{\text{even}}\rangle = -\sum_{q} \tfrac{1}{2} f(q)\{a^{\dagger}_{K_1 q} a^{\dagger}_{K_1,-q} - a^{\dagger}_{K_2 q} a^{\dagger}_{K_2,-q}\}\,|0\rangle. \tag{7.34a}$$

This is a linear combination of states of two K_1-mesons or of two K_2-mesons. The K_1 can decay into two pions with a short lifetime and the K_2 cannot decay into two pions and has a sufficiently long lifetime so its decay is generally not observed in a buble chamber. Thus the state (7.34a) is one in which there is a probability of 50% that *both* kaons decay into pions and a probability of 50% that *neither* kaon decays into two pions. The probability that one kaon decays into two pions and the other lives long and does not decay is zero.

On the other hand if the state (7.33) is one of odd parity and odd angular momentum it has the form

$$|\psi_{\text{odd}}\rangle = \sum_{q} \tfrac{1}{2} f(q)\{a^{\dagger}_{K_1 q} a^{\dagger}_{K_2,-q} - a^{\dagger}_{K_2 q} a^{\dagger}_{K_1,-q}\}\,|0\rangle = \sum_{q} f(q)\{a^{\dagger}_{K_1 q} a^{\dagger}_{K_2,-q}\}\,|0\rangle. \tag{7.34b}$$

This state consists of one K_1-meson and one K_2-meson. The K_1 decays into two pions with a short lifetime; the K_2 does not decay into two pions. Thus one and only one of the kaons in the state (7.34b) decays into two pions.

We see that it is possible to determine whether the parity and angular momentum of the two-kaon state is even or odd by observing the decay into two pions. If neither or both mesons decay into two pions the state has even angular momentum and parity. If only one decays into two pions and the other does not the state has odd angular momentum and parity. This method has been used in the determination of the parity and angular momentum of the ϕ-meson and in antiproton annihilation at rest.

Note the simplification in eqs. (7.32) to (7.34) resulting from the use of second-quantized notation rather than the Schrödinger notation. The latter would use labels for the two particles and would have two possible states for the physical state (7.32): one where particle number one is a K^0 and particle number 2 is $\overline{K}^0$ and vice versa. If one starts with a symmetrized wave function one then carries along twice as many terms as in the treatment using the second-quantized notation.

7.6 The Einstein–Rosen–Podolsky paradox in two-kaon systems

The decays of two-kaon systems provide an interesting example of one of the paradoxes of quantum mechanics which disturbed people in the early days of the theory; the paradox of Einstein, Podolsky and Rosen. Consider two kaons emitted in the $+z$ and $-z$ directions from the decay of an odd-parity state at rest. If we detect them by observing their decays, half of them will be K_1 and half will be K_2. If we look simultaneously at the two kaons coming from a single decay, one will be K_1 and the other K_2. There will be no case where both are K_1 or both are K_2. On the other hand, if we detect them by measuring their interaction with matter, half of them will be K^0 and half will be $\overline{K}^0$. If we examine the two kaons coming from a single decay one will always be K^0 and the other $\overline{K}^0$. They will never both be K^0 or both be $\overline{K}^0$.

It is instructive to consider the analogous case of simultaneous emission of two photons in such a state that they always have opposite polarizations from one another no matter how they are measured. If the photons emitted in a given direction are separated into horizontally and vertically polarized components, half are horizontal and half are vertical. If the horizontal or vertical polarization of two photons coming out simultaneously is measured one is horizontal and the other is vertical, they are never both horizontal or both vertical. However, the same occurs if one measures polarization at $+45°$ and $-45°$. Half of the photons are polarized at $+45°$ and half at $-45°$. Whenever two are measured simultaneously, they are always opposite, if one is $+45°$ the other is $-45°$.

Consider a kaon detector observing the kaons emitted in the $+z$ direction. No matter whether we measure the decays or the strong interactions we find that the kaon beam is 'unpolarized'. It is an equal statistical mixture of the two states K_1 and K_2, or K^0 and $\overline{K}^0$. There is always a 50–50 mixture regardless of the basis chosen for the measurement. This is analogous to an unpolarized light beam.

Now, consider a coincidence measurement of the decays of kaons emitted in the $+z$ and $-z$ directions. The measurement of the kaon in the $-z$ direction automatically separates the kaon beam emerging in the $+z$ direction into its two components, the K_1 and the K_2-components. Those in coincidence with K_1-mesons observed in the $-z$ direction constitute a beam of pure K_2-states and vice versa. In the photon analogy the beam of unpolarized photons is separated into components of horizontal and vertical polarization by measuring the horizontal and vertical polarization of the other beam emitted in the opposite direction.

So far everything is quite reasonable even from a classical point of view. The kaon beam consists half of K_1 and half of K_2-mesons. Each kaon can be assumed to be in a well-defined state, either K_1 or K_2, and two mesons emitted simultaneously are always in opposite states. Therefore, by measuring the meson in the $-z$ direction we simply find out information about its partner emitted at the same time in the $+z$ direction. The same is true for the photon analog. The photons can be assumed to be emitted in definite states of horizontal or vertical polarization, always in pairs with opposite polarization. The measurement of one beam then gives information about the other.

However, suppose that instead of measuring the decays, we measure the strong interactions of the kaons and determine whether each kaon detected in the $-z$ direction is a K^0 or $\overline{K}^0$. We then separate the unpolarized kaon beams observed in the $+z$ direction into two components which are K^0 and $\overline{K}^0$ respectively; those in coincidence with K^0's in the $-z$ direction are a pure $\overline{K}^0$ beam and vice versa. In the photon analogy, this would be separating the outgoing unpolarized beam in the $+z$ direction into components polarized at $+45°$ and $-45°$ by means of a measurement on the other beam. Here again there is a simple classical interpretation of the result. The kaons are emitted in definite states, either K^0 and $\overline{K}^0$, and the two emitted together are always in opposite states. Thus, a given kaon observed in the $+z$ direction is always in a definite state and the measurement on its partner in the $-z$ direction merely gives us the information what state it is.

However, when we combine the two types of measurements discussed above, we can no longer say that the kaon is emitted in a definite state. A kaon emitted in the $+z$ direction can be either K_1 or K_2, but it can also be either K^0 or $\overline{K}^0$. If we measure the *decay* of its companion emitted in the $-z$ direction, it will be definitely a K_1 or definitely a K_2, depending on the result of the measurement in the $-z$ direction. However, if we measure the *interactions with matter* at $-z$, the kaon observed at $+z$ is either definitely a K^0 or a $\overline{K}^0$ depending on what is observed at $-z$. Thus, even when the two kaons are far apart, one at $+z$ and the other at $-z$ and there is no interaction between them, we are still able to influence the state of the kaon at $+z$ *by our choice of what we measure at* $-z$, decays or strong interactions. In the analogous photon case, a given photon emerging at $+z$ is either in a state of definite horizontal or definite vertical polarization if we measure horizontal and vertical polarization of its partner emitted in the $-z$ direction. However, it is definitely polarized at $+45°$ or at $-45°$ if we choose instead to measure polarizations of $\pm 45°$ for the photon emitted in the $-z$ direction.

In considering single-kaon states, the analog with polarized light is helpful because a given state of polarization can always be expressed as a linear combination of polarization states in a different basis. A measurement of 45° polarization on a photon which is known to be horizontally polarized changes its state and forces it to become polarized either at $+45°$ or $-45°$, and a measurement of the strong interactions of a kaon known to be K_2 changes its state and forces it to become K^0 or $\overline{K}^0$. The case of the two-kaon system is formally the same as that of the one-kaon system but is physically quite different because a change in basis involves simultaneously the states of two kaons separated in space by a large distance. In the two-photon analog a measurement of 45° polarization not only forces the photon being measured to be in a state of either $\pm 45°$ but also similarly affects the other photon which is far away. A measurement of the interactions of one kaon not only forces it to be either K^0 or $\overline{K}^0$ but also similarly affects the other kaon which is far away. The description with a quantum-mechanical wave function thus includes coherent correlations over macroscopic distances. These correlations exist in quantum mechanics and are observed experimentally.

This case provides an example of quantum-mechanical interference even further removed from our intuitive picture of interfering waves arriving from different paths. However, formally it is simply another version of the 'two slit' experiment. The wave function (7.32) contains a term with a K^0 emitted in the $+z$ direction and a $\overline{K}^0$ emitted in the $-z$ direction. It also contains a term with the K^0 in the $-z$ direction and the $\overline{K}^0$ in the $+z$ direction. If we now look at the decays of these kaons and determine whether or not they are in the states K_1 or K_2, we find that there are contributions from both of the terms mentioned above. The two contributions are analogous to two 'paths' in the two-slit experiment. For the case of an odd-parity state (7.34b) the two contributions 'interfere constructively' for the K_1K_2 amplitude and 'interfere destructively' for the K_1K_1 and K_2K_2-amplitudes. Thus we have probability amplitudes for the results of simultaneous measurement on two particles separated by macroscopic distances, and these amplitudes have contributions from different 'paths', the $K^0(z)\overline{K}^0(-z)$ path and the $K^0(-z)\overline{K}^0(z)$ path.

7.7 The kaon mass shift

The interaction causing the kaon to decay into two pions also produces a change in the mass of the kaon. This is expected since the addition of a perturbation to a given Hamiltonian changes its energy eigenvalues. Such a shift in the energy of a decaying state is not generally observable experimentally, since it involves measuring the energy in the *absence* of the interaction as well as the physical energy. In the case of the neutral kaons, such a mass shift is measurable because there are two states K_1 and K_2, which are degenerate in the absence of this interaction. In the approximation where we consider only the 2π decay mode and neglect other decay modes, the interaction (7.27a) affects only the K_1 and not the K_2. The corresponding shift in the mass of the K_1 thus appears as a mass difference between the K_1 and K_2.

Let us attempt to calculate the mass shift by ordinary stationary perturbation theory. The unperturbed state consists of a kaon at rest and we calculate the change in the energy of the state as a result of the perturbation (7.12). Since the expectation value of the perturbation (7.12) is zero in a state with only one kaon present, there is no contribution in first order. The contribution in second order is given by the usual second-order perturbation theory formula

$$\Delta E = \sum_{q'} \frac{\langle K_0|V|\pi_{q'}\pi_{-q'}\rangle\langle\pi_{q'}\pi_{-q'}|V|K_0\rangle}{M_K c^2 - 2E_{\pi q'}} \tag{7.35a}$$

$$\Delta E = \sum_{q'} \frac{|g(0, q')|^2}{M_K c^2 - 2E_{\pi q'}}. \tag{7.35b}$$

This second-order result can be interpreted as due to the kaon decaying virtually into two pions and coming back again into a kaon, as indicated by the diagram fig. 7.1.

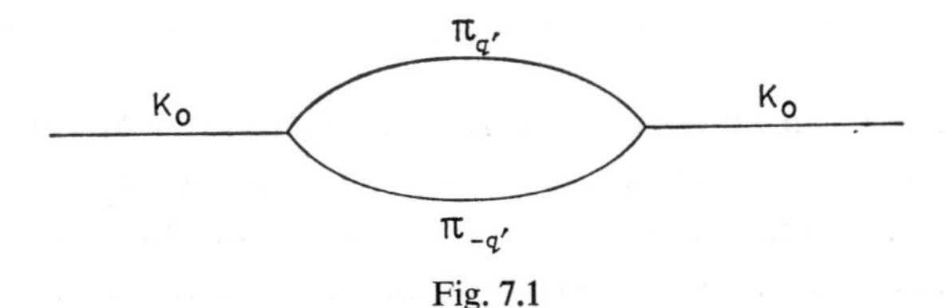

Fig. 7.1

Diagrams of this type, first introduced by Feynman, are exceedingly useful for labeling and keeping track of different contributions in perturbation theory.

The result (7.35) is not very useful as it stands since the sum on the index q' clearly diverges. There are two sources of divergence: (1) The denominator becomes zero for that value of q' where the energy is exactly conserved in the intermediate state. (2) The contributions for large values of q' increase without limit unless the function $g(0, q')$ cuts off at high values of q'. For the case of a local interaction, where g is independent of q', the sum (7.35) diverges at high values of q' since the density of states increases as q'^2 whereas the energy in the denominator only goes as q'. These divergences at low q (infrared) and high q (ultraviolet) illustrate the following general properties of such divergences in quantum field theory. Infrared divergences are not serious and can be handled in the same calculation by better approximation methods. Ultraviolet divergences are serious and indicate a profound difficulty in the theory.

The vanishing of the denominator in eq. (7.35) is not serious. It arises from the use of non-degenerate stationary state perturbation theory and the neglect of the degeneracy in the unperturbed Hamiltonian. The vanishing denominator simply indicates that there exists a state which is degenerate with the initial state and into which the kaon can decay and conserve energy. A proper treatment discussed below modifies the result to give the principal value of the expression (7.35), expressed as an integral over the variable q'.

The second divergence is much more serious and is still not adequately understood. Such ultraviolet divergences appear in many problems in quantum field theory as the result of contributions from very high momenta in intermediate states. So far no satisfactory means has been found to make such integrals finite although the method known as renormalization allows finite values to be obtained for experimentally measurable quantities even though infinite integrals appear in the theory. An arbitrary cut-off of the interaction at high momenta is sometimes introduced in calculations to obtain finite results, but no satisfactory method of determining the cut-off momentum has been found.

7.8 The neutral-kaon mass difference in time-dependent perturbation theory

The calculation of the mass shift is most conveniently done by time-dependent perturbation theory. Eq. (7.35) indicates that the shift is second order in the interaction V and requires second-order perturbation theory. Rather than extending the first-order treatment of section 7.2 to second order, we introduce a different formalism which is more suitable for higher-order calculations. The approximate solution of a set of simultaneous linear differential equations is very good for pedagogical purposes. Each of the coefficients α and $\beta_{q'}$ in section 7.2 has a well-defined physical meaning. The study of the behavior of these coefficients as functions of time gives a physical picture and provides physical insight. However, the extension of this approach to higher orders rapidly becomes impractical because the number of equations, coefficients and terms in a given equation becomes so large that it is impossible to handle them.

A more compact and tractable formalism is obtained by an approach which begins by converting the Schrödinger equation to an integral equation. We use this approach now in considering the kaon mass difference.

We begin by the following formal manipulations of the time-dependent Schrödinger equation

$$(H_0 + V)\psi = \mathrm{i}\hbar \frac{\partial \psi}{\partial t}. \qquad (7.36a)$$

Multiplying by $\mathrm{e}^{\mathrm{i}H_0 t/\hbar}$, we obtain

$$\mathrm{e}^{\mathrm{i}H_0 t/\hbar}[H_0 + V]\psi = \mathrm{i}\hbar\,\mathrm{e}^{\mathrm{i}H_0 t/\hbar}\frac{\partial \psi}{\partial t} = \mathrm{i}\hbar \frac{\partial}{\partial t}[\mathrm{e}^{\mathrm{i}H_0 t/\hbar}\psi] + H_0 \mathrm{e}^{\mathrm{i}H_0 t/\hbar}\psi. \qquad (7.36b)$$

This can be simplified to give

$$\mathrm{e}^{\mathrm{i}H_0 t/\hbar}V\psi = \mathrm{e}^{\mathrm{i}H_0 t/\hbar}V\mathrm{e}^{-\mathrm{i}H_0 t/\hbar}\mathrm{e}^{\mathrm{i}H_0 t/\hbar}\psi = \mathrm{i}\hbar \frac{\partial}{\partial t}[\mathrm{e}^{\mathrm{i}H_0 t/\hbar}\psi]. \qquad (7.36c)$$

Eq. (7.36c) can be written

$$V(t)\psi_{\mathrm{I}} = \mathrm{i}\hbar \frac{\partial}{\partial t}\psi_{\mathrm{I}} \qquad (7.37a)$$

where

$$\psi_{\mathrm{I}} = \mathrm{e}^{\mathrm{i}H_0 t/\hbar}\psi \qquad (7.37b)$$

$$V(t) = \mathrm{e}^{\mathrm{i}H_0 t/\hbar}V\mathrm{e}^{-\mathrm{i}H_0 t/\hbar}. \qquad (7.37c)$$

The function ψ_{I} is called the wave function in the interaction representation, Eq. (7.37a) is the Schrödinger equation in this representation.

The simplification introduced by the use of the interaction representation can be seen by examining the first-order solutions in section 7.2 in the interaction representation. We see that the additional exponential factor in eq. (7.37b) simplifies the wave function by exactly canceling the time-dependent exponential phase factors appearing in eqs. (7.16a) and (7.16b).

Formal integration of eq. (7.37a) gives an integral equation for ψ_{I}

$$\psi_{\mathrm{I}}(t) = \psi_{\mathrm{I}}(t_0) + \frac{1}{i\hbar}\int_{t_0}^{t} V(t_1)\psi_{\mathrm{I}}(t_1)\,\mathrm{d}t_1. \tag{7.38a}$$

If the interaction V is small, the integral equation (7.38a) can be solved by iteration to obtain a solution expressed as a power series in the interaction

$$\psi_{\mathrm{I}}(t) = \psi_{\mathrm{I}}(t_0) + \frac{1}{i\hbar}\int_{t_0}^{t} V(t_1)\psi_{\mathrm{I}}(t_0)\,\mathrm{d}t_1$$

$$+ \left(\frac{1}{i\hbar}\right)^2 \int_{t_0}^{t} V(t_1)\,\mathrm{d}t_1 \int_{t_0}^{t_1} V(t_2)\psi_{\mathrm{I}}(t_0)\,\mathrm{d}t_2 + \dots . \tag{7.38b}$$

Let us now apply eq. (7.38b) to the case of the decay of a kaon at rest. It is convenient to set $t_0 = 0$. Thus

$$\psi_{\mathrm{I}}(t_0) \equiv \psi_{\mathrm{I}}(0) = |\mathrm{K}_{r0}\rangle = a^{\dagger}_{\mathrm{K}_{r0}}|0\rangle \tag{7.39}$$

where we have introduced the additional index r to specify the kaon state in the two-dimensional subspace of K_1 and K_2. At later times the wave function contains a one-kaon component and also a two-pion component as in eq. (7.8a). If we use a more general interaction V including other decay modes, these also appear in the wave function.

The mass shift of the kaon states is given by the time behavior of the one-kaon component in the wave function. We therefore take the scalar product of the total wave function (7.38b) and a one-kaon state of zero momentum $|\mathrm{K}_{s0}\rangle$

$$\langle \mathrm{K}_{s0}|\psi_{\mathrm{I}}(t)\rangle = \langle \mathrm{K}_{s0}|\mathrm{K}_{r0}\rangle$$

$$+ \left(\frac{1}{i\hbar}\right)^2 \int_0^{t} \mathrm{d}t_1 \int_0^{t_1} \mathrm{d}t_2 \langle \mathrm{K}_{s0}|V(t_1)V(t_2)|\mathrm{K}_{r0}\rangle + \dots \tag{7.40a}$$

The linear term in V does not contribute since V has no matrix elements connecting two one-kaon states. The matrix elements appearing on the right-hand side of eq. (7.40a) can be rewritten

$$\langle \mathrm{K}_{s0}|V(t_1)V(t_2)|\mathrm{K}_{r0}\rangle = \sum_i \langle \mathrm{K}_{s0}|V|i\rangle\langle i|V|\mathrm{K}_{r0}\rangle \mathrm{e}^{-i(E_i - E_{\mathrm{K}0})(t_1 - t_2)/\hbar}, \tag{7.40b}$$

where we sum over all intermediate states $|i\rangle$ which are eigenfunctions of H_0 with eigenvalue E_i. The time dependence has been taken out of the matrix element by use of the definition (7.37c).

We now integrate eq. (7.40b) over t_2.

$$\frac{1}{i\hbar}\int_0^{t_1} dt_2\langle \mathrm{K}_{s0}|V(t_1)V(t_2)|\mathrm{K}_{r0}\rangle = \sum_i \langle \mathrm{K}_{s0}|V|i\rangle\langle i|V|\mathrm{K}_{r0}\rangle \left\{\frac{1-\mathrm{e}^{-\mathrm{i}(E_i-E_{\mathrm{K}0})t_1/\hbar}}{E_i - E_{\mathrm{K}0}}\right\}. \quad (7.41\mathrm{a})$$

The sum over the intermediate states can be simplified by writing it as an integral over the energy E_i

$$\frac{1}{i\hbar}\int_0^{t_1} dt_2\langle \mathrm{K}_{s0}|V(t_1)V(t_2)|\mathrm{K}_{r0}\rangle = \sum_i \int_{-\infty}^{\infty} \langle \mathrm{K}_{s0}|V|i\rangle\langle i|V|\mathrm{K}_{r0}\rangle \left\{\frac{1-\mathrm{e}^{-\mathrm{i}(E_i-E_{\mathrm{K}0})t_1/\hbar}}{E_i - E_{\mathrm{K}0}}\right\} \varrho(E_i)\,\mathrm{d}E_i \quad (7.41\mathrm{b})$$

where $\varrho(E_i)$ is the density of intermediate states at the energy E_i for a given decay mode and the sum is over all decay modes having the energy E_i. The limits of the integral are taken as $\pm\infty$, even though there are no states contributing with energies less than the rest energy of two pions. The expression is still formally correct since the density of states $\varrho(E_i)$ is zero for these low energies.

The integral (7.41b) is simplified by considering it as a contour integral in the complex plane. We can deform the contour to avoid the point $E_i = E_{\mathrm{K}0}$ by going around it in a small semicircle in the lower half plane as shown in fig. 7.2, since the integrand is analytic in this region. Although each of the terms in the bracket has a pole at $E_i = E_{\mathrm{K}0}$, there is no pole in the complete integrand. We can now consider each term in the bracket separately because the contour of fig. 7.2 avoids the pole. The integral of the second term over

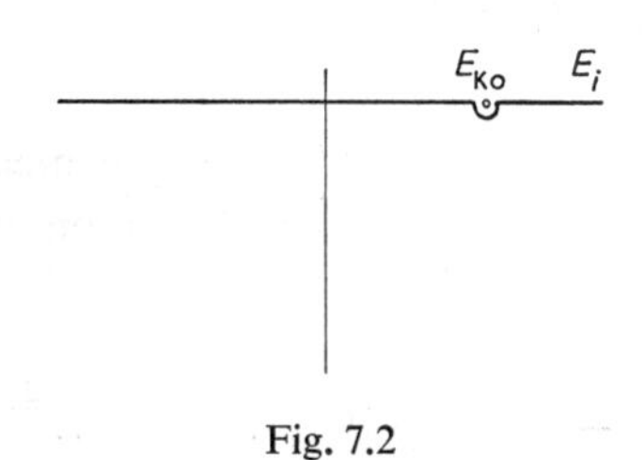

Fig. 7.2

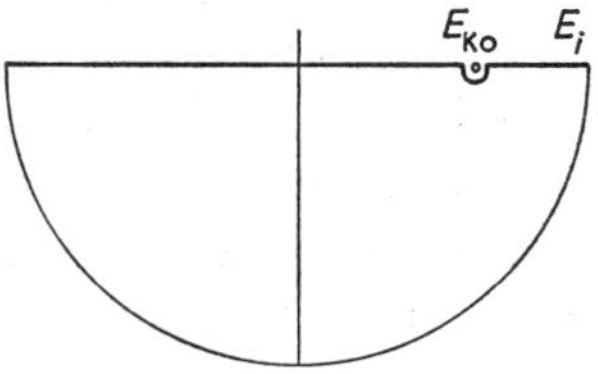

Fig. 7.3

this contour is simplified by closing the contour with a large semicircle of radius ΔE in the lower half plane, as shown in fig. 7.3. Since t is positive, the exponential is negative for negative imaginary values of E_i, and the integral over the large semicircle is negligible if $t_1 \gg \hbar/\Delta E$. Thus the integral (7.41) over the contour of fig. 7.2 is equal to the integral over the closed contour fig. 7.3. This vanishes by Cauchy's theorem if the integrand is analytic. If ΔE is chosen to keep all singularities outside the integrand, a minimum value of t_1 is determined for which our treatment applies. If V is regular, the singularity in $\varrho(E_i)$ at $E_i = 0$ requires $t_1 \gg \hbar/E_{\mathrm{K0}}$, which is not a serious restriction. Singularities in V would be interpreted in the spirit of the discussion following eq. (7.22). Thus

$$\int_{\mathrm{C}} \langle \mathrm{K}_{s0}|V|i\rangle\langle i|V|\mathrm{K}_{r0}\rangle \left\{ \frac{\mathrm{e}^{-\mathrm{i}(E_i - E_{\mathrm{K0}})t_1/\hbar}}{E_i - E_{\mathrm{K0}}} \right\} \times \varrho(E_i)\mathrm{d}E_i \approx 0 \quad \text{if} \quad t_1 \gg \Delta E/\hbar. \tag{7.42}$$

The integral (7.42) is the only contribution to the expression (7.41b) which depends explicitly on the time variable t_1. Since it vanishes the expression (7.41b) is time-independent and is equal to the integral over the contour C of the first term in the integrand. The contour integral over the small semicircle in the lower half plane is easily evaluated. The remaining portion of the contour is the entire real axis except for a small region around the pole. The integral over this portion is just the principal value of the integral on the real axis. Thus we can write

$$\sum_i \int_{\mathrm{C}} \frac{\langle \mathrm{K}_{s0}|V|i\rangle\langle i|V|\mathrm{K}_{r0}\rangle}{E_{\mathrm{K0}} - E_i} \varrho(E_i)\mathrm{d}E_i = \langle \mathrm{K}_{s0}|M|\mathrm{K}_{r0}\rangle - \tfrac{1}{2}\mathrm{i}\langle \mathrm{K}_{s0}|\Gamma|\mathrm{K}_{r0}\rangle \tag{7.43a}$$

where

$$\langle \mathrm{K}_{s0}|M|\mathrm{K}_{r0}\rangle = \mathscr{P} \sum_i \int_{-\infty}^{\infty} \frac{\langle \mathrm{K}_{s0}|V|i\rangle\langle i|V|\mathrm{K}_{r0}\rangle}{E_{\mathrm{K0}} - E_i} \varrho(E_i)\mathrm{d}E_i \tag{7.43b}$$

and

$$\langle \mathrm{K}_{s0}|\Gamma|\mathrm{K}_{r0}\rangle = 2\pi \sum_f \langle \mathrm{K}_{s0}|V|f\rangle\langle f|V|\mathrm{K}_{r0}\rangle\varrho(E_f) \tag{7.43c}$$

and $|f\rangle$ denotes the final state for the actual decay which conserves energy.

We have inserted summations over the states $|i\rangle$ and $|f\rangle$ in eqs. (7.43) to include the case where there is more than one decay mode. It is convenient to define a transition matrix

$$\langle \mathrm{K}_{s0}|T|\mathrm{K}_{r0}\rangle = \langle \mathrm{K}_{s0}|M|\mathrm{K}_{r0}\rangle - \tfrac{1}{2}\mathrm{i}\langle \mathrm{K}_{s0}|\Gamma|\mathrm{K}_{r0}\rangle. \tag{7.44a}$$

The matrix T is not Hermitean, M and $-\frac{1}{2}i\Gamma$ are the Hermitean and anti-Hermitean parts. By eqs. (7.41), (7.42) and (7.43)

$$\langle K_{s0}|T|K_{r0}\rangle = \frac{1}{i\hbar}\int_0^{t_1} dt_2 \langle K_{s0}|V(t_1)V(t_2)|K_{r0}\rangle \quad \text{if} \quad t_1 \gg \hbar/\Delta E. \tag{7.44b}$$

This result (7.44b) can now be substituted into eq. (7.40a). Since the expression (7.44b) is time-independent, the time integration is trivial. Thus

$$\langle K_{s0}|\psi_I(t)\rangle = \langle K_{s0}|K_{r0}\rangle - \frac{it}{\hbar}\langle K_{s0}|T|K_{r0}\rangle. \tag{7.45}$$

For a particular initial state $|K_r\rangle$ which has a normal exponential decay, we expect the expression (7.40a) to have the form

$$\langle K_{r0}|\psi_I(t)\rangle = e^{-\gamma t/2\hbar} e^{-i\delta mc^2 t/\hbar} \tag{7.46a}$$

where $\gamma/\hbar$ is the decay constant or the transition probability per unit time and δm is the mass shift. Expanding eq. (7.46a) to first order in t shows that it has a form similar to eq. (7.45),

$$\langle K_{r0}|\psi_I(t)\rangle \approx 1 - \frac{it}{\hbar}[\delta mc^2 - \tfrac{1}{2}i\gamma]. \tag{7.46b}$$

If the state $|K_{r0}\rangle$ is a simultaneous eigenvector of the matrices M and Γ, eq. (7.45) reduces to eq. (7.46b) with δmc^2 the eigenvalue of M and γ the eigenvalue of Γ. The value of γ is seen from eq. (7.43c) to be in exact agreement with the expression (7.20) for the transition probability per unit time given by first-order perturbation theory. The expression for the mass shift is seen from eq. (7.43b) to be just the result (7.35) obtained naively from second-order stationary non-degenerate perturbation theory, with the infrared divergence removed by taking the principal value.

Since the matrices M and Γ are 2×2, they have two eigenvectors and two eigenvalues. For the simple interaction V (7.24) which has only a single decay mode, the state K_2 (7.25b) is completely decoupled from the decay mode and the matrix T is trivial. For a general interaction which conserves CP the states K_1 and K_2 defined by eq. (7.30) must be simultaneous eigenvectors of M and Γ because they are shown by eq. (7.31) to be eigenvectors of CP with opposite eigenvalues. Thus they cannot be connected by matrix elements of operators which conserve CP.

The two states K_1 and K_2 have different eigenvalues for M and Γ, i.e. different mass shifts and lifetimes. If the initial state $|K_{r0}\rangle$ is a linear combination of K_1 and K_2, it will not have a simple exponential decay; its K_1 and K_2-components will decay with different lifetimes. Furthermore if K_{s0} is the state orthogonal to K_{r0}, $\langle K_{s0}|\psi_I(t)\rangle$ will be zero at $t=0$, but will not remain zero. It is only the eigenstates K_1 and K_2 of the matrix T which have a simple exponential decay, and which do not make transitions into one another as a function of time.

If the matrices M and Γ (7.43) do not commute, they cannot be simultaneously diagonalized. Since CP is known to be violated, the matrices M and Γ may not commute. In that case the two states which do not make transitions to one another are the eigenvectors of the non-Hermitean matrix T, eq. (7.44). These states will not necessarily be orthogonal.

7.9 The exponential decay and the width of the kaon

The expression (7.45) holds only to first order in t. To get an expression which shows the exponential decay explicitly we consider the exact integral equation (7.38a) and try a solution having the form

$$\psi_I(t) = \sum_s \alpha_s(t)|K_{s0}\rangle + \sum_i \beta_i(t)|i\rangle \tag{7.47a}$$

where the coefficients $\alpha_s(t)$ and $\beta_i(t)$ satisfy the initial conditions

$$\begin{aligned} \alpha_s(0) &= \langle K_{s0}|K_{r0}\rangle \\ \beta_i(0) &= 0, \end{aligned} \tag{7.47b}$$

and $|K_{r0}\rangle$ is the initial state of the kaon. The coefficients $\alpha_s(t)$ and $\beta_i(t)$ are determined directly from eq. (7.47a) and the integral equation (7.38a).

$$\alpha_s(t) = \langle K_{s0}|\psi_I(t)\rangle = \langle K_{s0}|K_{r0}\rangle + \frac{1}{i\hbar}\int_0^t dt_1 \sum_i \langle K_{s0}|V(t_1)|i\rangle\beta_i(t_1) \tag{7.48a}$$

$$\beta_i(t_1) = \langle i|\psi_I(t_1)\rangle = \frac{1}{i\hbar}\sum_{s'}\int_0^{t_1} dt_2\langle i|V(t_2)|K_{s'0}\rangle\alpha_{s'}(t_2). \tag{7.48b}$$

Substituting eq. (7.48b) into eq. (7.48a) and summing over the states $|i\rangle$ by closure, we obtain

$$\alpha_s(t) = \langle K_{s0}|K_{r0}\rangle - \frac{1}{\hbar^2}\sum_{s'}\int_0^t dt_1\int_0^{t_1} dt_2\langle K_{s0}|V(t_1)V(t_2)|K_{s'0}\rangle\alpha_{s'}(t_2). \tag{7.49a}$$

This expression can be simplified by reversing the order of integration over t_1 and t_2

$$\alpha_s(t) = \langle \mathrm{K}_{s0}|\mathrm{K}_{r0}\rangle - \frac{1}{\hbar^2}\sum_{s'}\int_0^t \mathrm{d}t_2\alpha_{s'}(t_2)\int_{t_2}^t \mathrm{d}t_1\langle \mathrm{K}_{s0}|V(t_1)V(t_2)|\mathrm{K}_{s'0}\rangle. \quad (7.49\mathrm{b})$$

The integral over t_1 is just equal to the transition matrix defined by eq. (7.44b). This can be seen by transforming the integration variables in eq. (7.49b) and noting that the integrand depends only on the difference $t_1 - t_2$ as indicated explicitly in eq. (7.40b).

$$\int_{t_2}^t \mathrm{d}t_1\langle \mathrm{K}_{s0}|V(t_1)V(t_2)|\mathrm{K}_{s'0}\rangle = \int_{t_2}^t \mathrm{d}t_1\langle \mathrm{K}_{s0}|V(t-t_2)V(t-t_1)|\mathrm{K}_{s'0}\rangle =$$

$$= \int_0^{t_\mathrm{a}} \mathrm{d}t_\mathrm{b}\langle \mathrm{K}_{s0}|V(t_\mathrm{a})V(t_\mathrm{b})|\mathrm{K}_{s'0}\rangle \quad (7.50)$$

where we have set $t_1 = t - t_\mathrm{b}$ and $t_2 = t - t_\mathrm{a}$. Thus

$$\alpha_s(t) = \langle \mathrm{K}_{s0}|\mathrm{K}_{r0}\rangle - \frac{\mathrm{i}}{\hbar}\int_0^t \mathrm{d}t_2 \sum_{s'} \alpha_{s'}(t_2)\langle \mathrm{K}_{s0}|T|\mathrm{K}_{s'0}\rangle. \quad (7.51\mathrm{a})$$

This is the natural generalization of eq. (7.45) and reduces to it when $\alpha_{s'}(t_2)$ is replaced by its unperturbed value.

Let the initial state $|\mathrm{K}_{r0}\rangle$ be an eigenvector of the matrix T with the eigenvalue $\varepsilon - \frac{1}{2}\mathrm{i}\gamma$. Then the solution of eq. (7.51a) is the exponentially decaying function

$$\alpha_s(t) = \delta_{sr}\mathrm{e}^{-(\frac{1}{2}\gamma + \mathrm{i}\varepsilon)t/\hbar}, \quad (7.51\mathrm{b})$$

exactly as expected from the first-order treatment, eq. (7.46).

There are two solutions of the form (7.51b), corresponding to the two eigenvectors of T. The corresponding values of ε and γ are the real and imaginary parts of the corresponding eigenvalues of T. For the case where M and Γ commute, these are just the eigenvalues of M and Γ.

We thus find solutions of the Schrödinger equation with exponential decay (7.51b) for states $|\mathrm{K}_{s0}\rangle$ which are eigenvectors of T. For small t this reduces to the approximate solution (7.45). The eigenvectors are denoted by $|\mathrm{K_S}\rangle$ and $|\mathrm{K_L}\rangle$ for the short and long-lived states. These states differ slightly from the CP eigenstates (7.30) when CP violation is taken into account.

Let us now examine the energy spectrum of the emitted pions from the decay. The expression (7.48b) for $\beta_i(t)$ can now be evaluated by substituting eq. (7.51b) and integrating

$$\beta_i(t) = \frac{\langle i|V|\mathrm{K}_{r0}\rangle[\mathrm{e}^{-\mathrm{i}(E_{\mathrm{K}0}-E_i+\varepsilon-\frac{1}{2}\mathrm{i}\gamma)t/\hbar} - 1]}{E_{\mathrm{K}0} - E_i + \varepsilon - \frac{1}{2}\mathrm{i}\gamma}. \tag{7.52a}$$

Note that this result is the same as the previous result (7.16b) except for the addition of $\varepsilon-\frac{1}{2}\mathrm{i}\gamma$ to the exponent and the denominator. Remember that the additional exponential factor in eq. (7.16b) disappears in the transformation to the interaction representation.

The probability that a decay mode with energy E_i is present is obtained by squaring this expression (7.52a). Neglecting the first term which decreases as $\exp(-\frac{1}{2}\gamma t)$ and is negligible after a long time, we obtain

$$|\beta_i(t)|^2 = \frac{|\langle i|V|\mathrm{K}_{r0}\rangle|^2}{(E_{\mathrm{K}0} - E_i + \varepsilon)^2 + \frac{1}{4}\gamma^2}. \tag{7.52b}$$

The energy spectrum of the decay products therefore is a resonance curve with a peak at the value $E_{\mathrm{K}0}+\varepsilon$ and a width at half maximum equal to γ, the total transition probability for unit time multiplied by $\hbar$. The quantity γ thus characterizes the rate of decay of the kaon and also the width of the energy spectrum of the pions. That the decay rate and the spectrum width are related follows from the uncertainty relation for energy and time,

$$\Delta E \Delta t \sim h. \tag{7.53}$$

Since the kaon exists only for a length of time of the order $1/\gamma$, the energy of the system is determined only with an accuracy $\hbar\gamma$; i.e. the energy level has a width of the order of $\hbar\gamma$.

The effect of the $\mathrm{K}_1-\mathrm{K}_2$ mass difference can be observed experimentally by measuring the behavior of a kaon beam as a function of time. Suppose at time $t=0$ a K^0-meson is produced in an elementary-particle reaction. The state of the system at time $t=0$ is then represented as

$$|\psi_{\mathrm{I}}(0)\rangle = |\mathrm{K}^0\rangle = \frac{1}{\sqrt{2}}\{|\mathrm{K}_1\rangle + |\mathrm{K}_2\rangle\}. \tag{7.54}$$

Let us now examine the development of this state as a function of time. For simplicity, we consider only the 2π-decay mode. Then the K_2-component does not decay, whereas the K_1-component decays exponentially with a decay constant γ. The projection of the state $\psi(t)$ on an arbitrary kaon state $|\mathrm{K}_s\rangle$ is then

$$\begin{aligned}\langle \mathrm{K}_s|\psi_{\mathrm{I}}(t)\rangle &= \frac{1}{\sqrt{2}}\{\mathrm{e}^{-\frac{1}{2}\gamma}\mathrm{e}^{-\mathrm{i}\delta m t/\hbar}\langle \mathrm{K}_s|\mathrm{K}_1\rangle + \langle \mathrm{K}_s|\mathrm{K}_2\rangle\} \\ &= \{\tfrac{1}{2}(1+\mathrm{e}^{-\frac{1}{2}\gamma t}\mathrm{e}^{-\mathrm{i}\delta m t/\hbar})\langle \mathrm{K}_s|\mathrm{K}^0\rangle + \tfrac{1}{2}(1-\mathrm{e}^{-\frac{1}{2}\gamma t}\mathrm{e}^{-\mathrm{i}\delta m t/\hbar})\langle \mathrm{K}_s|\overline{\mathrm{K}}^0\rangle\}.\end{aligned} \tag{7.55}$$

The effect of the mass difference is to introduce an additional phase factor between the K_1 and K_2-components which varies periodically with time. Thus the relative amounts of K^0 and $\overline{K}^0$-states present after a time t depend upon this mass difference. The relative amounts of K^0 and $\overline{K}^0$ are measurable by allowing the kaon beam to pass through matter which absorbs the $\overline{K}^0$-component much more strongly than the K^0 component.

7.10 Parity, charge conjugation and time reversal

In the above treatment we have included the restrictions imposed on the kaon decay interaction by conservation of momentum and angular momentum and by Lorentz invariance. We now consider invariance under space inversion, charge conjugation and time reversal. Since our interaction is a function of pion and kaon creation operators, we first examine the behavior of these operators under P and T.

The space inversion operator P is defined for one-particle Schrödinger wave functions by the relation:

$$P\psi(x, y, z) = \psi(-x, -y, -z). \tag{7.56a}$$

It follows immediately that:

$$P^2 = 1. \tag{7.56b}$$

In a description using the notation of second quantization, we wish to describe the action of the operator P on a wave function written as the product of several operators acting on the vacuum,

$$P\,ABCD|0\rangle = (PAP)(PBP)(PCP)(PDP)P|0\rangle \tag{7.57a}$$

where A, B, C, and D are operators. Since the vacuum is invariant under space inversion

$$P|0\rangle = |0\rangle. \tag{7.57b}$$

Thus

$$P\,ABCD|0\rangle = (PAP)(PBP)(PCP)(PDP)|0\rangle. \tag{7.57c}$$

From the relations (7.56) the operator P is both Hermitean and unitary. The transformation

$$A \to PAP \tag{7.58}$$

is thus a unitary transformation describing space inversion on the operator A. The application of the operator P on a wave function in the second-quantized notation is seen from eq. (7.57c) to be expressed as the implementation of the transformation (7.58) on all operators.

If the operator A is a creation operator for a particle in a momentum eigenstate, the transformed operator must be a creation operator for a particle with the sign of the momentum reversed. In the Schrödinger representation

$$P e^{i\boldsymbol{q}\cdot\boldsymbol{x}} = e^{-i\boldsymbol{q}\cdot\boldsymbol{x}}. \tag{7.59a}$$

Thus in the second-quantized notation

$$P a_{\boldsymbol{q}}^{\dagger} P = P_{\text{int}} a_{-\boldsymbol{q}}^{\dagger} \tag{7.59b}$$

where P_{int} is a phase factor which remains to be determined. Such a phase factor does not arise in the Schrödinger description (7.59a) where the number of particles is a constant of the motion. In processes where particles are created and destroyed and in which parity is conserved, each particle can carry an intrinsic parity in addition to the parity defined by its orbital motion through eq. (7.56a). This intrinsic parity

$$P_{\text{int}} = \pm 1 \tag{7.60}$$

appears as a phase factor in the transformation of the creation operators for the particle in a second-quantized description.

Since the intrinsic parity of pions and kaons is negative:

$$P a_{\mathrm{K}\boldsymbol{q}}^{\dagger} P = -a_{\mathrm{K},-\boldsymbol{q}}^{\dagger} \tag{7.61a}$$

$$P a_{\pi\boldsymbol{q}}^{\dagger} P = -a_{\pi,-\boldsymbol{q}}^{\dagger}. \tag{7.61b}$$

We can simplify the treatment of space inversion by using rotational invariance and noting that a momentum vector is reversed by a 180° rotation about any axis perpendicular to it.

If R is a 180° rotation about any axis perpendicular to $\boldsymbol{q}$,

$$R a_{\mathrm{K}\boldsymbol{q}}^{\dagger} R = a_{\mathrm{K},-\boldsymbol{q}}^{\dagger} \tag{7.62a}$$

$$R a_{\pi\boldsymbol{q}}^{\dagger} R = a_{\pi,-\boldsymbol{q}}^{\dagger}. \tag{7.62b}$$

Then

$$P R a_{\mathrm{K}\boldsymbol{q}}^{\dagger} R P = -a_{\mathrm{K},\boldsymbol{q}}^{\dagger} \tag{7.63a}$$

$$P R a_{\pi\boldsymbol{q}}^{\dagger} R P = -a_{\pi,\boldsymbol{q}}^{\dagger}. \tag{7.63b}$$

The operator PR is a 180° rotation followed by a space inversion. Since the inversion and rotation commute

$$PR = RP. \tag{7.64}$$

PR is just a reflection in the plane normal to the axis of R. Any interaction which is invariant under rotations (conserves angular momentum) is also invariant under PR if it is invariant under space inversion. Thus we can use the operator PR to test space inversion invariance instead of P. The advantage of PR is that it does not change momenta perpendicular to the axis defined by R. In the decay of a kaon at rest into two or three particles, all momenta lie in a plane. Thus it is always possible to choose R as a 180° rotation about the normal to the plane. The operator PR thus leaves *all* relevant momenta invariant.

Charge conjugation is defined as the transformation of a particle into its antiparticle without changing the other degrees of freedom. The $\overline{\mathrm{K}}^0$ is the antiparticle of K^0. The π^+ is the antiparticle of the π^-, and the π^0 is its own antiparticle. Thus,

$$Ca^{\dagger}_{\mathrm{K}^0}C = a^{\dagger}_{\overline{\mathrm{K}}^0} \tag{7.65a}$$

$$Ca^{\dagger}_{\pi^+}C = a^{\dagger}_{\pi^-} \tag{7.65b}$$

$$Ca^{\dagger}_{\pi^0}C = a^{\dagger}_{\pi^0}. \tag{7.65c}$$

The phases of (7.65a) and (7.65b) are determined to be positive by convention, since the relative phases of the K^0 and $\overline{\mathrm{K}}^0$ states and of the π^+ and π^- states are arbitrary. The phase of (7.65c) is an experimentally observable quantity. It cannot be changed by adopting a different phase convention because the same particle π^0 appears on both sides of the equation. The π^0 must be an eigenstate of C and the eigenvalue can be measured in processes where π^0's are created or destroyed, and in which C is conserved. One example is the decay $\pi^0 \to 2\gamma$, which proves that the eigenvalue is positive, i.e. the π^0 is even under charge conjugation.

In our treatment we consider for simplicity only neutral pions and drop the charge superscript. Charged pions appear only as a $\pi^+\pi^-$ pair in neutral kaon decays and charge conjugation interchanges the two charges. If the two pions are in a state which is symmetric under permutations, the operation of C on the charged pion state is the same as on the corresponding neutral pion state. This is true for two-pion decays of kaons and also for those three-pion decays where all the pions have zero relative angular momentum. Thus, our treatment for neutral pions also applies to charged pions in these two cases.

We now consider time reversal. Suppose that $\psi(t)$ is a solution of the time-dependent Schrödinger equation

$$H\psi(t) = \mathrm{i}\hbar \frac{\partial \psi(t)}{\partial t}. \tag{7.66a}$$

Under the substitution $t \to -t$, which describes a reversal of the direction of time, this equation becomes

$$H\psi(-t) = -i\hbar \frac{\partial\psi(-t)}{\partial t}. \tag{7.66b}$$

The Schrödinger equation (7.66) is thus not invariant under time reversal.

Although $\psi(-t)$ is not a solution of the Schrödinger equation (7.66a) we can ask if there exists some solution of the Schrödinger equation (7.66a) *which is physically equivalent* to $\psi(-t)$. By physically equivalent we mean that it gives the same results for all observable quantities. A function which differs from $\psi(-t)$ only by phase factors might satisfy this condition as the phase factors all cancel in the calculation of probabilities and observable matrix elements of Hermitean operators. Consider the complex conjugate of eq. (7.66b)

$$H^*\psi^*(-t) = i\hbar \frac{\partial\psi^*(-t)}{\partial t}. \tag{7.66c}$$

The function $\psi^*(-t)$ is physically equivalent to the function $\psi(-t)$ in giving the same probabilities and the same expectation values of Hermitean operators. Furthermore, the function $\psi^*(-t)$ is a solution of the Schrödinger equation (7.66a) if the Hamiltonian H is *real*. We can therefore define a time reversal operator T, by the relation

$$T\psi(t) = \psi^*(-t) \tag{7.67a}$$

$$TAT^{-1} = A^* \tag{7.67b}$$

where $\psi(t)$ is any wave function and A is any operator which does not depend explicitly on the time.

If the Hamiltonian in a particular problem is invariant under the transformation defined by the operator T we say that it is invariant under time reversal. This does not mean that the Schrödinger equation (7.66a) is invariant under time reversal but only that it has a solution which is physically equivalent to the time reversed solution. If H is not invariant under the transformation defined by the operator T, this does not necessarily mean that time reversal invariance is violated, because there is still the possibility that the Schrödinger equation (7.66a) has another solution which differs only from the solution (7.67) by phase factors and is physically equivalent to it. For our purposes however, it is sufficient to consider the operator T, defined by eq. (7.67).

Because of the complex conjugation T is not a linear operator but an antilinear operator.

$$T[a\psi_1 + b\psi_2] = a^*T\psi_1 + b^*T\psi_2. \tag{7.68}$$

Antilinear operators have many peculiar properties. For example, the phase of the eigenvalue of such an operator is not defined – multiplying the eigenvector by a complex constant changes the phase of the eigenvalue.

To find how T acts on creation operators for particles in plane-wave states, we note that

$$Te^{i\boldsymbol{q}\cdot\boldsymbol{x}} = (e^{i\boldsymbol{q}\cdot\boldsymbol{x}})^* = e^{-i\boldsymbol{q}\cdot\boldsymbol{x}}. \tag{7.69}$$

Thus time reversal reverses the sign of a momentum vector as one would expect, and

$$Ta^\dagger_{\pi\boldsymbol{q}}|0\rangle = a^\dagger_{\pi,-\boldsymbol{q}}|0\rangle \tag{7.70a}$$

$$Ta^\dagger_{K\boldsymbol{q}}|0\rangle = a^\dagger_{K,-\boldsymbol{q}}|0\rangle. \tag{7.70b}$$

Thus

$$Ta^\dagger_{\pi\boldsymbol{q}}T^{-1} = a^\dagger_{\pi,-\boldsymbol{q}} \tag{7.70c}$$

$$Ta^\dagger_{K\boldsymbol{q}}T^{-1} = a^\dagger_{K,-\boldsymbol{q}}. \tag{7.70d}$$

There is no 'intrinsic' phase factor in this case, as for parity because T is an antilinear operator.

We can again make use of the operator R to define an operator TR which does not change the momenta in the plane defined by R. Then

$$TRa^\dagger_{\pi\boldsymbol{q}}RT^{-1} = a^\dagger_{\pi\boldsymbol{q}} \tag{7.71a}$$

$$TRa^\dagger_{K\boldsymbol{q}}RT^{-1} = a^\dagger_{K\boldsymbol{q}}. \tag{7.71b}$$

7.11 *CPT* in kaon decays

We now examine the behavior under P, C and T of the kaon decay interaction. We extend the interaction (7.24) by including terms which describe kaon decays into three pions as well as into two pions.

$$V = g_1 a^\dagger_\pi a^\dagger_\pi a_{K^0} + g_2 a^\dagger_\pi a^\dagger_\pi a_{\bar{K}^0} + g_3 a^\dagger_\pi a^\dagger_\pi a^\dagger_\pi a_{K^0} + g_4 a^\dagger_\pi a^\dagger_\pi a^\dagger_\pi a_{\bar{K}^0} + \text{h.c.} \tag{7.72}$$

We consider only neutral pions, and suppress the momentum variables for convenience, even though they are no longer trivial in the three-pion decay as in the two-pion decay. The momenta of the three pions in the final state are not uniquely determined by energy and momentum conservation; the final state exhibits a continuous energy spectrum. If the momentum labels are added to the pion creation operators then g_3 and g_4 are non-trivial functions of the continuous variable describing the momentum spectrum in the final state. However, we shall test for P, C and T invariance by using the operators PR, C and TR which do not change momenta.

From rotational invariance, we note that

$$RVR = V. \tag{7.73}$$

From eqs. (7.63), (7.72) and (7.73) we obtain

$$\begin{aligned} PVP &= PR\,V\,RP \\ &= -g_1 a_\pi^\dagger a_\pi^\dagger a_{K^0} - g_2 a_\pi^\dagger a_\pi^\dagger a_{\bar{K}^0} + g_3 a_\pi^\dagger a_\pi^\dagger a_\pi^\dagger a_{K^0} + g_4 a_\pi^\dagger a_\pi^\dagger a_\pi^\dagger a_{\bar{K}^0} + \text{h.c.} \end{aligned} \tag{7.74a}$$

From eq. (7.65)

$$CVC = g_1 a_\pi^\dagger a_\pi^\dagger a_{\bar{K}^0} + g_2 a_\pi^\dagger a_\pi^\dagger a_{K^0} + g_3 a_\pi^\dagger a_\pi^\dagger a_\pi^\dagger a_{\bar{K}^0} + g_4 a_\pi^\dagger a_\pi^\dagger a_\pi^\dagger a_{K^0} + \text{h.c.} \tag{7.74b}$$

From eqs. (7.67) and (7.71),

$$\begin{aligned} TVT^{-1} &= TR\,V\,RT^{-1} \\ &= g_1^* a_\pi^\dagger a_\pi^\dagger a_{K^0} + g_2^* a_\pi^\dagger a_\pi^\dagger a_{\bar{K}^0} + g_3^* a_\pi^\dagger a_\pi^\dagger a_\pi^\dagger a_{K^0} + g_4^* a_\pi^\dagger a_\pi^\dagger a_\pi^\dagger a_{\bar{K}^0} + \text{h.c.} \end{aligned} \tag{7.74c}$$

Since the operators PR, C and TR do not change momenta, these equations hold as well, term by term, when the momentum dependence is included in the interaction (7.72). For each set of terms describing decays into a given plane, we can define R for that plane. Thus eq. (7.73) holds for that set and eqs. (7.74) also apply.

Space inversion thus reverses the sign of the two-pion decay terms which have an odd number of boson operators and does not change the three-pion decay terms which have an even number. Charge conjugation interchanges K^0 and $\bar{K}^0$ everywhere. Time reversal changes each coefficient g_i into its complex conjugate g_i^*. The transformations are thus equivalent to the following changes in the coefficients g_i.

$$PVP \Rightarrow g_1 \to -g_1, \quad g_2 \to -g_2, \quad g_3 \to g_3, \quad g_4 \to g_4 \tag{7.75a}$$

$$CVC \Rightarrow g_1 \to g_2, \quad g_2 \to g_1, \quad g_3 \to g_4, \quad g_4 \to g_3 \tag{7.75b}$$

$$TVT^{-1} \Rightarrow g_1 \to g_1^*, \quad g_2 \to g_2^*, \quad g_3 \to g_3^*, \quad g_4 \to g_4^*. \tag{7.75c}$$

We can also combine eqs. (7.75) to obtain the results of the transformations CP and CPT. We first consider the product of the operators C and P. From the relations (7.61) and (7.65), C and P commute in their action on pion and kaon operators. (This is not true in general because fermions and charge conjugate antifermions have opposite parity.) For the boson case, we can write

$$(CP)^{-1} = PC. \tag{7.76a}$$

Similarly

$$(CPT)^{-1} = T^{-1}PC. \tag{7.76b}$$

Thus

$$CP\,V\,(CP)^{-1} \Rightarrow g_1 \leftrightarrow -g_2; \quad g_3 \leftrightarrow g_4 \tag{7.77a}$$

$$CPT\,V\,(CPT)^{-1} \Rightarrow g_1 \leftrightarrow -g_2^*; \quad g_3 \leftrightarrow g_4^* \tag{7.77b}$$

The requirement that the interaction be invariant under any of these transformations restricts the values of the coefficients g_i to be unaffected by the appropriate transformation (7.75).

Parity conservation requires V to be invariant under the transformation (7.75a)

$$PVP = P \Rightarrow g_1 = g_2 = 0. \tag{7.78a}$$

The two-pion decays thus violate parity conservation, and must vanish if parity is conserved. The three-pion decays satisfy parity conservation and are not restricted by the requirement. The observation of both 2π and 3π decays is evidence that parity is not conserved in the decay.

If the interaction is invariant under charge conjugation,

$$CVC = V \Rightarrow g_1 = g_2; \quad g_3 = g_4. \tag{7.78b}$$

Both two and three-pion decays are allowed, but the interaction coefficients are such that only the K_2-state, eq. (7.30b), is allowed to decay at all and the decay of the K_1-state is forbidden. This is obvious, since the π^0 is an eigenstate of C with the eigenvalue $+1$. The K_1-state which is odd under C cannot decay into any number of π^0's if C is conserved.

The observation that both K_1 and K_2-states decay into pions is thus evidence that C invariance is violated.

We now consider CP invariance

$$CP\,V\,(CP)^{-1} = V \Rightarrow g_1 = -g_2; \quad g_3 = g_4. \tag{7.78c}$$

Both two-pion and three-pion decays are allowed, but only the K_1-state decays into two pions and only the K_2-state decays into three pions. Since we have already seen that parity cannot be conserved in the two-pion decay it is not surprising that invariance under C and under CP are incompatible for the case of the two-pion decay.

If CP invariance is valid, and the two-pion decay is observed, the conditions (7.78a) and (7.78b) on g_1 and g_2 are badly violated because there is 100% C and P violation in this decay. In this case the K_1 is coupled only to the two-pion decay mode, and the K_2 is coupled only to the three-pion decay mode. The decays of the K_1 and K_2 can be treated independently by time-dependent perturbation theory and the decay rates or lifetimes for the two states obtained from the appropriate golden rule formula. In the formalism of section 7.8, the matrices M and Γ are both diagonal in the K_1-K_2 basis. Until 1964, this description was believed to be in agreement with experiment. The two states K_1 and K_2 were observed experimentally with lifetimes of about 10^{-10} and 5×10^{-8} seconds respectively. The former decays into two pions and the latter was observed to decay only into three pions and into other three and four-body decay modes including leptons.

In 1964 the long-lived K_2 was found also to decay into two pions, rather weakly with a probability of about 10^{-3}. This indicated a small violation of CP invariance in the decay process. For this case the treatment of section 7.9 is still valid, but the matrix T is non-trivial. The states K_S and K_L which have exponential decays are the eigenvectors of the matrix T, eq. (7.51a). Since the CP violation is small, these states are very close to the CP eigenstates (7.30).

Note that the particular state K_2, defined by eq. (7.25b) is *still* decoupled from the 2π decay mode, since we did not assume CP invariance at that point. If a kaon is in this state at given time t, there can be no 2π decays at that time. However, the state of the kaon will change, as a result of the 3π decays, and after a finite time, a component of the K_1-state (7.25a) will be present which can decay into two pions. The exact time behavior can be computed by expressing the initial state in the basis K_S and K_L, eigenvectors of the matrix T, eq. (7.51a).

Time reversal invariance and CPT invariance require the following

$$TVT^{-1} = V \Rightarrow g_1 = g_1^*; \quad g_2 = g_2^*; \quad g_3 = g_3^*; \quad g_4 = g_4^* \tag{7.79a}$$

$$CPT\,V\,(CPT)^{-1} = V \Rightarrow g_1 = -g_2^*; \quad g_3 = g_4^*. \tag{7.79b}$$

If CP were conserved, T and CPT invariance would both be possible by requiring all the g_i to be real. However, if CP is violated, then either T or CPT must be violated. There is no conclusive experimental evidence on this point. However, theorists tend to favor CPT conservation and T violation, because CPT invariance is more fundamental on theoretical grounds; it follows from certain general assumptions in field theory.

Note that if T is conserved and CP and CPT are violated, either $|g_1| \neq |g_2|$ or $|g_3| \neq |g_4|$. We then find that the mass matrix (7.43b) in the K^0–$\overline{K}^0$ basis has *different diagonal* terms (the K^0 and $\overline{K}^0$ masses are unequal). Since the difference between the eigenvalues of a 2×2 matrix is always *greater* than the difference between the diagonal elements, the observed K_1–K_2 mass difference gives an upper limit on the K^0–$\overline{K}^0$ mass difference and therefore an upper limit on the degree of CPT violation in kaon decays.

PROBLEMS

1. In the isospin formalism the pion is described as an isospin triplet with isospin $T=1$.

Let $a^{\dagger}_{\pi m \boldsymbol{q}}$ represent a creation operator for a pion of momentum $\hbar\boldsymbol{q}$ and charge indicated by the isospin quantum number m. For π^{+}, π^{0} and π^{-}-mesons $m=+1$, 0 and -1 respectively. The quantum number m represents the eigenvalue of the 'z-component' of the isospin and isospins can be combined in the same way as angular momenta.

(a) What values of the total isospin T can occur in states of a two-pion system?

(b) Using the second-quantized notation write down a wave function describing a two-pion state with total momentum zero and total charge zero and which is an eigenfunction of the total isospin T with eigenvalue $T=0$. Let the relative motion of the two pions be described by an arbitrary function $f(\boldsymbol{q}')$.

(c) Repeat part (b) for all other eigenvalues of T which can occur in the two-pion system.

(d) Using the commutation rules for the creation operators and the symmetry properties of the Clebsch–Gordan coefficients, find restrictions on the parity and the angular momentum of the state of the two-pion system of parts (b) and (c).

2. The ϱ, ω, η, and π-mesons have the following properties:

ϱ; spin = 1, odd parity, isospin $T=1$
ω; spin = 1, odd parity, isospin $T=0$
η; spin = 0, odd parity, isospin $T=0$
π; spin = 0, odd parity, isospin $T=1$

(a) Which of the ϱ, ω, and η can decay into two pions and conserve angular momentum and isospin?

(b) Which of the ϱ, ω, and η can decay into two pions conserving angular momentum and parity?

(c) Which of the ϱ, ω, and η can decay into two pions conserving angular momentum, parity and isospin?

(d) Let $a^{\dagger}_{\eta \boldsymbol{q}}$ represent a creation operator for an η-meson with momentum $\hbar\boldsymbol{q}$. Since the η is a neutral meson which is an isospin singlet ($T=0$), no isospin index is necessary. Write down an interaction describing the decay of an η-meson into two pions, which conserves momentum and isospin. Make the interaction Hermitean. What conservation law is violated by this interaction?

3. Let $a^{\dagger}_{\pi m q \theta \phi}$ describe creation operators for a pion with a momentum whose magnitude is $\hbar q$ and whose direction is given by the spherical coordinates θ, ϕ.

(a) Write a wave function for a two-pion system having total momentum zero and total angular momentum L.

(b) Let $a^{\dagger}_{\varrho m s 0}$ represent the creation operator for a ϱ-meson in the isospin state m and in the ordinary spin state s with total momentum zero. The allowed values of s are 1, 0, and -1 since the total spin of the ϱ-meson is 1. Write down an interaction describing the decay of a ϱ-meson at rest into two pions and conserving momentum, angular momentum and isospin.

4. Consider a beam of mesons which is initially a pure K_2-beam (all of the K_1-component has been allowed to decay). Let this meson beam pass through a thin layer of material which absorbs 30% of the $\overline{K}^0$-component. After a time t_1 let the beam pass through another identical absorber. Determine the intensity of the K_1-component after passing through the second absorber as a function of the time interval t_1. Show how this effect could be used experimentally to measure the K_1–K_2 mass difference and indicate briefly what the experimental setup should be.

5. A kaon beam is passed through two slabs of matter separated by a distance d which is of the order of several centimeters (fig. 7.4).

If the beam is initially in the long-lived neutral kaon state, K_L, it can be converted to the short-lived state K_S by interaction with nuclei in either of the two slabs. The amplitude for the K_S final state consists of two components, corresponding to whether the $K_L \to K_S$ transition takes place in the first slab or in the second slab. The relative phase of the two components depends upon the distance d and the K_L–K_S mass difference. Interference experiments are actually used to measure the mass difference.

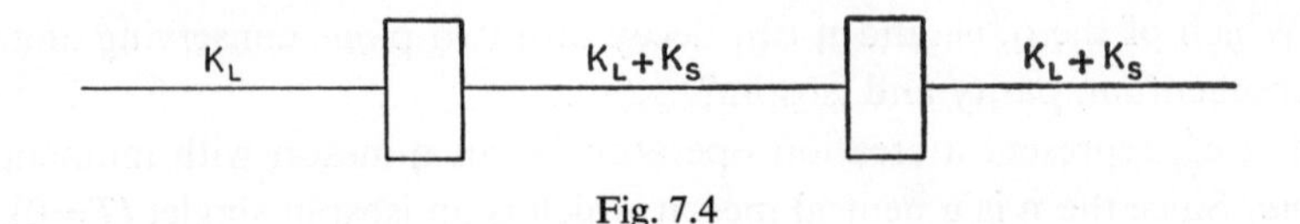

Fig. 7.4

The two components are coherent and can interfere only if it is impossible to determine in which of the two slabs the transition took place. However, because of the K_L–K_S mass difference the transition involves a transfer of momentum and energy between the kaon and the slab. By measuring the momenta of the slabs before and after the pasage of the kaon, one can determine which slab received the momentum transfer.

Show by analogy with the Mössbauer effect that the momentum transfer argument given above does not affect the coherence, and that interference should be observed.

6. Consider a Hamiltonian with one kind of kaon and one kind of pion, with the decay interaction:

$$V = \sum_{\boldsymbol{q}'\boldsymbol{q}''} g(\boldsymbol{q}', \boldsymbol{q}'') a^{\dagger}_{\pi(\boldsymbol{q}'-\boldsymbol{q}'')} a^{\dagger}_{\pi(\boldsymbol{q}'+\boldsymbol{q}'')} a_{\mathrm{K},2\boldsymbol{q}'} + \text{h.c.}$$

where $g(0, \boldsymbol{q}'')$ depends only on the magnitude of the vector $\boldsymbol{q}''$, and not on its direction.

Write down some eigenfunctions of the total Hamiltonian H, which describe stationary states. Hint: look for eigenfunctions of the unperturbed Hamiltonian H_0 which are unaffected by the interaction V, and are therefore also eigenfunctions of the total Hamiltonian. What is the physical interpretation of these eigenfunctions?

7. Consider the Hamiltonian describing two kinds of neutral kaons with the possibility of decay into both two-pion and three-pion states, i.e. with the interaction:

$$V = g_1 a^{\dagger}_{\pi} a^{\dagger}_{\pi} a_{\mathrm{K}^0} + g_2 a^{\dagger}_{\pi} a^{\dagger}_{\pi} a_{\overline{\mathrm{K}}^0} + g_3 a^{\dagger}_{\pi} a^{\dagger}_{\pi} a^{\dagger}_{\pi} a_{\mathrm{K}^0} + g_4 a^{\dagger}_{\pi} a^{\dagger}_{\pi} a^{\dagger}_{\pi} a_{\overline{\mathrm{K}}^0} + \text{h.c.}$$

We have suppressed the momentum variables and assume that there is only one state for each kind of particle.

Let a general solution of the time-dependent Schrödinger equation be:

$$|\psi(t)\rangle = \alpha(t) a^{\dagger}_{\mathrm{K}^0}|0\rangle + \beta(t) a^{\dagger}_{\overline{\mathrm{K}}^0}|0\rangle + \gamma(t) a^{\dagger}_{\pi} a^{\dagger}_{\pi}|0\rangle + \delta(t) a^{\dagger}_{\pi} a^{\dagger}_{\pi} a^{\dagger}_{\pi}|0\rangle.$$

Discuss the existence of a 'time-reversed' solution for the same Schrödinger equation. This is a solution in which the probabilities of finding a K^0, $\overline{\mathrm{K}}^0$, two pions or three pions vary with time in the exact opposite way as the above wave function (as if a movie film showing the above wave function were run backwards). The time reversed wave function:

$$|\psi_{\mathrm{TR}}(t)\rangle = \alpha'(t) a^{\dagger}_{\mathrm{K}^0}|0\rangle + \beta'(t) a_{\overline{\mathrm{K}}^0}|0\rangle + \gamma'(t) a^{\dagger}_{\pi} a^{\dagger}_{\pi}|0\rangle + \delta'(t) a^{\dagger}_{\pi} a^{\dagger}_{\pi} a^{\dagger}_{\pi}|0\rangle$$

is defined by the conditions:

$$|\alpha'(t)|^2 = |\alpha(-t)|^2; \qquad |\beta'(t)|^2 = |\beta(-t)|^2;$$

$$|\gamma'(t)|^2 = |\gamma(-t)|^2; \qquad |\delta'(t)|^2 = |\delta'(-t)|^2.$$

Show that if $g_3 = g_4 = 0$; i.e. the only allowed decay mode for kaons is the two-pion decay mode, then a time-reversed solution always exists.

What are the conditions on g_1, g_2, g_3 and g_4 in order that a time-reversed solution exists?

8. Write down a phenomenological Hamiltonian describing pions and photons which includes the decay of a pion into two photons. Write down an expression for the lifetime of the pion in terms of the parameters appearing in the Hamiltonian.

9. Consider the treatment of kaon decay into two pions as described in sections 7.1 and 7.2 applied to the case where there already are pions present in the initial state.

Discuss how the formal treatment of eqs. (7.8)–(7.20) is changed if the vacuum state in eq. (7.8a) is replaced by a multipion state:

$$|0\rangle \to \prod_{q_x > 0} (a^\dagger_{\pi q})^{n_q} |0\rangle \tag{1}$$

where there are n_q pions present in the initial state with momentum $\hbar \boldsymbol{q}$, and all pions have a *positive* x-component of $\boldsymbol{q}$. In particular:

(a) After assuming the intial conditions (7.8b) and (7.8c), show how eqs. (7.9), (7.15), (7.16), (7.17), (7.19) and (7.20) must be changed if the transformation (1) above is introduced into the wave function (7.8a).

(b) What is the physical interpretation of the change in the transition probability per unit time as given by eq. (7.20) when the pions are present in the initial state?

(c) Does the change discussed in question (b) have any appreciable effect on kaon decays which can be observed in the laboratory under conditions which are feasible with present-day techniques?

(d) Can you think of a different decay process, where the analog of the situation with many pions present in the initial state is of great importance?

(e) How would the discussion be changed if there were pions in the initial state which had all possible values of $\boldsymbol{q}$, including those with negative values of q_x?

10. Consider the emission of a gamma ray by a nucleus in an excited state using a treatment analogous to the treatment of kaon decay. Let the total energy of a particular nucleus be E_g when it is in its ground state and at rest. Let its energy be E_e when it is in a particular excited state and is at rest. Neglect all other states of internal motion of the nucleus. Let $a_{ek}^{\dagger}$ and $a_{gk}^{\dagger}$ be creation operators for the nucleus in its excited state and its ground state respectively in states having a total momentum $\hbar k$. Let $b_k^{\dagger}$ be the creation operator for a photon of momentum $\hbar k$. Neglect the polarization degree of freedom for the photon and the spin of the nuclear states, and assume that there is only one state for each having a momentum $\hbar \boldsymbol{k}$.

(a) Write in the second-quantized notation the Hamiltonian for a system containing a number of nuclei in the excited state, a number of nuclei in the ground state and a number of photons. Assume no interactions between the nuclei and the photons.

(b) Write an interaction describing a transition of the nucleus from the excited state to the ground state with the emission of a photon. Make the interaction Hermitean but write the most general interaction possible (without restrictions due to conservation laws). What other process is described by this interaction in addition to the transition of the nucleus from the excited state to the ground state with the emission of a photon?

(c) What restriction is imposed on the interaction of part (b) by the requirement of momentum conservation? How should energy conservation be handled in this treatment?

(d) Consider a state in which one nucleus is present in the excited state with total momentum zero at time $t=0$, and there are no nuclei present in the ground state and no photons. Write the correct time-dependent wave function for this state to first order in the perturbation. What is the transition probability per unit time for the emission of a photon?

(e) How would the result for the transition probability in part (d) be changed if the nucleus were initially in a state of total momentum $\hbar \boldsymbol{k}$ corresponding to a velocity very much smaller than that of light: What further restrictions do these considerations place on the parameters of the interaction in part (b) for the case where the nuclear velocities are kept well in the nonrelativistic region?

(f) Suppose that the nucleus instead of being free is bound to the origin of our coordinate system by a harmonic oscillator potential $V=\frac{1}{2}M\omega^2\boldsymbol{x}^2$ where $\boldsymbol{x}$ is the coordinate of the center-of-mass of the nucleus. Write this potential in the second-quantized notation using the symbol $\langle k|V|k'\rangle$ for the matrix elements of the potential in the momentum representation. Assume that the same potential acts on the nucleus whether it is in the excited state or in the ground state.

(g) Consider the system described by the Hamiltonian of questions b and c with the added potential of question f. Simplify the Hamiltonian by a transformation to a new set of states described by the creation operators $a_{en}^{\dagger}$ and $a_{gn}^{\dagger}$. These operators describe the creation of a nucleus in the excited or ground state moving in the nth state of excitation in the harmonic oscillator potential, rather than as a plane wave.

(h) How does the energy spectrum of the gamma rays emitted for the case described in part (g) differ from that in the absence of the harmonic oscillator potential? Neglect the natural line width.

SCATTERING THEORY FOR PEDESTRIANS

Chapter 8 presents the basic physical features and mathematical techniques underlying scattering theory in a one-dimensional example. In one dimension there are only two directions of scattering, forward and backward, rather than a continuous infinity of scattering angles as in three dimensions. The Hamiltonian has only a twofold instead of an infinite degeneracy. Thus, one-dimensional scattering is described at a given energy in a two-dimensional vector space rather than the infinite-dimensional space required for three-dimensional scattering, and the S- and T-matrices are 2×2 rather than infinite. These 2×2 matrices are sufficiently non-trivial to allow most of the physical content of three-dimensional scattering theory to be presented without the mathematical complications of infinite matrices. For example, in three dimensions the rotational symmetry of the potential which 'looks the same from all directions' enables the separation of the Schrödinger equation into an infinite number of partial waves. These are decoupled from one another because they behave differently under rotations. In one dimension reflection symmetry of a potential which looks the same from *all two* directions enables the separation of the Schrödinger equation into *two* independent partial waves. These are decoupled because they behave differently under reflections.

The treatment begins with partial wave analysis, phase shifts and definition of the scattering amplitude and the S-matrix in the two-dimensional vector space expressed in terms of a two-valued angular variable. The treatment is generalized to multichannel scattering by introducing an additional degree of freedom, charge, in the example of kaon–nucleon scattering with possible charge exchange. Isospin symmetry is used to define a complete set of partial waves and phase shifts. A series of specific problems involving delta potentials are solved exactly to demonstrate the use of the phase shift method and also many interesting properties of scattering amplitudes. Analytic properties are discussed, in particular, the relation between poles in the scattering amplitude and the existence of physical bound states and resonances and the general relation between analytic properties and conservation of probability. The single and double delta potentials are solved in a straightforward manner. A simple example with inelastic scattering is illustrated by the example of a scatterer which has two internal states. An exact solution with a delta potential shows the relation between inelastic scattering and complex phase shifts. Two poles found in the scattering amplitude correspond to bound states in the delta potential with the scatterer in the ground and excited states. The second bound state is stable if the binding energy is greater than the excitation energy of the excited state of the scatterer. Otherwise it is a resonance and decays into the ground state of the scatterer and a free particle. The associated pole develops an imaginary part whose magnitude gives a value for the width of the resonance which is exactly equal to the width given by the golden rule formula of time-dependent perturbation theory. Another case of resonance scattering presented is pion–pion scattering in one dimension due to the interaction used to describe kaon decay in ch. 7. The second-quantized Hamiltonian is transformed to give a Schrödinger equation for a particle with a reduced mass in a delta potential. Solution by the phase shift method gives the expected Lorentzian resonance curve in the scattering amplitude.

The presentation of formal perturbation theory in scattering emphasizes the particular difficulties arising in scattering perturbation theory which are absent in stationary perturbation theory. Ambiguities and paradoxes result from the use of non-normalizable wave functions and from the lack of a unique answer to the question 'which solution of the exact scattering problem corresponds to a given solution of the unperturbed Schrödinger equation'. This ambiguity is removed by specifying that the desired solution must satisfy the boundary condition of having only outgoing waves in addition to the unperturbed incident wave. The formal treatment then introduces the T-matrix, Green's functions and the Born series.

Section 8.13, 8.14 and 8.15 present a self-contained unorthodox treatment of motion in a one-dimensional periodic potential with solid state applications useful in elementary courses in solid state physics. The band structure of the energy spectrum is obtained simply and without approximations for the case of a general reflection-symmetric potential by the use of the scattering phase shifts. The essential advantage over conventional treatments results from parametrization of all the relevant properties of the scattering potential by the phase shifts. The condition for the existence of a solution of the Schrödinger equation is then expressed entirely in terms of cosines of various variables. Forbidden energy bands occur whenever this condition requires a cosine to be greater than unity. Section 8.15 generalizes this problem by allowing the potential to be not strictly periodic and treating the displacement of each lattice point from equilibrium as a dynamical variable. This gives a one-dimensional model of a crystal which describes Bloch electrons and phonons after a transformation to normal coordinates and has an electron–phonon interaction similar to the phenomenological interactions of ch. 4. Electron–electron and electron–phonon scattering in this model show the peculiarities of crystal momentum conservation.

CHAPTER 8

ONE-DIMENSIONAL SCATTERING IN QUANTUM MECHANICS

8.1 The one-dimensional scattering problem

In a scattering experiment a beam of particles is scattered by a target, and the scattered particles are detected. The interaction between the beam and target takes place in a small volume of space. The region where the beam is prepared and the region where the scattered beam is detected are both outside the region of interaction between the beam and the target. Thus both in the initial and the final states the particles in the beam behave as free particles. Scattering processes are transitions from one free-particle state to another as a result of an interaction which takes place within a small volume. These features are all illustrated in the following one-dimensional example.

Consider a single particle of mass m and momentum p moving freely in one dimension. The Hamiltonian is:

$$H_0 = p^2/2m. \tag{8.1}$$

The momentum p commutes with H_0. Thus H_0 and p can be simultaneously diagonalized. The eigenfunctions are plane waves:

$$\psi_k = \mathrm{e}^{ikx} \tag{8.2a}$$

$$p\psi_k = \hbar k \psi_k \tag{8.2b}$$

$$H_0\psi_k = \frac{(\hbar k)^2}{2m}\psi_k. \tag{8.2c}$$

The energy spectrum is continuous and doubly degenerate since the eigenvalue (8.2c) depends only on the magnitude of k and not on the sign. Any linear combination of the degenerate eigenfunctions ψ_k and ψ_{-k} is also an eigenfunction, e.g.

$$A\psi_k + B\psi_{-k} = A\mathrm{e}^{ikx} + B\mathrm{e}^{-ikx} \tag{8.3a}$$

or

$$\Phi_k = \sin(kx + \delta). \tag{8.3b}$$

where A, B and δ are constants.

The momentum eigenfunctions (8.2) describe travelling waves. The function (8.3a) is a linear combination of two travelling waves in opposite directions. The function (8.3b) describes a standing wave.

The Hamiltonian (8.1) is invariant under a reflection about the origin. The parity P thus commutes with the Hamiltonian and a complete set of simultaneous eigenfunctions of H and P can be found. The even and odd parity eigenfunctions are respectively

$$\psi_{ko} = \cos kx \tag{8.4a}$$

$$\psi_{k1} = \sin kx \tag{8.4b}$$

$$P\psi_{ko} = \psi_{ko} \tag{8.5a}$$

$$P\psi_{k1} = -\psi_{k1}. \tag{8.5b}$$

Let us now add to the Hamiltonian (8.1) a potential V, which is confined to a finite region bounded by the value $|x| = X$.

$$H = \frac{p^2}{2m} + V(x) \tag{8.6a}$$

$$V(x) = 0 \quad \text{for} \quad |x| > X. \tag{8.6b}$$

The eigenvalue spectrum for $E \geqq 0$ is not changed by the added potential. It is still continuous and doubly degenerate. The form of these eigenfunctions is also not changed in the region outside of the potential. One can find eigenfunctions which behave like any of the free-particle eigenfunctions (8.2a), (8.3) or (8.4b) for $x > X$, and similarly for $x < -X$. However we do not know the connection between the wave functions in the positive and negative domains. A wave function which has the form (8.2a) for $x > X$, must be some linear combination (8.3a) in the region $x < -X$, but we do not know a priori which linear combination. This depends upon the potential V.

Let $\psi^{(+)}(x)$ be an eigenfunction of H which has the form (8.2a) of a single plane wave for $x > X$:

$$\psi^{(+)}(x) = S\mathrm{e}^{ikx} \quad \text{for} \quad x > X \tag{8.7a}$$

where S is a numerical coefficient. Then for $x < -X$, this eigenfunction has the form (8.3a),

$$\psi^{(+)}(x) = \mathrm{e}^{ikx} + R\mathrm{e}^{-ikx} \quad \text{for} \quad x < -X, \tag{8.7b}$$

where R is a numerical coefficient, and we have chosen the normalization so that the coefficient of the first term is one. This eigenfunction (8.7) has a very simple physical interpretation. For $x < -X$, there are waves travelling in both directions, whereas for $x > X$, there is only an 'outgoing' wave moving to the right. The first term on the right-hand side of (8.7b) can be interpreted as an incident wave, the second term as a reflected wave and the wave function (8.7a) as a transmitted wave. Then R and S are the reflection and transmission coefficients for the potential V. They can be determined by the explicit solution of the Schrödinger equation, including the region of the potential.

8.2 Reflection and rotation symmetry and phase shifts

Suppose that the potential is invariant under reflections,

$$V(x) = V(-x) \tag{8.8a}$$

$$[P, V] = 0. \tag{8.8b}$$

Then the Hamiltonian (8.6) and the parity operator P can be simultaneously diagonalized to give even and odd standing-wave solutions. These can be written in the convenient form

$$\psi_0 = \cos(kx + \delta_0) \quad (x > X); \qquad \psi_0 = \cos(kx - \delta_0) \quad (x < -X) \tag{8.9a}$$

$$\psi_1 = \sin(kx + \delta_1) \quad (x > X); \qquad \psi_1 = \sin(kx - \delta_1) \quad (x < -X). \tag{8.9b}$$

These states differ from the corresponding free-particle parity eigenstates (8.4) by the 'phase shifts', δ_0 and δ_1. The values of these phase shifts depend upon the potential V and are obtained by the explicit solution of the Schrödinger equation.

The particular linear combination of the parity eigenstates (8.9) which has the form (8.7) is easily constructed

$$\psi^{(+)} = e^{+i\delta_0}\psi_0 + ie^{+i\delta_1}\psi_1 = \tfrac{1}{2}(e^{2i\delta_0} + e^{2i\delta_1})e^{ikx} \qquad (x > X) \tag{8.10a}$$

$$= e^{ikx} + \tfrac{1}{2}(e^{2i\delta_0} - e^{2i\delta_1})e^{-ikx} \qquad (x < -X) \tag{8.10b}$$

Thus:

$$S = \tfrac{1}{2}(e^{2i\delta_0} + e^{2i\delta_1}) = \tfrac{1}{2}[(e^{2i\delta_0} - 1) + (e^{2i\delta_1} - 1)] + 1$$

$$= 1 + \sum_{l=0,1} ie^{i\delta_l} \sin \delta_l \tag{8.11a}$$

$$R = \tfrac{1}{2}(e^{2i\delta_0} - e^{2i\delta_1}) = \tfrac{1}{2}[(e^{2i\delta_0} - 1) - (e^{2i\delta_1} - 1)]$$

$$= \sum_{l=0,1} i(-1)^l e^{i\delta_l} \sin \delta_l. \tag{8.11b}$$

The transmission and reflection coefficients are determined completely by the values of the phase shifts of the even and odd solutions.

Let us now express these results in a language which is more easily generalized to the physical three-dimensional case. The one-dimensional reflection symmetry of the potential, eq. (8.8a) can also be called invariance under a 180° rotation about an axis perpendicular to the x-axis. The natural generalization to three dimensions is full rotational invariance under an arbitrary rotation. Since the most convenient coordinates for discussing rotational invariance in three dimensions are spherical polar coordinates, our one-dimensional results will be more easily generalized to three dimensions if we express them in terms of 'one-dimensional polar variables'. We therefore define

$$r = |x| \tag{8.12a}$$

$$\theta = 0 \quad \text{if} \quad x > 0; \qquad \theta = \pi \quad \text{if} \quad x < 0. \tag{8.12b}$$

In the one-dimensional case θ has only two values, 0 and π, for the forward and backward directions respectively. However the dependence of the wave functions and the scattering process on this two-valued angular variable already gives considerable insight into the angular dependence for the three-dimensional case.

In three-dimensional scattering problems, a combination of cartesian and polar coordinates is often used, to write a wave function as a linear combination of an incident plane wave and an outgoing spherical wave. The wave function (8.7) can be rewritten as a single equation in this form, using our one-dimensional polar variables (8.12)

$$\psi^{(+)}(x) = \mathrm{e}^{ikx} + g(\theta)\mathrm{e}^{ikr}; \qquad (r > X) \tag{8.13a}$$

where

$$g(0) = S - 1 \tag{8.13b}$$

$$g(\pi) = R. \tag{8.13c}$$

The first term e^{ikx} in eq. (8.13a) is present not only for $x < -X$ but also for $x > X$. It describes, therefore, not only the incoming incident wave but also an outgoing wave which would be the complete solution of the Schrödinger equation in the absence of the potential. The two terms in the wave function (8.13a) thus describe an unperturbed incident wave which is the complete solution in the absence of the potential and a scattered wave which is entirely due to the potential. The function $g(\theta)$ describes the 'angular dependence' of the scattered amplitude.

The separation of the wave function into unperturbed and scattered waves differs from the separation in eqs. (8.7) into incoming and outgoing waves, where the outgoing waves include the continuation of the incident wave after it has passed the potential. This difference between the two descriptions appears as an additional term in the expression (8.13b) relating the forward scattering amplitude $g(0)$ to the transmission coefficient S. There is no such additional term in the relation (8.13c) between the backward scattering amplitude $g(\pi)$ and the reflection coefficient R.

These two alternative descriptions are both useful in the treatment of scattering phenomena. The division into incoming and outgoing waves is convenient for the discussion of conservation of probability, which requires that the current carried by the incoming waves be equal to the current carried by the outgoing waves. The division into an unperturbed wave and a scattered wave is useful for the treatment in perturbation theory where one begins with the unperturbed wave as the zero-order solution and calculates the scattered wave by a method of successive approximations.

8.3 Conservation of probability and the optical theorem

From conservation of probability the currents carried by the two outgoing waves must be equal to the current from the incoming wave. Since all waves in this elastic scattering process have the same wave number and velocity, the currents are all proportional to the densities, with the same proportionality factor. Thus the sum of the densities of the two outgoing waves must be equal to the density of the incoming wave.

$$|R|^2 + |S|^2 = 1. \tag{8.14}$$

Note that the solution (8.11) satisfies this condition.

The total scattered intensity is the sum of the intensities of the forward and backward scattered waves (in the three-dimensional case it would be the integral of the scattered intensity over all angles).

$$|g(0)|^2 + |g(\pi)|^2 = |R|^2 + |S-1|^2 = 2\,\mathrm{Re}(1-S) = -2\,\mathrm{Re}\; g(0), \tag{8.15}$$

where we have used eq. (8.14).

The function $g(\theta)$ is dimensionless and its square defines a scattering probability. In three dimensions, the scattering amplitude is naturally defined to give it the dimensions of length and its square defines the scattering cross section. The extra factor with dimensions of length arises naturally in three-dimensions because the free-particle solution corresponding to an outgoing wave is e^{ikr}/r rather than e^{ikr}. The three-dimensional analog of eq. (8.13a) is

$$\psi^{(+)}(x) = \mathrm{e}^{ikx} + f(\theta)\mathrm{e}^{ikr}/r \tag{8.16a}$$

where $f(\theta)$ is the scattering amplitude having the dimensions of length. Let us write

$$f(\theta)\frac{e^{ikr}}{r} = g(\theta)\frac{e^{ikr}}{ikr}. \tag{8.16b}$$

This relates $f(\theta)$ to a dimensionless amplitude $g(\theta)$ which is the natural generalization of the function $g(\theta)$ appearing in the one-dimensional case. We therefore define for the one-dimensional scattering amplitude

$$f(\theta) \equiv \frac{1}{ik} g(\theta). \tag{8.16c}$$

Substituting eq. (8.16c) into eq. (8.15), we obtain:

$$\sum_{l=0,\pi} |f(\theta)|^2 = -2k^2 \operatorname{Re}[ikf(0)] = 2k^{-1} \operatorname{Im}[f(0)]. \tag{8.17}$$

This relation showing that the total scattered intensity is proportional to the imaginary part of the forward scattering amplitude is called the 'optical theorem'. In the three-dimensional case the numerical factor is 4π instead of 2. This non-linear relation has a left-hand side quadratic in the scattering amplitudes and a linear right-hand side. The non-linearity arises because the scale of the wave functions has already been set by normalizing the coefficient of the incident wave to unity in the right-hand side of eq. (8.14). The occurrence of the imaginary part of an amplitude on the right-hand side of eq. (8.17) does not indicate a physical significance to the absolute phase in a wave function. This is a relative phase because the absolute phase of the wave function has been fixed by choosing the coefficient of the incident wave to be real.

The scattering amplitude $f(\theta)$ is very simply expressed in terms of the phase shifts by using eqs. (8.11), (8.13) and (8.16):

$$f(\theta) = k^{-1} \sum_{l=0,1} e^{il\theta} e^{i\delta_l} \sin \delta_l. \tag{8.18}$$

The generalization of eq. (8.18) to three dimensions is intuitively obvious. The scattering amplitude $f(\theta)$ is a function of continuous angular variables describing the scattering in any direction rather than only forward and backward. The parity symmetry of the potential becomes a rotational symmetry, expressing the invariance of the potential with respect to all changes of direction in space, rather than only the change from forward to backward. The conserved quantity corresponding to rotational invariance is angular momentum. The two parity eigenstates (8.9) are thus replaced by an infinite discrete series of angular momentum eigenstates each having its own phase shift. The expansion of a scattering wave function into angular momentum eigenstates is called a partial wave expansion. In three dimensions the scattering amplitude (8.18) is also expressed as the sum of the contributions of the partial waves with each contribution expressed as a function of the corresponding phase shift.

8.4 The *S*-matrix

The wave function (8.13) can be expressed as the sum of incoming and outgoing waves by writing the first term in polar coordinates as well as the second term.

$$\psi_0^{(+)} = \delta_{\theta\pi} \mathrm{e}^{-ikr} + [g(\theta) + \delta_{\theta 0}]\mathrm{e}^{ikr}; \qquad r > X \tag{8.19a}$$

where the subscript zero on the wave function indicates that its incident wave is in the forward direction. Since our potential is invariant under reflections we can construct another solution of the Schrödinger equation from eq. (8.19a) by performing a reflection on this wave function. In polar coordinates a reflection replaces θ by $\pi - \theta$

$$\psi_\pi^{(+)} = \delta_{\theta 0} \mathrm{e}^{-ikr} + [g(\pi - \theta) + \delta_{\theta\pi}]\mathrm{e}^{ikr}; \qquad r > X, \tag{8.19b}$$

where the subscript π indicates that the incident wave is in the backward direction. Eqs. (8.19a) and (8.19b) can be combined in the form

$$\psi_{\theta'}^{(+)} = \delta_{\theta(\pi-\theta')} \mathrm{e}^{-ikr} + [g(\theta' - \theta) + \delta_{\theta\theta'}]\mathrm{e}^{ikr}; \qquad r > X. \tag{8.19c}$$

Any linear combination of eqs. (8.19a) and (8.19b) is also a solution of the Schrödinger equation. Since any function of the two-valued variable θ can be expressed as a linear combination of $\delta_{\theta\pi}$ and $\delta_{\theta 0}$, we can construct solutions with an incoming wave e^{-ikr} multiplied by an arbitrary function of θ. Let $\phi_1(\theta)$ and $\phi_2(\theta)$ be any two orthonormal functions of θ in the two-dimensional vector space defined by the values $\theta = 0$ and $\theta = \pi$. Then we can construct the two corresponding solutions by combining the solutions (8.19a) and (8.19b):

$$\psi_\alpha^{(+)} = \phi_\alpha(0)\psi_0^{(+)} + \phi_\alpha(\pi)\psi_\pi^{(+)} = \sum_{\theta'=0,\pi} \phi_\alpha(\theta')\psi_{\theta'}^{(+)} \tag{8.20a}$$

$$\psi_\alpha^{(+)} = \phi_\alpha(\pi - \theta)\mathrm{e}^{-ikr} + \sum_{\beta=1,2} S_{\alpha\beta}\phi_\beta(\theta)\mathrm{e}^{ikr}; \qquad r > X, \quad \alpha = 1, 2 \tag{8.20b}$$

where

$$S_{\alpha\beta} = \sum_{\theta,\theta'} \phi_\beta^*(\theta)[g(\theta' - \theta) + \delta_{\theta\theta'}]\phi_\alpha(\theta'). \tag{8.20c}$$

The matrix $S_{\alpha\beta}$ is called the S-matrix and gives the amplitude of the outgoing wave of type β corresponding to an incoming wave of type α.

Conservation of probability requires that the total intensity of outgoing waves be equal to the intensity of incoming waves for any linear combination $\sum U_\alpha \psi_\alpha^{(+)}$ $(\alpha=1, 2)$ of the states (8.20b). Since the functions ϕ_α and ϕ_β are orthonormal, equating intensities of outgoing and incoming waves gives

$$\sum_{\alpha\beta\gamma} U_\gamma^* S_{\gamma\beta}^* U_\alpha S_{\alpha\beta} = \sum_\alpha U_\alpha^* U_\alpha. \tag{8.21a}$$

Since this must hold for all values of the coefficients U_α,

$$\sum_{\beta=1,2} S_{\alpha\beta} S_{\gamma\beta}^* = \delta_{\alpha\gamma}. \tag{8.21b}$$

Thus the S-matrix is unitary.

If there is no potential, the outgoing wave is the same as the incoming wave and the S-matrix is seen from eq. (8.20b) to be equal to the unit matrix.

Using the unitarity relation (8.21b) we can construct another set of two corresponding solutions

$$\begin{aligned} \psi_\gamma^{(-)} &= \sum_{\alpha=1,2} \psi_\alpha^{(+)} S_{\alpha\gamma}^* \\ &= \sum_{\alpha=1,2} \phi_\alpha(\pi-\theta) S_{\alpha\gamma}^* e^{-ikr} + \phi_\gamma(\theta) e^{ikr}; \quad r > X, \quad \alpha = 1, 2. \end{aligned} \tag{8.22a}$$

These solutions have a single outgoing wave and a sum of incoming waves, rather than a single incoming wave and a sum of outgoing waves. Note that a function having these properties can also be generated from any solution (8.20b) by interchanging θ and $\pi-\theta$ and taking the complex conjugate

$$[\psi_\alpha^{(+)}(\pi-\theta)]^* = \phi_\alpha^*(\theta) e^{ikr} + \sum_{\beta=1,2} S_{\alpha\beta}^* \phi_\beta^*(\pi-\theta) e^{-ikr}. \tag{8.22b}$$

Substituting the functions (8.22b) into the Schrödinger equation shows that they are solutions if the potential is *real*, i.e. if the potential is invariant under time reversal. In that case the solutions (8.22a) and (8.22b) must describe the same physical states, differing only by phases. This gives conditions on the S-matrix imposed by time reversal invariance. For the case where the phases of the basic states $\phi_\alpha(\theta)$ are chosen to be real, e.g. the states (8.19), time reversal invariance requires the S-matrix to be *symmetric*. This agrees with an intuitive picture of time reversal which would require the transition probability from state α to state β to be the same as that from β to α.

The S-matrix can be diagonalized for a reflection-symmetric potential by choosing the parity eigenstates (8.9) as our basic states. In polar coordinates, these are

$$\psi_0 = \cos(kr + \delta_0) = \tfrac{1}{2}e^{-i\delta_0}[e^{-ikr} + e^{2i\delta_0}e^{ikr}]; \qquad r > X \tag{8.23a}$$

$$\psi_1 = e^{i\theta}\sin(kr + \delta_1) = \tfrac{1}{2}ie^{-i\delta_1}e^{i\theta}[e^{-ikr} - e^{2i\delta_1}e^{ikr}]; \qquad r > X. \tag{8.23b}$$

These two can be combined with new normalization and phase factors in the form

$$\psi_l = e^{il\theta}[e^{-ikr} + (-1)^l e^{2i\delta_l}e^{ikr}]. \tag{8.23c}$$

This can also be written in a form resembling eq. (8.20b)

$$\psi_l = -[e^{il(\pi-\theta)}e^{-ikr} + e^{2i\delta_l}e^{i\theta}e^{ikr}]. \tag{8.23d}$$

By comparison with eqs. (8.20b) we see that

$$\phi_l(\theta) = e^{il\theta} \tag{8.24a}$$

and the S-matrix is

$$S_{ll'} = e^{2i\delta_l}\delta_{ll'}. \tag{8.24b}$$

A knowledge of the S-matrix gives a complete description of the scattering process. The S-matrix gives the scattered waves for all possible incident waves. In the general case where there are inelastic scattering processes as well as elastic, the S-matrix relates all possible states which are coupled together by the scattering process, the indices α and β take on values for all possible 'channels' rather than just the two values for the forward and backward channels. There are some schools of thought in particle physics which see the S-matrix as the most basic and fundamental quantity in particle physics, since the elements of the S-matrix are measured in scattering experiments rather than the Hamiltonian or other dynamical variables like fields.

8.5 KN charge exchange and multichannel scattering

As an example of multichannel scattering processes we can consider particles having additional internal degrees of freedom, such as electric charge. This introduces the possibility of inelastic processes, such as charge exchange scattering in addition to elastic scattering. Consider the scattering of a kaon by a potential due to a nucleon held fixed at the origin. The motion of the nucleon is neglected, but it can be either a proton or a neutron and exchange charge with the kaon which can be either a K^+ or K^0. If the initial state is a K^+ and a neutron, a charge exchange scattering can occur to a final state which is a K^0 and a proton. The n–p and K^0–K^+ mass differences are neglected so that charge exchange occurs with no change in energy or momentum. We assume that the potential is invariant under reflections and also conserves isospin.

Consider systems of electric charge $Q = +1$; i.e. the K^+n and K^0p systems for which charge exchange is possible. We do not consider the K^+p and K^0n systems, as only elastic scattering is allowed by charge conservation, and the treatment is exactly the same as the previous example, eq. (8.6). For positive energies the $Q = +1$ states have a continuous spectrum and a four-fold degeneracy. The two-fold degeneracy of the uncharged example eq. (8.6) is doubled because there are K^+n and K^0p states for each eigenfunction of the Hamiltonian (8.6). In the regions $x > X$ and $x < -X$ the eigenfunctions of the KN Hamiltonian are the same as for free particles and any basis can be chosen to describe the four degenerate states. However we do not know the relation between the wave functions between the positive and negative domains. To describe a scattering problem with a K^+ beam on a proton target, we need a wave function with the following properties:

In the region $x < -X$, it has an incident K^+n wave with momentum $+k$ and reflected K^+n and K^0p waves with momentum $-k$. In the region $x > X$ it has transmitted K^+n and K^0p waves with momentum $+k$ but no waves with momentum $-k$. Thus it has an incident K^+ wave coming from $x = -\infty$ on a proton target and outgoing waves in both directions which can be either K^+n or K^0p.

Our formalism is easily generalized to describe these multichannel wave functions. We introduce an additional index q to label the charge channel, $q = +$ for the K^+n state and $q = 0$ for K^0p. Then eq. (8.13a) can be written

$$\psi_q^{(+)} = U_q e^{ikx} + \sum_s g_{qs}(\theta) U_s e^{ikr}; \qquad r > X \tag{8.25a}$$

where U_q describes the charge of the state, and $g_{qs}(\theta)$ is a 2×2 matrix in the charge space. Similarly eq. (8.20b) becomes

$$\psi_{q\alpha}^{(+)} = U_q \phi_\alpha(\pi - \theta) e^{-ikr} + \sum_{\beta=1,2} S_{q\alpha, s\beta} U_s \phi_\beta(\theta) e^{ikr}; \qquad r > X. \tag{8.25b}$$

The S-matrix now has additional labels for the charge channel, and is given by the corresponding generalization of eq. (8.20c)

$$S_{q\alpha, s\beta} = \sum_{\theta, \theta'} \phi_\beta^*(\theta) [g_{qs}(\theta' - \theta) + \delta_{\theta\theta'} \delta_{qs}] \phi_\alpha(\theta'). \tag{8.25c}$$

Since isospin is conserved in the interaction, it is convenient to define isospin eigenfunctions.

$$U^{(1)} = \frac{1}{\sqrt{2}}(U_+ + U_0) \tag{8.26a}$$

$$U^{(0)} = \frac{1}{\sqrt{2}}(U_+ - U_0). \tag{8.26b}$$

We can then define simultaneous eigenfunctions of isospin and parity by generalizing eq. (8.23d)

$$\psi_l^{(T)} = -U^{(T)}[\mathrm{e}^{il(\pi-\theta)}\mathrm{e}^{-ikr} + \mathrm{e}^{2\mathrm{i}\delta_l{}^T}\mathrm{e}^{il\theta}\mathrm{e}^{ikr}]. \tag{8.27a}$$

The four isospin and parity eigenfunctions are decoupled and each is characterized by its individual phase shift. The S-matrix is then

$$S_{Tl,T'l'} = \mathrm{e}^{2\mathrm{i}\delta_l{}^T}\delta_{TT'}\delta_{ll'}. \tag{8.27b}$$

The scattering amplitudes $g_{qr}(\theta)$ can then be expressed in terms of these phase shifts.

For the case of incident K^+, $q=+$ and

$$g_{++}(\theta) = \tfrac{1}{2}\sum_{l,T} \mathrm{i}\mathrm{e}^{il\theta}\mathrm{e}^{\mathrm{i}\delta_l{}^T}\sin\delta_l^T \tag{8.28a}$$

$$g_{+0}(\theta) = \tfrac{1}{2}\sum_{l,T} \mathrm{i}\mathrm{e}^{il\theta}(-1)^{T-1}\mathrm{e}^{\mathrm{i}\delta_l{}^T}\sin\delta_l^T. \tag{8.28b}$$

The phase shift method thus allows the dependence on isospin and scattering angle to be unscrambled by the use of the isospin and parity eigenstates. In the general case any symmetry in the scattering problem allows a separation of the scattering amplitude into terms which are decoupled from one another because they correspond to different eigenvalues of a conserved quantity. One thus has partial waves for each set of eigenvalues and can express the corresponding partial-wave amplitudes in terms of the relevant phase shifts.

In a general collision problem there may be many possible channels for the reaction instead of only two, K^+n and K^0p, as in this simple example. For example in the scattering of protons on ^{12}C there can be reactions in which there are outgoing neutrons, deuterons or alpha particles each leaving a different residual nucleus which can be either in its ground state or in some excited state. We can generalize the notation of eq. (8.25) for the wave function to apply to the case of an arbitrary number of channels. Let q describe all the 'internal' degrees of freedom of the incident and scattered particles. In a case like proton scattering on ^{12}C the function U_q could describe not only the labels of the two particles but also the structure of the particular state of the residual nucleus. The function $g_{qs}(\theta)$ is then an $n\times n$ matrix where n is the number of channels, and the S-matrix is correspondingly enlarged. If there is no symmetry which allows us to separate the problem into uncoupled channels, then the phase shift method does not give a complete solution and the multichannel equation must be solved by other means. This would be the case in the proton–^{12}C example.

8.6 The delta potential

Let us now consider the solution of the scattering problem for a short-range potential which exists only in a region very small compared to the wavelength of the scattered particle

$$kX \ll 1. \tag{8.29}$$

The Schrödinger equation

$$\left[-\frac{\hbar^2}{2m}\frac{\mathrm{d}^2}{\mathrm{d}x^2}+V\right]\psi = E\psi = \frac{\hbar^2 k^2}{2m}\psi \tag{8.30a}$$

is commonly rewritten

$$\left(\frac{\mathrm{d}^2}{\mathrm{d}x^2}+k^2\right)\psi = \frac{2mV}{\hbar^2}\psi = U\psi \tag{8.30b}$$

where

$$U(x) \equiv \frac{2m}{\hbar^2}V(x). \tag{8.30c}$$

Eq. (8.30b) can be integrated between the points $-X$ and $+X$ where the potential vanishes.

$$\frac{\mathrm{d}\psi(+X)}{\mathrm{d}x}-\frac{\mathrm{d}\psi(-X)}{\mathrm{d}x}+k^2\int_{-X}^{X}\psi\,\mathrm{d}x = \int_{-X}^{X}U\psi\,\mathrm{d}x. \tag{8.31}$$

For a short-range potential satisfying eq. (8.29) the wave function does not change very much in the region of the potential, and eq. (8.31) gives the change in the derivative of the wave function from one side of the potential to the other. The right-hand side is just the product of the wave function and the integral of the potential. To simplify the calculations we assume a 'zero range' potential.

$$U(x) = -U_0\delta(x). \tag{8.32}$$

We now let $X \to 0$ in eq. (8.31). Then for the even-parity solution (8.9a):

$$\frac{\mathrm{d}\psi_0(0+)}{\mathrm{d}x}-\frac{\mathrm{d}\psi_0(0-)}{\mathrm{d}x} = -2k\sin\delta_0 = -U_0\psi_0(0) = U_0\cos\delta_0. \tag{8.33}$$

The phase shift for the even-parity wave is thus

$$\tan\delta_0 = \frac{U_0}{2k} \tag{8.34a}$$

and the S-matrix element is

$$e^{2i\delta_0} = \frac{1 + i \tan \delta_0}{1 - i \tan \delta_0} = \frac{1 + i(U_0/2k)}{1 - i(U_0/2k)} = \frac{2k + iU_0}{2k - iU_0}. \qquad (8.34b)$$

The odd-parity solution (8.34b) vanishes at the origin. Its phase shift thus vanishes for the zero-range potential.

$$\delta_1 = 0 \qquad (8.35a)$$

$$e^{2i\delta_1} = 1. \qquad (8.35b)$$

Substituting (8.34) and (8.35) into eq. (8.18) we obtain the scattering amplitude:

$$f(\theta) = \frac{1}{2ik}\,[e^{2i\delta_0} - 1] = \frac{U_0}{k[2k - iU_0]} \qquad (8.36a)$$

$$g(\theta) = i[U_0/(2k - iU_0)]. \qquad (8.36b)$$

We now have the complete solution to the scattering problem. The intensities of the forward and backward scattered waves are given by equation (8.36).

Since $f(\theta)$ is independent of θ the forward and backward scattered amplitudes are equal. This is expected since only the even-parity wave is scattered by this potential and the odd-parity wave is not. Since both even and odd-parity waves have equal amplitudes forward and backward but opposite relative phase, there can be a difference between the forward and backward scattering amplitudes only when both waves are present. This can also be seen directly by examining eq. (8.23). The same result holds in three dimensions. Any difference between forward and backward scattering requires interference between even and odd-parity waves.

The expression (8.36b) substituted into the solution (8.13a) defines a solution to the Schrödinger differential equation (8.30a) for all values of k. It is instructive to examine the solutions for complex values of k as well as real values. If k is complex, the imaginary part of k contributes a real part to the exponents in eq. (8.36a). Either the incoming or outgoing wave has a positive real exponent corresponding to an amplitude which increases without limit at large distances. Such solutions do not represent physical states; they are excluded by the boundary conditions used with the Schrödinger equation. However, the scattering amplitude (8.36) and the S-matrix elements (8.34b) have a pole at the value $k = \frac{1}{2}iU_0$. Thus as k approaches this value the amplitude of the scattered wave in (8.13a) increases without limit. If we normalize this solution to keep the amplitude of the scattered wave constant, the first term approaches zero and vanishes at the pole. Thus at the value $k=\frac{1}{2}iU_0$, eq. (8.13a) gives a solution in which there is only the outgoing wave and no incoming wave. For positive U_0, k is pure imaginary and positive, the exponent is real and negative and the amplitude decreases to zero at large distances thus satisfying the boundary condition. Thus the imaginary value $k=\frac{1}{2}iU_0$ defines a physically admissible solution of the Schrödinger equation. Since the amplitude decreases to zero at large distances this solution corresponds to a bound state.

$$\psi_B(x) = e^{-\frac{1}{2}U_0 r}. \tag{8.37a}$$

This bound-state wave function satisfies the Schrödinger equation:

$$H\psi_B(x) = E\psi_B(x) = -\hbar^2 \frac{U_0^2}{8m} \psi_B(x). \tag{8.37b}$$

The wave function (8.37) has a very peculiar feature: its entire amplitude is outside the range of the potential. A measurement of the position of the particle therefore always finds it in a place where there is no force on it; yet the particle remains bound. Although this peculiarity is exaggerated by the unphysical delta function potential, a similar situation can exist with potentials of finite range and is related to wave–particle duality in quantum mechanics. If the range of the potential is small compared to the wavelength of the particle, the wave cannot be confined to the region of the potential. However, solutions may exist which have the form of standing waves with a tail that extends out beyond the potential. Thus there can be a very large probability of finding the particle in a region where there is no force on it. The binding is a result of the wave nature of the particle which spreads it over a finite region including the potential.

8.7 Analytical properties of scattering amplitudes

Interesting physical information is obtained by examining the properties of the scattering amplitude or the S-matrix considered as an analytic function in the complex plane. A pole in this function can reveal the existence of a bound state and give its energy.

The argument is general and applies to any scattering problem and not only to the delta potential. A pole imples the existence of a solution of a Schrödinger equation having the form of the second term in eq. (8.13a) with the incident wave missing. If such a solution exists for the time-independent Schrödinger equation, a solution of the time-dependent Schrödinger equation can be obtained by multiplying this solution by the factor $e^{-iEt/\hbar}$.

$$\psi(t) = g(\theta)e^{ikr}e^{-iEt/\hbar}, \qquad R > X, \tag{8.38a}$$

where the energy E is a complex number given by:

$$E = \hbar^2 k^2/2m. \tag{8.38b}$$

When the pole occurs at a purely imaginary value of k, as in eq. (8.36), the energy is real and negative and the solution (8.38a) corresponds to a bound state. If the pole occurs at a complex value of k with a positive imaginary part, a solution of the time-dependent Schrödinger equation of the form (8.38a) can also be written in which the wave function decreases exponentially to zero at large distances. If

$$k = k_1 + ik_2, \tag{8.39a}$$

then

$$E = \frac{\hbar^2}{2m}[k_1^2 - k_2^2 + 2ik_1k_2], \tag{8.39b}$$

and

$$\psi(t) = g(\theta)e^{ikr}e^{-i[k_1{}^2-k_2{}^2]t/m}e^{-k_2r}e^{\hbar k_1k_2t/m}. \tag{8.39c}$$

The function (8.39c) satisfies all the proper boundary conditions for a Schrödinger wave function at any fixed value of the time $t=t_0$. It is well behaved and decreases to zero exponentially at large distances. This function can therefore represent a state of the physical system at the time $t=t_0$. However, since the Schrödinger equation is first order in time derivatives, a solution is completely specified for all time by giving its value at one time $t=t_0$. Since the function $\psi(t)$ defined by eq. (8.39c) is a solution of the Schrödinger equation, it is the unique solution corresponding to the particular initial conditions chosen. However, if $k_1>0$ the function (8.39c) *increases* exponentially with time and indicates that the scattering amplitude in some particular channel increases without limit. This contradicts conservation of probability and implies that no solutions with $k_1>0$ can exist. Thus the S-matrix considered as an analytic function of the energy E is not allowed to have any poles in the upper complex half-plane; i.e. with a positive imaginary part. A pole in the lower half-plane, $k_1<0$ is perfectly admissible since this corresponds to an exponentially decreasing probability in a given channel. This implies that the probability must increase in some other channel to give overall conservation of probability. However, an unlimited exponential increase of probability with time in any one channel cannot be compensated by decreases in other channels and is therefore not allowed.

Any scattering amplitude or S-matrix obtained from solving a Schrödinger equation exactly automatically has these analytic properties since conservation of probability is automatically incorporated in the Schrödinger equation. However, these analytic properties can be used in cases where we do not know the Schrödinger equation or have not solved it. The fact that a function is known to be analytic in a certain region of the complex plane allows us to use Cauchy's theorem in order to evaluate integrals of this function. The values of direct physical interest are only on the real axis. Thus one can consider contour integrals along the real axis from $-\infty$ to $+\infty$ which are closed by a circle in the upper half-plane. Relations are thus obtained between integrals of the scattering amplitude or of functions of the scattering amplitude on the real axis and the assumed asymptotic behavior at infinite energy which determines the contribution of the integral on the large circle. Such relations are called dispersion relations and are of particular importance in particle physics.

8.8 The double delta potential

The simple delta function potential lacks one important feature which characterizes physical potentials, namely, a characteristic range. Most physical potentials act in a finite region and there are physical scattering processes when the wavelength of the incident particles is short compared to the size of the region in which the potential acts. There is also the possibility of resonances when a number of half wavelengths just fit in the region of the potential and can set up standing waves. These features are absent in the zero range delta potential.

A simple modification of the delta potential which introduces a finite range, is a potential of two delta functions separated by a finite distance. To keep the reflection symmetry we place the two delta functions at the points $\frac{1}{2}a$ and $-\frac{1}{2}a$

$$U(x) = \frac{2m}{\hbar^2} V(x) = -\tfrac{1}{2}U_0[\delta(x - \tfrac{1}{2}a) + \delta(x + \tfrac{1}{2}a)]. \qquad (8.40)$$

We have defined the strength of each delta potential to be $\frac{1}{2}U_0$, to keep the same overall strength as for the single delta potential. The potential (8.40) thus reduces to the single delta potential (8.32) if $a=0$.

To solve the Schrödinger equation we first integrate it between the points $\frac{1}{2}a-\varepsilon$ and $\frac{1}{2}a+\varepsilon$, where ε is very small,

$$\frac{\mathrm{d}\psi}{\mathrm{d}x}(\tfrac{1}{2}a+\varepsilon)-\frac{\mathrm{d}\psi}{\mathrm{d}x}(\tfrac{1}{2}a-\varepsilon)+k^2\int\limits_{\frac{1}{2}a-\varepsilon}^{\frac{1}{2}a+\varepsilon}\psi\,\mathrm{d}x=\int\limits_{\frac{1}{2}a-\varepsilon}^{\frac{1}{2}a+\varepsilon}U\psi\,\mathrm{d}x=-\tfrac{1}{2}U_0\psi(\tfrac{1}{2}a). \quad (8.41a)$$

After dividing by $\psi(\frac{1}{2}a)$ and neglecting the integral on the left-hand side which is of order ε,

$$\frac{1}{\psi}\frac{\mathrm{d}\psi}{\mathrm{d}x}(\tfrac{1}{2}a+\varepsilon)-\frac{1}{\psi}\frac{\mathrm{d}\psi}{\mathrm{d}x}(\tfrac{1}{2}a-\varepsilon)=-\tfrac{1}{2}U_0. \quad (8.41b)$$

Since ψ is continuous at $\frac{1}{2}a$, we can use the same value at $\frac{1}{2}a\pm\varepsilon$. The expression (8.41b) is particularly useful because the logarithmic derivative is independent of the normalization.

We now use eq. (8.41b) to relate the solutions on the two sides of the delta function. The exact solutions of the Schrödinger equation for the interval $-\frac{1}{2}a\leqq x\leqq+\frac{1}{2}a$ with even and odd parity are just the corresponding solutions for a free particle.

$$\psi_0(x)=\cos kx;\qquad \frac{1}{\psi_0}\frac{\mathrm{d}\psi_0}{\mathrm{d}x}=-k\tan kx;\qquad -\tfrac{1}{2}a\leqq x\leqq\tfrac{1}{2}a \quad (8.42a)$$

$$\psi_1(x)=\sin kx;\qquad \frac{1}{\psi_1}\frac{\mathrm{d}\psi_1}{\mathrm{d}x}=+k\cot kx;\qquad -\tfrac{1}{2}a\leqq x\leqq\tfrac{1}{2}a. \quad (8.42b)$$

Outside the region of the potential, the parity eigenfunctions are expressed as usual in terms of phase shifts.

$$\psi_0(x)=\cos(kx\pm\delta_0);\qquad \frac{1}{\psi_0}\frac{\mathrm{d}\psi_0}{\mathrm{d}x}=-k\tan(kx\pm\delta_0);\qquad \pm x>\tfrac{1}{2}a \quad (8.43a)$$

$$\psi_1(x)=\sin(kx\pm\delta_1);\qquad \frac{1}{\psi_1}\frac{\mathrm{d}\psi_1}{\mathrm{d}x}=+k\cot(kx\pm\delta_1);\qquad \pm x>\tfrac{1}{2}a. \quad (8.43b)$$

The values of the phase shifts are determined by matching the two solutions (8.42) and (8.43) at the point $x=\frac{1}{2}a$ using the condition (8.41b)

$$-k\tan(\tfrac{1}{2}ka+\delta_0)+k\tan(\tfrac{1}{2}ka)=-\tfrac{1}{2}U_0 \quad (8.44a)$$

$$k\cot(\tfrac{1}{2}ka+\delta_1)-k\cot(\tfrac{1}{2}ka)=-\tfrac{1}{2}U_0. \quad (8.44b)$$

Solving eqs. (8.44) for δ_0 and δ_1 gives

$$\cot\delta_0 = \tan\tfrac{1}{2}ka + \frac{2k}{U_0}\sec^2\tfrac{1}{2}ka = \frac{4k/U_0 + \sin ka}{1+\cos ka} \tag{8.45a}$$

$$\cot\delta_1 = -\cot\tfrac{1}{2}ka + \frac{2k}{U_0}\csc^2\tfrac{1}{2}ka = \frac{4k/U_0 - \sin ka}{1-\cos ka}. \tag{8.45b}$$

Eq. (8.45) show the following interesting features:

1. When $ka=2n\pi$, eqs. (8.45) reduce to the values (8.34) and (8.35) for a single delta potential. This includes the case $a=0$, when the potentials are equivalent and also all cases where there are an integral number of wavelengths between the two potentials.

2. $\delta_0=0$ when $ka=(2n+1)\pi$, $\delta_1=0$ when $ka=2n\pi$, since the denominators of eqs. (8.45a) and (8.45b) vanish at these points. This can also be seen by looking at eq. (8.41a). Whenever $\psi(\frac{1}{2}a)=0$, the right-hand side of eq. (8.41a) vanishes, there is no discontinuity in the derivative at the point $\frac{1}{2}a$, and the solutions must be identical to those for a free particle in the absence of a potential; i.e. there is no phase shift. This occurs for the even-parity solution whenever an odd number of half wavelengths fit exactly in the interval between $\pm\frac{1}{2}a$ and for the odd-parity solution for an even number of half wavelengths. The even and odd-parity phase shifts thus show an oscillatory behavior as a function of k with periodic zeros.

3. For very small values of k eqs. (8.45) reduce to the form

$$\tan\delta_0 = \frac{U_0}{2k[1+\frac{1}{2}aU_0]} \qquad k\to 0 \tag{8.46a}$$

$$\tan\delta_1 = \frac{kU_0a^2}{8[1-\frac{1}{2}aU_0]} \qquad k\to 0. \tag{8.46b}$$

The odd-parity phase shift goes to zero and the even-parity phase shift has a form similar to that for a single delta function potential (8.34a).

4. For very large values of k eqs. (8.45) become

$$\tan\delta_0 = \frac{U_0}{4k}(1+\cos ka) \qquad k\to\infty \tag{8.47a}$$

$$\tan\delta_1 = \frac{U_0}{4k}(1-\cos ka) \qquad k\to\infty. \tag{8.47b}$$

The tangents of both phase shifts oscillate between zero and the value $U_0/2k$ for a single delta function potential.

The most interesting region of values for k is the intermediate region where the numerators of eqs. (8.45a) and (8.54b) can vanish. The scattering amplitude is given by the expression

$$f(\theta) = \frac{1}{k} \sum_{l=0,1} e^{il\theta} e^{i\delta_l} \sin \delta_l = \frac{1}{k} \sum_{l=0,1} \frac{e^{il\theta}}{\cot \delta_l - i}. \tag{8.48}$$

The scattering amplitude thus has a maximum or resonance whenever $\cot \delta_0 = 0$ or $\cot \delta_1 = 0$. From eqs. (8.45)

$$\cot \delta_0 = 0 \quad \text{when} \quad \sin ka = -4k/U_0 \tag{8.49a}$$

$$\cot \delta_1 = 0 \quad \text{when} \quad \sin ka = 4k/U_0. \tag{8.49b}$$

If $U_0 a$ is large, eqs. (8.49) have many solutions and there are many such resonances. In the vicinity of a resonance where $\delta_l = 0$, we can expand $\cot \delta_l$ as a power series in the energy and keep only the linear term

$$\cot \delta_l = \frac{2}{\Gamma}(E - E_0) \tag{8.50a}$$

where E_0 is the value of the energy where $\cot \delta_l = 0$ and

$$\frac{2}{\Gamma} = \frac{d}{dE}(\cot \delta_l) \quad \text{at} \quad E = E_0. \tag{8.50b}$$

In the vicinity of E_0 eq. (8.48) for the scattering amplitude can be written

$$f_l(\theta) = \frac{1}{k} \frac{\frac{1}{2} e^{il\theta} \Gamma}{(E - E_0) + \frac{1}{2} i\Gamma} \tag{8.51}$$

where $f_l(\theta)$ is the contribution to the scattering amplitude of the particular partial wave having the resonance. The expression (8.51) has the typical form of a resonance curve where Γ is the width of the resonance at half maximum.

The scattering amplitude (8.48) has poles when $\cot \delta_l = i$. From eqs. (8.45)

$$\cot \delta_0 - i = \frac{(4k/U_0 + \sin ka) - i(1 + \cos ka)}{(1 + \cos ka)}$$

$$= -i \frac{[1 + e^{ika} + 4ik/U_0]}{(1 + \cos ka)} \tag{8.52a}$$

$$\cot \delta_1 - i = \frac{(4k/U_0 - \sin ka) - i(1 - \cos ka)}{(1 - \cos ka)}$$

$$= -i \frac{[1 - e^{ika} + 4ik/U_0]}{(1 - \cos ka)}. \tag{8.52b}$$

A pole in the scattering amplitude thus occurs at a pure imaginary value of k, $ik = -\lambda$ if

$$(1 + e^{-\lambda a}) - 4\lambda/U_0 = 0 \quad \text{for} \quad \delta_0 \tag{8.53a}$$

$$(1 - e^{-\lambda a}) - 4\lambda/U_0 = 0 \quad \text{for} \quad \delta_1. \tag{8.53b}$$

Eq. (8.53a) shows that there is always one such bound state of even parity. This can be seen graphically by drawing the functions $1+e^{-\lambda a}$ and $4\lambda/U_0$ and noting that they must intersect at one point. Eq. (8.53b) shows that for large values of U_0 there is one odd-parity bound state whereas for small values of U_0 there will be no bound state. This is seen graphically by drawing the curves $1-e^{-\lambda a}$ and $4\lambda/U_0$. The critical value of U_0 is the one for which both curves have the same slope at $\lambda=0$,

$$U_0 = 4/a. \tag{8.54}$$

A bound state is found for values of U_0 greater than this value (8.54). This can also be verified by solving the Schrödinger equation directly for the bound states.

8.9 A delta potential with an excited state

Another instructive example obtainable by simple modification of the delta potential is the scattering by a system with an excited state which can be excited during the scattering. This corresponds to the physical problems of scattering of a particle by an atom, nucleus or molecule with the possibility of inelastic scattering with the excitation of the scatterer. We assume that the scatterer is very heavy and fixed at the origin and has only two states, the ground state and an excited state. The only scatterer degree of freedom which need be considered is the internal one which specifies whether it is in the ground state or the excited state. The wave function describing this system then is a function of the coordinate x of the particle being scattered and the internal degrees of freedom of the scatterer. The most general such wave function can be written as a linear combination of two terms, one having the scatterer in the ground state and the other having the scatterer in the excited state.

$$|\psi\rangle = a_g^\dagger|0\rangle\psi_g(x) + a_e^\dagger|0\rangle\psi_e(x) \tag{8.55}$$

where $a_g^\dagger$ and $a_e^\dagger$ create the scatterer in the ground and excited states respectively.

The Hamiltonian for this system includes the kinetic energy of the particle and the energy of the excited and ground states of the scatterer. For the interaction we choose a delta function potential with two terms, an elastic scattering term which does not change the state of the scatterer and an inelastic scattering term which induces transitions between the ground and the excited states.

$$H = \frac{p^2}{2m} - \frac{\hbar^2}{2m}\,\delta(x)[U_0\{a_g^\dagger a_g + a_e^\dagger a_e\} + U_1\{a_g^\dagger a_e + a_e^\dagger a_g\}] + E_e a_e^\dagger a_e + E_g a_g^\dagger a_g. \tag{8.56}$$

The Schrödinger equation for the system is easily solved by the standard techniques used with delta function potentials. For all values of x except at the origin the solution is just a solution for a free particle with no interaction. The derivative of the solution is discontinuous at the origin. The discontinuity is given in terms of the strength of the potential by integrating the Schrödinger equation between $x = -\varepsilon$ and $x = +\varepsilon$. In the limit $\varepsilon \to 0$ the only terms which contribute are the discontinuity in the derivative and the integral over the delta function potential. For the Hamiltonian (8.56) we obtain

$$-\frac{\hbar^2}{2m}\frac{d\psi}{dx}\bigg|_{-\varepsilon}^{+\varepsilon} - \frac{\hbar^2}{2m}[U_0\{a_g^\dagger a_g + a_e^\dagger a_e\} + U_1\{a_g^\dagger a_e + a_e^\dagger a_g\}]\psi(0) = 0. \tag{8.57}$$

Substituting the wave function (8.55) into eq. (8.57) we obtain

$$a_g^\dagger|0\rangle \frac{d\psi_g}{dx}\bigg|_{-\varepsilon}^{+\varepsilon} + a_e^\dagger|0\rangle \frac{d\psi_e}{dx}\bigg|_{-\varepsilon}^{+\varepsilon} + [U_0\psi_g(0) + U_1\psi_e(0)]a_g^\dagger|0\rangle + [U_0\psi_e(0) + U_1\psi_g(0)]a_e^\dagger|0\rangle = 0. \tag{8.58}$$

Since the ground and excited states of the scatterer are orthogonal, the terms involving these two states must vanish separately, thus

$$\frac{d\psi_g}{dx}\bigg|_{-\varepsilon}^{+\varepsilon} + U_0\psi_g(0) + U_1\psi_e(0) = 0 \tag{8.59a}$$

$$\frac{d\psi_e}{dx}\bigg|_{-\varepsilon}^{+\varepsilon} + U_0\psi_e(0) + U_1\psi_g(0) = 0. \tag{8.59b}$$

These equations (8.59) are sufficient to determine the exact solutions, once we have specified the desired boundary conditions. Consider a scattering problem in which the scatterer is initially in its ground state. Since the interaction conserves parity and the odd-parity solution is always unaffected by a delta function potential at the origin, we consider the even-parity solution. This has the usual form with a phase shift for the function $\psi_g(x)$. For $\psi_e(x)$ however, we wish a solution which has only outgoing waves and no incoming waves. Thus we set

$$\psi_g(x) = \alpha \cos(k|x| + \delta_0) \tag{8.60a}$$

$$\psi_e(x) = \beta e^{ik_e|x|} \tag{8.60b}$$

where α and β are coefficients to be determined and

$$k^2 = \frac{2m(E - E_g)}{\hbar} \tag{8.60c}$$

$$k_e^2 = \frac{2m(E - E_e)}{\hbar} = k^2 - \frac{2m}{\hbar} [E_e - E_g]. \tag{8.60d}$$

Substituting eqs. (8.60) into eqs. (8.59) we obtain

$$-2k\alpha \sin \delta_0 + U_0\alpha \cos \delta_0 + U_1\beta = 0 \tag{8.61a}$$

$$2ik_e\beta + U_0\beta + U_1\alpha \cos \delta_0 = 0. \tag{8.61b}$$

Solving these equations for β/α we obtain

$$\frac{\beta}{\alpha} = -\frac{U_1 \cos \delta_0}{U_0 + 2ik_e} = \frac{2k \sin \delta_0 - U_0 \cos \delta_0}{U_1}. \tag{8.62a}$$

Solving eq. (8.62a) for the phase shift δ_0 we obtain

$$\begin{aligned} \tan \delta_0 &= \frac{U_0}{2k} - \frac{U_1^2}{2k(U_0 + 2ik_e)} \\ &= \frac{U_0}{2k}\left[1 - \frac{U_1^2}{U_0^2 + 4k_e^2}\right] + \frac{ik_eU_1^2}{k[U_0^2 + 4k_e^2]}. \end{aligned} \tag{8.62b}$$

Eq. (8.62b) shows that the phase shift can be complex. The meaning of a complex phase shift is easily seen from the expressions (8.11) for the amplitudes of the reflected wave R and the transmitted wave S as functions of the phase shift. For complex phase shifts the total intensity of the reflected and transmitted waves is given by

$$|R|^2 + |S|^2 = \tfrac{1}{2}[e^{2i(\delta_0 - \delta_0^*)} + e^{2i(\delta_1 - \delta_1^*)}] = \tfrac{1}{2}[e^{-4\,\mathrm{Im}\,\delta_0} + e^{-4\,\mathrm{Im}\,\delta_1}] \leq 1. \tag{8.63}$$

Thus, if the phase shifts have an imaginary part the total intensity of the outgoing waves is less than the intensity of the incoming waves. This is perfectly reasonable for the case where there can be inelastic scattering. The elastic scattered wave does not necessarily have the full intensity of the incoming wave.

The expression (8.62b) has exactly the right properties to describe the inelastic scattering. If the energy of the incident beam is too low to excite the excited state, then k_e as given by eq. (8.60d) is imaginary, the phase shift as given by eq. (8.62b) is purely real, and there is no loss of intensity in the scattered beam. Once the energy is high enough to excite the scatterer, k_e is real. It is defined in eq. (8.60b) to be positive in order to describe outgoing waves. Thus the imaginary part of the phase shift given by eqs. (8.62b) is positive so that the elastic scattered intensity as given by eq. (8.63) is less than one.

Let us now examine the poles in the scattering amplitude. These occur when

$$\tan \delta_0 = \frac{U_0}{2k} - \frac{U_1^2}{2k(U_0 + 2ik_e)} = -i. \tag{8.64}$$

This can be reduced to the form

$$(U_0 + 2ik)(U_0 + 2ik_e) = U_1^2. \tag{8.65}$$

In the limit $U_1 = 0$ where there is no inelastic scattering, there are two solutions to eq. (8.65) both giving pure imaginary values of k and k_e. Let us write

$$ik = -\lambda \tag{8.66a}$$

$$ik_e = -\lambda_e. \tag{8.66b}$$

The values of λ and λ_e for the case $U_1 = 0$ are just the poles expected for a delta function potential. They correspond to the bound states of the particle, together with either the ground state or the excited state of the scatterer.

For the case where U_1 is finite but small, it is convenient to rewrite eq. (8.65) in the form

$$\lambda = \frac{U_0}{2} - \frac{U_1^2}{2(U_0 - 2\lambda_e)} \tag{8.67a}$$

$$\lambda_e = \frac{U_0}{2} - \frac{U_1^2}{2(U_0 - 2\lambda)} = \frac{U_0}{2} - \frac{U_1^2 U_0}{2[U_0^2 - 4\lambda^2]} - \frac{U_1^2 \lambda}{[U_0^2 - 4\lambda^2]} \tag{8.67b}$$

where λ and λ_e are related by

$$\lambda^2 = \lambda_e^2 - \frac{2m}{\hbar}(E_e - E_g). \tag{8.67c}$$

We now note the following interesting feature. If the binding energy $\frac{1}{2}U_0$ is larger than the excitation energy $E_e - E_g$ of the scatterer, then the two bound states are stable. However, if the binding energy is less than the excitation energy of the excited state, the bound state of the particle with the excited state of the scatterer can decay to the ground state of the scatterer and a free particle. This is clearly seen from eqs. (8.76b) and (8.67c). If λ_e is chosen to correspond to the bound state of the particle with the scatterer in the excited state, then eq. (8.67c) shows that λ is real if the binding energy is greater than the excitation energy but will be imaginary if the binding energy is less than the excitation energy. If λ is real, then λ_e is given by eq. (8.67b) also to be real, and corresponds to a bound state. If however, λ is imaginary, then λ_e given by eq. (8.67b) has an imaginary part and corresponds to a decaying state or a resonance. For this case we should substitute eq. (8.66) back in (8.67b) to obtain

$$\lambda_e = \frac{U_0}{2} - \frac{U_1^2 U_0}{2[U_0^2 + 4k^2]} + \mathrm{i}\,\frac{U_1^2 k}{U_0^2 + 4k^2}. \tag{8.68a}$$

The resonance energy is then given by

$$E_r = \frac{\hbar^2}{2m}\lambda_e^2$$

$$= \frac{\hbar^2}{2m}\left[\left(\frac{U_0}{2} - \frac{U_1^2 U_0}{2[U_0^2 + 4k^2]}\right)^2 - \frac{U_1^4 k^2}{(U_0^2 + 4k^2)^2}\right] + \tfrac{1}{2}\mathrm{i}\Gamma \tag{8.86b}$$

where

$$\Gamma = \frac{\hbar^2 k U_0 U_1^2}{2m[U_0^2 + 4k^2]}\left[1 - \frac{U_1^2}{U_0^2 + 4k^2}\right]. \tag{8.68c}$$

The decay of the resonant state can also be treated by time-dependent perturbation theory. We describe the system with unperturbed wave functions for which $U_1 = 0$, but U_0 is included in the unperturbed Hamiltonian to give the bound states. In the absence of the perturbation U_1 both bound states are stable. In first-order time-dependent perturbation theory the decay rate of the excited bound state is given by the golden rule formula and the width is therefore

$$\Gamma = \hbar W_{i\to f} = 2\pi|\langle f|V|i\rangle|^2 \varrho(E_f). \tag{8.69}$$

The interaction matrix element is

$$\langle f|V|i\rangle = \frac{\hbar^2}{2m} U_1 \lambda_e^{\frac{1}{2}} \cos\delta_0 \tag{8.70a}$$

where the last two factors come from the normalization of the bound-state wave function and the value of the continuum wave function at the origin. The density of final states in one dimension is given by

$$\varrho(E_f) = \frac{1}{\hbar}\frac{\mathrm{d}p}{\mathrm{d}E} = \frac{m}{2\pi\hbar^2 k}. \tag{8.70b}$$

Substituting eqs. (8.70) into (8.69), and inserting the unperturbed values for λ_e and $\cos\delta_0$ gives exactly the first term of eq. (8.68c). This is to be expected since the second term is of higher order in the perturbation U_1 and the perturbation theory result is good only to first order.

8.10 Kaon decay and scattering problems

Another instructive example is pion–pion scattering in one dimension resulting from the interaction between kaons and pions responsible for the kaon decay. The interaction can change an initial two-pion state into a one-kaon state and then into another two-pion state.

Consider the Hamiltonian used in the treatment of kaon decay in ch. 7,

$$H = H_K + H_\pi + V \tag{8.71a}$$

where

$$H_K = \int E_{Kq} a^\dagger_{Kq} a_{Kq}\,\mathrm{d}q \tag{8.71b}$$

$$H_\pi = \int E_{\pi q} a^\dagger_{\pi q} a_{\pi q}\,\mathrm{d}q \tag{8.71c}$$

and we consider the momentum variable q in a one-dimensional space. For the interaction V we use the local interaction (7.13a)

$$V = g\int \mathrm{d}x\, a^\dagger_\pi(x) a^\dagger_\pi(x) a_K(x) + \text{h.c.} \tag{8.72}$$

Our treatment of pion–pion scattering is not relevant to the scattering of physical pions, whose additional strong interactions completely overwhelm the weak interaction (7.13a). Furthermore, we shall be making non-relativistic approximations which are not valid for pions in this energy region. The treatment would apply to the case where the scattering of two nonrelativistic particles is dominated by a metastable state which decays into these two particles, i.e. a resonance.

We wish to describe pion–pion scattering in the center-of-mass system; i.e. in a state of total momentum zero. The only kaon state which is coupled to this system is the zero-momentum kaon state because of momentum conservation. To handle the local interaction most conveniently the pion creation operators are expressed in configuration space as a function of two coordinates y and z. We therefore look for a solution of the Schrödinger equation having the form

$$|\psi\rangle = \left| \int \alpha(y - z) a_\pi^\dagger(y) a_\pi^\dagger(z) \mathrm{d}y \mathrm{d}z + \beta a_{\mathrm{K}0}^\dagger \right| |0\rangle \tag{8.73}$$

where the coefficients $\alpha(y-z)$ and β are to be determined by substituting in the Schrödinger equation. For a zero-momentum state α depends only on the relative coordinate $(y-z)$ of two pions.

We now express the Hamiltonian (8.71) in terms of the variables appearing in the wave function (8.73), namely, the pion coordinates and the kaon momentum. The interaction V is expressed as function of coordinates. We therefore Fourier-transform the kaon operator to momentum space. We also write the pion operators at points y and z, as in the wave function in (8.73) and introduce a factor $\delta(y-z)$

$$V = (2\pi)^{-\frac{1}{2}} g \int \mathrm{d}y \int \mathrm{d}z \int \mathrm{d}q \, a_\pi^\dagger(y) a_\pi^\dagger(z) a_{\mathrm{K}q} \mathrm{e}^{\frac{1}{2}iq(y+z)} \delta(y - z) + \text{h.c.} \tag{8.74}$$

The kaon Hamiltonian (8.71b) is already in the desired form, but the pion Hamiltonian (8.71c) must be expressed in terms of configuration space variables. To simplify this calculation we assume the non-relativistic relation

$$E_{\pi q} = mc^2 + \frac{\hbar^2 q^2}{2m}. \tag{8.75}$$

This approximation allows the use of the non-relativistic Schrödinger equation in configuration space. The treatment is easily generalized to the case where the pions are described by the Klein–Gordon equation.

The pion Hamiltonian (8.71c) can now be written in configuration space

$$H_\pi = \frac{1}{2\pi} \int \mathrm{d}x \int \mathrm{d}x' \int \mathrm{d}q \left[mc^2 + \frac{\hbar^2 q^2}{2m} \right] \mathrm{e}^{-iqx} a_\pi^\dagger(x) \mathrm{e}^{iqx'} a_\pi(x'). \tag{8.76a}$$

This can be rewritten

$$H_\pi = \frac{1}{2\pi} \int \mathrm{d}x \, a_\pi^\dagger(x) \int \mathrm{d}x' \int \mathrm{d}q \left[mc^2 - \frac{\hbar^2}{2m} \frac{\mathrm{d}^2}{\mathrm{d}x^2} \right] \mathrm{e}^{iq(x'-x)} a_\pi(x'). \tag{8.76b}$$

Performing the integrations over q and x', we obtain

$$H_\pi = \int \mathrm{d}x \, a_\pi^\dagger(x) \int \mathrm{d}x' \left[mc^2 - \frac{\hbar^2}{2m} \frac{\mathrm{d}^2}{\mathrm{d}x^2} \right] \delta(x' - x) a_\pi(x') \tag{8.77a}$$

$$H_\pi = \int \mathrm{d}x \, a_\pi^\dagger(x) \left[mc^2 - \frac{\hbar^2}{2m} \frac{\mathrm{d}^2}{\mathrm{d}x^2} \right] a_\pi(x). \tag{8.77b}$$

We now substitute the wave function (8.73) into the Schrödinger equation. We first calculate $H_\pi|\psi\rangle$ using eqs. (8.77b) and (8.73).

$$H_\pi|\psi\rangle = \int \mathrm{d}x\, a_\pi^\dagger(x)\left[mc^2 - \frac{\hbar^2}{2m}\frac{\mathrm{d}^2}{\mathrm{d}x^2}\right] \times a_\pi(x)\int \mathrm{d}y \int \mathrm{d}z\, \alpha(y-z)a_\pi^\dagger(y)a_\pi^\dagger(z)|0\rangle. \tag{8.78a}$$

Moving the annihilation operator to the right and using boson commutation relations we obtain

$$H_\pi|\psi\rangle = \int \mathrm{d}x\, a_\pi^\dagger(x)\left[mc^2 - \frac{\hbar^2}{2m}\frac{\mathrm{d}^2}{\mathrm{d}x^2}\right] \times \int \mathrm{d}y \int \mathrm{d}z\, \alpha(y-z)[\delta(x-y)a_\pi^\dagger(z) + \delta(x-z)a_\pi^\dagger(y)]|0\rangle. \tag{8.78b}$$

The dummy integration variables y and z can be interchanged in the last term on the right-hand side of eq. (8.78b) and the two terms combined

$$H_\pi|\psi\rangle = \int \mathrm{d}x\, a_\pi^\dagger(x)\left[mc^2 - \frac{\hbar^2}{2m}\frac{\mathrm{d}^2}{\mathrm{d}x^2}\right] \times \int \mathrm{d}y \int \mathrm{d}z[\alpha(y-z) + \alpha(z-y)]\delta(x-y)a_\pi^\dagger(z)|0\rangle. \tag{8.78c}$$

Eq. (8.78c) illustrates the symmetry requirements of the pion Bose statistics. If $\alpha(y-z)$ is an odd function the two-pion component in the wave function (8.73) vanishes. Only the even part of $\alpha(y-z)$ has any physical significance. We therefore assume that $\alpha(y-z)$ is an even function. The integration of eq. (8.78c) over the variable y is trivial because of the $\delta(x-y)$.

$$H_\pi|\psi\rangle = \int \mathrm{d}x\, a_\pi^\dagger(x)\left[2mc^2 - \frac{\hbar^2}{m}\frac{\mathrm{d}^2}{\mathrm{d}x^2}\right]\int \mathrm{d}z\, \alpha(x-z)a_\pi^\dagger(z)|0\rangle. \tag{8.79}$$

To evaluate $V|\psi\rangle$ we note that

$$\begin{aligned} Va_{\mathrm{K}0}^\dagger|0\rangle &= (2\pi)^{-\frac{1}{2}}g \int \mathrm{d}y \int \mathrm{d}z \int \mathrm{d}q\, a_\pi^\dagger(y)a_\pi^\dagger(z)\mathrm{e}^{\frac{1}{2}\mathrm{i}q(y+z)}\delta(q)\delta(y-z)|0\rangle \\ &= (2\pi)^{-\frac{1}{2}}g \int \mathrm{d}y \int \mathrm{d}z\, a_\pi^\dagger(y)a_\pi^\dagger(z)\delta(y-z)|0\rangle \end{aligned} \tag{8.80a}$$

$$\begin{aligned} V\int \mathrm{d}y \int \mathrm{d}z\, \alpha(y-z)a_\pi^\dagger(y)a_\pi^\dagger(z)|0\rangle &= \\ &= 2(2\pi)^{-\frac{1}{2}}g^* \int \mathrm{d}y \int \mathrm{d}z \int \mathrm{d}y' \int \mathrm{d}z' \int \mathrm{d}q\, \delta(y-y')\delta(z-z')\alpha(y-z)\delta(y'-z') \\ &\times \mathrm{e}^{-\frac{1}{2}\mathrm{i}q(y+z)}\, a_{\mathrm{K}q}^\dagger|0\rangle \\ &= 2(2\pi)^{-\frac{1}{2}}g^* \int \mathrm{d}z \int \mathrm{d}q\, \alpha(0)\mathrm{e}^{-\mathrm{i}qz}\, a_{\mathrm{K}q}^\dagger|0\rangle = 2(2\pi)^{\frac{1}{2}}\alpha(0)g^* a_{\mathrm{K}0}^\dagger|0\rangle. \end{aligned} \tag{8.80b}$$

Thus

$$V|\psi\rangle = (2\pi)^{-\frac{1}{2}}\beta g \int dy \int dz\, a_\pi^\dagger(y) a_\pi^\dagger(z) \delta(y-z)|0\rangle$$
$$+ 2\alpha(0) g^*(2\pi)^{\frac{1}{2}} a_{K0}^\dagger|0\rangle. \tag{8.81}$$

We also have

$$H_K|\psi\rangle = \beta m_K c^2 a_{K0}^\dagger|0\rangle. \tag{8.82}$$

Using eqs. (8.79), (8.81) and (8.82), we can now write the Schrödinger equation for the solution (8.73).

$$(H-E)|\psi\rangle = 0 = \int dy \int dz\, [a_\pi^\dagger(y) a_\pi^\dagger(z)]|0\rangle$$
$$\times \left\{ \left[2mc^2 - \frac{\hbar^2}{m}\frac{d^2}{dy^2} - E \right] \alpha(y-z) + (2\pi)^{-\frac{1}{2}}\beta g \delta(y-z) \right\}$$
$$+ a_{K0}^\dagger|0\rangle\{(m_K c^2 - E)\beta + 2\alpha(0) g^*(2\pi)^{\frac{1}{2}}\}. \tag{8.83}$$

Equating separately the coefficients of the kaon and pion states we obtain the equations for the coefficients β and $\alpha(y-z)$

$$(m_K c^2 - E)\beta + 2\alpha(0) g^*(2\pi)^{\frac{1}{2}} = 0 \tag{8.84a}$$

$$\left[2mc^2 - \frac{\hbar^2}{m}\frac{d^2}{dy^2} - E \right] \alpha(y-z) + (2\pi)^{-\frac{1}{2}}\beta g \delta(y-z) = 0. \tag{8.84b}$$

Eliminating β between these two equations we obtain

$$\left[2mc^2 - \frac{\hbar^2}{m}\frac{d^2}{dx^2} - E \right] \alpha(x) = \frac{2|g|^2}{(m_K c^2 - E)} \delta(x)\alpha(0) \tag{8.85a}$$

where

$$x = y - z \tag{8.85b}$$

is the relative coordinate.

Eq. (8.85a) is just a Schrödinger equation for a particle of mass $\frac{1}{2}m$ (the reduced mass) in a delta function potential.

$$\left[\frac{d^2}{dx^2} + k^2 \right] \alpha = -U_0 \delta(x) \tag{8.86a}$$

where

$$k^2 = \frac{m}{\hbar^2}[E - 2mc^2] \tag{8.86b}$$

and

$$U_0 = \frac{2m|g|^2}{\hbar^2 [m_K c^2 - E]}. \tag{8.86c}$$

We have thus derived a one-body Schrödinger equation starting from the second-quantized Hamiltonian (8.71a). This is just the reverse of second quantization which begins with a one-body Schrödinger equation and develops a formalism capable of dealing with many particles and creation and annihilation. One might say that we have 'un-second-quantized' the Hamiltonian (8.71a) to get the one-body eq. (8.86).

From the solution of eq. (8.86) in section 8.6, we can write

$$\alpha(x) = \cos(k|x| + \delta_0) \tag{8.87a}$$

where the phase shift δ_0 is given by eq. (8.34a)

$$\tan \delta_0 = \frac{U_0}{2k} = \frac{m|g|^2}{\hbar^2 k[m_K c^2 - E]}. \tag{8.87b}$$

The scattering amplitude is given by eq. (8.36a)

$$f(\theta) = \frac{1}{k[2k/U_0 - \mathrm{i}]} = \frac{m|g|^2}{k\{\hbar^2 k[mc^2 - E] - \mathrm{i}m|g|^2\}}. \tag{8.88a}$$

This can be written as a Lorentzian resonance curve

$$f(\theta) = \frac{1}{k} \frac{\frac{1}{2}\Gamma}{(m_K c^2 - E) - \frac{1}{2}\mathrm{i}\Gamma} \tag{8.88b}$$

where the width Γ is

$$\Gamma = \frac{2m|g|^2}{\hbar^2 k}. \tag{8.88c}$$

This is just the value for the width obtained from the golden rule formula for the transition probability per unit time:

$$\frac{\Gamma}{\hbar} = \frac{2\pi}{\hbar} |g|^2 \varrho(E) \tag{8.89a}$$

where the density of states in one dimension in a box of unit volume is

$$\varrho(E) = \frac{1}{h} \frac{\mathrm{d}p}{\mathrm{d}E} = \frac{2m}{hp} = \frac{2m}{2\pi\hbar^2 k}. \tag{8.89b}$$

Thus the pion–pion scattering amplitude is given by a resonance curve whose width is just that calculated for the decaying state by time-dependent perturbation theory. Note, however, that the result (8.88c) is *exact*, as we have made no approximation in the solution of the Schrödinger equation for the given Hamiltonian.

8.11 Perturbation theory and the *T*-matrix

In one-dimensional scattering problems partial-wave analysis (8.18) expresses the exact solution in terms of two phase shifts. In three dimensions there are an infinite number of phase shifts and the method is practical only if a general closed form is found for all phase shifts or if it is possible to neglect the contributions from all partial waves above some value of l.

Another approach uses perturbation theory to solve the Schrödinger equation by a series of successive approximations. However, a straightforward application of the perturbation methods commonly used to treat bound-state problems leads to difficulties. Consider for example, the matrix elements of the potential V in a scattering problem between the exact solution (8.13) of the scattering problem and an unperturbed plane-wave state. Such matrix elements arise in perturbation theory and it is important to be able to evaluate them properly. From the Schrödinger equation,

$$\langle e^{\pm ikx}|V|\psi^{(+)}(x)\rangle = \langle e^{\pm ikx}|H - H_0|\psi^{(+)}(x)\rangle \tag{8.90a}$$

Since $\psi^{(+)}(x)$ is an eigenfunction of H with the eigenvalue $E=(\hbar k)^2/2m$

$$H|\psi^{(+)}(x)\rangle = E|\psi^{(+)}(x)\rangle = \frac{\hbar^2 k^2}{2m}|\psi^{(+)}(x)\rangle. \tag{8.90b}$$

Since $e^{\pm ikx}$ is an eigenfunction of the Hermitean operator H_0 with the same eigenvalue E

$$\langle e^{\pm ikx}|H_0 = E\langle e^{\pm ikx}| = \frac{\hbar^2 k^2}{2m}\langle e^{\pm ikx}|. \tag{8.90c}$$

Thus:

$$\langle e^{\pm ikx}|V|\psi^{(+)}(x)\rangle = \langle e^{\pm ikx}|\psi^{(+)}(x)\rangle(E - E) = 0. \tag{8.91}$$

This result is clearly incorrect. What is wrong? Try and figure this out before reading further!

Another example of this trouble appears in the following equation obtained by algebraic manipulation of the Schrödinger equation.

$$\Psi = \frac{1}{E - H_0} V\Psi. \tag{8.92}$$

This expression might be used for perturbation theory. With a zero-order wave function on the right-hand side, the first approximation is calculated from eq. (8.92) and higher approximations are obtained by iteration. However, there are eigenfunctions of H_0 with the eigenvalue E and the operator $1/(E-H_0)$ blows up for these states.

These difficulties arise from the use of wave functions which are plane waves extending over all space. These wave functions are not square integrable and their normalization integrals are infinite. One way to avoid these difficulties is by normalizing wave functions in a box of finite volume. This procedure gives a discrete energy level spectrum with the allowed values of k determined by the boundary conditions on the eigenfunctions (8.9) at the boundary of the box. Since these allowed values of k depend upon the phase shift, the eigenvalues E of the exact Hamiltonian H are different from those of the unperturbed Hamiltonian H_0. The expression (8.90) can be evaluated without inconsistencies and there are no vanishing denominators in eq. (8.92). However, the boundary conditions choose different values of k for the even and odd-parity states and remove the two-fold degeneracy of the eigenfunctions. The parity eigenfunctions provide a unique basis and there are no eigenfunctions describing a single incident wave and scattered waves. This is expected since the boundary conditions transform outgoing scattered waves into incoming waves which are re-scattered. Functions normalized in a box are thus not convenient for scattering calculations.

The use of a time-dependent description with wave packets confined to a finite region of space leads to a consistent formalism and corresponds more closely to physical scattering experiments performed in the laboratory. However, it is simpler to use the time-independent plane-wave states and the stationary eigenfunctions (8.13), with the normalization properly formulated.

We can avoid the above difficulties by replacing the usual orthonormality conditions for a discrete spectrum by the relation

$$\int_{-\infty}^{\infty} e^{-ik'x} e^{ik''x} dx = 2\pi\delta(k' - k''). \tag{8.93}$$

The Dirac delta function is a peculiar object which is zero everywhere except at the point $k' - k''$ where it is infinite. However this singular function is useful in calculations. It never appears in a result for a physically measurable quantity but only in integrands at intermediate stages. The singularity always disappears in an integration necessary to obtain the physical results. For example in using results calculated with plane waves to obtain physical results described by wave packets, the construction of a wave packet from plane-wave states involves just such an integral over the variables appearing in the argument of the delta function.

There are two kinds of contributions to matrix elements calculated with these plane-wave states: (1) Singular expressions like (8.93) where the infinite volume of space contributes to the integral and the result contains a delta function, (2) Non-singular integrals which are finite and vary smoothly with k.

These have contributions only from a finite volume, like the left-hand side of eq. (8.90a) or contributions from large distances which decrease sufficiently rapidly to make the integral converge. Both the singular and non-singular contributions give reasonable answers when integrated over the variable k.

We now resolve the paradox of eq. (8.91) and evaluate properly the integral defined by the right-hand side of eq. (8.90a). Using eq. (8.90b) *but not eq.* (8.90c), and substituting eq. (8.1) we obtain

$$\langle e^{\pm ikx}|V|\psi^{(+)}(x)\rangle = \int_{-X}^{+X} e^{\mp ikx}(H - H_0)\psi^{(+)}(x)\,dx$$

$$= \frac{\hbar^2}{2m}\int_{-X}^{+X} e^{\mp ikx}\left(k^2 + \frac{d^2}{dx^2}\right)\psi^{(+)}(x)\,dx. \quad (8.94a)$$

We must now be careful about allowing differential operators to operate to the left, as in eq. (8.90c) for these non-normalizable states. The standard method for making the differential operator in the expression (8.94a) operate to the left is integration by parts. Two integrations shift the differential operator to operate on the plane wave $e^{\pm ikx}$ and gives a contribution which exactly cancels the k^2-term to give zero in agreement with the naive result (8.91). However in each partial integration there is also the integrated part. This vanishes if the wave functions go to zero sufficiently rapidly at large distances, but plane-wave functions manifestly do not vanish at large distances. Therefore the integrated part is finite and in this case gives the entire result

$$\langle e^{\pm ikx}|V|\psi^{(+)}(x)\rangle = \frac{\hbar^2}{2m}\, e^{\mp ikx}\left(\frac{d\psi^{(+)}}{dx} \pm ik\psi^{(+)}(x)\right)\Bigg|_{-X}^{X}. \quad (8.94b)$$

Substituting eq. (8.13a) for $\psi^{(+)}(x)$ shows that the unperturbed plane wave gives an equal contribution at both limits and does not contribute to the expression (8.94b). The only contribution comes from the particular scattered wave which is in the same direction as the plane wave on the left-hand side of the matrix element. Using the form (8.19c) for the wave function, we obtain

$$\langle e^{ikx\cos\alpha}|V|\psi_\beta^{(+)}(x)\rangle = \frac{ik\hbar^2}{m}\, g(\alpha - \beta) \quad (8.95)$$

where $\alpha = 0$ or π and $\beta = 0$ or π.

The matrix elements of V between all possible scattering states (8.19) and all possible plane-wave states are now defined. However this is not the usual expression for the matrix element of an operator since it is defined between states expressed in different bases.

The result (8.95) can be expressed in the single basis of the unperturbed plane-wave states by defining a new operator. The T-matrix is defined for this purpose by the relation

$$V\psi_\alpha^{(+)} = T\phi_\alpha \tag{8.96a}$$

where ϕ_α is the unperturbed plane-wave state corresponding to the exact solution $\psi_\alpha^{(+)}$

$$\phi_\alpha = e^{ikx\cos\alpha}. \tag{8.96b}$$

The matrix elements of T are directly obtained from eq. (8.95)

$$\langle\phi_\beta|T|\phi_\alpha\rangle = \frac{ik\hbar^2}{m} g(\alpha - \beta). \tag{8.97}$$

The T-matrix gives a complete description of the scattering since all the information about the scattering process is contained in the elements of the T-matrix. The S and T-matrices can be related by comparing eqs. (8.97) and (8.20):

$$S_{\alpha\beta} = \delta_{\alpha\beta} + \frac{m}{ik\hbar^2} T_{\alpha\beta}. \tag{8.98}$$

The T-matrix, as defined by eq. (8.98) has elements formally defined also between states having different energies, while the S-matrix as defined by eq. (8.20) connects only states of the same energy. Eqs. (8.97) and (8.98) apply only to matrix elements between states of the same energy. The T-matrix vanishes if there is no potential; i.e. no scattering, whereas the S-matrix becomes the unit matrix in that case. In the three-dimensional case the numerical factor (8.98) in the relation between S and T-matrices is different owing to the different dimensions of the various quantities.

An other expression for the S-matrix is obtained by evaluating matrix elements between the states (8.20b) with scattered outgoing waves and (8.22a) with scattered incoming waves. These have a singular part which comes from the region outside the potential.

$$\begin{aligned}\langle\psi_\gamma^{(-)}|\psi_\alpha^{(+)}\rangle = \sum_{\theta=0,\pi} \int_0^\infty \mathrm{d}r[\sum_{\beta=1,2} \phi_\beta^*(\pi-\theta)S_{\beta\gamma}e^{ikr} + \phi_\gamma^*(\theta)e^{-ikr}] \\ \times [\phi_\alpha(\pi-\theta)e^{-ikr} + \sum_{\beta=1,2} S_{\alpha\beta}\phi_\beta(\theta)e^{ikr}] = S_{\alpha\gamma} \times 2\int_0^\infty \mathrm{d}r.\end{aligned} \tag{8.99}$$

The divergent integral can be expressed as a delta function by using states $\psi_\alpha^{(+)}$ and $\psi_\gamma^{(-)}$ with slightly different values of k and applying eq. (8.93).

Thus the S-matrix element for the transition $\alpha \rightarrow \gamma$ is given by the matrix element between the solution with outgoing scattered waves and the one with incoming scattered waves with unperturbed states α and γ respectively.

8.12 Green's functions and the Born series

We now return to the formulation of perturbation theory for the scattering problem and the impossibility of using eq. (8.92) as it stands. Perturbation theory gives an approximation to the exact solution of the Schrödinger equation starting from a plane wave as the unperturbed solution. However, we do not want *any* exact solution of the Schrödinger equation. We want a solution having the form (8.20b) with outgoing waves added to the unperturbed wave. We do not want solutions like (8.22a) containing *incoming* waves added to the unperturbed wave. In the absence of the potential when $g(\theta)=0$ and $S_{\alpha\beta}=\delta_{\alpha\beta}$ eqs. (8.20b) and (8.22a) become identical. The same is true for any linear combination of (8.20b) and (8.22a). Thus *a given unperturbed solution* of the free-particle Schrödinger equation can correspond to several different solutions of the Schrödinger equation when the interaction is present, not only the desired one in which the interaction produces additional outgoing waves but also others in which it produces incoming waves. A one-to-one correspondence does not exist.

Perturbation theory is thus ambiguous if it simply attempts to find *a* solution of the perturbed Schrödinger equation which corresponds to a particular solution of the unperturbed Schrödinger equation. This ambiguity is reflected in the presence of the denominator in eq. (8.92) which can vanish and therefore does not define a perturbation expansion. To obtain a useful perturbation formulation we must re-define our problem as finding a solution of the perturbed Schrödinger equation which not only corresponds to a particular unperturbed state but is also restricted to having only additional outgoing waves. This problem does not arise for the case where the wave functions are normalized in a box. There a one-to-one correspondence does exist between perturbed and unperturbed solutions and eq. (8.92) has no vanishing denominators.

The desired perturbation formulation can be developed with the aid of the solution for the delta function potential already considered. Substituting the potential (8.32) and the solution (8.36) into the Schrödinger equation (8.30) we obtain

$$\frac{\hbar^2}{2m}\left[\frac{\mathrm{d}^2}{\mathrm{d}x^2} + k^2 + U_0\delta(x)\right]\left[\mathrm{e}^{ikx} + \frac{\mathrm{i}U_0}{2k - \mathrm{i}U_0}\,\mathrm{e}^{ikr}\right] = 0. \quad (8.100\mathrm{a})$$

This can be rewritten

$$\frac{\hbar^2}{2m}\left[\frac{d^2}{dx^2}+k^2\right]e^{ikr}=\frac{i\hbar^2 k}{m}\,\delta(x). \tag{8.100b}$$

This expression can be used for a delta function potential at any point x', not necessarily at the origin, by substituting $x-x'$ for x in equation (8.100b)

$$\frac{\hbar^2}{2m}\left[\frac{d^2}{dx^2}+k^2\right]e^{ik|x-x'|}=\frac{i\hbar^2 k}{m}\,\delta(x-x'). \tag{8.100c}$$

It is now convenient to define the Green's functions

$$G^{\pm}(x,x')=\frac{m}{i\hbar^2 k}\,e^{\pm ik|x-x'|}. \tag{8.101a}$$

From eq. (8.100c) the functions $G^{\pm}(x,x')$ are seen to satisfy the equation

$$\frac{\hbar^2}{2m}\left[\frac{d^2}{dx^2}+k^2\right]G^{\pm}(x,x')=\delta(x-x'). \tag{8.101b}$$

With an intuitive picture of a potential $V(x)$ built up of a continuum of delta function potentials at each point x', with strength $V(x')$, we can write the Schrödinger equation in the form:

$$\frac{\hbar^2}{2m}\left[\frac{d^2}{dx^2}+k^2\right]\psi(x)=V(x)\psi(x)=\int_{-\infty}^{+\infty}V(x')\delta(x-x')\psi(x')\,dx'. \tag{8.102a}$$

We can now write formal solutions of the Schrödinger equation by using the Green's functions (8.101b) which are solutions for the delta function potential. The function

$$\psi_\alpha^{(\pm)}(x)=\phi_\alpha+\int_{-\infty}^{+\infty}G^{\pm}(x,x')V(x')\psi_\alpha^{(\pm)}(x')\,dx' \tag{8.102b}$$

satisfies the Schrödinger equation (8.30b) if ϕ_α is any solution of the free-particle Schrödinger equation. This is sometimes described by considering eq. (8.30b) as an inhomogeneous equation. Adding any solution of the homogeneous equation to a given solution of the inhomogenous equation gives another solution.

However, eq. (8.30b) is not really inhomogenous as the 'inhomogeneous term' contains a factor ψ. Thus eq. (8.102b) is not really a *solution* of the equation, as ψ appears on the right-hand side as well as on the left. For a delta function potential, eq. (8.102b) gives the solution, because ψ at one point is a single number, determined later by normalization. For any other potential, $\psi(x)$ appears as an unknown function whose value can only be found by solving the whole equation. Thus in deriving eq. (8.102b) we have converted the Schrödinger equation to an integral equation which still must be solved to get $\psi(x)$.

This integral equation (8.102b) differs from the Schrödinger equation in a very useful way. All solutions of the integral equation are also solutions of the Schrödinger equation. However, not all solutions of the Schrödinger equation are solutions of the integral equation. We have chosen eq. (8.102b) to give only those solutions which have the desired properties at large values of x. They consist of an unperturbed wave having the given form ϕ_α plus outgoing scattered waves and no incoming waves if $G^+(x, x')$ is used and the reverse if $G^-(x, x')$ is used. This has been achieved by choosing the first term on the right-hand side of eq. (8.102b) to be the desired unperturbed wave and by choosing the Green's function in the integral to be that solution of eq. (8.101b) which has only outgoing or only incoming waves. Other solutions of the Schrödinger equation could be obtained by replacing the first term in eq. (8.102b) by any solution of the free-particle Schrödinger equation and by replacing the Green's function $G^\pm(x, x')$ by any solution of eq. (8.101b).

Eq. (8.102b) is a possible starting point for a perturbation treatment. We have removed the ambiguity implied by eq. (8.92) which cannot choose between the solutions (8.20b) and (8.22a). Consider the expansion of the wave function (8.102) in a perturbation theory. We consider $\psi^{(+)}(x)$, choose the zero order term ϕ_α to be unperturbed wave e^{ikx}, and define ψ_n to be the nth-order term in the perturbation series

$$\psi^{(+)}(x) = e^{ikx} + \psi_1 + \psi_2 + \dots \psi_n + \dots. \tag{8.103}$$

The entire perturbation series can be obtained from eq. (8.102b) by iteration. The substitution of the zero-order term into the integral on the right-hand side of eq. (8.102b) gives on the left-hand side the sum of the zero-order term and a first-order term which is proportional to the potential. The substitution of this function into the integral on the right-hand side gives back the unperturbed solution plus the same first-order solution and a second-order term which is proportional to the square of the potential. Each iteration gives the next higher term in the perturbation expansion. The approximate solution in each order has the desired asymptotic properties, the unperturbed plane wave plus outgoing scattered waves. This iteration is called the Born series for the scattering amplitude. The first term which is the result of first-order perturbation theory is called the Born approximation. This result is exactly the same as would be obtained by using the golden rule and first-order time-dependent perturbation theory.

The above one-dimensional discussion is presented as a simplification of the physical three-dimensional problems in order to give an intuitive picture with a minimum of complicated mathematics. In this connection it is worth mentioning some differences between the one and three-dimensional cases which follow from purely dimensional considerations. The analogs of the Green's function equations (8.100c) and (8.101) involve a delta function in three-dimensional space and outgoing or incoming waves in three dimensions which have the form $e^{\pm ikr}/r$. The delta function in a one-dimensional space has dimensions $1/r$ while the free-particle plane wave is dimensionless. In three dimensions, the delta function has dimensions $1/r^3$ and the free-particle solution has dimension $1/r$. The factor k on the right-hand side of eq. (8.100c) is required by dimensional considerations for the one-dimensional case but not for the three-dimensional case. Thus the momentum dependence in the definition of the one-dimensional Green's function (8.101) disappears in the three-dimensional case.

We now return to the formal perturbation expression (8.92) and consider a simple modification which allows us to remove its difficulties and ambiguities and to obtain a formal expression equivalent to the integral equation (8.102).

We write the product $V\psi$ in the form indicated by eq. (8.102a) and express the delta function as a Fourier transform

$$V(x)\psi(x) = \int_{-\infty}^{\infty} dx'\,\delta(x-x')V(x')\psi(x')$$

$$= \frac{1}{2\pi}\int_{-\infty}^{\infty} dk' \int_{-\infty}^{\infty} dx'\,e^{ik'(x-x')}V(x')\psi(x'). \tag{8.104}$$

We can write

$$\frac{1}{E-H_0}\,e^{ik'x} = \frac{2m}{\hbar^2(\kappa^2-k'^2)}\,e^{ik'x} \tag{8.105a}$$

where

$$\kappa^2 = \frac{2m}{\hbar^2}E. \tag{8.105b}$$

Substituting eqs. (8.104) and (8.105) into the right-hand side of eq. (8.92) we obtain

$$\frac{1}{E - H_0} V(x)\psi(x) = \frac{m}{\pi\hbar^2} \int_{-\infty}^{+\infty} \mathrm{d}x' \int_{-\infty}^{+\infty} \mathrm{d}k' \frac{\mathrm{e}^{\mathrm{i}k'(x-x')}}{(\kappa - k')(\kappa + k')} V(x')\psi(x'). \qquad (8.106a)$$

The difficulty due to the singular dominator $E - H_0$ is still present and appears in the denominators of the integrand on the right-hand side. However, if we add to E a small imaginary part, which we later allow to go to zero, we can evaluate the integral. The integrand remains finite along the path of integration as soon as the poles are moved off the real axis. Since the integrand is analytic in the complex k'-plane except for poles at $k' = \pm\kappa$, the integral over k' is conveniently evaluated by contour integration, closing the contour with the addition of a semicircle at infinity. The contribution of the semicircle to the integral vanishes if the semicircle is in the upper half plane for positive values of $x - x'$ and in the lower half plane for negative values of $x - x'$. The contour integrals are then evaluated by taking the residues at the two poles $k' = \kappa$ and $k' = -\kappa$. One of these is in the upper half plane and contributes to the integral for positive $x - x'$. The other is in the lower half plane and contributes to the integral for negative $x - x'$. Which of the two poles $\pm\kappa$ is in the upper and which is in the lower half plane depends on the sign of the imaginary part of κ and not on its magnitude. We thus obtain two possible values for the integral

$$\frac{1}{E - H_0} V(x)\psi(x) = \frac{m}{\mathrm{i}\hbar^2} \int_{-\infty}^{\infty} \mathrm{d}x' \frac{\mathrm{e}^{\mathrm{i}\kappa|x-x'|}}{\kappa} V(x')\psi(x') \quad \text{if} \quad \operatorname{Im}\kappa > 0 \qquad (8.106b)$$

$$\frac{1}{E - H_0} V(x)\psi(x) = \frac{m}{\mathrm{i}\hbar^2} \int_{-\infty}^{\infty} \mathrm{d}x' \frac{\mathrm{e}^{-\mathrm{i}\kappa|x-x'|}}{\kappa} V(x')\psi(x') \quad \text{if} \quad \operatorname{Im}\kappa < 0. \qquad (8.106c)$$

The integrands on the right-hand sides of eq. (8.106b) and (8.106c) contain just the Green's functions $G^{\pm}(x, x')$ defined by eq. (8.101a). Thus eq. (8.106) can be rewritten:

$$\lim_{\varepsilon\to 0} \frac{1}{E - H_0 \pm i\varepsilon} V(x)\psi(x) = \int_{-\infty}^{\infty} dx' G^{\pm}(x, x')V(x')\psi(x'). \quad (8.107)$$

Substituting this result into eq. (8.102b), we obtain

$$\psi_\alpha^{(\pm)}(x) = \phi_\alpha(x) + \lim_{\varepsilon\to 0} \frac{1}{E - H_0 \pm i\varepsilon} \psi_\alpha^{(\pm)}(x). \quad (8.108)$$

This is the formal operator representation of the integral equation (8.102b).

We have 'fixed up' eq. (8.92) by adding a small imaginary part to E, which is allowed to approach zero, and adding a 'solution of the homogeneous equation' $(E - H_0)\psi = 0$. The sign of the imaginary part determines whether the solution with incoming or outgoing scattered waves is chosen.

Eq. (8.108) can be used to obtain an integral equation for the T-matrix. Multiplying the equation for $\psi_\alpha^{(+)}$ by V and using the definition (8.96a) of T, we obtain

$$V\psi_\alpha^{(+)}(x) = T\phi_\alpha(x) = V\phi_\alpha(x) + V \frac{1}{E - H_0 + i\varepsilon} T\phi_\alpha(x) \quad (8.109)$$

where we have used the notation common in scattering theory, of implying the limit $\varepsilon \to 0$ without writing it explicitly.

The integral equation for the matrix elements of T is obtained by taking the scalar product of eq. (8.109) with any unperturbed state ϕ_β. We thus obtain the operator equation

$$T = V + V \frac{1}{E - H_0 + i\varepsilon} T. \quad (8.110)$$

The T-matrix is defined by eq. (8.96a) to have matrix elements between states which do not have the same energy as well as states which have the same energy. Only the matrix elements between states which have the same energy are given by eq. (8.97) and related to physically observable scattering amplitudes. However, the operator relation (8.110) must involve also matrix elements of T between states of different energies, since V has such matrix elements. We have seen that the evaluation of the integrand, eq. (8.106) involves other values of k'.

Eq. (8.110) also provides a basis for a perturbation calculation of the T-matrix. The unperturbed value of T is $T = 0$. Substitution of this value on the right-hand side of eq. (8.110) gives the first-order result $T = V$, which is just the Born approximation. Continued iteration gives the higher-order results.

The relation between eqs. (8.108) and (8.102b) is characteristic of two approaches to quantum mechanics. The analytical approach, which began with Schrödinger, uses differential and integral equations and other analytical mathematical tools. The algebraic approach, which began with Heisenberg, uses matrices and vectors and operators in Hilbert space, and other algebraic mathematical tools. The physical results obtained by the two methods are always equivalent. Once a result is obtained with one approach, it is easy to translate the work into the other language, and obtain the same result. However, the best method to use in a given case depends upon the problem and on the taste of the individual who is trying to solve it. It is good for a student to be familiar with both approaches. Then, when he encounters a difficulty in one approach, he can translate the problem into the other and perhaps see it more clearly. In the present example, the difficulty was encountered in the algebraic statement, eq. (8.92), clarified by going to the analytical approach to obtain eq. (8.102b) and then 'translated' back into algebra to obtain the equivalent expression (8.107).

8.13 A particle in a one-dimensional periodic potential

The phase shifts used in scattering problems can also be used to treat the problem of a particle moving in a one-dimensional periodic potential. Let us modify the scattering problem (8.6) by adding an infinite number of potentials, identical to the potential $V(x)$, but centered at periodic intervals with period $a \geqq 2X$. This new periodic potential is

$$V_{\mathrm{p}}(x) = \sum_{n=-\infty}^{n=+\infty} V(x - na) \tag{8.111a}$$

and the Hamiltonian is

$$H = \frac{p^2}{2m} + V_{\mathrm{p}}(x). \tag{8.111b}$$

Such a Hamiltonian is often used to describe the motion of electrons in a crystal, where the ions produce a periodically varying electric field.

The Hamiltonian H is invariant under the translation $x \to x+a$. Thus if $\psi(x)$ is an eigenfunction of H, $\psi(x+a)$ is also an eigenfunction with the same eigenvalue.

$$H\psi(x) = E\psi(x), \tag{8.112a}$$

$$H\psi(x + a) = E\psi(x + a). \tag{8.112b}$$

If the energy value E is non-degenerate, the two solutions (8.112a) and (8.112b) describe the same state and can differ only by a factor. This factor must be of modulus unity since $\psi(x+a)$ and $\psi(x)$ have the same normalization. Thus

$$\psi(x + a) = e^{i\phi}\psi(x). \tag{8.113}$$

A function satisfying this condition can be written in the following form

$$\psi_k(x) = e^{ikx}u_k(x) \tag{8.114a}$$

where

$$\phi = ka \tag{8.114b}$$

and $u_k(x)$ is a periodic function with period a,

$$u_k(x + a) = u_k(x). \tag{8.114c}$$

In this form (8.114) the wave function is seen to be the product of a free-particle wave function and a periodic function which expresses the effect of the potential. These wave functions are called Bloch waves and are used in the description of conduction electrons in a lattice.

If the potential is an even function of x, i.e. invariant under a reflection about the origin,

$$V(x) = V(-x). \tag{8.115}$$

The Hamiltonian (8.111b) and the parity P can therefore be simultaneously diagonalized. From our experience with the scattering problem, we expect a two-fold degeneracy of eigenfunctions,* and that one can choose parity eigenfunctions which describe standing waves, or the Bloch wave functions (8.114a) which describe traveling waves.

We can now investigate the properties of these Bloch waves by using an approach similar to the one used in scattering problems, namely by using phase shifts to describe the effect of the potential on the wave functions. If we know a Bloch wave solution in the interval $-\frac{1}{2}a \leqq x \leqq \frac{1}{2}a$, and we know the phase factor ϕ or the Bloch wave number k, then we can determine the solution everywhere by using the periodicity condition (8.113) or the equivalent relation (8.114). We therefore only have to solve the Schrödinger equation in the region $|x| \leqq \frac{1}{2}a$. But for $|x| \leqq \frac{1}{2}a$ the periodic potential is identical to that of the scattering problem, eq. (8.6), and the solutions of the scattering problem can be used for the periodic potential.

* If the eigenvalue E is degenerate, the eigenfunctions are not required to have the form (8.113) or (8.114). However, we can define an operator which performs the translation $x \to x + a$, and which commutes with the Hamiltonian (8.111b). This is discussed in detail in ch. 14. Thus, it is always possible to choose a basis of eigenfunctions of H which are also eigenfunctions of the translation operator and have the form (8.114).

If we substitute a solution of the scattering problem for the region $|x|\leqq\frac{1}{2}a$ into the periodicity condition (8.113) we obtain a solution of the Schrödinger differential equation for the periodic potential, but it will generally be discontinuous at the points $x=(n+\frac{1}{2}a)$. To construct an acceptable solution, we must find a solution of the Schrödinger equation in the region $|x|\leqq\frac{1}{2}a$ which also satisfies the periodicity condition (8.113) as boundary conditions on the function and on its derivative at the points $x=\pm\frac{1}{2}a$. Such a function will be continuous and have a continuous derivative when it is extended outside the region $|x|\leqq\frac{1}{2}a$ by use of the periodicity condition (8.113).

The periodic boundary condition is conveniently formulated by writing the Bloch wave solution as a linear combination of parity eigenfunctions.

$$\psi(x) = \psi_0(x) + \psi_1(x) \tag{8.116a}$$

where $\psi_0(x)$ and $\psi_1(x)$ are eigenfunctions of H with even and odd parity respectively and are not normalized. Then

$$\psi(-x) = \psi_0(x) - \psi_1(x). \tag{8.116b}$$

The derivative of this solution is

$$\psi'(x) = \psi_0'(x) + \psi_1'(x), \tag{8.117a}$$

$$\psi'(-x) = -\psi_0'(x) + \psi_1'(x). \tag{8.117b}$$

We now impose the periodicity condition (8.113) on $\psi(x)$ and $\psi'(x)$ at the points $x=\pm\frac{1}{2}a$

$$\frac{\psi(+\frac{1}{2}a)}{\psi(-\frac{1}{2}a)} = \frac{\psi_0 + \psi_1}{\psi_0 - \psi_1} = \mathrm{e}^{ika} \tag{8.118a}$$

$$\frac{\psi'(+\frac{1}{2}a)}{\psi'(-\frac{1}{2}a)} = \frac{\psi_0' + \psi_1'}{-\psi_0' + \psi_1'} = \mathrm{e}^{ika}, \tag{8.118b}$$

where we have used the shortened notation ψ_0, ψ_0', ψ_1 and ψ_1' to denote the values of these functions at $x=\frac{1}{2}a$. Combining these equations gives the relations

$$\frac{\psi_0 + \psi_1}{\psi_0 - \psi_1} - \frac{\psi_0' + \psi_1'}{-\psi_0' + \psi_1'} = 0 = \psi_1\psi_1' - \psi_0\psi_0' \tag{8.119a}$$

$$\cos ka = \frac{\mathrm{e}^{ika} + \mathrm{e}^{-ika}}{2} = \frac{1}{2}\left[\frac{\psi_0 + \psi_1}{\psi_0 - \psi_1} + \frac{\psi_1' - \psi_0'}{\psi_1' + \psi_0'}\right] = \frac{\psi_0\psi_1' + \psi_1\psi_0'}{\psi_0\psi_1' - \psi_1\psi_0'}, \tag{8.119b}$$

where we have used eq. (8.119a) to simplify eq. (8.119b).

Eqs. (8.119) express the boundary conditions on the solution of the Schrödinger equation in the region $|x| \leqq \frac{1}{2}a$ in terms of the even and odd-parity eigenfunctions. These can be taken from the scattering solution, since they are the same as those for the periodic potential in this interval. Thus from eq. (8.9), which is valid for $|x| \geqq \frac{1}{2}a$ and therefore for $|x| = \frac{1}{2}a$,

$$\psi_0 = A \cos(\tfrac{1}{2}ka + \delta_0) \tag{8.120a}$$

$$\psi_1 = B \sin(\tfrac{1}{2}ka + \delta_1) \tag{8.120b}$$

$$\psi_0' = -\kappa A \sin(\tfrac{1}{2}ka + \delta_0) \tag{8.120c}$$

$$\psi_1' = \kappa B \cos(\tfrac{1}{2}ka + \delta_1) \tag{8.120d}$$

where A and B are coefficients to be determined and

$$\kappa = (2mE/\hbar)^{\frac{1}{2}}. \tag{8.120e}$$

The wave number κ is to be distinguished from the Bloch wave number k defined by eq. (8.114). For a free particle $\kappa = k$. In the presence of a periodic potential, the boundary conditions (8.119) determine the relation between κ and k; i.e. between the energy and the momentum.

We can now substitute the values (8.120) into the boundary conditions (8.119). We first note that

$$2\psi_0\psi_0' = -\kappa a A^2 \sin(\kappa a + 2\delta_0) \tag{8.121a}$$

$$2\psi_1\psi_1' = \kappa B^2 \sin(\kappa a + 2\delta_1) \tag{8.121b}$$

$$2\psi_0\psi_1' = \kappa AB \cos(\kappa a + \delta_0 + \delta_1) + \cos(\delta_0 - \delta_1) \tag{8.121c}$$

$$2\psi_0'\psi_1 = \kappa AB \cos(\kappa a + \delta_0 + \delta_1) - \cos(\delta_0 - \delta_1). \tag{8.121d}$$

Substituting eqs. (8.121) into eqs. (8.119) we obtain

$$B^2 \sin(\kappa a + 2\delta_1) + A^2 \sin(\kappa a + 2\delta_0) = 0. \tag{8.122a}$$

$$\cos ka = \frac{\cos(\kappa a + \delta_0 + \delta_1)}{\cos(\delta_0 - \delta_1)}. \tag{8.122b}$$

This can be rewritten

$$\cos(\kappa a + \delta_0 + \delta_1) = \cos \kappa a \cos(\delta_0 - \delta_1). \tag{8.122c}$$

The left-hand sides of both equations (8.122b) and (8.122c) are cosine functions restricted to be less than unity. The right-hand side of eq. (8.122c) is a product of two cosine functions with the same restrictions. Thus for any given Bloch wave number k there always exists a value of κ from eq. (8.122c) and a solution of the Schrödinger equation. All values are thus allowed for the Bloch wave number k which has a continuous spectrum analogous to the free electron momentum.

The right-hand side of eq. (8.122b) is the quotient of two cosine functions. It is greater than unity for some values of κ and eq. (8.122b) then has no solution for k. The energy spectrum thus has a band structure with discontinuities and forbidden regions whenever the right-hand side of eq. (8.122b) becomes greater than unity. If the value of κ or of the energy is plotted against the Bloch wave number k the curve has discontinuities in κ or E at a fixed value of k.

8.14 Some properties of the band spectrum

Eqs. (8.122) have been obtained from general symmetry arguments without assumptions about the detailed shape of the potential $V(x)$, because the boundary conditions (8.199) depend on the solutions of the Schrödinger equation in the interior only through two parameters, the logarithmic derivatives of the even and odd-parity solutions at $x=\frac{1}{2}a$. These parameters are directly related to the scattering phase shifts. Thus the conditions (8.122) which determine the band structure are expressed completely in terms of these phase shifts and do not depend upon other details of the solution of the Schrödinger equation or of the potential.

We now use these results to determine specific properties of the band spectrum. The function $\cos(\kappa a+\delta_0+\delta_1)$ oscillates between -1 and $+1$. At the extrema of these oscillations when it is equal to ± 1, eq. (8.122b) has no solution for k if $\delta_0 \neq \delta_1$. The forbidden regions around these extrema are seen from eq. (8.122b) to be defined by the condition

$$-|\delta_0 - \delta_1| < (\kappa a + \delta_0 + \delta_1 - n\pi) < |\delta_0 - \delta_1| \qquad (8.123a)$$

where n is an integer chosen to minimize the absolute value of $\kappa a+\delta_0+$ $+\delta_1-n\pi$.

At the boundaries of the forbidden regions (8.123a) eqs. (8.122) show that the solutions have the following properties

$$B=0 \quad \text{and} \quad ka = n\pi = (\kappa a + \delta_1 + \delta_0) + (\delta_0 - \delta_1)$$
$$\text{when} \quad \kappa a = n\pi - 2\delta_0 \qquad (8.231b)$$

$$A=0 \quad \text{and} \quad ka = n\pi = (\kappa a + \delta_1 + \delta_0) - (\delta_0 - \delta_1)$$
$$\text{when} \quad \kappa a = n\pi - 2\delta_1. \qquad (8.123c)$$

The forbidden regions thus occur whenever $ka = n\pi$ and appear as jump in the value of κa between the values $n\pi - 2\delta_0$ and $n\pi - 2\delta_1$. At the limits of the forbidden bands the Bloch waves are also parity eigenstates having odd parity when $\kappa a = n\pi - 2\delta_0$ and even parity when $\kappa a = n\pi - 2\delta_1$.

When $\kappa a + \delta_0 + \delta_1 = (n + \frac{1}{2})\pi$, eq. (8.122c) vanishes. Thus

$$ka = (n + \tfrac{1}{2})\pi = \kappa a + \delta_0 + \delta_1 \quad \text{when} \quad \kappa a = (n + \tfrac{1}{2})\pi - \delta_0 - \delta_1. \tag{8.123d}$$

In the neighborhood of this point, eq. (8.122c) gives the inequality

$$|\kappa a + \delta_0 + \delta_1 - (n + \tfrac{1}{2})\pi| \leqq |ka - (n + \tfrac{1}{2})\pi|. \tag{8.123e}$$

The energy spectrum of the Bloch waves thus consists of alternating allowed and forbidden bands. The forbidden bands occur at the Bloch wave numbers $ka = n\pi$ where there are an integral number of half wavelengths in a lattice period. A plot of $\kappa a + \delta_0 + \delta_1$ vs. ka thus has the form shown in fig. 8.1. The points at $ka = (n + \frac{1}{2})\pi$ lie on the dotted line $\kappa a + \delta_0 + \delta_1 = ka$, as indicated by eq. (8.123d), but the slope of the curve is less than unity as indicated by the inequality (8.123e). At $ka = n\pi$, the curve jumps from a point $|\delta_0 - \delta_1|$ below the dotted line to a point $|\delta_0 - \delta_1|$ above. The forbidden values of $\kappa + \delta_0 + \delta_1$, lie between these two values.

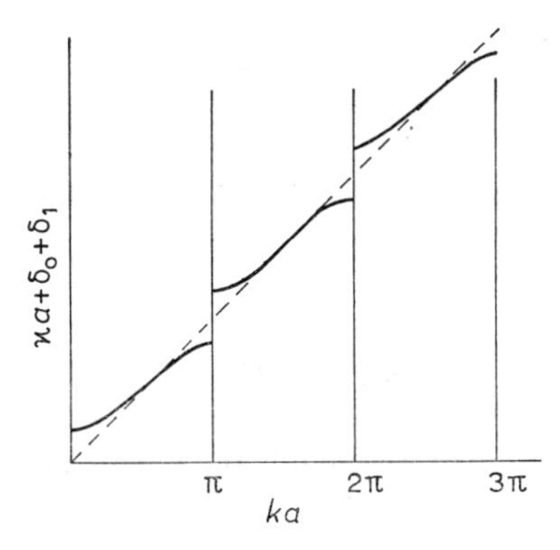

Fig. 8.1. Typical band spectrum for a periodic potential.

This band structure has been shown to be responsible for the classification of solids as conductors and insulators. Conduction occurs when it is easy to move electrons. In the absence of an applied electric field, there is no net current as the numbers of electrons moving in opposite directions are equal. An applied field can disturb this balance by putting more electrons in states moving in the direction of the field than in the reverse direction, provided that energy levels with a very small excitation energy are available for these electrons. In a system of free electrons there is a continuous spectrum and levels are always available at a very small excitation above the highest filled level. However, in a periodic potential with a band spectrum no very small excitations are possible if the electrons exactly fill a set of allowed energy bands. The lowest excitations then involve raising the energy of electrons across the forbidden region. This is not possible with a weak external field. Thus solids in which the number of electrons just fills the set of allowed band are insulators; those in which the highest band is only partly filled are conductors.

8.15 Lattice dynamics. Bloch electrons and phonons

In a real crystal the potential that an electron sees is not strictly periodic. The ions may be displaced slightly from their equilibrium positions in the lattice. We can take this into account by modifying the potential (8.111a) to allow for small displacements ξ_n of each ion from its equilibrium position

$$V(x, \xi_n) = \sum_n V(x - na - \xi_n) \approx V_p(x) - \sum_n \xi_n \frac{\partial V(x - na)}{\partial x}. \tag{8.124}$$

Since these displacements are small we can expand the potential in a Taylor's series and keep only first order terms.

The displacements ξ_n are not given parameters but dynamical variables whose motion should be described in our Hamiltonian. We thus generalize the Hamiltonian to include three parts describing the electron, the ions and the interaction between the electron and ions.

$$H = H_e + H_I + H_{int}. \tag{8.125a}$$

In the electron Hamiltonian we include the kinetic energy and the first term of the right-hand side of eq. (8.124).

$$H_e = \frac{p^2}{2m} + V_p(x). \tag{8.125b}$$

The ion Hamiltonian consists of the sum of the kinetic energies of all the ions and the potentials of the interatomic forces. Since the displacements ξ_n are small, we expand the interatomic potential in powers of ξ_n and keep only terms to second order. The constant term is removed by changing the zero of the energy. Linear terms are absent since the variables ξ_n are measured from their equilibrium position. Thus,

$$H_{\mathrm{I}} = \sum_n \frac{\pi_n^2}{2M} + \sum_{nn'} C_{nn'}\xi_n\xi_{n'} \tag{8.125c}$$

where π_n is the momentum canonically conjugate to ξ_n, M is the ion mass, and $C_{nn'}$ is the appropriate coefficient in the expansion of the interionic potential. Since the potential depends only upon the distance between the interacting ions and not on the position of the individual ions in the lattice, $C_{nn'}$ depends only on the difference $n-n'$.

The electron–ion interaction contains only one term in the approximation we are considering; namely the second term on the right-hand side of eq. (8.124)

$$H_{\mathrm{int}} = -\sum_n \xi_n \frac{\partial V}{\partial x}(x - na). \tag{8.125d}$$

Although this term was initially introduced to describe the effect of the displacement of the ion on the potential seen by the electron, the same interaction also affects the motion of the ion. The electric force between an electron and the positive ions affects the motion of the ions as well as the motion of the electron.

Although the periodic symmetry in the electron motion has been broken, there is an overall periodic symmetry for the whole lattice. We now define the transformation on the dynamical variables induced by a translation by a lattice period a

$$x \to x + a, \qquad p \to p \tag{8.126a}$$

$$\xi_n \to \xi_{n-1}, \qquad \pi_n \to \pi_{n-1}. \tag{8.126b}$$

The Hamiltonian (8.125) is seen to be invariant under this transformation since $C_{nn'}$ depends only on $n-n'$.

In order to proceed further, we consider a finite number of ions and use periodic boundary conditions in order to keep the invariance under the transformation (8.126b). This is common in solid state physics and field theory. We thus let n vary in steps of unity from $-N$ to $+N$ and set $\xi_N \equiv \xi_{-N}$ and $\xi_{n\pm 2N} = \xi_n$.

We have already seen how the periodic symmetry of the electron Hamiltonian H_{e} is used to find its eigenfunctions. The same symmetry is useful in treating the ion Hamiltonian H_{I}. Here we wish to transform the variables ξ_n to normal mode variables by diagonalizing the matrix $C_{nn'}$ with a linear transformation in the vector space of the dynamical variables ξ_n. The electron Hamiltonian was diagonalized by a transformation on vectors in Hilbert space, not in a space of dynamical variables. However, the symmetry is used the same way in both cases. Since the matrix to be diagonalized is invariant under a particular transformation, we look for eigenvectors of the matrix which are also simultaneous eigenvectors of this transformation.

As we have already seen in the example in Hilbert space, the eigenvectors of translations are just the Fourier transforms of the coordinates undergoing the translation, i.e. plane waves. We can therefore define our normal-mode variables immediately

$$a_q = \tfrac{1}{2}(N\hbar\alpha_q)^{-\frac{1}{2}} \sum_{n=-N}^{+N-1} (\xi_n + \mathrm{i}\alpha_q\pi_n)\mathrm{e}^{-\mathrm{i}qna} \tag{8.127a}$$

$$a_q^\dagger = \tfrac{1}{2}(N\hbar\alpha_q)^{-\frac{1}{2}} \sum_{n=-N}^{+N-1} (\xi_n - \mathrm{i}\alpha_q\pi_n)\mathrm{e}^{\mathrm{i}qna} \tag{8.127b}$$

where α_q is a constant to be determined which depends on the index q. We have chosen those linear combinations of the coordinates ξ_n and momenta π_n which make the normal-mode variables satisfy the commutation relations of boson annihilation and creation operators. The values of q are chosen to satisfy the periodic boundary conditions.

$$q = \frac{n'\pi}{Na}; \qquad -N \leqq n' \leqq N-1 \tag{8.127c}$$

$$a_{q+2\pi/a} \equiv a_q. \tag{8.127d}$$

We choose α_q to be an even function of q

$$\alpha_q = \alpha_{-q}, \tag{8.127e}$$

and note the orthogonality relation

$$\sum_{q=-\pi/a}^{q=(1-N^{-1})\pi/a} \mathrm{e}^{-\mathrm{i}q(n-n')a} = 2N\delta_{nn'}. \tag{8.127f}$$

The variables (8.127) are seen to be eigenfunctions of the transformation (8.126b) with the eigenvalues $\mathrm{e}^{\pm \mathrm{i}qa}$. Using the orthogonality relation (8.127f) we see that they also satisfy the boson commutation relation:

$$[a_q, a_{q'}^\dagger] = \delta_{qq'}. \tag{8.127g}$$

When the ion Hamiltonian (8.125c) is expressed in terms of the normal-mode variables it reduces to the form:

$$H_\mathrm{I} = \sum_q \tfrac{1}{2}\hbar\omega_q(a_q^\dagger a_q + a_q a_q^\dagger) \tag{8.128a}$$

where

$$\omega_q = 2 \sum_r \alpha_q C_{0r} e^{iqra} = (M\alpha_q)^{-1} \tag{8.128b}$$

$$\alpha_q = [2M \sum_r C_{0r} e^{iqra}]^{-\frac{1}{2}} \tag{8.128c}$$

and

$$C_{0r} = C_{n,n+r}. \tag{8.128d}$$

This is just the Hamiltonian for a system of free bosons, with the relation between frequency and wave number given by eq. (8.128b).

We can now express the interaction (8.125d) in terms of the normal coordinates by writing the potential as a Fourier transform.

$$\frac{\partial V}{\partial x}(x - na) \equiv \sum_q (N\hbar\alpha_q)^{-\frac{1}{2}} v_q e^{iq(x-na)}. \tag{8.129a}$$

We then have

$$H_{\text{int}} = -\sum_q \sum_n (N\hbar\alpha_q)^{-\frac{1}{2}} v_q e^{iqx} \xi_n e^{-iqna} = -\sum_q v_q e^{iqx}(a_q + a^\dagger_{-q}). \tag{8.129b}$$

The interaction (8.129b) thus describes a change in the electron momentum by an amount $\hbar q$, accompanied by either the absorption of a boson of wave vector q, or the creation of a boson of wave vector $-q$.

We now see the physical significance of the periodic symmetry, expressed by the invariance of the Hamiltonian under the generalized discrete translation (8.126). The Bloch wave number k or momentum of the electron is no longer a constant of the motion, but is changed by the interaction (8.129b). However, an overall 'momentum conservation' when the phonon variables are considered can be shown as follows.

The operator e^{iqx} in eq. (8.129b) describes a momentum transfer of $\hbar q$ to the electron and has a simple interpretation if the electron is in a momentum eigenstate. However, the eigenfunctions of the electron Hamiltonian (8.125b) are the Bloch waves (8.114) and not ordinary plane waves. These are not momentum eigenstates since an electron moving through any potential which is not constant has its momentum changed as it moves through the potential. Although the wave vector k associated with the Bloch wave function formally resembles the wave vector of a plane wave, the difference between this momentum and ordinary momentum is illustrated by calculating the matrix element of the operator e^{iqx} between two Bloch wave functions.

$$\langle k' | e^{iqx} | k \rangle = \int_{-\infty}^{\infty} u^*_{k'}(x) e^{-ik'x} e^{iqx} e^{ikx} u_k(x) \, dx. \tag{8.130a}$$

If the functions $u_k(x)$ were constant, as in the case of plane waves, the integral in eq. (8.130a) would give a single delta function corresponding to momentum conservation. Since the functions $u_k(x)$ are not constant but are periodic with period a, they can be expanded in a Fourier series in this interval. Thus the integral (8.130a) gives a delta function contribution whenever the difference $k'-k-q$ is equal to an allowed value for a Fourier component in the expansion of $u_k(x)$. Thus

$$\langle k'|\mathrm{e}^{iqx}|k\rangle = \sum_n 2\pi\delta(k' - k - q - 2n\pi/a) \int_{-\frac{1}{2}a}^{\frac{1}{2}a} u_{k'}^*(x)\mathrm{e}^{-i2n\pi x/a} u_k(x)\,\mathrm{d}x. \tag{8.130b}$$

We assume that the functions $u_k(x)$ are normalized in the interval from $-\frac{1}{2}a$ to $\frac{1}{2}a$. They are not required to be orthogonal for different values of k, because the orthogonality condition applies to the entire wave function including the factor e^{ikx}.

The Bloch wave number or crystal momentum is thus conserved modulo $2n\pi/a$. This is reasonable, since the Hamiltonian is not invariant under the continuous group of translations, but only under the discrete translation by a lattice vector a, and the Bloch wave number is only defined modulo $2n\pi/a$. Translation symmetry is discussed further in ch. 14.

We can easily generalize the Hamiltonian H to apply to many electrons instead of a single electron, if we disregard the direct Coulomb interaction between the electrons. This is justified in the Hartree–Fock approximation (see ch. 9) where we only consider the average interaction of a given electron with all other electrons, and neglect the fluctuations. The average interaction is included by changing the parameters of the periodic potential. If $c_k^\dagger$ denotes the creation operator of the Bloch wave vector k, we can rewrite eqs. (8.125b) and (8.129).

$$H_\mathrm{e} = \sum_k E_k c_k^\dagger c_k \tag{8.131a}$$

$$H_\mathrm{int} = \sum_{knq} v_{knq} c_{k+q+2n\pi/a}^\dagger c_k (a_q + a_{-q}^\dagger) \tag{8.11b}$$

where

$$E_k = \frac{\hbar^2}{2m}[\kappa(k)]^2 \tag{8.131c}$$

$$v_{knq} = \langle k + q + 2n\pi/a|\mathrm{e}^{iqx}|k\rangle v_q \tag{8.131d}$$

and $\kappa(k)$ is given by eq. (8.122c).

The Hamiltonian defined by eqs. (8.131) and (8.128) is formally the same as the phenomenological Hamiltonians constructed to treat boson emission, absorption and decay processes, such as eqs. (4.2) and (7.12). However, we have here constructed the Hamiltonian explicitly from the electron–ion and ion–ion interactions. We have derived, rather than postulated the existence of lattice vibration (phonon) creation and annihilation operators (8.129) which satisfy boson commutation rules. We have also derived interaction terms describing the emission and absorption of phonons by an electron, and can express the parameters of this interaction directly in terms of the electron–ion interaction.

This one-dimensional electron–phonon Hamiltonian provides a model which can be used to treat many interesting physical processes. For example:

Phonon–electron scattering is a second-order process in perturbation theory between an initial state $c_k^\dagger a_q^\dagger|0\rangle$ and a final state $c_{k'}^\dagger a_{q'}^\dagger|0\rangle$ each containing one electron and one phonon, via intermediate states $c_{k''}^\dagger|0\rangle$ or $c_{k''}^\dagger a_q^\dagger a_{q'}^\dagger|0\rangle$ containing either none or two phonons.

Electron–electron scattering via a phonon exchange is a second-order process in perturbation theory between an initial state $c_k^\dagger c_{k'}^\dagger|0\rangle$ and a final state $c_{k''}^\dagger c_{k'''}^\dagger|0\rangle$ each containing two electrons via intermediate states $c_{k''}^\dagger c_{k'}^\dagger a_q^\dagger|0\rangle$ containing two electrons and a phonon. Thus there is an effective electron–electron interaction via the lattice analogous to the ordinary electromagnetic interaction which arises from the exchange of *photons*. This electron–electron interaction mediated by the lattice can give rise to the pairing correlations which result in superconductivity.

The polaron is an electron which creates a potential well around itself by polarizing the lattice, repelling negative ions and attracting positive ions. It then oscillates in this well while at the same time moving through the lattice and carrying the well along with it. This can be treated by various methods.

THE MANY-BODY PROBLEM FOR PEDESTRIANS (PART 2)

Chapters 9, 10 and 11 constitute the more advanced portion of the monograph on many-body problems. The general discussions in ch. 9 and the treatment of the Thomas–Fermi and Hartree–Fock approximations are still presented at an elementary level and can be read by a student who has read the material in ch. 5, without requiring the treatment of composite systems, time-dependent perturbation theory and scattering in chs. 6, 7 and 8. Except for the end of section 9.7 which mentions the *T*-matrix, the more expanded treatment of pairing correlations in ch. 10 is still sufficiently elementary to be understandable to a student who has read ch. 6 but not yet learned time-dependent perturbation theory or scattering.

The treatment of elementary excitations in ch. 11 begins with a discussion of stationary states, giant resonances and elementary excitations which are also understandable at an elementary level, including the random-phase approximation applied both to a nuclear and a solid state problem. The discussion of sum rules and time-dependent formulations in sections 11.9 and 11.10 are closely analogous to the time-dependent formulations discussed in ch. 3 for the case of the Mössbauer effect.

Section 11.11 is an introduction to the more sophisticated formalism using spectral functions and Green's functions without entering into their actual applications. This material which represents the application to many-body problems of techniques borrowed from field theory is presented also to familiarize the student with them before he encounters them in field theory and in the perturbation theory of ch. 12. A more extended treatment of the application of methods of field theory to the many-body problem and superconductivity is given by J. Robert Schrieffer, Theory of Superconductivity (Benjamin, New York 1964).

CHAPTER 9

THE MANY-BODY PROBLEM IN QUANTUM MECHANICS

9.1 Introduction

Many-body problems arise in many branches of physics. A number of general methods for treating these problems have been developed. Before examining these problems and methods, let us first make a few general observations, which may be useful as a guide through the bewildering mass of confusion in the literature on the many-body problem.

1. *No one can solve them.* The main feature of all many-body problems is that no one can solve them exactly. The three-body problem is already well known to be insoluble exactly. Some kind of approximation must be made to get an answer for any specific case. The validity of the approximation depends upon the particular features of the physical problem. A general formulation apparently applicable to all many-body problems cannot indicate which approximation should be used in any specific case. The student should beware of elegant formalisms which appear to solve all many-body problems. Elegance cannot solve an insoluble problem.

2. *What do we want? Certainly not the exact solution!* The exact solution of the many-body problem is not only impossible; it is not wanted. The exact wave function for a many-particle system is so complicated and contains so much undesired information that it would be a major undertaking to obtain the small amount of desired information from it. Approximation methods generally used simplify the problem at the expense of undesired information. No one is interested in a system of successive approximations which approaches the exact solution of the Schrödinger equation. Only certain features of the system are of physical interest, and the formalism used hopefully gives a good description of these features, while very bad descriptions of other features or no description at all. To the extent that wave functions are used in these methods, they can be very bad wave functions,

with very little resemblance to the exact wave functions of the system. If they are used as a basis for perturbation theory, the perturbation treatment might even diverge. However, in each case, one finds reasons for believing that the treatment is valid for those quantities of physical interest.

The treatment of the quantum-mechanical many-body problem is thus qualitatively different from that of the few-body problem, such as the hydrogen or helium atom. The treatment of simple atoms aims at an exact solution of the problem, either in closed form or as a series of successive approximations. In using approximation methods one believes that one can get as close as desired to the exact solution by using the same method and doing more work. In the many-body problem, the exact solution is so complicated that the usual approaches would not converge to the exact solution or to anything remotely resembling it. On the other hand, the exact solution of a many-body problem is really irrelevant since it includes a large mass of information about the system which although measurable in principle is never measured in practice. In this respect, the quantum-mechanical many-body problem resembles the corresponding problem in classical statistical mechanics. Although it is possible in principle to solve Newton's equations of motion and obtain the classical trajectories of all the particles in the system for all times, only a small part of this information is relevant to anything that is measurable in practice in the laboratory.

Classical statistical mechanics uses a statistical description in which measurable macroscopic quantities such as temperature, pressure and entropy play an important role. An incomplete description of the system is considered to be sufficient if these measurable quantities and their behavior are described correctly. In the quantum-mechanical many-body problem, the aim is really the same as in the corresponding classical case. An accurate description is desired only of certain features of the problem which correspond to quantities that are measured in the laboratory. However, the formalism used is often identical to that for exact solution of the many-body problem. In fact, many treatments of the quantum-mechanical many-body problem *give the misleading impression that they are indeed methods for obtaining approximations to the exact solution of the Schrödinger equation.*

The wave functions used are very poor approximations if evaluated by the standards of the corresponding few-body problems. The overlap integral of one of these approximate wave functions with the exact wave function is certainly very close to zero. Any perturbation treatment starting with the approximate wave function in zero order would not converge to give the exact wave function. The approximate wave functions used are considered to be good if they lead to good approximate values for quantities of physical interest. Since these represent a very small portion of the total amount of information contained in the wave function, most of the information is irrelevant for practical purposes. It is not really important whether or not this irrelevant information is correct.

3. *How many particles*? There are two quite different types of many-body problems: (1) macroscopic many-body systems, (2) atomic and nuclear systems. Macroscopic systems contain a very large number of particles, $N \approx 10^{23}$. Effects which are of order N^{-1} or even $N^{-\frac{1}{3}}$ are clearly small. Surface phenomena can be neglected (unless one is particularly interested in studying surfaces). Atoms and nuclei contain just enough particles to be frustrating, of the order of 10 or 100. They have more than the hydrogen atom or the deuteron but any neglect of terms of order N^{-1} is subject to suspicion, and $N^{-\frac{1}{3}}$ is of order unity. Surface effects are important; 216 billard balls arranged in a cube have 152 balls on the surface and only 64 in the interior. It is thus surprising that the same approximation methods should seem to be useful in treating both macroscopic and nuclear systems. These approximations may turn out to be valid for entirely different reasons in the two cases.

4. *What is the problem*? Some many-body problems require the solution of a well-defined Schrödinger equation; in others, the exact form of the Hamiltonian is not known. In some atomic or solid state problems, the motion of the system is studied under the influence of well-defined forces, such as electromagnetic forces. In other problems, such as the structure of complex nuclei, the forces are not known, and a model Hamiltonian is often used, such as the shell model. In these models, parameters appear whose values are not determined a priori. They can be varied to fit experimental data when comparing with experiment the results of a calculation based upon the model. The physical meaning of these parameters is not very clear if approximations are made in the calculation. Approximations which are not justified from first principles, may work in practice if parameters in the model Hamiltonian are varied, this can occur when the terms neglected in the approximation are not small enough to be negligible, but happen to affect the result in the same way as changing parameters in the Hamiltonian. One can then talk about an 'effective' Hamiltonian or interaction, which somehow includes effects which are not treated directly.

5. *Discrete and continuous spectra.* In atoms and nuclei the ground state and a number of low-lying bound states are discrete and their energy levels are

directly measurable. These states are eigenfunctions of the Hamiltonian. The properties of these states could in principle be obtained by solving the Schrödinger equation to the desired degree of approximation and examining the properties of the eigenfunctions.

At higher excitation in atoms and nuclei and at nearly all excitations in solids there are continuous spectra of unbound states. The exact eigenfunctions of the Hamiltonian are never measured and are of no particular interest. The states of physical interest are produced experimentally by excitation of the ground state or are low-lying states present in thermal equilibrium at a finite temperatures. Experimental excitations never produce a small number of exact continuum eigenstates but rather coherent wave packets from the continuum of eigenstates. Such wave packets often have very simple properties which depend on the particular mode of excitation as well as upon the energy levels of the system. The properties of such elementary excitations play an important role in theoretical and experimental investigations of many-body systems.

6. *Elementary excitations*. The excitation spectrum of a many-particle system is very complex because of a large number of degrees of freedom. Nearly all of these excitations are not of any physical interest. Consider, for example, all of the possible ways of exciting twelve million particles in a system of 10^{23}, or even all possible ways of exciting 57 particles in a nucleus containing 122. Most experimental measurements on many-particle systems involve many simple kinds of excitation such as those produced by radiation or by bombardment with particles. The properties of these elementary excitations are therefore of particular interest. The aim of many treatments of the many-body problem is to study these elementary excitations with a minimum amount of complication from other effects of the large number of degrees of freedom in the system.

In studying elementary excitations one is interested in the change of the state of the system but not in a detailed description of the system before and after the excitation. Complicated properties of this system which are unchanged by the excitation are not of interest, and can either be ignored or approximated very crudely or badly. In the Green's function formalism and in the formalism of linearizing equations of motion, one deals mainly with the operators which generate these elementary excitations and avoids, in so far as possible, the use of many-body wave functions.

The quantum-mechanical many-body problem is thus similar to classical statistical mechanics in that only a small amount of the information provided by a detailed dynamical theory is of physical interest. The formalisms of the two cases differ because wave functions commonly used in the quantum-mechanical problem contain all the detailed information but allow simplifications by approximations which may give an extremely poor description of those quantities which are not of direct physical interest. These approximations are often exceedingly difficult to justify in any rigorous fashion, since the terms neglected are not necessarily small in the usual sense. It is only their effect on results which are of particular physical interest which need be small in order for the approximation to be valid.

9.2 The non-interacting Fermi gas

Most treatments of many-fermion systems begin with the free Fermi gas, in which all interactions are neglected. The Hamiltonian is

$$H = \sum_{k\sigma} \varepsilon_k a_{k\sigma}^\dagger a_{k\sigma} \tag{9.1}$$

where $a_{k\sigma}^\dagger$ is a creation operator for a plane-wave state with wave number k and spin σ, and ε_k is the kinetic energy of a particle with wave number k. The ground state of the system is obtained by filling all the lowest lying single-particle energy levels until all the particles are used up. The value of k at the top of this distribution is called the Fermi momentum and is denoted by k_F. The Fermi ground state wave function is thus

$$|F\rangle = \prod_{\substack{|k| \leqq k_\mathrm{F} \\ \sigma}} a_{k\sigma}^\dagger |0\rangle. \tag{9.2a}$$

This state can also be completely defined by the relations

$$a_{\boldsymbol{k}}^\dagger |F\rangle = 0 \quad \text{for} \quad |k| \leqq k_\mathrm{F} \tag{9.2b}$$

$$a_{\boldsymbol{k}} |F\rangle = 0 \quad \text{for} \quad |k| > k_\mathrm{F}. \tag{9.2c}$$

Simple excited states are obtained by exciting a single particle to a level above k_F

$$a_{\boldsymbol{m}} a_{\boldsymbol{k}} |F\rangle \qquad (|m| > k_\mathrm{F};\, |k| \leqq k_\mathrm{F}). \tag{9.3a}$$

The excitation energy of this state is

$$E_{mk} = \varepsilon_m - \varepsilon_k. \tag{9.3b}$$

In the treatment of many-fermion systems with second quantization it is customary to consider the Fermi state as one normally uses the vacuum state. Other wave functions are represented by combinations of creation and annihilation operators acting on the Fermi state. One then considers $a_{\boldsymbol{m}}^\dagger$ in eq. (9.3a) as creating a particle above the Fermi sea and the operator a_k as annihilating a particle or creating a 'hole' in the Fermi sea. The single-particle energies are often normalized by subtracting a constant from the Hamiltonian H to make the energy zero at the Fermi momentum k_F. The eigenvalues of the Hamiltonian then give the excitation energies above the ground state. The energy $\varepsilon_m - \varepsilon_F$ can be considered as that required to create a particle above the Fermi sea and $\varepsilon_F - \varepsilon_k$ as the energy required to create a hole in the Fermi sea. The energy required for the excitation of the state (9.3) can be written as the sum of the particle energy and the hole energy

$$E_{mk} = (\varepsilon_m - \varepsilon_F) + (\varepsilon_F - \varepsilon_k). \tag{9.3c}$$

Most states of physical interest are simple elementary excitations of the system involving a small number of excited particles and holes. The energy of such a state is just the sum of the energies of the individual particles and individual holes. This can be expressed formally by defining the hole creation operator

$$b_k^\dagger = a_k \quad \text{for} \quad |m| \leqq k_F. \tag{9.4a}$$

The relations (9.2b) and (9.2c) which define the Fermi ground state can now be written

$$b_k|F\rangle = 0 \quad \text{for} \quad |k| \leqq k_F \tag{9.4b}$$

$$a_k|F\rangle = 0 \quad \text{for} \quad |k| > k_F. \tag{9.4c}$$

The set of operators $b_k^\dagger$ for $|k| \leqq k_F$ and $a_k^\dagger$ for $|k| > k_F$, are hole and particle excitation operators. They can be considered together on a common footing and called 'quasiparticle' excitation operators; namely operators which create excitations behaving like single particles. For this set of quasiparticle operators, eqs. (9.4b) and (9.4c) show that the Fermi ground state $|F\rangle$ behaves like the vacuum.

In terms of these quasiparticle creation operators, the Hamiltonian for a non-interacting Fermi gas can now be written

$$H = \sum_{k \leqq k_F} (\varepsilon_F - \varepsilon_k) b_k^\dagger b_k + \sum_{k > k_F} (\varepsilon_k - \varepsilon_F) a_k^\dagger a_k \tag{9.5}$$

where the energy of the ground state has been normalized to zero.

The energy spectrum of the elementary excitations is a continuum for a system of infinite volume. For finite volume it is a very closely spaced discrete spectrum with the spacing depending on the volume of the system. There is no energy gap between the ground state and the first simple elementary excitations of one or two particles.

The Fermi ground state is an eigenfunction of all the number operators n_k with the eigenvalue 1 below the Fermi surface and 0 above. This is represented graphically by the plot of fig. 9.1.

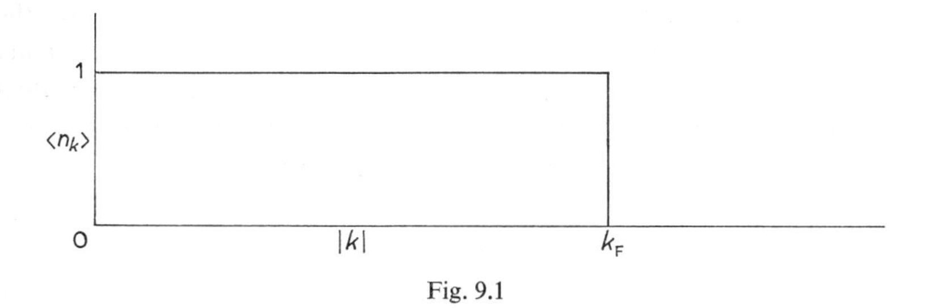

Fig. 9.1

In momentum space the Fermi surface for a non-interacting Fermi gas is a sphere with radius $\hbar k_F$. For a system in a finite volume with periodic boundary conditions the allowed momentum values appear in momentum space as points in a cubic lattice with a lattice spacing such that the volume occupied by each point is h. We consider fermions with spin $\frac{1}{2}$, the only case of physical interest for the many-fermion problem. There are thus two states at each lattice point and the density of states per unit volume in momentum space is

$$n_p \mathrm{d}^3 p = \frac{2\Omega}{\hbar^3} \mathrm{d}^3 p \tag{9.6a}$$

where Ω is the volume of the system. The total number of states in the Fermi sphere of radius $\hbar k_F$ is

$$N = \frac{\Omega k_F^3}{3\pi^2} = \frac{\Omega(2m\varepsilon_F)^{\frac{3}{2}}}{3\pi^2 \hbar^3} \tag{9.6b}$$

where ε_F is the Fermi energy. The number of particles within the Fermi sphere is proportional to the volume Ω of the box; i.e. the density in configuration space is independent of the size of the box.

9.3 Non-interacting fermions in a slowly varying potential The Thomas–Fermi model

Consider a system of noninteracting fermions in a constant potential V. The potential does not affect the allowed values of the momentum but only the relation between energy and momentum. The number of particles per unit volume below the Fermi surface is given by

$$\varrho(\boldsymbol{r}) = \frac{N}{\Omega} = \frac{[2m(\varepsilon_F - V)]^{\frac{3}{2}}}{3\pi^2\hbar^3}. \tag{9.7}$$

Now let the potential vary as a function of position in space, but sufficiently slowly so that we can still consider a box at each point in space in which the variation of the potential can be neglected. Then eq. (9.7) applies where V and $\varrho(\boldsymbol{r})$ are now functions of position. The Fermi ground state is constructed as above by filling the lowest levels in order until all the particles are used up. If we picture this system as consisting of many boxes all filled simultaneously, they will all be filled to the same level. Thus ε_F must be a constant, independent of position, whose value is determined by the total number of particles. This description is applicable, for example, to the case of an electron gas in a solid in the presence of an external potential which varies slowly over macroscopic distances.

This description of a many-fermion system in a slowly varying external potential has been used in the Thomas–Fermi model for many-electron atoms. Consider eq. (9.7) for a system of electrons moving in an electrostatic potential. If the potential V is given, eq. (9.7) gives the electron density. If there are no other charged particles in the region this gives the total electric charge density. However, the potential V is not a given function but depends upon the electron density distribution. This dependence of the potential on the charge density is expressed by Poisson's equation

$$\nabla^2 V = -4\pi e^2 \varrho. \tag{9.8}$$

Poisson's equation from classical electrostatics gives the potential V if the charge density ϱ is known, and boundary conditions are specified. Eq. (9.7) from quantum mechanics gives the electron density ϱ if the potential V is known. The simultaneous solution of eqs. (9.7) and (9.8) can give a self-consistent description of the motion of electrons in a potential which includes the potential produced by the electron charge density itself.

For an atom or an ion eqs. (9.7) and (9.8) are assumed to hold over all space except at the origin where there is a point nuclear charge. The value of the nuclear charge appears as a boundary condition on the behavior of the potential near the origin. The remaining boundary condition is provided by the behavior of the potential at large distances; namely that of a Coulomb potential corresponding to a charge at the origin equal to the total charge of the system. For a neutral atom the total charge is zero and the potential should approach zero at large distances faster than a Coulomb potential.

Combining eqs. (9.7) and (9.8) to eliminate ϱ and assuming spherical symmetry, we obtain

$$\frac{1}{r}\frac{\mathrm{d}^2(rV)}{\mathrm{d}r^2} = -\frac{4e^2[2m(\varepsilon_F - V)]^{\frac{3}{2}}}{3\pi\hbar^3}. \tag{9.9a}$$

For a nucleus of charge Z at the origin the boundary condition is

$$rV = -Ze^2 \quad \text{at} \quad r = 0. \tag{9.9b}$$

The non-linear second-order differential equation (9.9a) can be rewritten

$$r^2\frac{\mathrm{d}^2}{\mathrm{d}r^2}\left[\frac{r(\varepsilon_F - V)}{Ze^2}\right] = \frac{4e^2}{3\pi\hbar^3}\left[\frac{2mr(\varepsilon_F - V)}{Ze^2}\right]^{\frac{3}{2}} r^{\frac{3}{2}}(Ze^2)^{\frac{1}{2}}. \tag{9.9c}$$

It is therefore convenient to introduce the new variables

$$\Phi = \frac{r(\varepsilon_F - V)}{Ze^2} \tag{9.10a}$$

$$X = r\frac{Z^{\frac{1}{3}}e^2m}{\hbar^2}\frac{2^{\frac{7}{3}}}{(3\pi)^{\frac{2}{3}}} = \frac{r}{a_0}\frac{Z^{\frac{1}{3}}}{0.885} \tag{9.10b}$$

where $a_0 = \hbar^2/me^2$ is the first Bohr radius. Eqs. (9.9c) and (9.9b) then have the simple form

$$X^{\frac{1}{2}}\frac{\mathrm{d}^2\Phi}{\mathrm{d}X^2} = \Phi^{\frac{3}{2}} \tag{9.11a}$$

$$\Phi(0) = 1. \tag{9.11b}$$

This equation (9.11) is now independent of Z. For a neutral atom the boundary condition as $X \to \infty$ requires the potential to go to zero faster than $1/X$ and is also independent of Z. Thus a single numerical solution of the eqs. (9.11) subject to these boundary conditions gives a universal solution for all neutral atoms. This gives a universal shape for the potential of all atoms and predicts that the size of atoms is proportional to $Z^{\frac{1}{3}}$.

Eq. (9.11) has been solved numerically and the solutions are given in the literature. They can be used to give approximate properties of atoms, and also to provide first approximations for other methods which give better results, such as the Hartree–Fock approximation.

The Thomas–Fermi model neglects two effects which can be important in a many-body system: (1) the variation of the potential, (2) the correlations between the particles. The particle density at any point is assumed to be that for a constant potential whose value is equal to the value of the potential at that point. This approximation will give rise to errors when the potential changes appreciably in a distance equal to the wavelength at the Fermi surface. In the case of atoms, the Fermi–Thomas model breaks down both at small distances and at large distances. At small distances, the potential becomes singular and varies very rapidly. At large distances the potential approaches the Fermi energy and the wavelength becomes very long. However, in the intermediate region, which is important for many atomic phenomena, the model is quite good.

The assumption that each particle moves in the same static potential $V(X)$ neglects the correlations between particles. The true Coulomb interaction between two electrons depends upon the positions of *both* electrons, and the potential seen by any one electron varies as the other electrons move. The charge density of the many-electron system is not a function of position, but is a dynamical variable, which is a function of the dynamical variables of the electrons. Thus Poisson's equation (9.8) must be considered as an operator equation, and the potential is a many-body operator depending upon the positions of all the electrons. By taking the average or expectation value of this potential, we reduce this many-body problem to a set of one-body problems which we can solve, but we lose the effects of the correlations between the motions of the different electrons.

9.4 The Hartree–Fock approximation

We now consider an improvement on the Thomas–Fermi model which takes the variation of the potential into account exactly, but still neglects correlations. Instead of using the density for a box, the Schrödinger equation is solved exactly for particles moving in the potential V. Thus an exact equation is obtained instead of eq. (9.7) which gives the charge density corresponding to filling the N lowest levels in the given potential. This equation can then be combined with an equation like the Poisson equation (9.8) which gives the potential produced by this distribution of particles. This can also be done for a general interaction, such as nuclear forces, where there is no simple relation between the density and the potential. The self-consistent solution of these two equations is called the Hartree–Fock approximation or self-consistent field.

The Hartree–Fock approximation is easily formulated using the variational principle. Let H be the Hamiltonian for the many-particle system and $|\xi\rangle$ be a trial wave function. The variational principle requires that for any variation $|\delta\xi\rangle$

$$\delta\langle H\rangle = \langle\xi + \delta\xi|H|\xi + \delta\xi\rangle - \langle\xi|H|\xi\rangle = 0 \tag{9.12a}$$

$$\langle\delta\xi|H|\xi\rangle + \langle\xi|H|\delta\xi\rangle + \langle\delta\xi|H|\delta\xi\rangle = 0. \tag{9.12b}$$

The third term in eq. (9.12b) is discarded as being of second order in $\delta\xi$. The first two terms must vanish individually since they are complex conjugates of one another and are varied independently by arbitrary variations $|\delta\xi\rangle$. The variational principle is therefore expressed by the requirement

$$\langle\delta\xi|H|\xi\rangle = 0 = \langle\xi|H|\delta\xi\rangle. \tag{9.13}$$

The state $|H\xi\rangle$ is thus required to be orthogonal to the state $|\delta\xi\rangle$ for all possible variations $|\delta\xi\rangle$ to first order in $|\delta\xi\rangle$.

We now look for the best possible wave function which has the form of independent single-particle motion in some external potential. In the second-quantized notation such a wave function is denoted by a simple product of creation operators acting on the vacuum.

$$|\xi\rangle = \prod_{i=1}^{N} a_{k_i}^{\dagger}|0\rangle \tag{9.14}$$

where the creation operators $a_{k_i}^{\dagger}$ are creation operators for a single particle in some set of orthonormal states corresponding to an unknown potential which is to be determined. The wave function (9.14) describes a state in which each one of a set of N single-particle states is occupied by a particle. The most general N-particle wave function can be expressed as a linear combination of all possible states of the form (9.14). In the Hartree–Fock approximation, we restrict our set of trial wave functions to those which can be written as a single term having the form (9.14) in some basis. In the Schrödinger representation such a wave function has the form of a determinant to give the required antisymmetrization. This wave function is called a Slater determinant. The variational principle is applied by varying the basis to minimize the energy. Since the basis is related to independent-particle motion in an effective external potential, the variation is equivalent to varying this potential.

Consider the variation of the wave function (9.14) under an infinitesimal change of basis,

$$a_{k_i}^{\dagger} \to a_{k_i}^{\dagger} + \delta_i^{\dagger} \tag{9.15a}$$

$$|\xi + \delta\xi\rangle = \prod_{i=1}^{N} (a_{k_i}^{\dagger} + \delta_i^{\dagger})|0\rangle \tag{9.15b}$$

where $\delta_i^{\dagger}$ represents an arbitrary set of creation operators multiplied by an infinitesimal constant.

The variation in the wave function to first order in $|\delta\xi\rangle$ is

$$|\delta\xi\rangle = \sum_{j=1}^{N} \pm\delta_j^\dagger \prod_{\substack{i=1 \\ i\neq j}}^{N} a_{k_i}^\dagger|0\rangle \tag{9.16}$$

where the $\pm$ depends on the number of anticommutators required to bring the factors into the particular order indicated by eq. (9.16). This sign is irrelevant for our considerations.

The variation (9.16) consists only of states in which all but one of the original set of states k_i are occupied and only one state is changed. An arbitrary variation includes all possible states of the many-particle system in which the occupation number of only one state is changed, while all the other occupied states remain occupied.

The subspace spanned by the variation thus includes all states which can be reached by operating on the trial wave function with an arbitrary single-particle operator. The variational principle (9.13) thus requires that the state $|H\xi\rangle$ be orthogonal to all states which are obtained by changing the state of only one particle in the trial wave function. The state $|H\xi\rangle$ must therefore be a linear combination of the original state $|\xi\rangle$ and those formed by the excitation of at least two particles. *A state $|\xi\rangle$ that satisfies the variational principle must have the property of having no single-particle excitations produced by the action of the Hamiltonian H.* Conversely, any state on which the Hamiltonian produces no single-particle excitations satisfies the variational principle in the subspace of states having the form (9.14).

The Hartree–Fock variational principle can also be expressed in terms of particles, holes and quasiparticles. Let us define hole creation operators for the N states in (9.14)

$$b_{k_i}^\dagger = a_{k_i} \tag{9.17a}$$

and let $a_m^\dagger$ represent creation operators for a set of one-particle states orthogonal to the set k_i and constituting a complete set together with them. The state (9.14) is then defined by the relations

$$b_{k_i}|\xi\rangle = 0 = a_m|\xi\rangle. \tag{9.17}$$

The state $|\xi\rangle$ is therefore the vacuum for the new set of creation operators $a_m^\dagger$ and $b_{k_i}^\dagger$, which create particle and hole excitations. We can call these quasiparticle excitations and define the state $|\xi\rangle$ as the quasiparticle vacuum. The Hartree–Fock variational principle then requires that the Hamiltonian H acting on the state $|\xi\rangle$ have no terms exciting only two quasiparticles (i.e. a particle–hole pair). The state $H|\xi\rangle$ contains only the state $|\xi\rangle$ itself and states containing at least four excited quasiparticles.

So far, there has been no indication of a 'self-consistent field'. We also have not indicated how the Hartree–Fock basis can be found in practice. We now illustrate this feature of the Hartree–Fock approximation. Let the Hamiltonian be the sum of the kinetic energy T of the particles, a one-body potential W and a two-body interaction V. In an atom, for example, W is the central field of the nucleus and V is the interaction between the electrons.

$$H = T + V = \sum_{kk'} \langle k'|T + W|k\rangle a^\dagger_{k'} a_k + \tfrac{1}{2} \sum_{kk'k''k'''} \langle k'k'''|V|k''k\rangle a^\dagger_{k'} a^\dagger_{k'''} a_{k''} a_k. \tag{9.18}$$

The variational principle requires that all matrix elements of H which excite only one particle from the state (9.14) must vanish when the state is expressed in the Hartree–Fock basis. Thus all the off-diagonal matrix elements of the one-body operator $T+W$ must be exactly cancelled by those matrix elements of the potential in which the state of only one particle is changed. Collecting all terms in (9.18) which annihilate a particle in a particular occupied state k_i, create a particle in an arbitrary particular state k', and do not change the states of any other particles, we obtain

$$\langle k'|T + W|k_i\rangle + \sum_{k_i''} \{\langle k'k_i''|V|k_i''k_i\rangle - \langle k_i''k'|V|k_i''k_i\rangle\} = 0 \quad \text{if} \quad k' \neq k_i \tag{9.19}$$

where the sum is over all occupied states k_i''. Two matrix elements of the potential appear which differ by an exchange of the particles in the final state. An additional factor of two appears since the sum in eq. (9.18) is over all the k's. Each term is counted twice because of the possibility of exchanging the two labels simultaneously in both the initial and final states. This double counting does not occur in eq. (9.19) and must be compensated by the addition of a factor two.

Let us define an effective single-particle potential by setting

$$\langle k'|U|k_i\rangle = \sum_{k_i''} \langle k'k_i''|V|k_i''k_i\rangle - \langle k_i''k'|V|k_i''k_i\rangle. \tag{9.20}$$

The first term on the right-hand side of eq. (9.20) is just the 'average field' of the particles, since it is the sum of the two-body matrix elements changing the state of one particle and summed over all the others. This is seen clearly in the Schrödinger representation

$$\sum_{k_i''} \langle k'k''|V|k''k_i\rangle = \sum_{k_i''} \int \mathrm{d}x' \int \mathrm{d}x''\, \phi^*_{k'}(x')\phi^*_{k_i''}(x'')V\phi_{k_i''}(x'')\phi_{k_i}(x'). \tag{9.21a}$$

If V depends only on the coordinates x', x'' and not on the momenta, we can write

$$\sum_{k_i''} \langle k'k_i''|V|k_i''k_i\rangle = \int \mathrm{d}x' \, \phi^*_{k'}(x')\{\int V(x', x'')\varrho(x'')\mathrm{d}x''\}\phi_{k_i}(x')$$

where

$$\varrho(x'') = \sum_{k_i''} \phi^*_{k_i''}(x'')\phi_{k_i''}(x'') \tag{9.21b}$$

is the probability density for finding a particle at the point x''. The matrix element (9.21) is thus that for a single particle in a single-particle potential obtained by taking the potential between this particle and another particle at a point x'' and integrating over all possible points x'' with a weighting of the probability density for finding another particle at x''. If V depends on momenta, the potential U will depend on the probability current density as well as the probability density.

The last term in eq. (9.20) is the exchange term. This corresponds in the Schrödinger picture to exciting a particle from a state k_i'' into a state k' and exciting a particle in the state k_i into the hole at k_i'' left by the previous excitation. The summation over k_i'' includes the state k_i which appears as if the particle's own field were included in the average field. However, this term is cancelled by the corresponding exchange term, and there is no error introduced by including $k_i''=k_i$ in the sum (9.20).

From eqs. (9.19) and (9.20) we can define a single-particle Hamiltonian

$$\langle k'|h|k_i\rangle = \langle k'|T+W|k_i\rangle + \langle k'|U|k_i\rangle = \varepsilon_{k_i}\delta_{k'k_i}. \tag{9.22}$$

The right-hand side of this equation expresses the Hartree–Fock condition (9.19) which requires that off-diagonal elements of the single-particle Hamiltonian h vanish between occupied and unoccupied states. There is no condition on the elements between two occupied states, since the wave function (9.6) is invariant under an arbitrary transformation of basis *which affects only the occupied states*. We choose a basis in which h is diagonal in the subspace of occupied states and denote the values of these diagonal elements by ε_{k_i}.

Eq. (9.22) has the form of a single-particle Schrödinger equation which defines the states $|k_i\rangle$ to be those which diagonalize the single-particle Hamiltonian h. The treatment so far gives no physical significance to the eigenvalues ε_{k_i} although it is tempting to interpret this quantity as the energy of a single particle moving in the average field. Eqs. (9.20) and (9.22) represent the 'self-consistent' form of the Hartree–Fock condition, and are analogous to eqs. (9.7) and (9.8) for the Thomas–Fermi model. Eq. (9.22) requires the single-particle states $|k_i\rangle$ to be eigenstates for single-particle motion in a potential $W+U$. Like eq. (9.7) it defines the density corresponding to a given potential, but it is an exact quantum-mechanical result. Eq. (9.20) defines the potential U produced by the two-body potential V for the N-particle system in the states defined by the single-particle equation (9.22), and is exactly equivalent to the Poisson equation (9.8) for the case of a Coulomb potential.

The value for the energy of the system obtained from the variational principle is not obtained by summing the eigenvalues of eq. (9.22) interpreted as single-particle energies. The energy is given by the expectation value of the Hamiltonian (9.18) with the many-body wave function (9.14),

$$E = \sum_{k_i} \langle k_i|T + W|k_i\rangle + \tfrac{1}{2} \sum_{k_i k_i''} \{\langle k_i k_i''|V|k_i'' k_i\rangle - \langle k_i'' k_i|V|k_i'' k_i\rangle\} \quad (9.23a)$$

$$E = \sum_{k_i} \langle k_i|T + W|k_i\rangle + \tfrac{1}{2}\langle k_i|U|k_i\rangle. \quad (9.23b)$$

The energy differs from the sum of the eigenvalues of eq. (9.22) because of the factor $\frac{1}{2}$ appearing in front of the matrix elements of the potential U. This factor can be interpreted as canceling the 'double counting' of the two-body potential between any two particles a and b. This potential appears twice; once in the average field determining the motion of particle a and once in the average field determining the motion of particle b. The sum of the potential energies of the average fields of all the particles is included in summing the eigenvalues of eq. (9.22) and counts the contribution of each two-body interaction twice. We have not used any of these intuitive arguments to obtain the result (9.23b) but only to interpret the result which follows rigorously from the variational principle.

In practice, eqs. (9.20) and (9.22) are solved self-consistently by an iteration procedure. A trial potential is guessed and used in eq. (9.22) to find the wave functions. These are then inserted into eq. (9.20) to find a better approximation to the potential. This process is then iterated until it has converged to the desired accuracy. The Thomas–Fermi potential is often used as the first approximation.

The Hartree–Fock approach can also be used for the description of excited states. However, the variational principle requires the effective single-particle potential defined by eq. (9.20) to have a summation over the states k_i'' *which are occupied in the particular excited state considered.* The potential is thus different for each excited state, as expected, since the average potential seen by any given particle depends upon the states of the other particles. However, it is very inconvenient to have a different single-particle basis for each state of a many-particle system. If single-particle states above the Fermi surface are defined with a potential which is different from that used to define the single-particle states below the Fermi surface the two sets of states are not orthogonal.

In the applications to excited states a common effective potential for all single-particle states is defined by eq. (9.20) with the sum taken over the same set of states k_i''; namely those occupied in the ground state of the many-particle system. The definition of the single-particle Hamiltonian (9.20) is extended to include the matrix elements between all single-particle states, and it is required to be diagonal also for states above the Fermi surface. Eq. (9.20) now defines a one-particle Schrödinger equation for a complete set of single-particle states in a given external potential. The potential is determined only by the properties of the ground state and only the ground state wave function for the many-particle system satisfies the variational principle. However, if the number of particles is large, one can expect that the potential should not change much if only a few particles are excited and the wave functions for states with only a small number of excited particles should still give a good approximation.

The Hartree–Fock approximation has been applied to a large variety of many-particle systems. In cases where the approximation is not adequate to describe the desired phenomena, additional approximation methods are used to include the terms omitted in the Hartree–Fock approximation; namely, the portion of the two-body interaction which excites two particles out of the ground state. This portion of the interaction, sometimes called the residual interaction, has been treated by perturbation theory and by fancier methods when perturbation theory is inadequate. These are discussed below.

The applications of the Hartree–Fock approximation can be divided into two general classes each characterized by a different symmetry. There are small systems like atoms and nuclei which have a well-defined center and for which rotational invariance and spherical symmetry about this center play an important role. There are large systems like solids and nuclear matter which are infinite for all practical purposes and for which a translational symmetry plays an important role.

9.5 Spherical symmetry and shell structure in atomic and nuclear systems

The two-body interactions in atoms and nuclei are rotationally invariant. One naturally looks for a Hartree–Fock potential which possesses spherical symmetry (the exceptional case of deformed nuclei is considered below). The energy levels for a single particle in such a spherical potential are characterized by the angular momentum j of the level and have a degeneracy of

$2j+1$. In atoms where the electrostatic interactions between the particles are spin-independent to a good approximation one can label the single-particle levels by the orbital angular momentum l and the spin s with a degeneracy $2(2l+1)$ for spin $\frac{1}{2}$ electrons.

The filling of these discrete degenerate energy levels gives rise to a shell structure in atoms and nuclei. A typical system consists of a 'core' of particles completely filling the lowest levels and a few additional 'valence' particles in an unfilled level. Many physical properties of atoms and nuclei can be obtained by considering the core of closed shells as inert and examining only the valence particles in detail. A closed shell has zero total angular momentum and all electromagnetic moments are zero. The angular momentum and the moments of the system are therefore determined completely by the valence particles. The elementary excitations of this system of physical interest also generally involve only valence particles, although there are some excitations of interest having a hole in a closed shell.

When there are several particles in an unfilled shell the Hartree–Fock approximation gives a very high degeneracy because of the large number of ways of distributing these particles among the degenerate states of the shell. The removal of this degeneracy by the residual interaction is calculated by degenerate perturbation theory; i.e., diagonalizing the residual interaction in the subspace of states degenerate in the Hartree–Fock approximation. Since the residual interaction is also rotationally invariant, the linear combinations of Hartree–Fock states which diagonalize the interaction should be eigenfunctions of the total angular momentum. The properties of angular momentum algebra and couplings have been extensively used in atomic and nuclear spectroscopy.

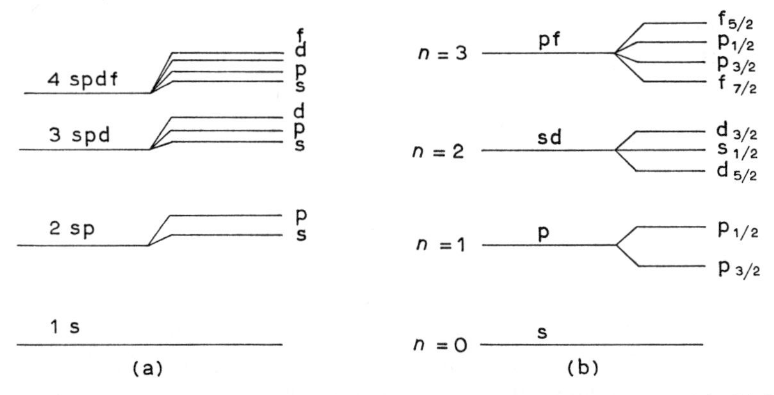

Fig. 9.2. Energy levels. (a) Coulomb potential, and screened atomic potential, (b) harmonic oscillator potential and nuclear potential.

We now examine some of the qualitative features of atomic and nuclear spectroscopy. The Coulomb and harmonic oscillator potentials are convenient starting points for atomic and nuclear spectroscopy respectively. Both have simple exact solutions and the orderings of their levels are shown in fig. 9.2.

The difference between the orderings of the levels in the two cases is easily understood from the difference in the shapes of the potentials. The Coulomb potential becomes very strong near the origin whereas the harmonic oscillator is quite flat in this region. The difference between the two potentials is felt more strongly by states whose wave functions are appreciable in the vicinity of the origin. For a state of orbital angular momentum l the repulsive centrifugal potential $l(l+1)/r^2$ is dominant at small distances and prevents the wave function from being appreciable in that region. Thus, the strong attraction of the Coulomb potential near the origin is felt most by levels of low l.

In comparing the levels of the Coulomb potential with the harmonic oscillator one sees that the excited levels of low angular momentum are systematically lower in the Coulomb potential than in the harmonic oscillator relative to the states of higher angular momentum. The ground state in both systems is an s-state. The first excited level contains a p-state, the next excited level contains a d-state. In general the *first* level having a given value of l occurs for both potentials at the lth excited level in both potentials. The second s-state is degenerate with the first p-state for the Coulomb potential but only with the first d-state in the harmonic oscillator potential. Similarly successive states of lower angular momentum are displaced downward in the Coulomb potential relative to their position in the oscillator potential.

In multi-electron atoms the Hartree–Fock potential deviates from the pure Coulomb potential by the contribution of the field produced by the negative electrons. The potential is thus reduced and this reduction is greater at large distances. This reduction in the attractive potential is then felt more strongly by states of high orbital angular momentum where a larger part of the wave function is in the region where the screening is effective. Thus the first excited level splits with the p-state lying higher than the s-state, etc. as shown in fig. 9.2a. In the nuclear case the harmonic oscillator potential is just a convenient idealization. The Hartree–Fock potential for a short-range force is expected to follow roughly the nuclear density and be more like a finite square well. It would be flatter at the origin than the harmonic oscillator and would stop at some constant value, rather than going upward to infinity like the harmonic oscillator. Both of these effects favor states of higher angular momentum. Thus the ordering of levels found in nuclei deviate from those of the harmonic oscillator by bringing down the levels of high orbital angular momentum. In the second excited state the d-level is brought below the s. A much larger effect is the spin–orbit coupling which splits the states having a given l into two levels having $j=l+\frac{1}{2}$ and $j=l-\frac{1}{2}$. The state of higher j always lies lower. This is illustrated in fig. 9.2b.

We can also make some qualitative observations on the ordering of the energy levels obtained from coupling the angular momenta of the valence particles in different ways. In atoms the residual interaction is spin-independent to a good approximation and is repulsive. The effect of the spin on the energy comes entirely from the symmetry properties of the wave functions determined by the Pauli principle. Electrons with parallel spins are prevented by the Pauli principle from coming close together in space whereas electrons with antiparallel spins do not have this restriction. Thus a spatial wave function for a system of electrons with parallel spins has a smaller Coulomb repulsion energy than a wave function for electrons with antiparallel spins. Thus, for a system of several electrons the state of maximum spin should lie lowest and successively lower values of spin should be higher. This is empirically true in atoms and is known as Hund's rule.

In the nuclear case the residual interaction is attractive and short-range. The lowest energy is obtained when the spatial wave functions of the nucleons overlap as much as possible, in contrast to the repulsive Coulomb case. Because of the large spin–orbit interaction the single-particle orbits are characterized by the total angular momentum j, and the projection m on the z-axis. For two nucleons the maximum overlap is obtained by coupling the two nucleons to the maximum total angular momentum $J=2j$. For the state with projection $M=J$ each nucleon is in the state $m=j$. Thus both nucleons are in exactly the same state and give the maximum overlap. Since nucleons are fermions the symmetric $J=2j$ is not allowed for two neutrons or two protons but only for a proton and a neutron. If we use the isospin formalism we see that the $J=2j$ state is allowed for isospin zero but not for isospin one.

For identical nucleons one cannot put both particles in the same state. However, the pairs of states with equal and opposite values of m, namely m and $-m$, have exactly the same spatial density distribution and differ only by a phase factor in the wave function. Thus for two identical nucleons the two-particle state with $M=0$ has a lower energy than states with other values of M. When we extend this argument to states which are constructed to be eigenfunctions of J as well as M we see that the $J=0$ state lies lowest for a pair of identical nucleons.

The experimental values of the ground state spins of nuclei having two valence nucleons outside the shells agree with these arguments. If the two nucleons are identical, $J=0$; if they are a neutron and a proton, $J=2j$. When several identical valence nucleons are present, they tend to combine in pairs having $J=0$ and the wave functions are simply described using the concept of seniority discussed in a simple manner in ch. 6 and considered in more detail in ch. 10.

When there are several valence neutrons and protons, a new effect is found; namely deformation. Consider for example, the filling of a shell having $j=\frac{7}{2}$. The lowest state for two nucleons is the $T=0$ neutron-proton state with $J=7$. The $M=7$ state has both particles in eigenstates of $m=\frac{7}{2}$. Let us now add another neutron and another proton to this state. If we put them into states having a definite m the lowest energy is obtained if they are both put in the state $m=-\frac{7}{2}$ as these wave functions have exactly the same probability density distribution as the state with $m=+\frac{7}{2}$. The density distribution for these four nucleons is then not spherically symmetric but has the oblate (disc shape) form of a wave function having $m=\pm j$. The interaction of these four valence nucleons with the other nucleons in the closed shell core produces a field which is not spherically symmetric and which tends to deform the core into an oblate shape. If this interaction is sufficiently strong a deformation of the core can be produced. This would appear in the Hartree–Fock approximation as the existence of a solution of the Hartree–Fock equations with a potential and a nucleon density not having spherical symmetry but having an oblate deformation.

On the other hand one might add the four nucleons as two neutrons in a $J=0$ state and two protons in a $J=0$ state, thus maximizing the interaction between the identical nucleons and retaining the spherical symmetry. This would give a nuclear wave function which is an eigenstate of seniority. The Hartree–Fock equations can have both spherical and deformed solutions, indicating that both the spherical and the deformed states give a local minimum in the energy with respect to infinitesimal changes in the wave function. Whether or not the ground state is spherical or deformed then depends upon the energies of the two minima.

Qualitative aspects of the deformation are often described by the use of a deformed harmonic oscillator potential; i.e. one in which the force constants are different in the z-direction from the x and y-directions. One can then plot the expectation value of the Hamiltonian in the ground state of the harmonic oscillator potential as a function of the deformation parameter which expresses the difference between the force constants in the different directions. There may be a minimum at spherical symmetry, then a maximum and a second minimum which can be higher or lower than the minimum for the spherical case. If the second minimum is higher than the first, the ground state should be spherical whereas if it is lower, it should be deformed. This is a highly simplified version of the Hartree–Fock argument, as we have restricted the use of the variational principle only to harmonic oscillator potentials and varied only the force constants and not the shape of the potential. When the general Hartree–Fock equations are used the qualitative picture is the same, but it is not possible to define a single parameter to characterize the variation of the wave function.

The occurrence of a Hartree–Fock solution which does not have spherical symmetry seems to be in contradiction with the rotational invariance of the Hamiltonian. The deformed Hartree–Fock solution is not an eigenfunction of the total angular momentum J, and we know that the exact eigenfunctions of the Hamiltonian must be eigenfunctions of J unless there are degeneracies. However, it is quite simple to restore the overall spherical symmetry to the Hartree–Fock solution. The physical picture is that a disc-shaped Hartree–Fock solution tells us that the equilibrium shape for the many-body system does have an oblate deformation. The spherical symmetry of a Hamiltonian then requires that there is no preferred direction for the orientation in space of this disc and that the wave function describing the ground state must have an equal probability for all possible orientations. In the Hartree–Fock approximation one sees that the existence of *one* deformed solution having a particular orientation in space implies the existence of an infinite number of solutions, all having the same energy, generated by rotating this particular solution through some angle. The desired wave function, which has an equal probability for all spatial orientations, is obtained by taking linear combinations of these states; i.e. an integral over the angle parameters describing the rotation.

9.6 Translational symmetry in solids and nuclear matter

Large systems are considered either to be infinite or to be quantized in a large box with periodic boundary conditions. If the interactions between particles depend only on the distance between the particles and possibly on momenta but not on the absolute position of either particle, the Hamiltonian is invariant under translations and momentum is conserved. For this case the plane-wave basis of the non-interacting Fermi gas is always a solution of the Hartree–Fock equations. This is easily seen by the following argument.

Consider the ground state of a non-interacting Fermi gas in which all states below the Fermi surface are filled and those above are occupied. This state is a solution of the Hartree–Fock equations for interacting fermions if the interaction Hamiltonian acting on this state does not produce any single particle–hole excitations but only excitations of two particles and two holes. A single particle–hole excitation removes a particle from within the Fermi sphere and places it in a state outside which has a different momentum from the initial state. Thus any single particle–hole excitation must change the momentum of the state. A Hamiltonian which conserves momentum can have non-zero matrix elements only connecting states having the same total momentum, and has no matrix elements between the Fermi ground state and any state which has only a single particle excited. Thus, the Fermi ground state is a solution of the Hartree–Fock equations for any translationally invariant Hamiltonian; i.e. for any interaction which conserves momentum.

The plane-wave basis is therefore used for many problems in solids and for nuclear matter. In considering the motion of electrons in a solid there is also the underlying crystal lattice structure which breaks translational invariance. If the crystal structure can be neglected, as in many treatments of the electron gas where it is replaced by a uniform background of positive charge, the plane-wave basis can be used. In cases where the presence of the lattice is taken into account there is invariance under the discrete translations by lattice vectors. For this case Bloch waves are solutions of the Hartree–Fock equations and the effective self-consistent potential is a periodic potential.

Note that an ionic crystal could also be considered as a many-body problem under the influence of translationally invariant ionic forces. Here one has a situation similar to that of the deformed nucleus. Although the plane-wave basis gives a solution of the Hartree–Fock equations, another solution exists with lower energy in which the particles execute small vibrations around the points in a crystal lattice. Although such a wave function is not an eigenfunction of the total momentum one could in principle construct momentum eigenfunctions by taking linear combinations of lattice wave functions with lattice points located at different places in space. In practice this is not done because it is not of physical interest.

Another type of large system which is of physical interest is a spin system. In many solids the spins of the nuclei, atoms or particular important electrons can be considered as degrees of freedom separated to a very good approximation from the other degrees of freedom of the system. For example, a simple model for a ferromagnet consists of a single electron located at each lattice point of a crystal lattice with an interaction between different spins. The dynamical variables of the problem are then the electron spin orientations at each lattice point. The Hamiltonian contains a term expressing the interaction of each spin with an external magnetic field and a term including the spin–spin interaction. It is a good approximation to disregard the other degrees of freedom; i.e. the spatial degrees of freedom of the electron and the motion of all the other particles in the crystal. For this case one can also define a Hartree–Fock approximation in which each spin moves in a field which is the sum of the external magnetic field and the average field produced by the interactions with the other spins.

9.7 Beyond Hartree–Fock. The residual interaction and correlations

The results of Hartree–Fock theory can be improved by taking into account the residual interaction. If the Hartree–Fock approach gives a good zero-order approximation one might expect that the effects of residual interaction are small and can be treated by perturbation theory. There are many cases where this is done. However, there are also other cases where perturbation theory in the residual interaction breaks down. This would indicate that the Hartree–Fock wave function is not a good approximation from the point of view of perturbation theory. Yet the results show that the Hartree–Fock wave function is in some sense a good approximation in giving good zero-order predictions of many physical properties. There are two types of situations of this kind.

One case where the naivest form of Hartree–Fock wave function is not good is the case of the atom or nucleus with several valence particles in the outer shell. Here we have seen that one must use degenerate perturbation theory and diagonalize the interaction in a subspace of states which are all degenerate in the Hartree–Fock approximation. Consider now the case where the atom or nucleus is in a weak magnetic field which removes the degeneracy of many of these Hartree–Fock states. To get a good approximate wave function one would still need to diagonalize the residual interaction together with the interaction with the external magnetic field in the subspace of states which would be degenerate in the absence of the magnetic field.

A similar situation can exist in solid state problems. The Hartree–Fock ground state may be non-degenerate but there are a large number of excited states in the Hartree–Fock approximation which are very close in energy to the ground state and mixed with the ground state by the residual interaction. The residual interaction cannot be treated as a simple perturbation on the non-degenerate Hartree–Fock ground state, requires something analogous to diagonalization in the subspace of nearly degenerate Hartree–Fock states. The resulting wave function would then be very different from the Hartree–Fock ground state, and would have a very small overlap with it. However, because the difference affects only particles in the states very close to the

Fermi surface, it would not appreciably affect the distribution of the number operator shown in fig. 9.1. Instead a short drop at the Fermi surface there would be a smooth transition over a small region where $\langle n_k \rangle$ would drop from one to zero. The Fermi wave function would then give a good approximation for the calculation of all properties of this system which are not sensitive to the precise shape of the Fermi surface even though it is a bad approximation from the point of view of perturbation theory.

The above argument shows that the Fermi ground state is unstable against the effects of the residual interaction because the levels are so closely spaced and there is always a small region in momentum space where the excitation energy can be much smaller than the matrix elements of the residual interaction. Because of the enormous number of states of this type in a macroscopic system, it is out of the question to diagonalize the residual interaction in a particular subspace as is done in atomic and nuclear shell models. Instead approximation methods have to be devised for treating such cases. Some of these are discussed in chs. 10 and 11.

The above examples describe the breakdown of perturbation theory in the residual interaction with a weak interaction, due to the near degeneracy of states at the Fermi surface. There is also a different kind of breakdown where the interaction is strong, but strong in a peculiar way so that it does not destroy the Hartree–Fock picture completely. The principal example of this type is the effect of the repulsive core in the nucleon–nucleon interaction in the many-body treatment of nuclear matter and finite nuclei. Consider for example the treatment of nuclear matter with a two-body interaction which contains a repulsive core that becomes infinite at a particular radius; i.e. an impenetrable sphere of finite radius surrounded by an attractive potential. For this case, since the interaction conserves momentum, the plane-wave basis is a solution of the Hartree–Fock equations. However, any attempt to calculate even the ground state energy in the Hartree–Fock approximation breaks down. Since the potential becomes infinite in a finite volume around each particle the matrix elements of the potential between plane-wave states are also infinite and no sensible results can be calculated. One can argue that the repulsive potential is not really infinite but some very large finite value. In this case a numerical answer would be obtained but would be ridiculously large and not have any physical significance.

The clue to the proper treatment of particles with infinite repulsive cores is seen in the two-body scattering problem for hard spheres. Here the infinite repulsive core causes no trouble. The solutions of the two-body Schrödinger equation give a wave function which vanishes whenever the distance between the two particles is within the repulsive core. The core does not give any infinite energy, it simply excludes the particles from those regions of space where they are too close together. One might think that the plane-wave Hartree–Fock wave function for the many-body system may still be good approximations to the exact wave function as long as no pair of particles is close enough together to be within the repulsive core. One might modify the wave function by operating on it with projection operators which automatically make the wave function vanish whenever any pair of particles is too close together. Such projection operators are of necessity highly singular two-body operators and therefore very difficult to use in precise calculations.

A fruitful approach to this type of problem was developed by Brueckner to use the properties of the T-matrix appearing in scattering theory. One property of the T-matrix is that its operation on the unperturbed wave function gives the same result as operating on the exact wave function with the potential

$$T\phi = V\psi \tag{9.24}$$

where V is the two-body interaction, ϕ is the unperturbed plane-wave function and ψ is the exact solution of the two-body scattering problem. If we consider this equation in configuration space we see that for small distances between the particles where V becomes infinite the exact wave function ψ must vanish and the product is not infinite. Since the unperturbed wave function ϕ does not vanish the operator T must be well behaved in this region.

The basic philosophy of the Brueckner approach is to solve the two-body scattering problem exactly and determine the T-matrix. One then formulates the many body problem using the unperturbed or Hartree–Fock wave function and replacing the singular two-body interaction V by the 'effective interaction' T. Thus the effects of the correlations are taken into account, not by improving the wave function but rather by modifying the residual interaction to an effective interaction which by eq. (9.24) gives the same result when used with unperturbed wave functions as the real interaction with the exact wave functions.

The application of this approach in practice is quite complicated. First, the two-body scattering problem is not that of two isolated free particles. Even if one neglects the interaction of the two particles under consideration with the remaining particles in the system one cannot neglect the Pauli principle which prevents the two particles from being scattered into states already occupied by the others. One must therefore solve a two-body scattering problem in the presence of a non-interacting Fermi sea. This is called the Bethe–Goldstone equation. The introduction of the two-body T-matrix into the many-body problem is also quite complicated. Explicit formalisms have been developed to do this in a consistent manner which allows for higher-order corrections. These methods are described in the literature of nuclear matter and the details are beyond the scope of this book. However, the essential physical basis behind the method is that described above.

PROBLEMS

1. Consider a system of many fermions of mass m moving in a one-dimensional harmonic oscillator potential, $V=\frac{1}{2}m\omega^2x^2$. Consider the ground state of the N-particle system and calculate the Fermi energy (the energy of the highest filled level) (a) using the exact harmonic oscillator wave functions, (b) using an approach like that of the Thomas–Fermi model, in which the density of particles at a point x is that for a box with a constant potential $V(x)$.

2. Let $|\mathrm{HF}\rangle$ be the solution of the Hartree–Fock equations for a many-fermion system whose Hamiltonian is H. Which of the following matrix elements must vanish?

$$\langle \mathrm{HF}|a_k^\dagger H a_m^\dagger|\mathrm{HF}\rangle$$
$$\langle \mathrm{HF}|a_k^\dagger a_m H|\mathrm{HF}\rangle$$
$$\langle \mathrm{HF}|a_m a_k^\dagger H|\mathrm{HF}\rangle$$
$$\langle \mathrm{HF}|a_k H a_m^\dagger|\mathrm{HF}\rangle$$
$$\langle \mathrm{HF}|H|\mathrm{HF}\rangle.$$

3. For what kind of two-body interaction is a set of plane-wave states automatically a solution of the Hartree–Fock equations. Explain.

4. Consider a simplified model of a ferromagnet, in which N spins of $\frac{1}{2}$ interact with one another and with an external magnetic field, described by the Hamiltonian

$$H = -g\sum_{i=1}^{N} h_i\sigma_{zi} - \sum_{\substack{i=1\\ i\neq j}}^{N}\sum_{j=1}^{N} G_{ij}\boldsymbol{\sigma}_i\cdot\boldsymbol{\sigma}_j$$

where h_i is the intensity of the external magnetic field (taken to be in the z-direction) at the position of the spin i. $\boldsymbol{\sigma}_i$ is the Pauli spin matrix describing the orientation of the spin i. G_{ij} is the strength parameter for the interaction between the spin i and the spin j.
(a) Let $a_{i+}^\dagger$ and $a_{i-}^\dagger$ represent creation operators for a spin at position i with $S_z=+\frac{1}{2}$ and $-\frac{1}{2}$ respectively. Rewrite the Hamiltonian in terms of these second-quantized operators.

(b) Write the most general trial wave function for the solution of this problem by the Hartree–Fock method (i.e. assuming that each spin moves independently in the average field of the others). Use the second-quantized notation of part (a).
(c) Write the Hartree–Fock equations for this problem.
(d) Show that the z-component of the total angular momentum

$$S_z = \tfrac{1}{2} \sum_{i=1}^{N} \sigma_{zi}$$

commutes with the Hamiltonian if $h_i = h$, independent of i (uniform external magnetic field).
(e) Show how the Hartree–Fock variational principle can be modified to obtain approximate solutions of the Schrödinger equation for states which are simultaneous eigenfunctions of H and S_z corresponding to a given eigenvalue S_0 for S_z. Use a Lagrange multiplier in the variational principle.

CHAPTER 10

PAIRING CORRELATIONS AND THE BCS THEORY

10.1 Introduction

In ch. 6 we have considered a many-fermion system with a particular type of interaction which tends to produce correlated or bound pairs. The ground state of this system consists entirely of correlated pairs each having the same total momentum. A finite energy is required to break a pair or even to move a pair to a state with a slightly different momentum, because of the effect of the Pauli principle. This energy gap is characteristic of superconducting or superfluid systems, in contrast to so-called 'normal' systems like the non-interacting Fermi gas. In normal systems one can find excited states of very low or infinitesimal excitation energy which differ from the ground state only by the excitation of one or two particles at the top of the Fermi sea.

In this simplified model, the kinetic energy of the individual fermions was neglected and the interaction chosen (6.21) was simplified to produce bound pairs having the flat, rectangular momentum spectrum of fig. 6.1b. These simplifications make the exact solution easily obtainable with quasi-spin operators. A more realistic treatment is obtained by examining a Hamiltonian including the kinetic energy and having a more general form for the interaction.

$$H = \sum_{k\sigma} \varepsilon_k a^\dagger_{k\sigma} a_{k\sigma} - \sum_{kk'} G_{kk'} a^\dagger_{k\uparrow} a^\dagger_{-k\downarrow} a_{-k'\downarrow} a_{k'\uparrow} \tag{10.1}$$

where ε_k is the kinetic energy of a particle in the state with wave number $\boldsymbol{k}$ and $G_{kk'}$ is a function specifying the strength of the interaction. $G_{kk'}$ is now a function of the momenta $\boldsymbol{k}$ and $\boldsymbol{k}'$ rather than being a constant as in eq. (6.21). The interaction term in this Hamiltonian tends to bind pairs of particles into bound states having a form more general than the flat momentum state. We have chosen for convenience an interaction which binds pairs into states with zero total momentum rather than some particular value $2K$ as in eq. (6.21). This 'BCS reduced Hamiltonian' was first proposed in the celebrated paper on superconductivity.

The BCS Hamiltonian cannot be solved exactly using the quasi-spin techniques of the simpler example. Approximate solutions are obtainable by methods developed by BCS and by Bogoliubov and Valatin. These solutions have been shown to become exact in the limit of infinite volume and have qualitative properties similar to the exact solution of the simple example.

10.2 A simplified BCS model (strong-coupling limit)

Before presenting the BCS solution let us consider a simplified case. We assume that $G_{\boldsymbol{kk}'}$ has a constant value over a region in the vicinity of k_{F}, and vanishes outside this region, and we neglect the variation in the kinetic energy ε_k in the region of the interaction and replace it by the mean value ε_{F}.

$$G_{\boldsymbol{kk}'} = G \quad \text{for} \quad k_{\mathrm{F}} - \Delta k \leqq |k| \leqq k_{\mathrm{F}} + \Delta k \tag{10.2a}$$

$$G_{\boldsymbol{kk}'} = 0 \quad \text{for} \quad \big||k| - k_{\mathrm{F}}\big| > \Delta k \tag{10.2b}$$

$$\varepsilon_k \approx \varepsilon_{\mathrm{F}} \quad \text{for} \quad k_{\mathrm{F}} - \Delta k \leqq |k| \leqq k_{\mathrm{F}} + \Delta k. \tag{10.2c}$$

This simplified problem can be solved with the quasi-spin operators of ch. 6. Below $k_{\mathrm{F}} - \Delta k$ there is no interaction and all the levels are occupied as in the non-interacting Fermi gas. Above $k_{\mathrm{F}} + \Delta k$ there is also no interaction and all the levels are empty as in the non-interacting case. In the region near k_{F} there is a constant interaction, and the variation in kinetic energy is neglected. This defines a problem exactly like the simplified model eq. (6.21) and can be solved exactly using quasi-spin operators defined in this region of $\boldsymbol{k}$. The solution has all levels in the region of interaction equally occupied and leads to a distribution as indicated in fig. 10.1.

We now have correlated or bound pairs of particles built up of states having equal and opposite momenta near the Fermi momentum. Using the results of ch. 6, the ground state energy and the excitation energies of simple excited states can be calculated for this model by substituting directly into

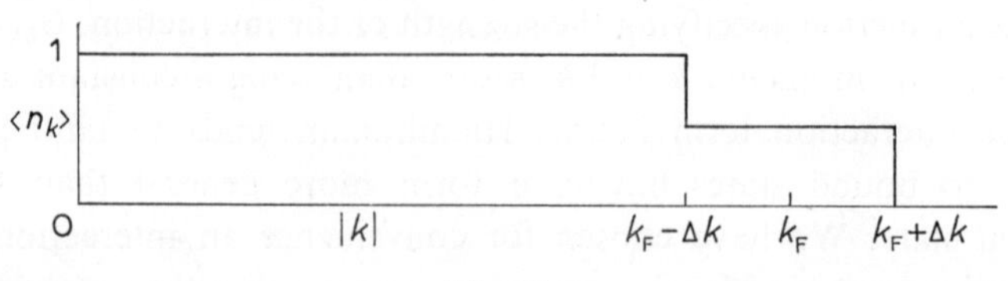

Fig. 10.1

eq. (6.35) for the interaction energy

$$V|m, v\rangle = -mG\Omega\left\{1 - \frac{(m - 1 + v)}{\Omega}\right\}|m, v\rangle. \tag{6.35}$$

For cases of practical interest the width of the interaction region in momentum space is small compared to the Fermi momentum and the density of states in this region can be considered to be constant to a good approximation. The number of interacting states above k_F is therefore equal to the number below k_F. The total number of states Ω for particles of a given spin in the interaction region is therefore just twice the number of states which would be completely occupied in the non-interacting Fermi gas. The number of particles in the interaction region is therefore exactly equal to Ω; i.e. half the number required to fill all the states completely.

The ground state can be seen from eq. (6.35) to have all the particles in the interaction region paired and no unpaired particles, i.e. $m=\frac{1}{2}\Omega$; $v=0$. The interaction energy in this state is

$$V_g = \frac{-G\Omega^2}{4}\left\{1 + \frac{2}{\Omega}\right\}. \tag{10.3a}$$

For an excited state with $m=\frac{1}{2}\Omega-\delta$, $v=v$, the excitation interaction energy is

$$V(\delta, v) - V_g = G\Omega\left\{\frac{v}{2} + \frac{(\delta + 1 - v)}{\Omega}\right\}. \tag{10.3b}$$

The kinetic energy for the ground state is just equal to that of the non-interacting Fermi gas in this simple model which neglects the variation in the kinetic energy in the interaction region. If we choose the zero of our energy scale as the energy of the non-interacting Fermi gas, the ground state energy is given by eq. (10.3a). Let us now examine the energies of the simplest excited states analogous to those involving the excitation of a single particle for the non-interacting Fermi gas. There are several types of excitations:

(a) Break-up of a pair with both particles remaining in the interaction region. The kinetic energy is the same as that of the ground state. The excitation energy is thus given by eq. (10.3b) with $\delta=1$, $v=2$.

(b) Break-up of a pair with one particle remaining in the interaction region and the other in a state $\boldsymbol{k}$ outside the interaction region. The interaction contribution to the excitation energy is given by eq. (10.3b) with $\delta=1$, $v=1$. The kinetic energy is increased over that of the ground state by the amount $\varepsilon_k-\varepsilon_F$.

(c) Excitation of a particle from a state $\boldsymbol{k}$ below the interaction region into the interaction region. The interaction contribution to the excitation energy is given by eq. (10.3b) with $\delta=0$, $v=1$. The kinetic energy is greater than that of the ground state by the amount $\varepsilon_F - \varepsilon_k$.

(d) Excitation of a particle in a state $\boldsymbol{k}$ below the interaction region to a state $\boldsymbol{k}'$ above the interaction region. The excitation energy for such a state is just the difference between the kinetic energies and is the same as for a non-interacting Fermi gas.

The excitation energy for each of the above states is thus

$$E_a - E_g = G\Omega \tag{10.4a}$$

$$E_b - E_g = G(\tfrac{1}{2}\Omega + 1) + \varepsilon_k - \varepsilon_F \tag{10.4b}$$

$$E_c - E_g = G\cdot\tfrac{1}{2}\Omega + \varepsilon_F - \varepsilon_k \tag{10.4c}$$

$$E_d - E_g = \varepsilon_{k'} - \varepsilon_k. \tag{10.4d}$$

These values show that there is no single-particle excitation of arbitrarily small energy as in the non-interacting Fermi gas. There is a minimum excitation energy or 'energy gap'. For excitation of particles within the interaction region the energy gap is $G\Omega$; for excitation of particles outside the interaction region, the energy gap is just the difference in the kinetic energy on opposite sides of the interaction region.

Another feature of these excitations is the blurring of particle and hole states in the interaction region. For example, the excited state b has one excited particle outside of the interaction region, but the interaction region, with one unpaired particle, is not a 'single hole' state in the sense of the non-interacting Fermi gas. Furthermore, removing a particle from the interaction region or adding an unpaired particle affects the motion and energies of all the other particles through the Pauli principle. Thus these excitations are not simple single-particle excitations in the sense of the non-interacting Fermi gas. Such 'quasiparticle' excitations carry the quantum numbers of a single particle, but may be either an ordinary particle or an ordinary hole or something more complicated, as in these examples where the change in the state of one particle also induces some change in the rest of the system. All the four types of excitations given above would be classified as two-quasiparticle excitations. The single-particle–single-hole excitations of a non-interacting Fermi gas would also be called two-quasiparticle excitations.

This simplified version of the BCS reduced Hamiltonian is sometimes called the strong-coupling limit because it represents the case where the interaction strength G (the coupling constant) is so strong that the variation of the kinetic energy in the interaction region can be neglected with respect to the interaction.

For the exact solution of the BCS Hamiltonian (10.1), where the variation of the kinetic energy in the interaction region is taken into account, one would expect the curve of $\langle n_k \rangle$vs. k to be a smoothed-out version of fig. 10.1, without the sharp steps at $k_F \pm \Delta k$, as indicated in fig. 10.2. The effect of the kinetic energy term should be to decrease the occupation of states above k_F and increase the occupation of those below. In the BCS treatment of this problem, a variational trial wave function is used which incorporates the features of this kind of distribution, namely, one exhibiting a 'spreading of the Fermi surface'. The variational parameters in the BCS wave function are essentially the expectation values $\langle n_k \rangle$. Thus the shape of the resulting distribution is determined by the variational principle.

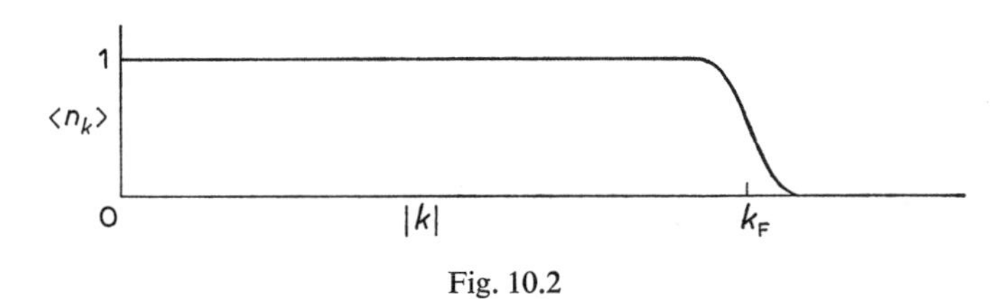

Fig. 10.2

10.3 The BCS wave function and the Bogoliubov–Valatin transformation

The BCS wave function is chosen to have the form

$$\text{BCS} = \prod_{\boldsymbol{k}} \{u_k + v_k a^{\dagger}_{\boldsymbol{k}\uparrow} a^{\dagger}_{-\boldsymbol{k}\downarrow}\}|0\rangle \tag{10.5a}$$

where u_k and v_k are real variational parameters satisfying the normalization condition

$$u_k^2 + v_k^2 = 1. \tag{10.5b}$$

By inspection, the expectation value for the number of particles in a particular state is given by

$$\langle \text{BCS}|n_{\boldsymbol{k}\uparrow}|\text{BCS}\rangle = \langle \text{BCS}|n_{-\boldsymbol{k}\downarrow}|\text{BCS}\rangle = |v_k^2|. \tag{10.5c}$$

Thus changing the values of the variational parameters v_k allows the BCS wave function to represent any arbitrary distribution of values of $\langle n_k \rangle$ and in particular those like fig. 10.2.

It is evident that the state (10.5) does not contain a definite number of particles. Each factor in the product consists of a term u_k which does not change the number of particles and a term proportional to v_k which adds two particles to the system. Thus the BCS wave function is a linear combination of states having many different numbers of particles. Although this may seem at first to be unreasonable, the BCS treatment actually gives a very good description of the ground state of the Hamiltonian (10.1) for large numbers of particles. One can consider the BCS wave function as describing average properties of an ensemble of systems having different numbers of particles analogous to the grand canonical ensemble in statistical mechanics. Since the number of particles in a superconductor is very large, the fluctuation in the number of particles in the BCS wave function is comparatively small and does not introduce any serious error.

The BCS wave function (10.5) is simply related to the 'bound pair' wave functions (6.7) used in the simple model. The wave function (6.7) for total momentum zero can be written

$$(D_0^\dagger)^m|0\rangle = \{\sum_{\boldsymbol{k}} (v_k/u_k)a^\dagger_{\boldsymbol{k}\uparrow}a^\dagger_{-\boldsymbol{k}\downarrow}\}^m|0\rangle \tag{10.6}$$

with the function g_k replaced by v_k/u_k. This wave function can be seen to be the component of the BCS wave function (10.5) having exactly $2m$ particles. The $2m$-particle component of (10.5) is obtained by taking all possible terms containing m factors $v_k a^\dagger_{\boldsymbol{k}\uparrow}a^\dagger_{-\boldsymbol{k}\downarrow}$ and factors u_k for all other values of $\boldsymbol{k}$. This is just the function (10.6) multiplied by a normalization factor $\prod_{\boldsymbol{k}} u_k$, since (10.6) also contains all possible products of m factors with the same coefficients. Note that all terms in (10.6) in which the same factor appears twice must vanish by the fermion anticommutation rules. Because the different $g_k = v_k/u_k$ are not all equal, as in the simpler quasi-spin model, calculations with this wave function are not simple. The BCS wave function which does not have a definite number of particles is much more easily used in calculations.

The BCS wave function can also be related to the Hartree–Fock formalism. The wave function (10.5) can be expressed as a 'vacuum' for a particular set of quasiparticle creation operators. Note that

$$(u_k a_{\boldsymbol{k}\uparrow} - v_k a^\dagger_{-\boldsymbol{k}\downarrow})|\mathrm{BCS}\rangle = 0 = (v_k a^\dagger_{\boldsymbol{k}} + u_k a_{-\boldsymbol{k}\downarrow})|\mathrm{BCS}\rangle. \tag{10.7}$$

Consider the operators

$$\alpha_{\boldsymbol{k}\uparrow} = u_k a_{\boldsymbol{k}\uparrow} - v_k a^\dagger_{-\boldsymbol{k}\downarrow} \tag{10.8a}$$

$$\alpha^\dagger_{\boldsymbol{k}\uparrow} = u_k a^\dagger_{\boldsymbol{k}\uparrow} - v_k a_{-\boldsymbol{k}\downarrow} \tag{10.8b}$$

$$\alpha_{\boldsymbol{k}\downarrow} = u_k a_{\boldsymbol{k}\downarrow} + v_k a^\dagger_{-\boldsymbol{k}\uparrow} \tag{10.8c}$$

$$\alpha^\dagger_{\boldsymbol{k}\downarrow} = u_k a^\dagger_{\boldsymbol{k}\downarrow} + v_k a_{-\boldsymbol{k}\uparrow}. \tag{10.8d}$$

These operators satisfy the same anticommutation rules as fermion creation and annihilation operators. Since the operators (10.8b) and (10.8d) add momentum $\hbar \boldsymbol{k}$ and $s_z = \pm\frac{1}{2}$, they can be considered as operators describing quasiparticles of momentum $\hbar \boldsymbol{k}$ and spin up and down respectively. These quasiparticles are neither pure particles nor pure holes but combinations of both. The BCS wave function is the quasiparticle vacuum:

$$\alpha_{\boldsymbol{k}\uparrow}|\text{BCS}\rangle = \alpha_{\boldsymbol{k}\downarrow}|\text{BCS}\rangle = 0. \tag{10.9}$$

The BCS wave function is thus a generalization of the Hartree–Fock wave function. It is a solution of the variational principle where the variation is the most general linear transformation of particle creation and annihilation operators, while the Hartree–Fock variation does not allow linear combinations of creation and annihilation operators. Such linear combinations cannot be allowed in a formulation where the number of particles is held constant. By analogy with the Hartree–Fock variational principle, the matrix elements of the Hamiltonian between the BCS quasiparticle vacuum and states containing only two quasiparticles would be expected to vanish. The only non-vanishing matrix elements would be the diagonal one and those exciting at least *four* quasiparticles. We shall see below that this is indeed the case.

The calculation of the variational principle can be done either in the original representation (10.5) or by the canonical transformation (10.8) to quasiparticle operators, due to Bogoliubov and Valatin. We use the original representation because the physical meaning of states and operators is more easily seen when expressed in terms of real particle states rather than those of quasiparticles.

The values of the parameters u_k and v_k are determined by minimizing the expectation value of the Hamiltonian (10.7). An additional constraint is required because varying these parameters can change the mean number of particles in the system. The procedure for keeping the mean number of particles at the desired value N_0 is to minimize the expectation value of the operator $H - \lambda N$ where N is the operator for the total number of particles and λ is a Lagrange multiplier, commonly called the chemical potential in statistical mechanics. The value of λ is determined by the condition that the expectation value of the operator N should be equal to N_0, the desired number of particles. The number operator is similar in form to the kinetic energy; thus the two terms are easily combined to give

$$H - \lambda N = \sum_{\boldsymbol{k}\sigma} (\varepsilon_k - \lambda) a^\dagger_{\boldsymbol{k}\sigma} a_{\boldsymbol{k}\sigma} - \sum_{\boldsymbol{k}\boldsymbol{k}'} G_{\boldsymbol{k}\boldsymbol{k}'} a^\dagger_{\boldsymbol{k}\uparrow} a^\dagger_{-\boldsymbol{k}\downarrow} a_{-\boldsymbol{k}'\downarrow} a_{\boldsymbol{k}'\uparrow}. \tag{10.10}$$

The chemical potential thus has the effect of changing the zero level of the single-particle energies.

The expectation value of (10.10) with the BCS wave function is easily calculated

$$\langle \mathrm{BCS}|H-\lambda N|\mathrm{BCS}\rangle = \sum_k 2(\varepsilon_k - \lambda)v_k^2 - \sum_{kk'} G_{\boldsymbol{kk}'}u_k v_k u_{k'} v_{k'}. \quad (10.11)$$

If this expectation value is minimized with respect to variation of v_k one obtains

$$2(\varepsilon_k - \lambda)u_k v_k = (\sum_{k'} G_{\boldsymbol{kk}'}u_{k'}v_{k'})(u_k^2 - v_k^2). \quad (10.12)$$

The requirement on the mean number of particles gives the condition

$$\sum_k 2v_k^2 = N_0. \quad (10.13)$$

The set of equations (10.12) and (10.13) are sufficient to determine the variational parameters v_k and the chemical potential λ. The explicit solution of these equations is rather complicated and in the general case is best done by numerical methods. Note that for an infinite system, the quantities u_k and v_k are continuous functions of the variable k and eq. (10.12) becomes an integral equation.

The solution simplifies in the special case of a constant value for the interaction matrix element $G_{\boldsymbol{kk}'}=G$ over a particular range of values of $\boldsymbol{k}$ as in eq. (10.12). For this case it is convenient to define

$$\Delta \equiv \sum_{k'} G_{\boldsymbol{kk}'}u_{k'}v_{k'} = G\sum_k u_k v_k. \quad (10.14)$$

Eq. (10.12) then becomes

$$2(\varepsilon_k - \lambda)u_k v_k = \Delta(u_k^2 - v_k^2). \quad (10.15)$$

Squaring eq. (10.15) and remembering that $u_k^2+v_k^2=1$, we obtain

$$4(\varepsilon_k - \lambda)^2 u_k^2 v_k^2 = (\varepsilon_k - \lambda)^2[1 - (u_k^2 - v_k^2)^2] = \Delta^2(u_k^2 - v_k^2)^2 \quad (10.16)$$

from which we obtain

$$u_k^2 - v_k^2 = \frac{\varepsilon_k - \lambda}{[(\varepsilon_k - \lambda)^2 + \Delta^2]^{\frac{1}{2}}} \quad (10.17a)$$

$$u_k^2 = \tfrac{1}{2}\left[1 + \frac{\varepsilon_k - \lambda}{[(\varepsilon_k - \lambda)^2 + \Delta^2]^{\frac{1}{2}}}\right] \quad (10.17b)$$

$$v_k^2 = \tfrac{1}{2}\left[1 - \frac{\varepsilon_k - \lambda}{[(\varepsilon_k - \lambda)^2 + \Delta^2]^{\frac{1}{2}}}\right]. \quad (10.17c)$$

Eqs. (10.17) give u_k and v_k in terms of the parameters λ and Δ which are as yet undetermined. However, some general features of the wave function are already evident. In particular, one can see that the plot of v_k^2 versus k has indeed the form of fig. 10.2. We note that v_k^2 is a monotonically decreasing function of ε_k and that when $\varepsilon_k=\lambda$, $v_k^2=\frac{1}{2}$; i.e. the probability of finding a particle in this state is just $\frac{1}{2}$. For $\varepsilon_k \gg \lambda$, $v_k^2=0$, whereas for $\varepsilon_k \ll \lambda$, $v_k^2=1$. Thus the curve of v_k^2 versus k drops from 1 to 0 in the vicinity of the energy λ. The transition is smooth rather than sharp as for a non-interacting Fermi gas and occurs over an energy region of the order of Δ. In order for the mean number of particles to come out correctly, the value of λ must clearly be somewhere in the neighborhood of ε_F, as can be seen from eq. (10.13).

10.4 The excitation spectrum and the energy gap

We can now also calculate the form of the quasiparticle excitation spectrum. We first consider the wave function

$$|\boldsymbol{k}\uparrow\rangle = \frac{1}{u_k} a_{\boldsymbol{k}\uparrow}^{\dagger}|\mathrm{BCS}\rangle. \tag{10.18}$$

Since the square of any fermion creation operator vanishes. the operator $a_{\boldsymbol{k}\uparrow}^{\dagger}$ eliminates the term $v_k a_{\boldsymbol{k}\uparrow}^{\dagger} a_{-\boldsymbol{k}\downarrow}^{\dagger}$ in the BCS wave function (10.5) and replaces the constant term u_k by the creation operator $a_{\boldsymbol{k}\uparrow}^{\dagger}$. The state (10.14) is thus one in which the state $\boldsymbol{k}\uparrow$ is certainly occupied by a particle whereas the state $-\boldsymbol{k}\downarrow$ is certainly empty, while the other states are occupied in the same way as in the BCS ground state. This state can be considered as representing a quasiparticle in the state $\boldsymbol{k}\uparrow$ added to the BCS ground state. It corresponds to adding a particle in this state to the system, thereby changing the state of the other particles by making the pair of states $\boldsymbol{k}\uparrow$ and $-\boldsymbol{k}\downarrow$ unavailable to them.

Let us now calculate the change in the expectation value of the Hamiltonian (10.11) when the BCS ground state is replaced by the single-quasiparticle state (10.18). The only terms which are changed in (10.11) are those involving the particular state $\boldsymbol{k}$ of the quasiparticle. The first term, the kinetic energy, is changed by replacing the total occupation of the states $(\boldsymbol{k}\uparrow, -\boldsymbol{k}\downarrow)$ by 1 instead of $2v_k^2$. The second term, the interaction, is changed by eliminating from the summation, over both indices, the terms involving the particular state $\boldsymbol{k}$, as the corresponding terms in the Hamiltonian (10.10) now give zero in acting on the state (10.18). Since one of the two paired states $(\boldsymbol{k}\uparrow, -k\downarrow)$ is occupied and the other is not, it is neither possible to add a pair of particles nor to remove a pair of particles from these states. The energy difference between the state (10.18) and the ground state is then

$$E_k = \langle \boldsymbol{k}\uparrow|H - \lambda N|\boldsymbol{k}\uparrow\rangle - \langle \text{BCS}|H - \lambda N|\text{BCS}\rangle$$

$$= (1 - 2v_k^2)(\varepsilon_k - \lambda) + 2\Delta u_k v_k. \tag{10.19}$$

Substituting eqs. (10.15) and (10.17) we obtain

$$E_k = \left[(u_k^2 - v_k^2)(\varepsilon_k - \lambda) + \frac{\Delta^2}{(\varepsilon_k - \lambda)}(u_k^2 - v_k^2)\right. \tag{10.20a}$$

$$E_k = \frac{(\varepsilon_k - \lambda)^2}{[(\varepsilon_k - \lambda)^2 + \Delta^2]^{\frac{1}{2}}} + \frac{\Delta^2}{[(\varepsilon_k - \lambda)^2 + \Delta^2]^{\frac{1}{2}}} \tag{10.20b}$$

$$E_k = [(\varepsilon_k - \lambda)^2 + \Delta^2]^{\frac{1}{2}}. \tag{10.20c}$$

If $\Delta = 0$ and $\lambda = \varepsilon_F$, this result is just that expected for a non-interacting Fermi gas, either for adding a particle in a state k above the Fermi surface, or removing a particle (adding a hole) at a state k below the Fermi surface. If $\Delta \neq 0$, the energy is always larger than the corresponding value for the non-interacting Fermi gas and approaches the finite value Δ as a minimum energy for $\varepsilon_k = \lambda$. The quantity Δ thus represents an energy gap in the quasi-particle spectrum.

The state (10.18) clearly cannot be an excited state of the system for which the wave function (10.5) is the ground state, since the state (10.18) is a linear combination of states having an odd number of particles whereas the state (10.5) is a linear combination of states having an even number of particles. To obtain excited states of the system, one must excite an even number of quasiparticles corresponding to the creation of an integral number of particle–hole pairs for the case of the non-interacting Fermi gas.

The form of the Hamiltonian and the wave functions show that the energies of quasiparticles are additive except for a small correction. Each additional quasiparticle changes the appropriate term in the kinetic energy and eliminates the corresponding terms from the interaction energy. In adding the energy of a quasiparticle in a state $\boldsymbol{k}'$ to that of the energy of a quasiparticle in a state $\boldsymbol{k}''$ to get the energy of a two-quasiparticle state, a small error is made by counting twice the one interaction term involving both $\boldsymbol{k}'$ and $\boldsymbol{k}''$. If the total number of states in the sum is very large, this is a very small error. Another small error is introduced by the change in the mean number of particles upon adding a quasiparticle in a state for which v_k^2 is not exactly equal to $\frac{1}{2}$. This is also small if the number of particles is very large.

The concept of particle or hole, useful for the non-interacting Fermi gas, is not strictly relevant here. The state (10.18) apparently produced by adding a particle to the ground state can also be produced by adding a hole. By inspection

$$|k\uparrow\rangle = \frac{1}{u_k} a^{\dagger}_{k\uparrow}|\mathrm{BCS}\rangle = \frac{1}{v_k} a_{-k\downarrow}|\mathrm{BCS}\rangle. \tag{10.21}$$

A quasiparticle is thus neither a particle nor a hole but combines features of both. A two-quasiparticle state is therefore analogous to a particle–hole excitation in the non-interacting Fermi gas. From the result (10.20) we see that if Δ is not equal to 0, there is an *energy gap* in the quasiparticle excitation spectrum.

Let us now evaluate the energy-gap parameter Δ. From eqs. (10.14) and (10.15)

$$\Delta = \frac{G}{2}\sum_k \left\{\frac{\Delta}{(\varepsilon_k - \lambda)}(u_k^2 - v_k^2)\right\} = \Delta \frac{G}{2}\sum_k \frac{1}{[(\varepsilon_k - \lambda)^2 + \Delta^2]^{\frac{1}{2}}}. \tag{10.22}$$

Eq. (10.18) clearly has the trivial solution $\Delta=0$. This gives no energy gap. From (10.13) $v_k^2=1$ for $\varepsilon_k<\lambda$ and $v_k^2=0$ for $\varepsilon_k>\lambda$. The wave function is therefore just that of the non-interacting Fermi gas, the so-called 'normal state', and λ must be equal to ε_F to give the proper number of particles.

The non-trivial 'superconducting' solution $\Delta\neq 0$ is given by the solution of the equation

$$1 = \frac{G}{2}\sum_k \frac{1}{[(\varepsilon_k - \lambda)^2 + \Delta^2]^{\frac{1}{2}}}. \tag{10.23}$$

The chemical potential λ is determined by the equation

$$\sum_k 2v_k^2 = \sum_k \left\{1 - \frac{\varepsilon_k - \lambda}{[(\varepsilon_k - \lambda)^2 + \Delta^2]^{\frac{1}{2}}}\right\} = N_0. \tag{10.24}$$

If the region of interaction is small compared to the Fermi energy, the density of states of a given spin per unit energy in the interaction region can be considered to be constant. Let us denote this density by ϱ and convert the sum in eq. (10.24) to an integral. Let us define a quantity ω such that the spreading of the Fermi surface is all contained within the region from $\lambda-\omega$ to $\lambda+\omega$; i.e. below $\lambda-\omega$, $v_k=1$ and above $\lambda+\omega$, $v_k=0$. Then

$$\sum_{\substack{k \\ \varepsilon<(\lambda-\omega)}} 2 + \int_{\lambda-\omega}^{\lambda+\omega} \varrho \left\{1 - \frac{\varepsilon - \lambda}{[(\varepsilon - \lambda)^2 + \Delta^2]^{\frac{1}{2}}}\right\} \mathrm{d}\varepsilon = N_0. \tag{10.25}$$

From the symmetry of the integral about the point $\varepsilon=\lambda$ the second term in the integral vanishes and the left-hand side of eq. (10.25) is just equal to the total number of electrons which can be place below the energy λ. Thus

$$\lambda = \varepsilon_F. \tag{10.26}$$

We now determine the energy-gap parameter Δ by converting the sum in eq. (10.23) to an integral and evaluating it. The limits of the integration are the limits of the region of interaction expressed in energy units. If we define the parameter ω such that the region of interaction is between $\lambda-\omega$ and $\lambda+\omega$ we obtain

$$1 = \tfrac{1}{2}G\varrho \int_{\lambda-\omega}^{\lambda+\omega} \frac{d\varepsilon}{[(\varepsilon-\lambda)^2 + \Delta^2]^{\frac{1}{2}}} = G\varrho \sinh^{-1}(\omega/\Delta). \tag{10.27}$$

Thus

$$\Delta = \frac{\omega}{\sinh(1/G\varrho)}. \tag{10.28}$$

For the case of the strong-coupling limit ($G\varrho \gg 1$),

$$\Delta \approx G\varrho\omega \approx \tfrac{1}{2}G\Omega \qquad (G \gg 1) \tag{10.29}$$

where $\Omega = 2\varrho\omega$ is the total number of states in the interaction region for electrons of a given spin as was defined in the simpler model. Since the energy gap is 2Δ for an excitation of two quasiparticles this agrees with our exact result (10.4a) for the strong-coupling limit.

The weak-coupling limit $G\varrho \ll 1$ is relevant to physical superconductors. For this case we obtain

$$\Delta \approx 2\omega e^{-(1/G\varrho)} \qquad (G\varrho \ll 1). \tag{10.30}$$

The results (10.28) and (10.30) are not analytic functions of G at $G=0$, but have essential singularities at this point. These results cannot be expanded in a power series in the parameter G in the vicinity of $G=0$. The treatment of the BCS Hamiltonian by perturbation theory with the interaction as a perturbation gives a result as a power series in G. Because of the non-analytic behavior of the solution such a perturbation expansion would not give the correct result.

The expression for the energy-gap parameter Δ has the following peculiar feature: It is proportional to a parameter ω having the dimensions of energy and representing the range of single-particle energies over which the interaction exists. However Δ is of a completely different order of magnitude from ω. The factor ω is multiplied by a dimensionless exponential factor which can be quite small. In physical superconductors Δ is two orders of magnitude smaller than ω. The value of the energy-gap parameter Δ is characterized by the critical temperature; i.e. the amount of thermal energy necessary to destroy the superconducting state. This is of the order of a few degrees Kelvin. On the other hand, the energy range ω over which the interaction takes place, is characterized by the energies of the lattice vibrations (phonons) corresponding to several hundred degrees Kelvin.

The BCS theory gives a very good description of many properties of superconductors including the persistent current, the Meissner effect and electromagnetic properties. The qualitative picture of a current of bound pairs is similar to that discussed in the context of the simple model in section 6.4.

10.5 Pairing correlations in nuclei

The treatment in section 6.4 gives a good description of pairing correlations and seniority in a single shell in the *j–j* coupling nuclear shell model. Because of spherical symmetry all of the single-particle states in the same shell have the same Hartree–Fock energy and it is only necessary to diagonalize the interaction. When several shells are considered the generalized seniority (6.39) provides a good description only in the strong-coupling limit; i.e. when the difference between the Hartree–Fock energies of the different shells can be neglected. For realistic cases where these energies must be taken into account one has a Hamiltonian similar to eq. (10.1) where ε_k is the Hartree–Fock energy and the labels of the single-particle states should be changed to a spherical basis so that the interaction has the form of eq. (6.39b). For this case the BCS treatment can be used with a BCS wave function (10 5). The treatment formally follows exactly that for the superconductor except that ε_k are the single-particle Hartree–Fock energies and the states are the shell model orbitals. The BCS equations (10.23) and (10.24) remain discrete sums over the relevant shell model states, and are solved numerically.

In the nuclear case eq. (10.23) does not have a solution unless the interaction strength G is greater than some critical value. This is easily seen by writing eq. (10.23) in the form

$$\frac{1}{G} = \frac{1}{2}\sum_{k}\frac{1}{[(\varepsilon_k - \lambda)^2 + \Delta^2]^{\frac{1}{2}}} \leqq \frac{1}{2}\sum_{k}\frac{1}{|\varepsilon_k - \lambda|}\,. \tag{10.31}$$

Since the summation on the right-hand side of eq. (10.31) is over a finite number of states the right-hand side of the inequality has a finite value and defines a lower limit for G for which a non-trivial solution to the equation exists. This limit does not exist in the case of the superconductor where the spectrum of single-particle states is continuous for all practical purposes and the sum on the right-hand side of eq. (10.31) is either infinite or very large and proportional to the size of the box. Thus one finds a solution (10.31) which is valid for arbitrarily small G. The BCS formalism applies also to deformed nuclei where the Hartree–Fock states are not spherically symmetric. In that case the single-particle energy ε_k in eq. (10.1) is the energy in the deformed self-consistent potential and depends on m as well as on j. The formal solution is exactly the same as in the spherical case with several shells or the superconductor.

The BCS formulation has been applied extensively to the structure of heavy nuclei where there is appreciable configuration mixing between different shells.

CHAPTER 11

ELEMENTARY EXCITATIONS IN MANY-BODY SYSTEMS

11.1 Two examples of many-body problems

The Hartree–Fock approximation neglects the residual two-body interaction and the correlations between the motions of different particles. As illustrations of the problems raised by correlations and the methods of treatment of the residual interaction, we consider two problems: a macroscopic problem for which the exact Hamiltonian is known and a nuclear problem in which a model Hamiltonian is used which contains parameters not known a priori. We shall see how the same formalism can be used in the treatment of both problems, and how this usage is justified by different considerations in the two cases.

Consider first an electron gas in a uniform background of positive charge to make the system electrically neutral. This is a simplified model for conduction electrons in a metal or for electrons in a plasma, in which the positive ions are replaced by a continuum of positive charge. For convenience the electrons are enclosed in a cubical box with periodic boundary conditions. The Hamiltonian consists of the kinetic energy and the Coulomb interaction between the charges.

$$H = \sum_i (p_i^2/2m) + \sum_{i>j} e^2/(|\boldsymbol{r}_i - \boldsymbol{r}_j|) - \sum_i V(\boldsymbol{r}_i), \tag{11.1}$$

where $V(\boldsymbol{r}_i)$ is the potential due to the uniform background of positive charge.

For a zero-order wave function we choose a single-particle basis of plane-wave states in the box, which are eigenfunctions of the kinetic energy. The ground state is the 'Fermi sphere', in which all states within a sphere in momentum space are occupied. This wave function is a solution of the Hartree–Fock equations, since the Coulomb interaction conserves momentum and has no matrix elements exciting only a single particle from the Fermi sphere. Many physical phenomena, particularly conductivity of metals, are well described by the simple wave functions which neglect the off-diagonal elements of the Coulomb interaction between electrons. However the residual interaction cannot be treated by simple perturbation theory. The second-order contribution to the ground-state energy turns out to be infinite, and higher-order contributions diverge even worse. We shall see below the reason for this divergence, and then understand how these 'bad' wave functions can still give results which are good approximations for certain physical quantities.

As a second example consider the nuclear shell model for the nucleus ^{16}O. The Hamiltonian includes a single-particle potential $V(\boldsymbol{r}_i)$ and a residual two-body interaction $v_{ij}(\boldsymbol{r}_i - \boldsymbol{r}_j)$

$$H = \sum_i (p_i^2/2m) + \sum_i V(\boldsymbol{r}_i) + \sum_{i>j} v_{ij}(|\boldsymbol{r}_i - \boldsymbol{r}_j|). \tag{11.2}$$

The electron gas treats a very large number of particles, this problem has only sixteen. The electron gas has a well-defined Hamiltonian; namely one involving the electrostatic Coulomb interaction; this Hamiltonian contains a potential $V(\boldsymbol{r}_i)$ which is not known at all from first principles, and a 'residual interaction' which is also not clearly defined. One might expect that the single-particle potential $V(\boldsymbol{r}_i)$ should be defined as the Hartree–Fock potential for the real two-body interaction. However, the real two-body interaction is not yet known, and there are difficulties in obtaining a Hartree–Fock solution for realistic potentials because of the strong repulsive core. The potential $V(\boldsymbol{r}_i)$ is usually an ad hoc potential chosen to fit certain experimental nuclear data.

The zero-order wave function considers individual particles moving in single-particle orbits determined by the potential $V(\boldsymbol{r}_i)$. The lowest orbit is 1s, the next is 1p. Since the protons and neutrons have two possible spin orientations, the 1s orbit can accommodate 4 nucleons and the 1p can accommodate 12. Thus the zero-order ground-state wave function for ^{16}O has all the 1s and 1p orbits filled; i.e. it is a 'double closed shell'. The lowest excited-state wave functions then have one particle excited from the p-shell into the next shell, including 2s and 1d orbits. These excitations can be considered as 'particle–hole' excitations, since they involve removing a particle from a filled shell (creating a hole) into an unfilled shell (creating an excited particle).

The 'particle–hole' excitations are of particular interest in both these problems because they are produced by the action of a single-particle operator on the ground state; e.g. by photon absorption or by electron scattering in the Born approximation. However, although the action of such a single-particle operator on the ground state is trivial if the ground-state wave function is of the single-particle type the action of such an operator is very complicated on a ground-state wave function which is *not* a simple single-particle state. There is good reason to believe that the simple single-particle description is not adequate for either of the two examples under consideration. For the electron gas this is evident from the failure of second-order perturbation theory. For the ^{16}O nucleus (where we have no a priori theory which is tractable) this is evident from experimental data. Thus in both cases, the problem arises of how to calculate the excitation spectrum produced by a particular type of single-particle excitation, such as photon absorption, using a zero-order wave function of the single-particle type which is known to be a *bad* wave function.

We now investigate both these cases in detail, remembering that we want the excitation spectrum for a particular set of single-particle operators, and that we do not care about other aspects of the system under consideration.

It is convenient to express the Hamiltonians for both these problems in second quantization. For the electron gas, we use the Hartree–Fock plane-wave basis and express the Coulomb interaction as a Fourier expansion

$$\sum_{i>j} \frac{e^2}{|\boldsymbol{r}_i - \boldsymbol{r}_j|} = \sum_{\boldsymbol{k}} \sum_{i>j} v_k \mathrm{e}^{\mathrm{i}\boldsymbol{k}\cdot(\boldsymbol{r}_i - \boldsymbol{r}_j)}. \tag{11.3a}$$

The expansion coefficient v_k is

$$v_k = 4\pi e^2/k^2\Omega \tag{11.3b}$$

where Ω is the volume of the box.

The expression (11.3a) can be transformed to a form more suitable for the change to second-quantized operators by making the sums on i and j independent. This can be done by adding and subtracting a term $i=j$, and introducing a factor $\frac{1}{2}$ to compensate for double counting with independent i and j.

$$\sum_{i>j} \frac{e^2}{|\boldsymbol{r}_i - \boldsymbol{r}_j|} = \frac{1}{2} \sum_{\boldsymbol{k}} v_k [(\sum_i \mathrm{e}^{\mathrm{i}\boldsymbol{k}\cdot\boldsymbol{r}_i})(\sum_j \mathrm{e}^{-\mathrm{i}\boldsymbol{k}\cdot\boldsymbol{r}_j}) - n], \tag{11.4}$$

where n is the total number of particles, and the term $-n$ cancels out the added $i=j$ term.

The contribution of the term $k=0$ in the sum, which appears to be singular, is just cancelled by the uniform background of positive charge. The Hamiltonian is now easily written in second-quantized form. Let $c^{\dagger}_{\boldsymbol{k}\sigma}$ represent a creation operator for an electron in a state of momentum $\hbar\boldsymbol{k}$ and spin σ ($\sigma = \pm 1$ for spin 'up' or 'down'). Then

$$H = \sum_{p} \frac{(\hbar p)^2}{2m} c^\dagger_{p\sigma} c_{p\sigma} + \tfrac{1}{2} \sum_{k \neq 0} v_k \{ c^\dagger_{(p+k)\sigma} c_{p\sigma} c^\dagger_{q\sigma} c_{(q+k)\sigma} \}. \tag{11.5}$$

The operators appearing in the interaction (11.4) are just the Fourier components of the electron density. The electron density operator for a system of electrons is represented as a sum of delta functions, infinite at every point where an electron is, and zero elsewhere,

$$\varrho(r) = \sum_i \delta(r - r_i) = \sum_k \sum_i \mathrm{e}^{ik \cdot r} \mathrm{e}^{-ik \cdot r_i} = \sum_k \varrho_k \mathrm{e}^{ik \cdot r} \tag{11.6a}$$

where the Fourier component ϱ_k of the electron density is

$$\varrho_k = \sum_i \mathrm{e}^{-ik \cdot r_i}. \tag{11.6b}$$

The Coulomb interaction can thus be written

$$\tfrac{1}{2} \sum_k v_k (\varrho^*_k \varrho_k - n). \tag{11.6c}$$

The Hamiltonian for the ^{16}O nucleus can be written more conveniently in the shell-model basis, where each single-particle state is specified by the radial quantum number n, orbital angular momentum l, total angular momentum j, and the projection m of the total angular momentum on the z-axis. For convenience we use a single index k to denote the four quantum numbers (n, l, j, m) and let $a^\dagger_k$ represent a creation operator for a nucleus in a state k. The Hamiltonian then assumes the form

$$H = \sum_k \varepsilon_k a^\dagger_k a_k + \tfrac{1}{2} \sum_{k'k''k'''k^{\mathrm{iv}}} \langle k^{\mathrm{iv}} k''' | V | k'' k' \rangle a^\dagger_{k^{\mathrm{iv}}} a^\dagger_{k'''} a_{k''} a_{k'} \tag{11.7}$$

where ε_k are the energies of the single-particle states in the shell-model potential and the coefficients in the interaction term are the matrix elements of the residual interaction. Note again the difference between the two Hamiltonians. The electron gas Hamiltonian is completely defined in terms of known physical constants. The nuclear Hamiltonian contains as parameters the single-particle energies ε_k and the interaction matrix elements which are not known from first principles. Their values must somehow be fed into the problem. We shall see that the choice of these parameters is often determined from experimental data and therefore is also influenced by the approximations made in solving the problem.

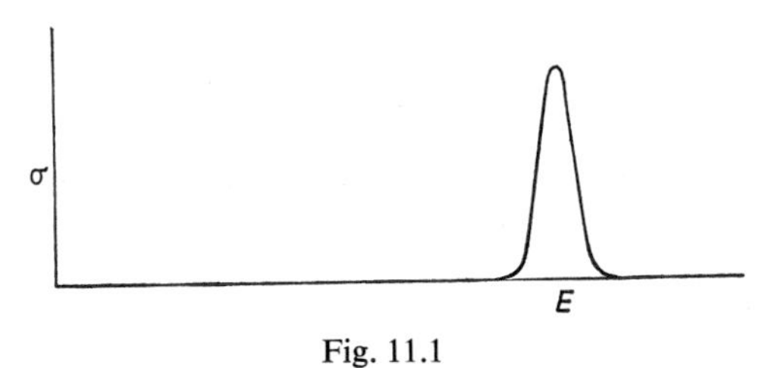

Fig. 11.1

11.2 Stationary states and giant resonances

Let us consider the absorption cross section for photons by the electron gas. Suppose, for example, that in a particular region, the cross section observed experimentally has a sharp peak as a function of energy, as shown in fig. 11.1. There are two possible explanations for such a peak. There can be a single state which is an eigenfunction of the Hamiltonian of the electron gas and which has a very strong transition matrix element for photon absorption from the ground state, while other states in the same energy region have very low matrix elements. One then says that all of the transition 'strength' in this region lies in this particular state. However, the transition strength can also be distributed among a large number of states which are eigenfunctions of the Hamiltonian, all of which lie very close to one another in energy. Such a concentration of transition strength in many states within a small energy region is often called a 'giant resonance'. If the peak is due to a single state, it should have a very narrow width, just the natural line width determined by the radiative transitions back to the ground state and by other possible transitions. If the peak is a giant resonance, the width is determined by the dynamics of the system, and a fine structure of the peak can be observed if the spacing between those stationary states of the system contributing to the giant resonance is greater than their natural line width.

A giant resonance can also be described as a single state which is not an eigenfunction of the Hamiltonian. It is a linear combination of eigenfunctions in a small energy range; i.e. a kind of 'wave packet' containing all the transition strength. The coefficient of each eigenfunction in the linear combination is proportional to the transition matrix element for that eigenfunction. Such a giant resonance state is not an approximate eigenfunction of the Hamiltonian in the sense of perturbation theory. It is a linear combination of many eigenfunctions, each of which has a small coefficient. However, it can be considered an approximate 'stationary state' of the system in the following sense. If the system is in this state at some time $t=0$, it will remain 'in this state' for a time determined by the width of the giant resonance. If the corresponding solution of the time-dependent Schrödinger equation is expressed as a linear combination of stationary states, each with a slightly different frequency, the state will remain in the giant resonance state to a good approximation until the different components in the expansion get 'out of phase'. The 'lifetime' of the state is just the reciprocal of the frequency spread in the expansion.

This can be expressed quantitatively as follows. Let the operator M describe the transition for absorption of a photon with a particular set of quantum numbers (e.g. electric dipole). The transition amplitude for the excitation of a particular eigenfunction $|E'\rangle$ of the Hamiltonian with energy E' is then proportional to the matrix element $\langle E'|M|i\rangle$, where $|i\rangle$ is the initial state.

Consider the state

$$|\psi(0)\rangle = N \sum_{E'} |E'\rangle\langle E'|M|i\rangle \tag{11.8a}$$

where N is a normalization constant. Then

$$\langle E'|\psi(0)\rangle = N\langle E'|M|i\rangle, \tag{11.8b}$$

and

$$\sum_{E'} \left|\langle E'|\psi(0)\rangle\right|^2 = \sum_{E'} N^2\left|\langle E'|M|i\rangle\right|^2 = 1. \tag{11.8c}$$

For any arbitrary state $|\xi\rangle$,

$$\langle \xi|M|i\rangle = \sum_{E'} \langle \xi|E'\rangle\langle E'|M|i\rangle = \frac{1}{N}\langle \xi|\psi(0)\rangle. \tag{11.9a}$$

From eqs. (11.8) we can write

$$\begin{aligned}\left|\langle\psi(0)|M|i\rangle\right|^2 &= \left|\sum_{E'} \langle\psi(0)|E'\rangle\langle E'|M|i\rangle\right|^2 \\ &= N^2\{\sum_{E'} \left|\langle E'|M|i\rangle\right|^2\}^2 = \sum_{E'} \left|\langle E'|M|i\rangle\right|^2.\end{aligned} \tag{11.9b}$$

The state $|\psi(0)\rangle$ is thus defined to contain all the transition strength. Eq. (11.9a) shows that the transition matrix element $\langle\xi|M|i\rangle$ is proportional to the overlap between the states $|\xi\rangle$ and $|\psi(0)\rangle$. Eq. (11.9b) shows that the square of the matrix element $\langle\psi(0)|M|i\rangle$ gives the total cross section summed over all states E.

If the system is in the state (11.8a) at time $t=0$, then the state at a later time t is

$$|\psi(t)\rangle = \sum_{E'} |E'\rangle\langle E'|\psi(0)\rangle \mathrm{e}^{-\mathrm{i}E't/\hbar}. \tag{11.10}$$

This state (11.10) has a mean energy and mean square width which is independent of time and given by

$$\bar{E} = \sum_{E'} E' \left|\langle E'|\psi(t)\rangle\right|^2 = \sum_{E'} E' \left|\langle E'|\psi(0)\rangle\right|^2 \tag{11.11a}$$

$$\overline{\Delta E^2} = \sum_{E'} (E' - \bar{E})^2 \left|\langle E'|\psi(t)\rangle\right|^2 = \sum_{E'} (E' - \bar{E})^2 \left|\langle E'|\psi(0)\rangle\right|^2. \tag{11.11b}$$

This state is a solution of the time-dependent Schrödinger equation which is the 'giant resonance' state at time $t=0$. To determine the 'lifetime' or 'decay' of the giant resonance state, we can calculate the overlap of the state (11.10) with the state $|\psi(0)\rangle$ as a function of the time t:

$$\begin{aligned} \langle\psi(t)|\psi(0)\rangle &= \sum_{E'} \left|\langle E'|\psi(0)\rangle\right|^2 e^{iE't/\hbar} \\ &= e^{i\bar{E}t/\hbar} \sum_{E'} \left|\langle E'|\psi(0)\rangle\right|^2 \\ &\quad \times \{1 + i(E' - \bar{E})t/\hbar - \tfrac{1}{2}(E' - \bar{E})^2 t^2/\hbar^2 + \ldots\} \qquad (11.12a) \\ &= e^{i\bar{E}t/\hbar}\{1 - \tfrac{1}{2}\overline{(\Delta E)^2}\, t^2/\hbar^2 + \ldots\}. \qquad (11.12b) \end{aligned}$$

The overlap is thus of magnitude unity for times t small compared with $\hbar[\overline{(\Delta E)^2}]^{-\frac{1}{2}}$, and decreases when t is of the same order as this quantity. Thus the lifetime of the state is related to the energy width of the packet as one expects from the uncertainty principle.

Suppose the giant resonance state $|\psi(0)\rangle$ is obtained as an approximate eigenstate of the Hamiltonian by some approximation method. According to the normal standards, such as that of perturbation theory, this is a very poor approximation to any particular exact eigenstate. However, if it is used to calculate the photon absorption cross section, it gives a peak at approximately the right energy, and with the right strength (area under peak). It does not give the correct line width, nor does it give any fine structure. However, the location of the peak and its strength are often all the information desired for comparison with experimental measurements, particularly if the latter do not have the resolution to determine fine structure or widths. In any case, good values for the location and strength of the peak are useful first approximations.

In many many-particle systems giant resonance excitations are produced by common experimental techniques. Photon absorption, electron scattering, and nuclear direct reactions often excite the system into a region where there is a continuum of energy levels. The particular type of excitation is expressed by a simple operator, such as a single-particle operator, acting on the initial state. If the system is described by a single-particle model, such as the Hartree–Fock approximation, these excitations often produce a number of discrete and distinct excited states. Although the single-particle description may not be very good at all, the deviations from this description may only affect these excitations by mixing the discrete excited states with a large number of neighboring states; i.e. by changing these single eigenstates into giant resonances. In such cases the single-particle description may be very useful in describing the location and strengths of the peaks, even though the eigenfunctions given in this description are far from the exact eigenfunctions of the system and may give an extremely poor description of inter-particle correlations not relevant to these giant resonances.

Excitations produced by a simple mechanism and leading to states of the system which are approximately stationary (i.e. they have a long lifetime or narrow width) are sometimes called 'elementary excitations'. The study of such elementary excitations, both theoretically and experimentally, has proved to be a valuable tool in exploring the structure of many-particle systems. Many approximate methods for calculation are aimed primarily at determining the properties of these elementary excitations, rather than at obtaining an exact or even good approximate solution of the Schrödinger equation.

11.3 An example of an elementary excitation

Consider electric quadrupole excitations of the nucleus ^{16}O. The operator for electric quadrupole absorption in the long-wavelength limit is just the electric quadrupole moment of the system, namely the sum of the quadrupole moments of the individual particles.

$$Q_M = \sum_i r_i^2 Y_M^2(\theta_i, \phi_i) \tag{11.13}$$

where Y_M^2 is a spherical harmonic.

If the ^{16}O nucleus is described by a harmonic oscillator shell model, the energy levels are equally spaced with a splitting $\hbar\omega$ and the lowest s and p-levels are filled. The operator Q_M consists of terms which each raise the energy of a given particle by $2\hbar\omega$, either from the $n=0$ s-level to the $n=2$ sd-shell, or from the $n=1$ p-shell to the $n=3$ pf-shell. The operator can also have diagonal matrix elements and matrix elements which lower the energy by two units but these vanish for the particular case of the ^{16}O ground state shell model wave function. Thus, in this approximation, the excitation spectrum produced by the absorption of electric-quadrupole photons would be a delta function at excitation $2\hbar\omega$.

We now note that the single-particle levels of the harmonic oscillator are split by spin–orbit forces and by deviations of the potential from a pure harmonic oscillator. Thus, the p-shell splits into $p_{\frac{3}{2}}$ and $p_{\frac{1}{2}}$, the sd-shell into $s_{\frac{1}{2}}$, $d_{\frac{3}{2}}$ and $d_{\frac{5}{2}}$ and the pf-shell into $p_{\frac{1}{2}}$, $p_{\frac{3}{2}}$, $f_{\frac{5}{2}}$ and $f_{\frac{7}{2}}$. There are thus three hole levels and seven particle levels participating in the quadrupole excitation. Fifteen of the 21 possible particle–hole configurations can be coupled to angular momentum two and have non-vanishing matrix elements of the quadrupole operator (11.12). Thus, when corrections to the harmonic oscillator single-particle levels are taken into account the quadrupole excitation is no longer a single line but is split into 15 components.

We next consider the effects of the residual two-body interactions, not included in the Hartree–Fock approximation. We first examine the matrix elements of these two-body interactions within the subset of excited particle–hole states with angular momentum and parity 2^+. Such a two-particle interaction has off-diagonal matrix elements connecting the different one-particle, one-hole states. We can obtain a better approximation by using degenerate perturbation theory and diagonalizing the two-body interaction in the subspace of these one-particle, one-hole states. This results in a shift of the energy levels and a change in the wave functions which shifts the strengths of the quadrupole excitation matrix elements among the set of 15 states.

We now note that excitations of $2\hbar\omega$ are also obtained in the harmonic oscillator shell model by exciting *two* particles, each to the next shell; i.e. from the p-shell to the sd-shell. A large number of such 2^+ states is obtained (about 100) and they are all mixed with the one-particle, one-hole excitations by the residual two-body interaction. The next step in a calculation using degenerate perturbation theory would be to diagonalize the two body interaction in the subspace of all states having excitation energies $2\hbar\omega$ in the simple harmonic oscillator shell model. However, the corrections to the quadrupole excitation spectrum as a result of this diagonalization are very different from that of the previous diagonalization. The quadrupole operator, being a single-particle operator, has no matrix elements between the ground state and the two-particle, two-hole states. All the 'quadrupole strength' is concentrated in the one-particle, one-hole states. Thus, the mixing by the interaction of a given one-particle, one-hole state with a number of neighboring two-particle, two-hole states only redistributes the quadrupole strength present in the particular one-particle, one-hole state among a number of neighboring levels, leaving the total strength constant. This contrasts very much with the mixing of the one-particle, one-hole states in the previous diagonalization, where each of the states had a non-vanishing quadrupole matrix element and the mixing effects could be coherent and shift laıge amounts of quadrupole strength from one state to another.

From this we see that diagonalizing the two-body interaction in the subspace of the one-particle, one-hole states is important to give a reasonable picture of the quadrupole excitation spectrum, but that the effect of admixtures of other states such as two-particle, two-hole states do not necessarily have a large effect. If the residual two-body interaction is very strong the approximation scheme breaks down completely. However there is a considerable range of strength for the residual two-body interaction where two-particle, two-hole states are appreciably mixed with one-particle, one-hole states in the exact eigenfunction of the Hamiltonian, but the only effect of these admixtures on the excitation spectrum is to split any given line into a number of lines in the same energy region with the same overall strength. In a 'poor resolution' experiment, this splitting would be unresolved and appear as a line broadening.

We now consider the effect of the residual interaction on the ground state. Since the Hamiltonian acting on the Hartree–Fock ground state produces states with two-particle, two-hole excitation, a better approximation to the ground state includes admixtures of states with excited particle–hole pairs. Similarly, for the excited state there are admixtures having additional particle–hole pairs. If we are interested only in the excitation spectrum produced by the quadrupole operator many effects of the additional particle–hole pairs cancel out since they give the same contribution to both the ground and the excited states.

There is one important effect of these 'ground state correlations' on the quadrupole matrix element. Suppose the ground state has a small admixture of a state containing two excited particle–hole pairs. The quadrupole operator A, acting on this component can 'de-excite' one of these pairs, leaving a wave function identical to one produced by a single-particle–hole excitation on the Hartree–Fock ground state. Such effects of the ground state correlation can be important, since they give a contribution to the transition matrix element which is *linear* in the perturbation.

We now see the features which are desired in an approximate treatment of quadrupole excitations. We do not necessarily want very good approximations to the exact ground-state and excited-state wave functions. We can diagonalize the first excited state in the one-particle, one-hole subspace but need not be too concerned by admixtures having larger numbers of particle–hole pairs. This is sometimes called the 'Tamm–Dancoff approximation'. We also need to be very concerned about the admixtures of higher configurations into the ground state, except for the possibility that the quadrupole operator can de-excite one of these admixed configurations to give the desired quadrupole state. This is sometimes called treating the ground state correlations in the random-phase approximation (RPA).

11.4 Operators generating excitations, collective and single-particle

Let us now consider the formal treatment of elementary excitations. Suppose that an operator A can be found in a particular many-body problem which satisfies the equation

$$[H, A]|g\rangle = \varepsilon A|g\rangle \tag{11.14a}$$

where $|g\rangle$ is the exact ground state wave function for the system, H is the Hamiltonian, and ε is a number. It follows immediately that the state $A|g\rangle$ is an eigenfunction of the Hamiltonian with an excitation energy ε above the ground state.

$$H(A|g\rangle) = (E_g + \varepsilon)(A|g\rangle). \tag{11.14b}$$

The operator A thus 'generates an excitation' of the system. Since the commutator $[H, A]$ is proportional to the time derivative of the Heisenberg operator A, eq. (11.13) can also be considered to be the Heisenberg equation of motion for the operator A. Thus one often says that *operators satisfying simple equations of motion generate excitations of the system.*

Two characteristic examples of operators satisfying an equation of the form (11.13) occur in the harmonic oscillator and in the non-interacting Fermi gas. In the harmonic oscillator, the creation operator satisfies an equation of type (11.13), whereas in the non-interacting Fermi gas an operator creating a particle above the Fermi surface and a hole below satisfies such an equation.

Consider first the harmonic oscillator. The Hamiltonian, written in terms of the creation and annihilation operators for oscillator quanta $a^\dagger$ and a, has the form

$$H_{\text{osc}} = \tfrac{1}{2}\hbar\omega(a^\dagger a + aa^\dagger). \tag{11.15a}$$

Thus

$$[H_{\text{osc}}, a^\dagger] = \hbar\omega a^\dagger. \tag{11.15b}$$

Eq. (11.15b) is of the form (11.13) but more general. It is an operator equation, valid for all states and matrix elements and not only when acting on the ground state. Eq. (11.15b) implies that an infinite set of states having equally spaced energy levels can be generated by repeated operation with the operator $a^\dagger$, starting from the ground state. Eq. (11.15b) is characteristic of all boson-like harmonic vibration modes found in many-particle systems. These are often called phonons. Usually in the many-particle system an operator is found which satisfies an equation like (11.15b) only approximately, because of anharmonic effects and couplings to other degrees of freedom.

For the non-interacting Fermi gas, we have

$$H_F = \sum_{\boldsymbol{k}} \varepsilon_k c_{\boldsymbol{k}}^\dagger c_{\boldsymbol{k}}, \tag{11.16a}$$

where ε_k is the single-particle energy, and

$$[H_F(c_{\boldsymbol{k}'}^\dagger c_{\boldsymbol{k}})] = (\varepsilon_{k'} - \varepsilon_k)(c_{\boldsymbol{k}'}^\dagger c_{\boldsymbol{k}}). \tag{11.16b}$$

Eq. (11.16b) seems at first to be general, like eq. (11.15b), valid regardless of the state on which it is operating. However, eq. (11.16b) is non-trivial only when operating on a many-particle state in which the single-particle state $\boldsymbol{k}$ is occupied and $\boldsymbol{k}'$ is empty; otherwise it reduces to $0=0$. Furthermore, once eq. (11.16b) operates on such a state the resulting state no longer satisfies this condition, and the relation cannot be used again. An excitation of this type is called a 'single-particle excitation'. It takes a single particle from one state $\boldsymbol{k}$ and moves it into another state $\boldsymbol{k}'$. The excitation energy is just the change in the single-particle energy. Once a particular particle is excited out of the state $\boldsymbol{k}$, it cannot be excited out of the state $\boldsymbol{k}$ again. Thus a particular single-particle excitation can only be used once, and does not lead to a set of equally spaced energy levels like a phonon or boson excitation.

An intermediate case between single-particle and boson excitations occurs in many-fermion systems. Suppose we consider the harmonic oscillator shell model for ^{16}O. Let $a_{nk}^\dagger$ represent operators creating particles in the shell labeled by the principal quantum number n, with additional quantum numbers k specifying the particular state in the shell. We neglect spin–orbit coupling and deviations from the harmonic potential, so that all states having the same value of n are degenerate, and have the energy $(n+\frac{3}{2})\hbar\omega$. The $n=0$ and $n=1$ levels are filled in ^{16}O and we consider excitations like the E2 excitations of section 11.3 from $n=0$ to $n=2$ and from $n=1$ to $n=3$. We can write

$$[H_{SM}, a_{2k''}^\dagger a_{0k'}] = 2\hbar\omega(a_{2k''}^\dagger a_{0k'}). \tag{11.17a}$$

$$[H_{SM}, a_{3k''}^\dagger a_{1k'}] = 2\hbar\omega(a_{3k''}^\dagger a_{1k'}). \tag{11.17b}$$

where H_{SM} is the shell model Hamiltonian, for all values of k' and k''. Any linear combination of the operators satisfying eq. (11.17) also satisfies eq.

(11.17). Thus, if we define

$$A \equiv \sum_{k'k''} C(k'k'')a^{\dagger}_{2k''}a_{0k'} + \sum_{k'k''} D(k'k'')a^{\dagger}_{3k''}a_{1k'} \tag{11.18a}$$

then

$$[H_{SM}, A] = 2\hbar\omega A. \tag{11.18b}$$

The electric quadrupole operator (11.13) is a special case of such an operator.

The individual operators satysfying eq. (11.17) can each only act once on a particular state; the square of each operator is zero. However, the linear combination A defined by eq. (11.18a) can act several times on a particular state, since $A^2 \neq 0$, thus producing a set of equally spaced energy levels analogous to those of the harmonic oscillator. The operator A generates an excitation which is a coherent linear combination of single-particle excitations. Such an excitation is often called a 'collective' excitation.

The square of each individual term in the sum (11.18) defining A vanishes. In the operator A^2, the non-vanishing terms are the 'cross terms' which do not have the same factor appearing twice. As one goes to higher powers of A, a point is reached where the power is greater than the number of terms in the sum (11.18a) and there is no term in which one factor does not appear twice. This power of A must vanish. Thus successive operation with powers of A can lead to a spectrum of equally spaced energy levels resembling that of the harmonic oscillator, but differing from it in cutting off at some finite excitation. One can say that the operator A has approximately the properties of a boson creation operator if there are a large number of terms in the sum (11.18), and if the degree of excitation of the system is small in comparison with the number of terms in the sum.

The relation (11.18) is rather trivial for the simple case of the harmonic oscillator shell model, where the degeneracy is so high that one can define a large number of linear combinations and still satisfy the equations. In physically interesting cases the degeneracy of the single-particle levels is removed by using a more realistic potential and adding residual two-body interactions to obtain a Hamiltonian of the form (11.7). An operator of the form (11.18a) may then still have useful properties under any of the following conditions:

(1) Certain particular linear combinations of the operators (11.18a) still satisfy eq. (11.18b) for the more realistic Hamiltonian, even though the individual operators (11.17) do not. The problem of finding those particular linear combinations is then equivalent to solving the Schrödinger equation for the excited states. The particular linear combinations satisfying eq. (11.18b) can then be considered as operators generating excitations which are either classified as 'single-particle' or 'collective' depending upon whether

there is one predominant term in the sum (11.18a) or whether there are many contributing terms of approximately equal magnitude. The term quasiparticle is sometimes used to describe these excitations. The single-particle type is called a quasi-fermion; the collective type a quasi-boson.

(2) A particular linear combination (11.18a) is an operator which produces the kind of excitation of physical interest. For example, in the investigation of nuclear excitation by absorption of electric dipole or quadrupole photons, the relevant electric multipole operator may be expressible in the form (11.18). (This is true in the long-wavelength limit with harmonic oscillator wave functions mentioned in section 11.3). For this case, with the simplified harmonic oscillator Hamiltonian, the multipole operator generates a single excited state of the system which has all of the transition strength. Since the multipole operator acting on the ground state is an eigenfunction of the Hamiltonian, any eigenfunction orthogonal to this 'multipole state' has a vanishing multipole matrix element connecting it to the ground state. If the multipole operator no longer satisfies eq. (11.18b) with a more realistic Hamiltonian, this means that the 'multipole state' is not an eigenfunction of the more realistic Hamiltonian. However, it may be a linear combination of many states having the form of a 'giant resonance'. In such a case, all the multipole transition strength is in this giant resonance, and the multipole operator can be considered as generating an 'elementary excitation' which is of physical interest. Another possibility would be for the multipole strength to be split between a small discrete number of giant resonances, each generated by some operator of the form (11.18b). The multipole operator would then be a linear combination of these operators.

One might look for elementary excitations of a many-particle system by trying to find operators satisfying simple equations of motion of the type (11.14a) in some approximation. The validity of the approximation should then be judged only with reference to results of physical interest. If the left-hand side of the equation (11.14a) is equal to the right-hand side plus correction terms which are neglected, it is not necessary that these additional terms be small. The state generated by A need not be an exact nor even an approximate eigenfunction; it can be a giant resonance. Thus additional terms in eq. (11.14a) can be discarded, even though they are not small, if their effect is only to convert the calculated excited state into a giant resonance by mixing it with neighboring states without changing relevant transition strengths. This is one example of an approximation which might be considered bad by conventional standards, but which does not seriously affect the results for quantities of physical interest.

11.5 Linearized equations of motion and the random-phase approximation

Suppose that we are looking for elementary excitations which can be produced by photon absorption in systems like the two examples under consideration, the electron gas and the nucleus ^{16}O. We are then interested in finding a single-particle operator which satisfies a simple equation of motion, like eq. (11.14a) and can generate such an excitation.

For both systems, the Hamiltonian has the general form

$$H = H_1 + H_2 \tag{11.19}$$

where H_1 is a single-particle operator and H_2 is a two-body operator. For any single-particle operator A, we have

$$[H, A] = C_1 + C_2, \tag{11.20a}$$

where

$$C_1 = [H_1, A] \tag{11.20b}$$

$$C_2 = [H_2, A]. \tag{11.20c}$$

The commutators C_1 and C_2 are one and two-body operators respectively. The commutator (11.20a) cannot be proportional to A, because A is a single-particle operator, and the commutator (11.20) includes a two-body part C_2 as well as a single-particle part. Thus a single-particle operator cannot satisfy eq. (11.14a) exactly.

An equation of the type (11.14a) can be obtained if the two-body part of C_2 can be neglected in some approximation. The two-body operator C_2, acting on the ground-state wave function can produce states having one particle excited, and states having two particles excited. If the part corresponding to two-particle excitations is neglected, then one can look for single-particle operators which satisfy eq. (11.14a) in this approximation. This approximation is sometimes called the random-phase approximation (RPA).

Another way to characterize the approximation is exhibited in the second-quantized formulation. Any single-particle operator is a linear combination of bilinear products of a creation and an annihilation operator; e.g. $a_m^\dagger a_n$. The commutator C_2, being a two-body operator is a linear combination of terms of fourth degree in these operators $a_m^\dagger a_n a_r^\dagger a_s$, or quadratic in the bilinear products. The equation of motion (11.20a) is called 'non-linear' because it includes a term C_2 which is quadratic in the kind of operator which appear linearly in A. The equation can be 'linearized' by assuming that operators $a_m^\dagger a_n$ are smaller if $m \neq n$ than if $m = n$. This might be justified by the following 'random phase' argument. In a sum of a large number of terms of this type, those for which $m \neq n$ are incoherent and add with 'random phases', while those with $m = n$ are just number operators which are positive definite and add in phase. It is difficult to justify this approximation for the general case, since the validity of the approximation depends upon the features of the individual problem. We shall look into this below for the two problems under investigation. However, if we make this approximation by keeping only those terms of type $a_m^\dagger a_n a_r^\dagger a_s$, for which two indices are equal, we have reduced the operator C_2 to one which only excites single particles. We have also linearized the equation of motion, by throwing away the terms of second order in the 'small quantities' $a_m^\dagger a_n$ with $m \neq n$.

Another reason for neglecting the two-particle excitations produced by C_2 is that these states cannot be reached from the ground state by the elementary excitation process under consideration. The inclusion of these terms would require an operator satisfying eq. (11.14a) to include a two-body part as well as a single-particle part, and thus a whole series of terms involving three-body, four-body, etc. excitations. However, even if these components in the exact solution of eq. (11.14a) are not small, they do not contribute to the transition matrix element for the elementary excitation. In this case the excited state obtained by solving the *linearized* equation of motion may be a giant resonance, containing all the excitation strength in a particular energy region, rather than a single state.

The arguments given above for neglecting the two-body part of C_2 are not meant to be convincing. They indicate why such an approximation may be plausible in some cases. The validity of the approximation must be investigated in each specific problem. Note also that the term 'single-particle' excitation has been used rather loosely. If the ground state is a pure single-particle state, then it is easy to define the set of states obtainable from the ground state by the excitation of a single particle. If the ground state is not a pure single-particle state and is furthermore *not known*, it is not easy to define 'that part of a two-body operator which excites only a single particle'. For the present, we ignore this point and define the single-particle part of a two-body operator *as if the ground state were a pure single-particle state.* This may seem inconsistent, particularly in those cases where the deviation of the real ground state from a single-particle state produces a significant contribution to the result. This point will be considered later on.

11.6 The electron gas in the random-phase approximation

Let us now solve the linearized equations of motion for the electron gas, just as an example. We first write down and solve the equations without attempting to justify the approximation.

It is convenient to write the Hamiltonian in the form

$$H = \sum_{k\sigma} \frac{(\hbar \boldsymbol{k})^2}{2m} c^\dagger_{\boldsymbol{k}\sigma} c_{\boldsymbol{k}\sigma} + \tfrac{1}{2} \sum_{\boldsymbol{k}\neq 0} v_k (\varrho_{-\boldsymbol{k}} \varrho_{\boldsymbol{k}} - n), \tag{11.21}$$

using the Fourier components of the electron density $\varrho_{\boldsymbol{k}}$ defined by eq. (11.6).

For any single-particle operator A, the commutator $[H, A]$ is the sum of a single-particle and a two-particle part, as in eq. (11.20). The two-particle part is

$$C_2 = \tfrac{1}{2} \sum_{\boldsymbol{k}\neq 0} v_k (\varrho_{-\boldsymbol{k}} [\varrho_{\boldsymbol{k}}, A] + [\varrho_{-\boldsymbol{k}}, A] \varrho_{\boldsymbol{k}}). \tag{11.22}$$

Since $\varrho_{\boldsymbol{k}}$ is a single-particle operator, the commutator C_2 can be linearized by keeping only the *diagonal* part of the commutator $[\varrho_{\pm\boldsymbol{k}}, A]$. These multiplied by the single-particle operator $\varrho_{\pm\boldsymbol{k}}$ give a single-particle operator, while the remaining parts of the commutators excite one particle and give a two-particle excitation when multiplied by $\varrho_{\pm\boldsymbol{k}}$. (Note that we are neglecting those terms in which the two factors each excite a single particle, but the result is only a single-particle excitation because the second factor moves another particle into the hole made by the first excitation. This is sometimes called the 'exchange' contribution and occurs only in systems of identical particles. If the two particles represented by r_i and r_j in eq. (11.4) were not identical these terms would not be single-particle excitations, but would be a single-particle excitation plus the exchange of two particles.

Let us choose for the operator A the most general single-particle operator which generates an excitation of a definite momentum $\hbar\boldsymbol{k}'$, without changing the spin orientation of the particles

$$A_{\boldsymbol{k}'} = \sum_{\boldsymbol{q}'\sigma'} g_{\boldsymbol{k}'\boldsymbol{q}'} c^\dagger_{(\boldsymbol{k}'+\boldsymbol{q}')\sigma'} c_{\boldsymbol{q}'\sigma'}. \tag{11.23}$$

where $g_{\boldsymbol{k}'\boldsymbol{q}'}$ are coefficients to be determined to make (11.23) a solution of the linearized equations. The commutator $[\varrho_{\boldsymbol{k}}, A_{\boldsymbol{k}'}]$ is

$$[\varrho_{\boldsymbol{k}}, A_{\boldsymbol{k}'}] = \sum_{\boldsymbol{q}\boldsymbol{q}'\sigma\sigma'} g_{\boldsymbol{k}'\boldsymbol{q}'} [(c^\dagger_{\boldsymbol{q}\sigma} c_{(\boldsymbol{k}+\boldsymbol{q})\sigma}), (c^\dagger_{(\boldsymbol{k}'+\boldsymbol{q}')\sigma'} c_{\boldsymbol{q}'\sigma'})]. \tag{11.24}$$

The diagonal part of this commutator comes from terms where $\boldsymbol{q}=\boldsymbol{q}'$, $\sigma=\sigma'$

and $\boldsymbol{k}=\boldsymbol{k}'$. Thus

$$\langle[\varrho_{\boldsymbol{k}}, A_{\boldsymbol{k}'}]\rangle = \delta_{\boldsymbol{k}\boldsymbol{k}'} \sum_{\boldsymbol{q}'\sigma'} g_{\boldsymbol{k}'\boldsymbol{q}'}(n_{\boldsymbol{q}'\sigma'} - n_{(\boldsymbol{k}'+\boldsymbol{q}')\sigma'}) \tag{11.25a}$$

$$= \delta_{\boldsymbol{k}\boldsymbol{k}'} \sum_{\boldsymbol{q}'\sigma'} (g_{\boldsymbol{k}'\boldsymbol{q}'} - g_{\boldsymbol{k}'(\boldsymbol{q}'-\boldsymbol{k}')})n_{\boldsymbol{q}'\sigma'} \tag{11.25b}$$

where $n_{\boldsymbol{q}'\sigma'}$ is a number operator, and we have obtained eq. (11.25b) from (11.25a) by changing the dummy index $\boldsymbol{q}'$ in the second sum.

The two terms in the commutator (11.22) are seen to make equal contributions to the diagonal part, since the sum is over all values of $\boldsymbol{k}$ and the second term is the same as the first with $\boldsymbol{k}$ replaced by $-\boldsymbol{k}$. The linearized equation of motion is thus

$$[H, A_{\boldsymbol{k}'}] = \sum_{\boldsymbol{q}'\sigma'} \left(\frac{\hbar^2(\boldsymbol{k}' + \boldsymbol{q}')^2}{2m} - \frac{\hbar^2 \boldsymbol{q}^2}{2m}\right) g_{\boldsymbol{k}'\boldsymbol{q}'} c^{\dagger}_{(\boldsymbol{k}'+\boldsymbol{q}')\sigma'} c_{\boldsymbol{q}'\sigma'}$$

$$+ v_{k'} c^{\dagger}_{(\boldsymbol{k}'+\boldsymbol{q}')\sigma'} c_{\boldsymbol{q}'\sigma'} \sum_{\boldsymbol{q}''\sigma''} (g_{\boldsymbol{k}'\boldsymbol{q}''} - g_{\boldsymbol{k}'(\boldsymbol{q}''-\boldsymbol{k}')} n_{\boldsymbol{q}''\sigma''}$$

$$= \hbar\omega_{k'} \sum_{\boldsymbol{q}'\sigma'} g_{\boldsymbol{k}'\boldsymbol{q}'} c^{\dagger}_{(\boldsymbol{k}'+\boldsymbol{q}')\sigma'} c_{\boldsymbol{q}'\sigma'}. \tag{11.26}$$

where $\hbar\omega_{k'}$ is the excitation energy of the desired excitation.

We can equate coefficients of the operators $c^{\dagger}_{(\boldsymbol{k}'+\boldsymbol{q}')\sigma'} c_{\boldsymbol{q}'\sigma'}$, since operators having different values of $\boldsymbol{q}'$ are linearly independent.

$$\left(\hbar\omega_{k'} - \frac{\hbar^2(\boldsymbol{k}' + \boldsymbol{q}')^2}{2m} + \frac{\hbar^2 \boldsymbol{q}'^2}{2m}\right) g_{\boldsymbol{k}'\boldsymbol{q}'}$$

$$= v_{k'} \sum_{\boldsymbol{q}''\sigma''} (g_{\boldsymbol{k}'\boldsymbol{q}''} - g_{\boldsymbol{k}'(\boldsymbol{q}''-\boldsymbol{k}')}) n_{\boldsymbol{q}''\sigma''}. \tag{11.27}$$

This is a set of simultaneous homogeneous linear equations for determining the coefficients $g_{\boldsymbol{k}'\boldsymbol{q}'}$.

The occupation number $n_{\boldsymbol{q}''\sigma''}$ is taken to be that for the unperturbed ground state; it can therefore be replaced by unity and the sum taken only over states within the Fermi sphere. Note that with this interpretation of the ground state, the only non-vanishing terms in (11.23) are those for which $\boldsymbol{q}'$ is inside the Fermi sphere and $\boldsymbol{k}'+\boldsymbol{q}'$ is outside. If $g_{\boldsymbol{k}'\boldsymbol{q}''}$ is the coefficient of a term which satisfies this criterion, $g_{\boldsymbol{k}'(\boldsymbol{q}''-\boldsymbol{k}')}$ is not, since $\boldsymbol{k}'+(\boldsymbol{q}''-\boldsymbol{k}')=\boldsymbol{q}'$ and is not outside the Fermi sphere. One might therefore expect that the second term of the right-hand side of eq. (11.27) should be discarded, since it comes from terms in (11.23) which vanish when acting on the unperturbed ground state. However, these terms may not vanish when acting on the *real* ground state, and they are therefore kept. They turn out to make an appreciable contribution. This is the inconsistency discussed above; the unperturbed ground state wave function is used to determine the single-particle part of the commutator, but is not assumed to be the real wave function.

Because of the particularly simple form of the set of homogeneous equations, they can be treated without solving an enormous determinant. Eq. (11.27) can be solved for $g_{\boldsymbol{k}'\boldsymbol{q}'}$ in terms of the sum on the right-hand side, and the g's can be eliminated by using this result to obtain an expression for this sum:

$$\sum_{|\boldsymbol{q}'|<k_F} (g_{\boldsymbol{k}'\boldsymbol{q}'} - g_{\boldsymbol{k}'(\boldsymbol{q}'-\boldsymbol{k}')}) = \sum_{|\boldsymbol{q}'|<k_F} \left(\frac{v_{k'}}{\hbar\omega_{k'} - \dfrac{\hbar^2(\boldsymbol{k}'+\boldsymbol{q}')^2}{2m} + \dfrac{\hbar^2 q'^2}{2m}} - \frac{v_{k'}}{\hbar\omega_{k'} - \dfrac{\hbar^2 q'^2}{2m} + \dfrac{\hbar^2(\boldsymbol{q}'-\boldsymbol{k}')^2}{2m}} \right) \sum_{|\boldsymbol{q}''|<k_F} (g_{\boldsymbol{k}'\boldsymbol{q}''} - g_{\boldsymbol{k}'(\boldsymbol{q}''-\boldsymbol{k}')}). \tag{11.28a}$$

Thus

$$1 = \sum_{\boldsymbol{q}'} \left(\frac{v_{k'}}{\hbar\omega_{k'} - \dfrac{\hbar^2(\boldsymbol{k}'+\boldsymbol{q}')^2}{2m} + \dfrac{\hbar^2 q^2}{2m}} - \frac{v_{k'}}{\hbar\omega_{k'} - \dfrac{\hbar^2 q'^2}{2m} + \dfrac{\hbar^2(\boldsymbol{q}'-\boldsymbol{k}')^2}{2m}} \right)$$
$$= \frac{4\pi e^2 \hbar^2}{m} \sum_{\boldsymbol{q}'} \left(\frac{1}{\hbar\omega_{k'} - \dfrac{\hbar^2 \boldsymbol{k}'\cdot\boldsymbol{q}'}{m} - \dfrac{\hbar^4 k'^4}{4m^2}} \right). \tag{11.28b}$$

This is an equation in which the only unknown is $\omega_{k'}$. It can therefore be solved to give the excitation energies. It gives the frequency of the excitation in terms of the wave number $\boldsymbol{k}'$ and is often called the Bohm–Pines dispersion formula.

General features of the solution of eq. (11.28b) can be obtained as follows: The right-hand side of eq. (11.28b) becomes infinite whenever one of the denominators vanishes. This occurs whenever $\omega_{k'}$ is equal to an unperturbed excitation energy; i.e. the energy required to take a particle from a state $\boldsymbol{q}'$ to a state $\boldsymbol{q}'+\boldsymbol{k}'$. The right-hand side plotted as a function of $\omega_{k'}$ thus has the form shown in fig. 11.2.

Eq. (11.28b) has a solution whenever the right-hand side is equal to unity; i.e. whenever the curve crosses the value unity. Since it passes from $+\infty$ to $-\infty$ between every adjacent pair of values of excitation of the unperturbed Fermi gas, there is an excitation of the interacting system between each pair of unperturbed excitation energies. From eq. (11.27) we seen that if $\omega_{k'}$ is close to an unperturbed excitation energy corresponding to a single-particle

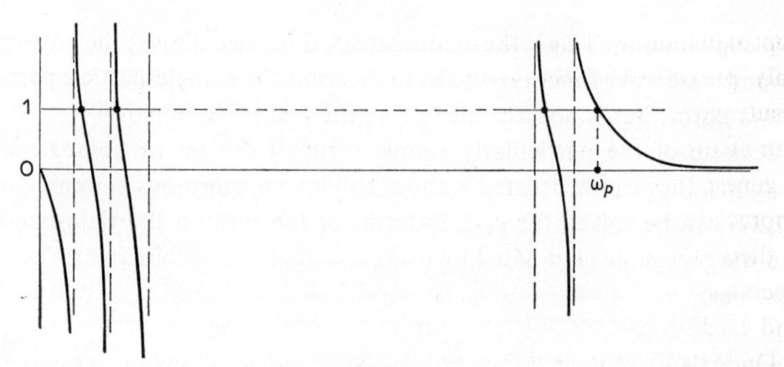

Fig. 11.2

transition from a state $\boldsymbol{q}'$, then $g_{\boldsymbol{k}'\boldsymbol{q}'}$ is very large compared to the other g's. This correspond to a single-particle excitation, as one term in eq. (11.23) predominates over all the others.

We also see that there is a solution for one large value of $\omega_{k'}$, greater than the maximum single-particle excitation energy. The value of $\omega_{k'}$ can be obtained for this case by changing the sum in eq. (11.28b) to an integral and evaluating the integral. Since $\hbar\omega_{k'}$ is higher than all the unperturbed excitation energies, there are no singularities in this integral. Eq. (11.28b) is easily solved approximately for the case of small $\boldsymbol{k}'$ by neglecting the terms in $\boldsymbol{k}'$. The sum then becomes $n/(\hbar\omega_{k'})^2$, where n is the number of particles, and we obtain

$$\omega_k^2 = \frac{4\pi e^2}{m} \frac{n}{\Omega}. \tag{11.29}$$

This is just the classical plasma frequency. The corresponding excitations are called 'plasmons' or plasma waves. They are density fluctuations of wave number $\boldsymbol{k}'$. Note that the plasma frequency depends upon n/Ω, the particle *density*, rather than on the number of particles or the volume.

The operator $A_{k'}$ which generates this plasmon excitation can be seen to be essentially the Fourier component $\varrho_{k'}$ of the density. In the limit $\boldsymbol{k}' \to 0$, the terms depending on $\boldsymbol{q}'$ in the left-hand side of eq. (11.27) can be neglected with respect to $\hbar\omega_{k'}$. The coefficient $g_{\boldsymbol{k}'\boldsymbol{q}'}$ is thus independent of $\boldsymbol{q}'$ in the limit $\boldsymbol{k}' \to 0$. Eq. (11.23) then gives

$$\lim_{\boldsymbol{k}' \to 0} A_{\boldsymbol{k}'} = \varrho_{\boldsymbol{k}'} \tag{11.30}$$

(one need not be disturbed by noting that the sum on the right-hand side of eq. (11.27) goes to zero as $\boldsymbol{k}' \to 0$ if $g_{\boldsymbol{k}'\boldsymbol{q}'}$ becomes independent of $\boldsymbol{k}'$, since the sum is multiplied by a factor $1/k^2$ in $v_{k'}$). The plasmon excitation is thus approximately a density fluctuation with wave number $\boldsymbol{k}'$. Note that the only properties of the ground state wave function used to obtain this result were the occupation numbers $n_{\boldsymbol{q}'\sigma'}$. Since exciting one particle–hole pair changes only two of these occupation numbers, the form of the operator A generating the plasma mode and the plasmon frequency $\omega_{k'}$ should not be appreciably affected if there are several particle–hole pairs initially excited or if there are already several plasmons present. Thus the plasmon creation operators $A_{\boldsymbol{k}'}$ should behave approximately like boson creation operators and excitations of several plasmons should be possible.

Once the picture of density waves behaving approximately like harmonic oscillations is accepted, it is clear that there must be zero-point plasma oscillations in the real ground state. These would appear in the ground state wave function as a mixture of states in which particle–hole pairs are excited out of the Fermi sphere. Such 'ground state correlations' are appreciable and account for the divergence of the perturbation calculation beginning with the unperturbed Fermi sphere.

In treatments where these density fluctuations are taken into account in calculating properties of the ground state, one finds that they effectively screen out the 'long-range part' of the Coulomb interaction which is responsible for the divergence, and leave a 'screened interaction' which no longer produces divergences at low $\boldsymbol{k}$ (long range). These treatments are beyond the scope of this book. However, one can see that this is reasonable from the following physical argument. Trouble should be expected in a model where the Coulomb energy is calculated under the assumption that the particles move independently with no correlations between them. If one electron is moved a small distance, the change in the potential produced is equivalent to the potential of an electric dipole (adding a positron at the original position of the electron and adding an electron at the new position). The potential due to a dipole decreases as $1/r^2$ with distance. However, the number of electrons in a spherical shell at radius r increases as r^2. Thus the change in the Coulomb energy between this electron and all those in a spherical shell of radius r and thickness dr is independent of r. The change in Coulomb energy of all the electrons within a sphere of radius r is then proportional to r and becomes infinite as r goes to infinity. This infinity is clearly unphysical, because the electrons are free to move and constitute a conductor on a macroscopic scale. The motion of a charge within a conductor does not produce a change in the electric field outside the conductor because the motion of the conduction electrons shields the exterior from the interior. If one considers that moving any one electron must induce motions of other electrons to screen out the change in the field at large distances, the infinity is eliminated. The motion of the electrons which produces the shielding is clearly a slowly varying function of the distance, and constitutes a wave packet of density fluctuations of long wave length. This is just the type of correlation which is introduced by proper consideration of plasmons, and is not present in the Hartree–Fock description.

Although the neglect of the ground state correlations causes serious divergences in the normal perturbation treatment, one can hope that their neglect in evaluating the linearized equations of motion (11.26) does not produce large errors. The effects of these correlations are taken into account by including in the definition of the operator $A_{\boldsymbol{k}'}$ (11.23) terms which vanish if there are no correlations; i.e. those which annihilate a particle outside the Fermi sphere and create one inside. In this formulation the assumed form of the ground state wave function plays a secondary role. The specific properties of the wave function enter mainly in the values of the occupation numbers $n_{\boldsymbol{q}''\sigma''}$ in eq. (11.27). Since a very large number of states $\boldsymbol{q}''$ contribute to the sum (11.27) for the case of the plasmon excitation, the result is insensitive to changes in a relatively small number of occupation numbers. Thus the real ground state wave function can be very different from the unperturbed one in the sense of perturbation theory without changing the value of the $\omega_{k'}$ appreciably when the correct occupation numbers are inserted in eq. (11.27).

11.7 The giant resonances in the nucleus ^{16}O

Let us now treat the nucleus ^{16}O in a similar manner. For this case our 'Fermi sphere' consists of the lowest s and p-shells, and we shall consider excitations produced by exciting a particle from the p-shell ($n=1$) into the next shell, the sd-shell ($n=2$). Since we do not have momentum eigenfunctions, but rather eigenfunctions of angular momentum and parity, we can talk about excitation of a particular *multipole* rather than of a particular wave number.

Suppose we consider the electric dipole excitation from the ground state, leading to excited states of angular momentum one and odd parity. We can then write the most general linear combination of operators which produce a 1^- state, and determine the coefficients by satisfying the linearized equations of motion. The procedure is directly analogous to that for the electron gas, but there are the following differences.

(1) Because angular momentum is not additive, like linear momentum, the simplest form of an operator producing the desired excitation is not a product of a creation and an annihilation operator, but a linear combination of such products with Clebsch–Gordan coefficients.

(2) The number of particles is not large. Only 12 particles (the four particles in the inner s-shell do not contribute) can be involved in this excitation.

Before we consider solving the linearized equations of motion, let us first examine how the ^{16}O problem would be treated by conventional perturbation theory. The unperturbed ground state is non-degenerate. However, the first excited 1^- states are all nearly degenerate, since they all produce excitations by taking some particle from the p-shell and putting it into the sd-shell. The first step in a conventional calculation is thus to remove the degeneracy by diagonalizing the interaction in the subspace of the nearly degenerate states. The next step is second-order perturbation theory, in which contributions to the energy are calculated from states which can be mixed in by the interaction to first order. Since the interaction is a two-body interaction, this means mixing into the ground state the excitation of two particle–hole pairs and into the excited state the excitation of three particle–hole pairs.

If we now examine the linearized equations of motion for the most general set of particle–hole operators involving the $n=1$ and the $n=2$ shells, we find that there are two sets of operators, those which excite a particle from the p-shell into the sd-shell, and those which take a particle from the sd-shell back into the p-shell. These correspond to the two types of operators found in the electron gas, those which excite a particle from inside the Fermi sphere to a state outside and those which do the reverse. Clearly the second type of operator gives zero when operating on the unperturbed ground state wave function for both the electron gas and ^{16}O. These operators are associated with the 'ground state correlations' and contribute only to the extent that the real ground state has appreciable deviations from the unperturbed ground state.

If we consider only the first type of operator, those which excite a particle from the p-shell to the sd-shell, we see that linearization of the equations of motion is exactly equivalent to the standard perturbation treatment of the shell model. In looking for the most general set of operators generating 1^- excitations by creating a particle in the sd-shell and annihilating one in the p-shell, one is examining just that set of unperturbed excited states used in the perturbation treatment. Linearizing the equations of motion means operating on these wave functions with the Hamiltonian, discarding all terms which do not have a single excited particle in the sd-shell and a single hole in the p-shell, and solving the resulting set of simultaneous homogeneous equations. This is identical with diagonalizing the Hamiltonian in the subspace of states having one excited particle in the sd-shell and one hole in the p-shell. Thus in the simplest approximation, ordinary shell model perturbation theory and the linearization of equations of motion, random-phase approximation, etc. are equivalent. The 'rediscovery of the standard shell model' by these fancier methods is sometimes called the Tamm–Dancoff approximation.

If we include the second type of operator, we are also including the possibility that excited particle–hole pairs are mixed into the ground state and that additional particle–hole pairs are present in the excited state. However, we are limited not only to two excited pairs in the ground state and three in the excited state, as in second-order perturbation theory. Thus the inclusion of the ground state correlations in the random-phase approximation is *not equivalent* to anything simple in ordinary perturbation theory. It includes some of the second-order contributions, possibly not all of them, and it also includes certain higher-order contributions. It is not clear off-hand, which approximation is better, RPA or ordinary perturbation theory, and it is necessary to examine specific cases to find out.

The contributions which are included in the various treatments are conveniently characterized by use of diagrams. The excitation of a particle–hole pair by an incoming photon is illustrated in fig. 11.3a. The wavy line indicates a photon, a line with an arrow going upward (forward in time) indicates a particle, a line with an arrow going downward (backward in time) indicates a hole. In fig. 11.3b a particular particle–hole pair is created by the photon, and is annihilated with the formation of another particle–hole pair as a result of the two-body interaction, indicated by a dashed line. Fig. 11.3c shows repeated annihilations and creations of particle–hole pairs. The shell model diagonalization of the Hamiltonian in the one-particle, one-hole subspace, is represented diagrammatically by the summation of all possible diagrams of all orders of the type shown in fig. 11.3c, since it includes all interactions coupling all states in which only one particle–hole pair is present at a time. Such diagrams are sometimes called 'forward-going bubble graphs'.

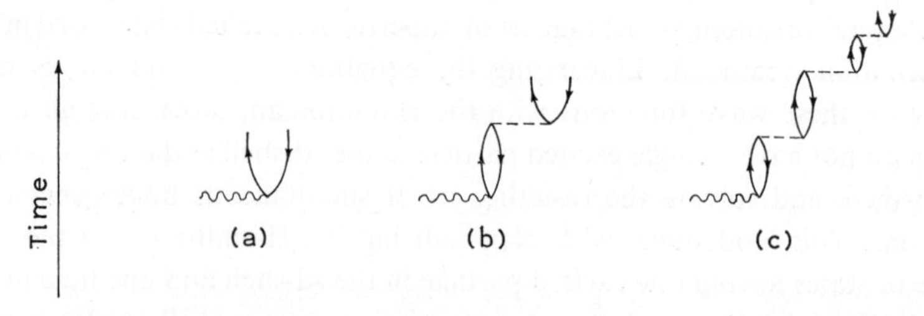

Fig. 11.3

The diagram in fig. 11.4a represents a typical 'ground state correlation' contribution. Two particle–hole pairs are present before the arrival of the photon, as a result of an interaction which creates them. The photon is absorbed with the annihilation of one pair; i.e. it is absorbed by a particle in an excited state which jumps down to fill the hole. In fig. 11.4b another ground state correlation diagram is shown in which two pairs are initially present and the photon is absorbed with the creation of a third pair which later annihilates one of the first two. This doubling-back and forth (forward and backward-going bubble graphs) is characteristic of the diagrams included in the random-phase approximation. A typical more complicated diagram is shown in fig. 11.4c.

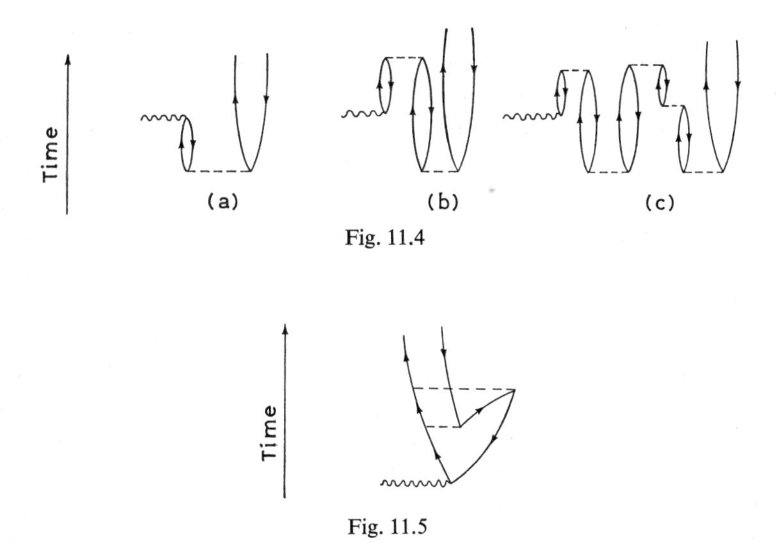

Fig. 11.4

Fig. 11.5

Fig. 11.5 illustrates a diagram which is not a 'bubble' and is not included in the random-phase approximation. It is a second-order diagram, since the interaction appears twice and corresponds to a mixture into the excited state of a component in which two particle–hole pairs are present. The two interactions involve the scattering of a particle and the creation or annihilation of a particle–hole pair. At each interaction there are *three* particle lines (arrow upward) and one hole (arrow downward) involved, rather than two particles and two holes. Thus these interactions are not simply described in terms of particle–hole pairs and are not included in RPA.

Let us now examine some of the general features of the solution in the random-phase approximation. We first define a complete set of linearly independent operators which create particle–hole excitations of a given angular momentum and parity, say 1^-, by annihilating a particle in the $n=1$ shell and creating one in the $n=2$ shell.

$$C_i = \sum_{k'k''} C_i(k', k'')a^\dagger_{2k''}a_{1k'}. \tag{11.31a}$$

The Hermitean conjugates of these operators

$$C^\dagger_i = \sum_{k'k''} C^*_i(k', k'')a^\dagger_{1k'}a_{2k''} \tag{11.31b}$$

create a particle in the $n=1$ shell and annihilate one in the $n=2$ shell. The operators C_i (11.31a) acting on the unperturbed ground state generate the set of excited states which appear in the shell model diagonalization. The operators $C^\dagger_i$ (11.31b) give zero when operating on the unperturbed ground state and are associated with ground state correlations.

The linearized equations of motion for the operators (11.31) have the form

$$[H, C_i]_{\text{lin}} = \sum_j a_{ji}C_j + \sum_j b^*_{ji}C^\dagger_j \tag{11.32a}$$

$$[H, C^\dagger_i]_{\text{lin}} = -\sum_j a^*_{ji}C^\dagger_j - \sum_j b_{ji}C_j, \tag{11.32b}$$

where a_{ji} and b_{ji} are coefficients, and eq. (11.32b) is obtained by taking the Hermitean conjugate of eq. (11.32a). To find an operator satisfying an e-quation of the form which generates an excitation, we substitute into eq. (11.14a) the most general linear combination of the operators (11.31)

$$C = \sum_i x_iC_i - \sum_i y_iC^\dagger_i \tag{11.33}$$

where the phase is chosen to give a result in the same form as found in the literature.

Substituting (11.33) into the linearized eq. (11.14a) and introducing the expressions (11.32) for the linearized commutators, we obtain

$$\sum_{ij} x_ia_{ji}C_j + \sum_{ij} x_ib^*_{ji}C^\dagger_j + \sum_{ij} y_ia^*_{ji}C^\dagger_j + \sum_{ij} y_ib_{ji}C_j = \varepsilon \sum_i (x_iC_i - y_iC^\dagger_i). \tag{11.34}$$

Equating coefficients of the individual linearly independent operators C_i and $C^\dagger_i$ leads to a set of homogeneous linear equations determining the coefficients x_i and y_i and the eigenvalues ε. These equations are written in a concise form if the coefficients a_{ji} and b_{ji} are represented by matrices A and B and the coefficients x_i and y_i by row matrices X and Y,

$$\begin{pmatrix} A & B \\ B^* & A^* \end{pmatrix}\begin{pmatrix} X \\ Y \end{pmatrix} = \varepsilon \begin{pmatrix} X \\ -Y \end{pmatrix} \tag{11.35a}$$

or

$$\begin{pmatrix} A & -B \\ B^* & -A^* \end{pmatrix} \begin{pmatrix} X \\ -Y \end{pmatrix} = \varepsilon \begin{pmatrix} X \\ -Y \end{pmatrix}. \tag{11.35b}$$

The problem of finding the excitations is thus reduced to diagonalization of a *non-Hermitean* matrix of twice the order of the matrix A. The latter is clearly the matrix which remains to be diagonalized if eq. (11.34) is assumed to operate on the unperturbed ground state and the contributions from all the $C_j^\dagger$ operators vanish. Thus including ground state correlations in the calculation in the random-phase approximation leads to the diagonalization of a matrix twice the size of the one diagonalized in the usual shell model calculation.

11.8 The time-dependent Hartree–Fock approximation

The neglect in the random-phase approximation of terms in C_2 which excite two particles out of the Hartree–Fock ground state recalls the general philosophy of the Hartree–Fock approximation which also neglects two-particle excitations. There is a close relation between the two approximations. The RPA can be formulated as a 'time-dependent Hartree–Fock' approximation, in which excited states are described as motion of independent particles in a time-dependent potential.

As a simple example of the time-dependent approach, consider the harmonic oscillator (11.15). Let us define the operator

$$S = a^\dagger e^{-i\omega t} + a e^{i\omega t}. \tag{11.36a}$$

Then if $\psi(t)$ is a solution of the time-dependent Schrödinger equation for the harmonic oscillator, it follows from the equation of motion (11.15b) that $e^{i\alpha S}\psi(t)$ is also a solution,

$$H_{\text{osc}} e^{i\alpha S}\psi(t) = i\hbar \frac{\partial}{\partial t} [e^{i\alpha S}\psi(t)], \tag{11.36b}$$

where α is any constant. If α is large, and $\psi(t)$ is the harmonic oscillator ground state, the motion described by the state (11.36b) is a classical harmonic vibration with frequency ω, a wave packet constructed from many eigenfunctions of the Hamiltonian.

Since this result depends only upon the equation of motion (11.15) and not on other properties of the oscillator, it holds for any quasi-boson excitation generated by an operator $A^\dagger$ satisfying the equation of motion (11.15b) with the Hamiltonian of the system. One can define a Hermitean operator S corresponding to this type of excitation and also a solution (11.36b) of the time-dependent Schrödinger equation describing periodic oscillations about the ground state.

Consider the case where the wave function $\psi(t)$ is a Hartree–Fock wave function and $A^\dagger$ is a single-particle operator. Then S is also a single-particle operator, and

$$\mathrm{e}^{\mathrm{i}\alpha S}|\mathrm{HF}\rangle = \mathrm{e}^{\mathrm{i}\alpha S}\prod_k c_k^\dagger|0\rangle = \prod_k [\mathrm{e}^{\mathrm{i}\alpha S} c_k^\dagger \mathrm{e}^{-\mathrm{i}\alpha S}]|0\rangle, \tag{11.37}$$

where $c_k^\dagger$ are creation operators for states k in the Hartree–Fock basis. The state (11.37) is thus expressible as a product of creation operators for single-particle states acting on the vacuum, but with a new basis of single-particle states obtained from the Hartree–Fock basis by a *time-dependent* unitary transformation.

Since the single-particle states of the Hartree–Fock basis are solutions of the one-particle Schrödinger equation in a self-consistent potential, the transformed single-particle states (11.37) are solutions of a transformed equation, in which the Hamiltonian is transformed by a time-dependent unitary transformation. Since S is a periodic function of time, the single-particle potential executes periodic vibrations about the Hartree–Fock value. This 'time-dependent Hartree–Fock' picture is very appropriate for describing nuclear multipole vibrations. If $A^\dagger$ is the quadrupole operator, the potential executes quadrupole oscillations about an equilibrium shape, describing a collective mode in which the particles move in an oscillating potential while the potential is just the average field produced by the particles themselves.

Detailed formulations of the time-dependent Hartree–Fock approach are available in the literature, and are beyond the scope of this book. The basic idea, however, is simply to find approximate solutions of the *time-dependent* Schrödinger equation which have the form (11.37); namely an extension of the Hartree–Fock wave function to include single-particle states which vary periodically with time.

11.9 Sum rules

Elementary excitations of the type discussed here very often satisfy sum rules which can be of considerable importance. For example the sum of the squares of the transition matrix elements for any operator M is easily seen to satisfy a sum rule:

$$\sum_{E'} |\langle g|M|E'\rangle|^2 = \sum_{E'} \langle g|M|E'\rangle\langle E'|M^\dagger|g\rangle$$
$$= \langle g||M|^2|g\rangle - |\langle g|M|g\rangle|^2 \tag{11.38}$$

where $|g\rangle$ is the exact ground state and the sum is over all excited states $|E'\rangle$ of the system.

Energy-weighted sum rules are often obtainable by commuting the operator M with the Hamiltonian. Consider the case where M is a single-particle operator depending only on the coordinates of the particles. This applies to electric multipole operators.

$$M = \sum_i M_i(x_i). \tag{11.39}$$

If the Hamiltonian consists only of the kinetic energy plus terms depending only on coordinates and not on the momenta, we have the commutators:

$$[H, M_i] = \left[\frac{p_i^2}{2m}, M_i\right] = -i\hbar[p_i M_i' + M_i' p_i]/2m \tag{11.40a}$$

$$[H, M_i], M_j^\dagger] = -2\hbar^2 |M_i'|^2 \delta_{ij}/m. \tag{11.40b}$$

Taking the matrix elements of eq. (11.40a) in the ground state and inserting a complete set of intermediate states in the commutator, we obtain:

$$\sum_{E'} \langle g|[H, M_i]|E'\rangle\langle E'|M_j^\dagger|g\rangle - \langle g|M_j^\dagger|E'\rangle\langle E'|[H, M_i]|g\rangle$$
$$= -2\sum_{E'}(E' - E_g)\langle g|M_i|E'\rangle\langle E'|M_j^\dagger|g\rangle$$
$$= -2\hbar^2\langle g||M_i'|^2|g\rangle\delta_{ij}/m. \tag{11.41a}$$

Summing over i and j, we obtain:

$$\sum_{E'}(E' - E_g)|\langle g|M|E'\rangle|^2 = \hbar^2\langle g|\sum_i |M_i'|^2|g\rangle/m. \tag{11.41b}$$

This energy-weighted sum rule relates the excitation matrix elements for the operator M to the expectation value of a function of the coordinates in the ground state. For electric dipole excitation, M_i is proportional to x_i, M_i' is a constant and the right-hand side of the sum rule is a constant. For electric quadrupole excitation, the sum rule relates excitation matrix elements to the ground state quadrupole moment and mean-square radius.

Sum rules have been derived and used in all branches of physics from the early applications to atomic spectroscopy to modern applications in particle physics. The two sum rules (11.39) and (11.41b) are typical examples. The square of a matrix element of an operator generating elementary excitations is summed over all possible states of excitation, possibly with weighting factors, to give a quantity which is experimentally measurable in principle (if all the relevant excited states can be reached). The expression is also transformed by the use of algebraic properties of the operators to a sum of products of matrix elements over all possible intermediate states. This sum is then evaluated by closure to give another observable experimental quantity.

In deriving the sum rules, no assumption is made about the form of the ground state wave function, and very few, if any, assumptions are made about the Hamiltonian. The derivation of the sum rule (11.38) involves no assumption and the derivation of (11.41b) assumes only the commutation relation (11.40a). Such sum rules are therefore useful in treating systems whose properties are not known exactly either at a theoretical or at an experimental level.

In nuclear physics, sum rules of the type (11.41b) can be used to check the validity of the assumed commutation relation (11.40a) without requiring a detailed knowledge of the nuclear Hamiltonian or the nuclear wave functions. Disagreement with the sum rule would indicate that the Hamiltonian contains other terms besides the kinetic energy which do not commute with the particle coordinates; i.e. either velocity-dependent or exchange forces.

In particle physics, sum rules are obtained by assuming commutation relations for the operators representing the electromagnetic and weak currents. These generate elementary excitations of strongly interacting systems produced by incident photons or neutrinos. This 'current algebra' has led to a number of successful results. Agreement of theoretical predictions with experiment provides evidence for the validity of the assumed commutation relations even though nothing at all is known about the wave functions or even the dynamical variables of the particles under consideration.

11.10 Time-dependent formulations

A number of elegant mathematical techniques have been developed for treating the elementary excitations of the system without requiring detailed knowledge of the wave functions. Many of these techniques, which were first developed for quantum field theory, can be presented on a more elementary level as mathematical techniques for studying the elementary excitations of many-particle systems. Some of these techniques were used in the treatment of the Mössbauer effect in ch. 2.

Consider the cross section as a function of energy for elementary excitations described by an operator M. The cross section can be expressed as follows:

$$\sigma(E)\mathrm{d}E = K(E)\sum_{E'} |\langle g|M|E'\rangle|^2 \delta(E - E')\mathrm{d}E \tag{11.42}$$

where $K(E)$ is a kinematic factor. Since the level spectrum is discrete, the excitation spectrum (11.42) is a series of delta functions at the energies corresponding to energies E' of excited states of the system with strengths proportional to the square of the matrix element of the operator M.

The right-hand side of the expression (11.42) can now be transformed as follows. We first write the delta function as a Fourier integral:

$$\sigma(E)\mathrm{d}E = \frac{K(E)\mathrm{d}E}{2\pi\hbar} \int_{-\infty}^{\infty} \mathrm{d}\tau \sum_{E'} \langle g|M|E'\rangle \mathrm{e}^{\mathrm{i}(E-E')\tau/\hbar} \langle E'|M^{\dagger}|g\rangle. \tag{11.43a}$$

Since E' is just the eigenvalue of the Hamiltonian in the state $|E'\rangle$, and E_g is the eigenvalue in the state $|g\rangle$

$$\sigma(E)\mathrm{d}E = \frac{K(E)\mathrm{d}E}{2\pi\hbar} \int_{-\infty}^{\infty} \mathrm{d}\tau\, \mathrm{e}^{\mathrm{i}(E-E_g)\tau/\hbar} \sum_{E'} \langle g|\mathrm{e}^{\mathrm{i}H\tau/\hbar} M \mathrm{e}^{-\mathrm{i}H\tau/\hbar}|E'\rangle\langle E'|M^{\dagger}|g\rangle. \tag{11.43b}$$

Thus

$$\sigma(E)\mathrm{d}E = \frac{K(E)\mathrm{d}E}{2\pi\hbar} \int_{-\infty}^{\infty} \mathrm{d}\tau\, \mathrm{e}^{\mathrm{i}(E-E_g)\tau/\hbar} \langle g|\mathrm{e}^{\mathrm{i}H\tau/\hbar} M \mathrm{e}^{-\mathrm{i}H\tau/\hbar} M^{\dagger}|g\rangle$$

$$= \frac{K(E)\mathrm{d}E}{2\pi\hbar} \int_{-\infty}^{\infty} \mathrm{d}\tau\, \mathrm{e}^{\mathrm{i}(E-E_g)\tau/\hbar} \langle g|M(\tau)M^{\dagger}(0)|g\rangle \tag{11.43c}$$

where $M(\tau)$ is the Heisenberg operator at the time τ. The Fourier transform of the excitation energy spectrum is thus expressed in terms of the matrix element of the product of two Heisenberg operators at different times.

Although the time variable was introduced in this derivation as a mathematical trick, the time dependence has a very simple physical interpretation in terms of the giant resonances discussed in section 11.2. The expression (11.12) for the decay of the giant resonance state can be transformed to a form similar to eq. (11.43c)

$$\langle\psi(0)|\psi(t)\rangle = \sum_{E'} |\langle E'|\psi(0)\rangle|^2 \mathrm{e}^{-\mathrm{i}Et/\hbar}$$

$$= N^2 \sum_{E'} \langle g|M|E'\rangle\langle E'|M^{\dagger}|g\rangle \mathrm{e}^{-\mathrm{i}E't/\hbar} \tag{11.44a}$$

$$\langle\psi(0)|\psi(t)\rangle = N^2 \mathrm{e}^{-\mathrm{i}E_g t/\hbar} \langle g|M(t)M^{\dagger}(0)|g\rangle. \tag{11.44b}$$

The physical interpretation of eqs. (11.43) and (11.44) is now clear. There are two complementary descriptions of the excitation spectrum produced by the action of the operator $M^\dagger$ on a state $|g\rangle$, using energy and time variables respectively. The decay of the state $M^\dagger|g\rangle$ produced by the excitation, expressed as a function of time is proportional to the Fourier transform of the excitation energy spectrum.

Elementary excitations concentrated in a small energy region with a finite width are described as states which decay with a definite lifetime, and often called 'quasiparticle' excitations since they 'almost behave like particles'. In the limit of infinite lifetime and zero width, the operator $M^\dagger$ could be considered as creating a particle added to the system. For the case of electric dipole or quadrupole excitation this operator creates particle–hole pairs. In the Hartree–Fock approximation, a single particle–hole pair is a stable excitation and behaves like two particles added to the system. A coherent linear combination of particle–hole pairs behaves like a boson describing a collective excitation which looks like a vibration of the whole system. Such an excitation with zero width would appear like the addition of a physical boson to the system. An excitation with finite width corresponds to a damped vibration and is a quasiparticle with a finite lifetime.

In many treatments the width and decay of the quasiparticle excitations are neglected and the state $M^\dagger|g\rangle$ is treated as an eigenfunction of the Hamiltonian. This only affects the excitation spectrum by replacing peaks with finite widths by delta functions. This approximation can be very useful in giving approximate properties of these excitations which are subject to verification by experiment.

11.11 Spectral functions and Green's functions

Operators which create or annihilate single particles in a particular state are of particular interest for the description of elementary excitations. Such excitations are observed directly in nuclear stripping or pickup reactions which add or remove single particles from a nucleus. Many excitations of physical interest are expressible as simple combinations of these excitations.

In the Hartree–Fock approximation, an operator $c_k^\dagger$ which creates a particle above the Fermi level generates an elementary excitation whose energy is just the single-particle energy. The operator satisfies the equation of motion

$$[H, c_k^\dagger] = \varepsilon_k c_k^\dagger; \qquad k > k_\mathrm{F} \tag{11.45a}$$

where k denotes the complete set of quantum numbers necessary to specify a Hartree–Fock state and $k > k_\mathrm{F}$ denotes that the state k is above the Fermi level. Similarly, an annihilation operator for an occupied state below the Fermi level generates an elementary excitation of the hole type and satisfies the equation:

$$[H, c_k] = \varepsilon_k c_k; \qquad k < k_F \tag{11.45b}$$

where ε_k is the hole energy.

The equations of motion (11.45) can also be used to define the Hartree–Fock approximation. If the commutators on the left-hand side are evaluated using the exact Hamiltonian, they contain linear terms in the creation and annihilation operators and also higher-order terms coming from the two-body interaction. Applying the standard linearization procedure to these equations of motion leads exactly to the Hartree–Fock equations.

In the Hartree–Fock approximation, the excitation spectrum produced by the operators (11.45) is trivial and consists of a single delta function at the appropriate energy. Beyond the Hartree–Fock approximation, where the residual interactions between particles are considered a more complicated excitation spectrum is obtained. For this case it is convenient to define the following functions by analogy to the expression (11.42)

$$\varrho^+(k, \omega) = \sum_{E'} |\langle E'|c_k^\dagger|g\rangle|^2 \delta(\omega + E_g + \mu - E') \tag{11.46a}$$

where the zero for the energy variable ω is chosen to make it an excitation energy. Since there are no negative-energy excitations from the ground state

$$\varrho^+(k, \omega) = 0 \quad \text{if} \quad \omega < 0. \tag{11.46b}$$

The chemical potential μ is introduced because the operator $c_k^\dagger$ changes the number of particles in the system. The state $|E'\rangle$ has $n+1$ particles if the state $|g\rangle$ has n particles. The chemical potential μ is the difference between the ground state energies of the $n+1$ and the n-particle systems.

$$\mu = E_g^{n+1} - E_g^n. \tag{11.47}$$

Thus $E_g + \mu$ is the ground state energy of the $(n+1)$-particle system and the argument of the delta function in eq. (11.46) vanishes when ω is equal to the excitation energy of the state $|E'\rangle$ above the ground state of the $(n+1)$-particle system.

Integrating eq. (11.46) over all values of ω, we obtain a sum rule:

$$\int_{-\infty}^{\infty} \varrho^+(k, \omega)\mathrm{d}\omega = \sum_{E'} \langle g|c_k|E'\rangle\langle E'|c_k^\dagger|g\rangle = \langle g|c_k c_k^\dagger|g\rangle = 1 - \langle g|n_k|g\rangle \tag{11.48}$$

where n_k is the number operator for the state k. For the Hartree–Fock approximation n_k is either 1 or 0 depending upon whether the state is occupied or unoccupied. In the presence of interactions, the ground state is no longer an eigenfunction of n_k and the sum rule depends on this property of the ground state. We can also define the analogous expression for the hole excitations.

$$\varrho^-(k, \omega) = \sum_{E'} |\langle E'|c_k|g\rangle|^2 \delta(-\omega + E_g - \mu - E'). \tag{11.49a}$$

Here we have defined the variable ω to be the negative of the excitation energy referred to the ground state energy $E_g - \mu$ of the $(n-1)$-particle system. Thus

$$\varrho^-(k, \omega) = 0 \quad \text{if} \quad \omega > 0. \tag{11.49b}$$

This function satisfies the sum rule:

$$\int_{-\infty}^{\infty} \varrho^-(k, \omega)\mathrm{d}\omega = \sum_{E'} \langle g|c_k^\dagger|E'\rangle\langle E'|c_k|g\rangle = \langle g|c_k^\dagger c_k|g\rangle = \langle g|n_k|g\rangle. \tag{11.50}$$

It is therefore convenient to define the function:

$$A(k, \omega) = \varrho^+(k, \omega) + \varrho^-(k, \omega). \tag{11.51a}$$

This function satisfies the sum rule:

$$\int_{-\infty}^{\infty} A(k, \omega)\mathrm{d}\omega = 1. \tag{11.51b}$$

The function $A(k, \omega)$ is called the spectral weight function and describes both particle-type excitations (11.46) and hole-type excitations (11.49) in a manner that allows us to separate the two components. Because $\varrho^+(k, \omega)$ vanishes for negative values of ω and $\varrho^-(k, \omega)$ vanishes for positive values, $A(k, \omega)$ describes particle excitations for positive frequencies and hole excitations for negative frequencies. Note that although the sum rules satisfied by ϱ^+ and ϱ^- individually depend upon the properties of the ground state, the sum rule satisfied by the spectral weight function $A(k, \omega)$ is independent of the ground state.

In the Hartree–Fock approximation only one of the quantities ϱ^+ or ϱ^- can be non-vanishing for a given value of k, depending upon whether k is above or below the Fermi surface. However, for the interacting system, levels on both sides are partially occupied. Thus both $c_k^\dagger$ and c_k give non-vanishing contributions when acting on the ground state of an interacting Fermi gas. The functions ϱ^+, ϱ^- and A can also be expressed in terms of Heisenberg operators by the same trick used to obtain eq. (11.43) from eq. (11.42).

$$\varrho^+(k,\omega) = \frac{1}{2\pi\hbar}\int_{-\infty}^{\infty} d\tau\, e^{i(\omega+\mu)\tau/\hbar}\langle g|c_k(\tau)c_k^\dagger(0)|g\rangle \tag{11.52a}$$

$$\varrho^-(k,\omega) = \frac{1}{2\pi\hbar}\int_{-\infty}^{\infty} d\tau\, e^{-i(\omega+\mu)\tau/\hbar}\langle g|c_k^\dagger(\tau)c_k(0)|g\rangle . \tag{11.52b}$$

The expression (11.52b) can be written in a form more like (11.52a) by the change of variable $\tau \to -\tau$ and noting that:

$$\langle g|c_k^\dagger(\tau)c_k(0)|g\rangle = \langle g|c_k^\dagger(0)c_k(-\tau)|g\rangle . \tag{11.52c}$$

Thus

$$\varrho^-(k,\omega) = \frac{1}{2\pi\hbar}\int_{-\infty}^{\infty} d\tau\, e^{i(\omega+\mu)\tau/\hbar}\langle g|c_k^\dagger(0)c_k(\tau)|g\rangle \tag{11.52d}$$

and

$$A(k,\omega) = \frac{1}{2\pi\hbar}\int_{-\infty}^{\infty} d\tau\, e^{i(\omega+\mu)\tau/\hbar}\langle g|\{c_k(\tau), c_k^\dagger(0)\}_+|g\rangle . \tag{11.52e}$$

The spectral weight function $A(k, \omega)$ satisfies a simple sum rule but is a highly singular function which is the sum of many delta functions. The same information about the elementary excitation spectrum is contained in the better behaved Green's function:

$$G(k,E) = \int_{-\infty}^{\infty} \frac{A(k,\omega)\,d\omega\, e^{iE\delta}}{E-\omega-\mu+i\omega\delta} = \int_{-\infty}^{\infty} \frac{[\varrho^+(k,\omega)+\varrho^-(k,\omega)]\,d\omega\, e^{iE\delta}}{E-\omega-\mu+i\omega\delta} \tag{11.53}$$

where δ is a positive infinitesimal quantity. The Green's function has a pole at all energy values corresponding to elementary excitations and the residue is proportional to the strength of the excitation as given by the value of the function $A(k, \omega)$. The only singularities in the Green's function are poles. These allow analytic manipulations which are not so easily accomplished for $A(k, \omega)$.

The Fourier transform of the expression (11.53) in the energy variable is:

$$G(k,\tau) = \frac{1}{2\pi}\int_{-\infty}^{\infty} e^{-iE\tau/\hbar}\, G(k,E)\, dE$$

$$= \frac{1}{2\pi}\int_{-\infty}^{\infty} d\omega \int_{-\infty}^{\infty} dE\, \frac{[\varrho^+(k,\omega)+\varrho^-(k,\omega)]}{E-\omega-\mu-i\omega\delta}\, e^{-iE(\tau-\delta)/\hbar} . \tag{11.54}$$

The integral over E can be evaluated by closing the contour with a semi-circle at infinity in the upper or lower half plane depending on the sign of τ. The integrand has a single pole which is in the upper or lower half plane depending upon the sign of ω. For $\tau \leqq 0$, the contour is closed in the upper half plane and the integrand has a contribution from a pole for negative values of ω. Since $\varrho^+(k, \omega)$ vanishes for negative ω, we obtain:

$$G(k, \tau) = \mathrm{i} \int_{-\infty}^{\infty} \mathrm{d}\omega \, \varrho^-(k, \omega) \mathrm{e}^{-\mathrm{i}(\omega+\mu)\tau/\hbar} \qquad (\tau \leqq 0)$$

$$= \mathrm{i} \int_{-\infty}^{\infty} \mathrm{d}\omega \int_{-\infty}^{\infty} \frac{\mathrm{d}\tau'}{2\pi} \mathrm{e}^{\mathrm{i}(\omega+\mu)(\tau-\tau')/\hbar} \langle g|c_k^\dagger(0) c_k(\tau)|g\rangle \qquad (\tau \leqq 0) \quad (11.55\mathrm{a})$$

where we have dropped the infinitesimal δ and substituted eq. (11.52b). Similarly, for $\tau > 0$ the contour is closed in the lower half plane, there is a contribution from a pole at positive values of ω and the contribution of ϱ^- vanishes:

$$G(k, \tau) = -\mathrm{i} \int_{-\infty}^{\infty} \mathrm{d}\omega \, \varrho^+(k, \omega) \mathrm{e}^{-\mathrm{i}(\omega+\mu)\tau/\hbar} \qquad (\tau > 0)$$

$$= -\mathrm{i} \int_{-\infty}^{\infty} \mathrm{d}\omega \int_{-\infty}^{\infty} \frac{\mathrm{d}\tau'}{2\pi} \mathrm{e}^{\mathrm{i}(\omega+\mu)(\tau-\tau')/\hbar} \langle g|c_k(\tau) c_k^\dagger(0)|g\rangle \qquad (\tau > 0). \quad (11.55\mathrm{b})$$

Eqs. (11.55) can be integrated over τ' and ω to obtain:

$$G(k, \tau) = \mathrm{i}\langle g|c_k^\dagger(0) c_k(\tau)|g\rangle \qquad (\tau \leqq 0) \qquad (11.56\mathrm{a})$$

$$G(k, \tau) = -\mathrm{i}\langle g|c_k(\tau) c_k^\dagger(0)|g\rangle \qquad (\tau > 0). \qquad (11.56\mathrm{b})$$

These two expressions can be combined by using the θ-function defined as:

$$\theta(\tau) = \begin{cases} 1 & (\tau > 0) \\ 0 & (\tau \leqq 0) \end{cases} \qquad (11.57\mathrm{a})$$

$$G(k, \tau) = -\mathrm{i}\langle g|c_k(\tau) c_k^\dagger(0)|g\rangle \theta(\tau) + \mathrm{i}\langle g|c_k^\dagger(0) c_k(\tau)|g\rangle [1 - \theta(\tau)]. \qquad (11.57\mathrm{b})$$

Another notation commonly used is:

$$G(k, \tau) = -\mathrm{i}\langle g|T\{c_k(\tau) c_k^\dagger(0)\}|g\rangle \qquad (11.57\mathrm{c})$$

where the time ordering symbol implies reordering the operators if necessary to put them in chronological order with the earliest times at the right, and including a minus sign for each interchange of anticommuting operators in the reordering.

If the single-particle basis k is a plane-wave basis, we can also consider the Fourier transform of the Green's function in configuration space:

$$G(\boldsymbol{k}, \tau) = \int e^{-i\boldsymbol{k}\cdot\boldsymbol{r}} G(\boldsymbol{r}, \tau) d^3\boldsymbol{r} \tag{11.58a}$$

where $G(\boldsymbol{r}, \tau)$ can be expressed in terms of the configuration space creation and annihilation operators:

$$\psi(\boldsymbol{r}, \tau) = \sum_{\boldsymbol{k}} c_{\boldsymbol{k}}(\tau) e^{i\boldsymbol{k}\cdot\boldsymbol{r}} \tag{11.59a}$$

$$\psi^{\dagger}(\boldsymbol{r}, \tau) = \sum_{\boldsymbol{k}} c_{\boldsymbol{k}}^{\dagger}(\tau) e^{i\boldsymbol{k}\cdot\boldsymbol{r}} \tag{11.59b}$$

$$G(\boldsymbol{r}, \tau) = -i\langle g|T\{\psi(\boldsymbol{r}, \tau)\psi^{\dagger}(0, 0)\}|g\rangle. \tag{11.59c}$$

This expression is often written in a more general form with an arbitrary origin for space and time:

$$G(\boldsymbol{r}_1 t_1; \boldsymbol{r}_2 t_2) = -i\langle g|T\{\psi(\boldsymbol{r}_1, t_1)\psi^{\dagger}(\boldsymbol{r}_2, t_2)\}|g\rangle. \tag{11.60}$$

The Green's function has many interesting properties. The function $G(\boldsymbol{k}, E)$ gives the excitation energies of all the elementary excitations induced by the operators $c_{\boldsymbol{k}}$ and $c_{\boldsymbol{k}}^{\dagger}$ and the strength of each excitation. If $\boldsymbol{r}_1 = \boldsymbol{r}_2$ and $t_1 = t_2$, the expression (11.60) is just the expectation value of the density of particles at time t. Similarly, from expression (11.56a), $G(\boldsymbol{k}, 0)$ gives the momentum distribution of particles in the ground state. The Green's function (11.60) describes creating a particle at the point $\boldsymbol{r}_2$ and time t_2 and annihilating the particle at the point $\boldsymbol{r}_1$ at the later time t_1. It can also describe creating a hole at the point $\boldsymbol{r}_1$ and time t_1 and annihilating the hole at the point $\boldsymbol{r}_2$ and a later time t_2. With this picture, the hole excitation is described also as creating a particle at the point $\boldsymbol{r}_2$ at the time t_2 and annihilating the particle at the point $\boldsymbol{r}_1$ at the earlier time t_1. Thus holes are represented as particles traveling backward in time. This picture is very useful in quantum field theory and in diagrammatic representations of various contributions in the many-body problems.

So far, we have written the spectral weight function and the Green's function and showed that they contain information of physical interest. However, in order for these functions to be useful, there must be some way of calculating them to some approximation from the Schrödinger equation without solving the Schrödinger equation for the ground state. This can be done by finding an equation of motion which is satisfied by the Green's function and which is obtained by differentiating the expression (11.57) with respect to τ. From eq. (11.57b) the derivative is seen to contain two kinds of terms, one from differentiating the argument of the Heisenberg operator and one from differentiating the θ-function to get a delta function.

$$\mathrm{i}\frac{\partial}{\partial\tau}G(k,\tau)=\langle g|T\left\{\frac{\partial c_k(\tau)}{\partial\tau}c_k^\dagger(0)\right\}|g\rangle+\delta(\tau), \tag{11.61}$$

since the matrix element multiplying $\delta(\tau)$ is just the anticommutator of the operators all at $\tau=0$, which is equal to unity by the anticommutation rules. To evaluate the time derivative of the annihilation operator, we note that:

$$\frac{\partial c_k(\tau)}{\partial\tau}=\frac{\mathrm{d}}{\mathrm{d}\tau}[\mathrm{e}^{\mathrm{i}H\tau/\hbar}c_k(0)\mathrm{e}^{-\mathrm{i}H\tau/\hbar}]=$$

$$=\frac{\mathrm{i}}{\hbar}\mathrm{e}^{\mathrm{i}H\tau/\hbar}[H,c_k(0)]\mathrm{e}^{-\mathrm{i}H\tau/\hbar}=\frac{1}{\hbar}[H,c_k(\tau)]. \tag{11.62}$$

Thus, if we know the Heisenberg equation of motion for the operator c_k we can obtain the equation of motion (11.61) for the Green's function. If we linearize the equation of motion (11.61) we obtain the Hartree–Fock approximation and a linear equation for $G(k,\tau)$ which can be immediately solved. Beyond the Hartree–Fock approximation, there are additional terms of higher order in the creation and annihilation operators on the right-hand side of the equation of motion (11.61). One method of treating these terms is to define and write equations of motion for higher-order Green's functions such as two-particle Green's functions, three-particle Green's functions etc., which involve the creation and subsequent annihilation of two and three particles. The equations for Green's functions of different orders are coupled, leading to an infinite hierarchy of equations. One approximation used is to truncate this hierarchy by neglecting correlations of more than two particles or more than three particles.

Another possible approach to the solution of the equation (11.61) is by perturbation theory, in which the Hamiltonian is written as a zero-order or Hartree–Fock part plus a perturbation. This method was developed first for quantum electrodynamics and is expressed most concisely in the diagrammatic notation of Feynman.

In a similar manner one can obtain the equation of motion for the space-time Green's function (11.60). Differentiating with respect to t_1, we obtain

$$i \frac{\partial}{\partial t_1} G(\boldsymbol{r}_1 t_1; \boldsymbol{r}_2 t_2) = \frac{i}{\hbar} \langle g | T\{[H, \psi(\boldsymbol{r}_1, t_1)]\} \psi^\dagger(\boldsymbol{r}_2, t_2) | g \rangle$$
$$+ \delta(\boldsymbol{r}_1 - \boldsymbol{r}_2)\delta(t_1 - t_2). \quad (11.63a)$$

This can be written

$$i \frac{\partial}{\partial t_1} G(\boldsymbol{r}_1 t_1; \boldsymbol{r}_2 t_2) = -\frac{i}{\hbar} [H(\boldsymbol{r}_1), G(\boldsymbol{r}_1 t_1; \boldsymbol{r}_2 t_2]$$
$$+ \delta(\boldsymbol{r}_1 - \boldsymbol{r}_2)(t_1 - t_2) \quad (11.63b)$$

where the operator $H(\boldsymbol{r}_1)$ operates only on operators in G at the time t_1. This form of the equation explains why these functions are called Green's functions.

PROBLEMS

Consider a simplified model of a ferromagnet, in which N spins of $\frac{1}{2}$ interact with one another and with an external magnetic field, described by the Hamiltonian

$$H = -g \sum_{i=1}^{N} h\sigma_{zi} - \sum_{\substack{i=1 \\ i \neq j}}^{N} \sum_{j=1}^{N} G_{ij} \boldsymbol{\sigma}_i \cdot \boldsymbol{\sigma}_j .$$

1. (a) Show that the total spin S^2 commutes with the Hamiltonian H.
(b) Write the Heisenberg equations of motion for the x- and y-components of the spin, S_x and S_y.
(c) What conclusions about the energy spectrum can you draw from the equations of motion for S_x and S_y?
(d) Show by the use of the variational principle that if the parameters g, h, and G_{ij} are all positive, the ground state wave function is an eigenfunction of S^2 and S_z with the eigenvalues $\frac{1}{2}N$, $+\frac{1}{2}N$. Denote this state by $|\frac{1}{2}N, +\frac{1}{2}N\rangle$.

2. Suppose that the spins are all in a one-dimensional lattice with equal spacing in the z-direction, and that there are periodic boundary conditions, $\boldsymbol{\sigma}_{i+N} = \boldsymbol{\sigma}_i$. Let g, h and G_{ij} all be positive. The state $|\frac{1}{2}N, +\frac{1}{2}N\rangle$ is the ground state as in problem 1d.

The Hamiltonian is invariant under the discrete translation $\boldsymbol{\sigma}_i \to \boldsymbol{\sigma}_{i+1}$.
(a) Is the ground state an eigenfunction of this translation operator? If so, give the eigenvalue.
(b) Write the equation of motion for the operator $\sum_{j=1}^{N} \mathrm{e}^{2\pi inj} \sigma_{j-}$, where n is any integer and $\sigma_{j-} = \sigma_{jx} - \mathrm{i}\sigma_{jy}$.

Show that this operator generates elementary excitations of the system above the ground state.
(c) Show that the state generated by the operator defined in problem 2b is degenerate with the ground state if $h=0$ and $n=0$.
(d) Are the states generated by the operator in problem 2b eigenfunctions of S^2?, of S_z?, of the translation operator $\boldsymbol{\sigma}_i \to \boldsymbol{\sigma}_{i+1}$? Give eigenvalues.

3. Calculate the ground state energy for the electron gas in the Hartree–Fock approximation and show that the second-order perturbation calculation of the energy leads to a divergent result.

4. Suppose that $|\langle E'|\psi(0)\rangle|^2 = C[(E'-\bar{E})^2 + (\frac{1}{2}\Gamma)^2]^{-1}$ in eq. (11.12a) where C and Γ are constants. Show that eqs. (11.11b) and (11.12b) cannot be used for this case, but that the conclusions relating the lifetime and width of the state are still valid.

$$\overline{\Delta E^2} = \sum_{E'} (E' - \bar{E})^2 |\langle E'|\psi(t)\rangle|^2 = \sum_{E'} (E' - \bar{E})^2 |\langle E'|\psi(0)\rangle|^2 \quad (11.11b)$$

$$\langle \psi(t)|\psi(0)\rangle = e^{i\bar{E}t/\hbar} \left\{ 1 - \tfrac{1}{2}\overline{(\Delta E)^2} \frac{t^2}{\hbar^2} + \ldots \right\}. \quad (11.12b)$$

5. For the ^{16}O nucleus, write down a complete set of linearly independent operators which create a particle–hole excitation with angular momentum and parity 1^-, by annihilating a particle in a $p_{\frac{1}{2}}$ or $p_{\frac{3}{2}}$ orbit and creating one in an $s_{\frac{1}{2}}$, $d_{\frac{3}{2}}$ or $d_{\frac{5}{2}}$ orbit. Do the same for octupole (3^-) excitations.

6. Consider the electron gas in a uniform background of positive charge and calculate the screening of the Coulomb potential in the Thomas–Fermi approximation. Suggestion: Calculate the potential due to a small positive point charge added to the electron gas by (a) modifying the Thomas–Fermi treatment in ch. 9 to include the effect of the uniform background of positive charge, (b) linearizing the non-linear differential equation by assuming that Z is small, (c) solving the linearized equation.

If this screened potential is used to calculate the second-order correction to the ground state energy, rather than the Coulomb potential, how would this affect the result of problem 3.

7. Write the linearized equations of motion describing quasiparticle excitations for the BCS reduced Hamiltonian in the simplified case where $G_{kk'} = G$. Calculate the energy of excitations by an operator $xa^{\dagger}_{k\uparrow} + ya_{-k\downarrow}$, where the parameters x and y are determined by solving the equations of motion. Use the BCS ground state wave function in the linearization procedure and note that the operators $a^{\dagger}_{k\uparrow}a^{\dagger}_{-k\downarrow}$ and $a_{k\uparrow}a_{-k\downarrow}$ have non-vanishing expectation values in this state.

FEYNMAN DIAGRAMS FOR PEDESTRIANS

Chapter 12 presents the modern formulation of time-dependent perturbation theory with Feynman diagrams in a simple non-relativistic form to illustrate the value of the method in nuclear and solid state problems as well as in relativistic quantum field theory. The essential new feature of the modern formalism is shown to be emphasis on the *particles* produced in the intermediate virtual transitions rather than on the intermediate *quantum states*. A single term in the new perturbation expansion represents a particular set of intermediate virtual particles and includes the contributions from all possible intermediate states arising from different time orderings of the transitions in which this same set of particles is created and absorbed. In high order where many intermediate particles are simultaneously present, many time orderings are possible for a given set of intermediate particles. Each time ordering appears in the old formalism as a separate intermediate state and contributes a separate term. Thus the new formulation greatly simplifies the calculation of high-order processes even in non-relativistic perturbation theory. Further simplification results from the representation of holes in many-body theory and antiparticles in quantum field theory as particles moving backward in time. This allows pair creation and annihilation processes to be included simply without additional terms.

In first-order perturbation theory the old-fashioned formulation used in ch. 7 gives a satisfactory description of the golden rule, and also gives an instructive picture of the time development, energy conservation and the uncertainty principle, while the new formulation offers no advantage. In second order there is only one intermediate state and the complications due to different time orderings of intermediate transitions do not arise. These complications first appear in third order; thus it is necessary to go to third order to see the advantages of the new formulation. In higher order the new formulation makes calculations feasible which would have been impractical in the old formulation and enables the summation of certain classes of diagrams to all orders.

To illustrate the method a third order 'bremsstrahlung' process is considered in detail in a theory defined by a phenomenological Hamiltonian containing fermions and bosons and an interaction which describes emission and absorption of single bosons by fermions. The different possible intermediate states are first considered and the sum of terms appearing in the old theory with different energy denominators is simplified algebraically to obtain a single term with one denominator. This new denominator describes the deviation from energy conservation associated with each intermediate particle rather than with each intermediate state. This 'deviation factor' is just the propagator for the particle in the new formalism and has the simple interpretation of describing the particle as 'off the energy shell' (or the mass shell in a relativistic theory). The third-order process is calculated explicitly in the new formalism and the Feynman rules for calculating processes from the Feynman diagrams are derived.

Section 12.4 'Fieldsmanship' shows how many of the problems arising in field theory can be seen in a simple way from the topological structure of the associated Feynman diagrams without performing detailed calculations. Many diagrams are shown to represent uninteresting quantities like the difference between properties of a physical state and an 'unphysical' unperturbed state in which all interactions are turned off. These diagrams can be eliminated formally by the process called renormalization; i.e. summing them and setting the sum equal to values of observed physical quantities. The polaron is presented as a simple example of quantum field theory and is treated by the analog of the Hartree–Fock method for the many-body problem. The classical limit is derived in which the presence of a large number of bosons around a fixed source behaves like a classical static field obeying a Poisson equation. The fieldsmanship and polaron sections can be read without reference to the preceding material in this chapter.

CHAPTER 12

FEYNMAN DIAGRAMS, PROPAGATORS AND FIELDS

12.1 Introduction

We now consider modern time-dependent perturbation theory first developed by Feynman for quantum electrodynamics and now in general use also in particle physics and solid state physics. The old-fashioned time-dependent perturbation theory described a transition from an initial state to a final state through various intermediate 'virtual' states. In intermediate states energy is not conserved and energy denominators appear in the transition matrix element. The energy denominator is the difference between the energy of the intermediate state and the energy of the initial or final state. Each intermediate state may contain several particles. In Feynman's approach, the emphasis is on the intermediate *particles* rather than the intermediate *states*. We illustrate this difference in a typical example.

Consider a system of interacting fermions and bosons with an interaction describing the emission or absorption of a boson by a fermion. This model applies to most cases of physical interest including quantum electrodynamics, electrons and phonons in solids, and interacting nucleons and mesons. Let $c_k^\dagger$ and $a_q^\dagger$ be creation operators for a fermion in a state k with energy E_k and a boson in a state q with energy ω_q. The quantum numbers k and q denote complete sets of quantum numbers including spins required to specify the one-particle states. They may be either plane waves or Hartree–Fock states.

Let the Hamiltonian be

$$H = H_0 + V \tag{12.1a}$$

where H_0 describes non-interacting fermions and bosons

$$H_0 = \sum_k E_k c_k^\dagger c_k + \sum_q \omega_q a_q^\dagger a_q \tag{12.1b}$$

and the interaction V has the general form

$$V = \sum_{jkq} g_{jkq} c_j^\dagger c_k a_q + g_{jkq}^* c_k^\dagger c_j a_q^\dagger \tag{12.1c}$$

where g_{jkq} are coefficients with arbitrary dependence on the indices j, k and q.

Let us now consider a 'bremsstrahlung' process in which two fermions in states j and k are scattered into states j' and k' and a boson in a state q is emitted. This process is third order in the interaction V, and is described in the old-fashioned perturbation treatment as the exchange of a boson between the two fermions plus the emission of an additional boson. There are many terms in the calculation corresponding to different time orderings of the emission of the final boson relative to the emission and absorption of the exchanged boson.

Consider two possible sets of intermediate states corresponding to the diagrams fig. 12.1a and fig. 12.1b.

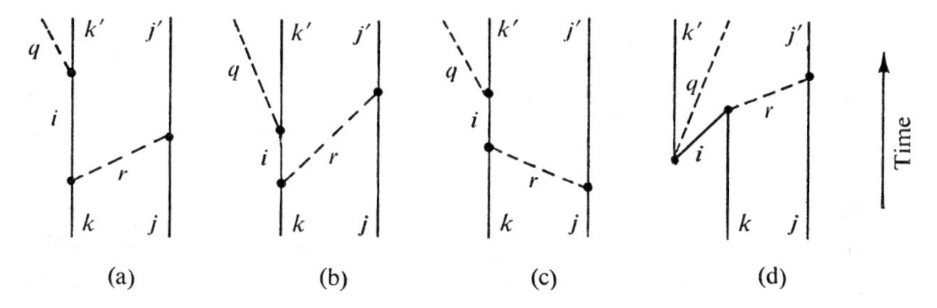

Fig. 12.1. In these diagrams the solid lines denote fermions, the dashed lines bosons, and the point where a dashed line joins a solid line represents the vertex or interaction in which the boson is emitted or absorbed.

In both diagrams the three steps in the transition are: (1) the fermion in the state k emits a boson in a state r and jumps into an intermediate state i; (2) the fermion in the state j absorbs the boson and jumps into the final state j' and (3) the fermion in the intermediate state i emits the final boson and jumps into the final state k'. The two diagrams fig. 12.1a and fig. 12.1b differ only in the time ordering of the last two transitions. The terms calculated in third-order perturbation theory for these two diagrams involve exactly the same interaction matrix elements but have different energy denominators.

12.2 Energy denominators and propagators

We now calculate these energy denominators explicitly by taking the difference in energy between each intermediate state and either the initial or the final state. Energy conservation is assumed, since time-dependent perturbation theory always requires it anyway in the final result. Thus

$$E_k + E_j = E_{k'} + E_{j'} + \omega_q. \tag{12.2}$$

The first intermediate state in fig. 12.1a differs from the initial state by having a fermion in a state i and a boson in a state r, rather than having a fermion in a state k. The energy denominator is

$$\Delta E_{a1} = E_i + \omega_r - E_k. \tag{12.3a}$$

The second intermediate state in fig. 12.1a differs from the final state in having a fermion in the state i instead of a boson in the state q and a fermion in the state k'. The energy denominator for this state is

$$\Delta E_{a2} = E_i - \omega_q - E_{k'}. \tag{12.3b}$$

Similarly for the diagram fig. 12.1b we have

$$\Delta E_{b1} = E_i + \omega_r - E_k \tag{12.3c}$$

$$\Delta E_{b2} = E_j + \omega_r - E_{j'}. \tag{12.3d}$$

The contribution of each diagram to the transition matrix element contains the product of the two energy denominators multiplied by interaction matrix elements and kinematic factors. Since the interaction matrix elements and the kinematic factors are exactly the same for the two diagrams of fig. 12.1a and fig. 12.1b we can write the sum of the contributions of these two diagrams to the transition matrix element T as the product of a common factor F and the sum of the two products of energy denominators

$$T = F\left\{\frac{1}{\Delta E_{a1}\Delta E_{a2}} + \frac{1}{\Delta E_{b1}\Delta E_{b2}}\right\} \tag{12.4a}$$

$$T = F\left\{\frac{1}{(E_i + \omega_r - E_k)(E_i - \omega_q - E_{k'})} + \frac{1}{(E_i + \omega_r - E_k)(E_j + \omega_r - E_{j'})}\right\}. \tag{12.4b}$$

Combining the two terms we obtain

$$T = F\left\{\frac{(E_j + \omega_r - E_{j'}) + (E_i - \omega_q - E_{k'})}{(E_i + \omega_r - E_k)(E_i - \omega_q - E_{k'})(E_j + \omega_r - E_{j'})}\right\}. \tag{12.4c}$$

This expression is simplified by substituting the energy conservation relation (12.2) into the numerator to obtain

$$T = F\left\{\frac{1}{(E_i - \omega_q - E_{k'})(E_j + \omega_r - E_{j'})}\right\}. \tag{12.4d}$$

The sum of the contributions from the two diagrams fig. 12.1a and fig. 12.1b thus has a form just as simple as the contribution of each individal diagram, a single term with a denominator which is the product of two factors.

The difference between the expression (12.4d) and the individual terms in the sum (12.4b) is conveniently characterized as follows: The diagrams 12.1a and 12.1b have two intermediate virtual states. They also have two intermediate virtual particles which exist only in intermediate states and not in the initial or final states; namely, the fermion in the state i and the boson in the state r. The energy denominators given by eqs. (12.3) characterize the two intermediate virtual *states*. The energy denominators in the expression (12.4d) characterize the two virtual *particles*.

The first factor in (12.4d) characterizes the non-conservation of energy in the absorption of the fermion i, the second factor characterizes the non-conservation of energy in the absorption of the boson r. These factors are common to the two diagrams 12.1a and 12.1b and are independent of the time orderings. Eq. (12.4d) suggests that we can associate some non-conservation of energy with *each intermediate virtual particle* rather than considering the many-particle state as a whole. These particles are considered as being 'off the energy shell'. With this approach we write a single term describing the sum of both contributions, fig. 12.1a and fig. 12.1b, by writing energy denominators for each virtual particle rather than for each virtual intermediate state. We thus have only a single term rather than two and need consider only a single diagram having a given topological structure. We need not worry about the different time orderings of interaction vertices obtained by distorting the diagrams without changing any quantum numbers on the particles. For higher order processes where more than two different time orderings are possible for each diagram the reduction in the number of terms by this procedure becomes much greater.

The diagram shown in fig. 12.1c resembles those of figs. 12.1a and 12.1b. It differs by the direction in time of the exchanged boson line. The energy denominators for this diagram are

$$\Delta E_{c1} = E_{j'} + \omega_r - E_j \tag{12.5a}$$

$$\Delta E_{c2} = E_i - \omega_q - E_{k'}. \tag{12.5b}$$

These energy denominators are similar to those of the expression (12.4d). If the kinematic and interaction factors are the same as for the diagrams 12.1a and 12.1b, the contributions of the three diagrams can be combined by adding the energy denominators,

$$\frac{1}{\Delta E_{a1}\Delta E_{a2}} + \frac{1}{\Delta E_{b1}\Delta E_{b2}} + \frac{1}{\Delta E_{c1}\Delta E_{c2}}$$

$$= \frac{2\omega_r}{(E_i - \omega_q - E_{k'})[\omega_r^2 - (E_j - E_{j'})^2]}. \tag{12.6}$$

In figs. 12.1a and b, the boson r is emitted by the fermion in the state k and absorbed by the one in the state j, while in fig. 12.1c it is emitted by the fermion in the state j and absorbed by the one in the state k. If the boson has an electric charge, e.g. a pion in nucleon–nucleon scattering by pion exchange, the bosons in fig. 12.1c must have the opposite electric charge from the one in figs. 12.1a and 12.1b. Thus fig. 12.1c describes the exchange of a boson which is the *antiparticle* of the boson exchanged in figs. 12.1a and 12.1b. In order for the result (12.6) to apply to the sum of the contributions of the three diagrams to the transition matrix element, the interaction matrix element must be the same for particle and antiparticle exchange with these quantum numbers.

The expression (12.6) again splits into two factors, one for each virtual particle. The factor for the boson r is not a simple energy denominator but something more complicated. This generalized energy denominator is called a propagator. A simple expression with a given factor or propagator for each virtual particle not only describes the sum of all diagrams with the same topology and different time orderings, but also includes the sum of particle and antiparticle exchanges. Furthermore, the quadratic denominator for the boson r in eq. (12.6) has a particularly interesting form. For a boson with momentum p_r and mass M,

$$\omega_r^2 = p_r^2 + M^2c^4 \tag{12.7a}$$

and

$$\omega_r^2 - (E_j - E_{j'})^2 = M^2c^4 - (E_r^2 - p_r^2), \tag{12.7b}$$

where

$$E_r^2 = (E_j - E_{j'})^2 \tag{12.7c}$$

is the square of the energy the boson would have if energy were conserved in the emission and absorption processes. The expression (12.7b) can be interpreted as the difference between the squared mass of a real boson and the squared mass of the virtual boson which is 'off its mass shell'. This expression is also relativistically invariant, and suitable for a relativistic theory.

In a similar way, the contribution of the diagram of fig. 12.1d can be included. For this case the direction in time of the fermion line has been reversed, and corresponds to an antiparticle or to a hole in many-body problems. The boson q is emitted with the creation of a particle–hole pair and the hole then annihilates with the initial fermion in the emission of the exchanged boson. The 'hole theory' of antiparticles is discussed in ch. 13.

12.3 The formal perturbation theory

The Feynman approach considers virtual particles, rather than virtual states, and defines a function called a propagator for each virtual particle rather than an energy denominator for each state. The propagator is defined so that a single diagram gives the contribution of many terms in old-fashioned 'intermediate state' perturbation theory, namely all terms which are related by changing the time ordering of the various emission and absorption processes, or by reversing the direction in time of lines, thereby changing particles into holes or antiparticles. The propagators can be defined relativistically, thus leading to a fully covariant relativistic theory.

The energy and momentum of virtual particles are defined so that energy and momentum are always conserved in all interactions. The energy and momentum of the virtual particle are determined entirely by the conservation laws and do not satisfy the relation between energy and momentum for a free physical particle. The virtual particle is 'off its mass shell' rather than 'off its energy shell' and the propagator describes the amount by which it is off the mass shell.

We now develop the formalism of this approach. We begin with the iterated solution (7.38b) of the time-dependent Schrödinger equation in the interaction representation.

$$\psi_{\mathrm{I}}(t) = \psi_{\mathrm{I}}(t_0) + \left(\frac{-\mathrm{i}}{\hbar}\right)\int_{t_0}^{t} V(t_1)\psi_{\mathrm{I}}(t_0)\,\mathrm{d}t_1$$

$$+\left(\frac{-\mathrm{i}}{\hbar}\right)^2 \int_{t_0}^{t} V(t_1)\,\mathrm{d}t_1 \int_{t_0}^{t_1} V(t_2)\,\mathrm{d}t_2\,\psi_{\mathrm{I}}(t_0)$$

$$+\left(\frac{-\mathrm{i}}{\hbar}\right)^3 \int_{t_0}^{t} V(t_1)\,\mathrm{d}t_1 \int_{t_0}^{t_1} V(t_2)\,\mathrm{d}t_2 \int_{t_0}^{t_2} V(t_3)\,\mathrm{d}t_3\,\psi_{\mathrm{I}}(t_0) + \ldots \tag{12.8}$$

In a scattering problem we are interested in an initial state $|i\rangle$ which is an eigenfunction of H_0, at a time in the remote past before the scattering process takes place. To describe the initial state, let us set $t_0 = -\infty$.

$$\psi_{\mathrm{I}}(t_0 = -\infty) = |i\rangle \tag{12.9a}$$

$$H_0|i\rangle = E_i|i\rangle. \tag{12.9b}$$

We are interested in the probability of a transition to a state $|f\rangle$ which is also an eigenfunction of H_0

$$H_0|f\rangle = E_f|f\rangle \tag{12.9c}$$

The element of the S-matrix between the states $|i\rangle$ and $|f\rangle$ is the scalar product of the state f with the exact solution of the Schrödinger equation at time $t = +\infty$ .Substituting eqs. (12.9) into the iterated solution (12.8) we obtain a perturbation series for the S-matrix element.

$$\langle f|S|i\rangle = \langle f|\psi_{\mathrm{I}}(+\infty)\rangle = \langle f|i\rangle + \left(\frac{-\mathrm{i}}{\hbar}\right)\int_{-\infty}^{+\infty} \mathrm{d}t_1 \,\langle f|V(t_1)|i\rangle$$

$$+\left(\frac{-\mathrm{i}}{\hbar}\right)^2 \int_{-\infty}^{\infty} \mathrm{d}t_1 \int_{-\infty}^{+\infty} \mathrm{d}t_2 \,\langle f|V(t_1)V(t_2)|i\rangle\theta(t_1 - t_2)$$

$$+\left(\frac{-\mathrm{i}}{\hbar}\right)^3 \int_{-\infty}^{+\infty} \mathrm{d}t_1 \int_{-\infty}^{+\infty} \mathrm{d}t_2 \int_{-\infty}^{+\infty} \mathrm{d}t_3 \langle f|V(t_1)V(t_2)V(t_3)|i\rangle\theta(t_1 - t_2)\theta(t_2 - t_3) + \dots$$

$$= \langle f|S^0|i\rangle + \langle f|S^{(1)}|i\rangle + \langle f|S^{(2)}|i\rangle + \langle f|S^{(3)}|i\rangle + \dots . \tag{12.10}$$

We have rewritten the second- and third-order terms in a more convenient form for later manipulations by changing the upper limits of all integrals to $+\infty$ and including the θ-functions in the integrand to remove the extraneous contributions.

For the bremsstrahlung problem defined above

$$|i\rangle = c_k^\dagger c_j^\dagger|0\rangle \tag{12.11a}$$

$$|f\rangle = a_q^\dagger c_{k'}^\dagger c_{j'}^\dagger|0\rangle. \tag{12.11b}$$

The first non-vanishing contribution in eq. (12.10) comes from the third-order term. The product of the three V-operators must contain the operators c_k and c_j to annihilate the particles in the initial state and $c_{k'}^\dagger c_{j'}^\dagger$ and $a_q^\dagger$ to create the particles in the final state. There remain two fermion operators and two boson operators unaccounted for in the nine operators appearing in a product of three V's. To give a non-vanishing matrix element, the fermion operators must create and annihilate the same fermion state and the boson operators must create and annihilate the same boson state. We denote the intermediate fermion state by i and the intermediate boson state by r, as in fig. 12.1.

We now consider all possible ways of distributing the nine operators $c_k\ c_j\ c_{k'}^\dagger\ c_{j'}^\dagger\ c_i^\dagger\ c_i\ a_q^\dagger\ a_r^\dagger$ and a_r in three interactions V. Let

$$V_{\mathrm{a}} = c_{k'}^\dagger c_i a_q^\dagger g_{ik'q}^* \tag{12.12a}$$

$$V_{\mathrm{b}} = c_i^\dagger c_k [a_r^\dagger g_{kir}^* + a_{-r} g_{ik-r}] \tag{12.12b}$$

$$V_{\mathrm{c}} = c_{j'}^\dagger c_j [a_r g_{j'jr} + a_{-r}^\dagger g_{jj'-r}^*] \tag{12.12c}$$

$$V_{\mathrm{b'}} = c_{k'}^\dagger c_i [a_r^\dagger g_{ik'r}^* + a_{-r} g_{k'i-r}] \tag{12.12d}$$

$$V_{\mathrm{a'}} = c_i^\dagger c_k a_q^\dagger g_{kiq}^* . \tag{12.12e}$$

We have introduced the label $-r$ for the case where r is a momentum, since annihilating a boson of momentum $-r$ gives the same momentum transfer as creating a boson with momentum r.

The products $V_{\mathrm{a}}V_{\mathrm{b}}V_{\mathrm{c}}$ and $V_{\mathrm{a'}}V_{\mathrm{b'}}V_{\mathrm{c}}$ have non-vanishing matrix elements between the states $|i\rangle$ and $|f\rangle$. Other products can be formed by interchanging the indices j and k and/or j' and k' in these products. These constitute all possible products which can be constructed which have non-vanishing matrix elements. We consider the product $V_{\mathrm{a}}V_{\mathrm{b}}V_{\mathrm{c}}$ and $V_{\mathrm{a'}}V_{\mathrm{b'}}V_{\mathrm{c}}$ and disregard for the present the products formed by permuting j and k and j' and k'. We can therefore substitute into the third-order term of eq. (12.10)

$$V(t_i) = V_{\mathrm{a}}(t_i) + V_{\mathrm{b}}(t_i) + V_{\mathrm{c}}(t_i) + V_{\mathrm{a'}}(t_i) + V_{\mathrm{b'}}(t_i). \tag{12.13}$$

There are twelve non-vanishing terms corresponding to six ways of distributing V_{a}, V_{b} and V_{c} among the times t_1, t_2 and t_3, and similarly for $V_{\mathrm{a'}}$, $V_{\mathrm{b'}}$, V_{c}. Since t_1, t_2 and t_3 are dummy variables, we can change them to new variables t_{a}, t_{b} and t_{c}. We change the variables differently for each term, so that the arguments of V_{a}, $V_{\mathrm{a'}}$, V_{b}, $V_{\mathrm{b'}}$ and V_{c} are always t_{a}, $t_{\mathrm{a'}}$, t_{b}, $t_{\mathrm{b'}}$ and t_{c} respectively. Thus for each term the θ-functions are different and choose a different domain for the variables t_{a}, t_{b}, t_{c}. In eqs. (12.10) the θ-functions choose $t_1 \geqq t_2 \geqq t_3$. In our change of variables, we have used all six ways of replacing t_1, t_2 and t_3 by t_{a}, t_{b} and t_{c}, the V's always have the same argument and they are always ordered 'chronologically' with the earlier time to the right. This can be conveniently written

$$\langle f|S^{(3)}|i\rangle = \left(-\frac{\mathrm{i}}{\hbar}\right)^3 \int_{-\infty}^{+\infty} \mathrm{d}t_{\mathrm{a}} \int_{-\infty}^{+\infty} \mathrm{d}t_{\mathrm{b}} \int_{-\infty}^{+\infty} \mathrm{d}t_{\mathrm{c}}$$

$$\times \langle f|T\{V_{\mathrm{a}}(t_{\mathrm{a}})V_{\mathrm{b}}(t_{\mathrm{b}})V_{\mathrm{c}}(t_{\mathrm{c}}) + V_{\mathrm{a'}}(t_{\mathrm{a}})V_{\mathrm{b'}}(t_{\mathrm{b}})V_{\mathrm{c}}(t_{\mathrm{c}})\}|i\rangle \tag{12.14}$$

where the time-ordered or T-product is defined to include re-ordering the factors in chronological order with the earliest time to the right. The six possible orderings give six integrals over different domains of t_a, t_b and t_c and correspond to just the domains defined by the θ-functions in each of the six terms obtained by substituting eq. (12.13) into eq. (12.10).

The expression (12.14) can be simplified by a change of time variables.

$$V_a(t_a)V_b(t_b)V_c(t_c) = e^{iH_0t_a/\hbar}V_a e^{iH_0(t_b-t_a)/\hbar}V_b e^{iH_0(t_c-t_b)/\hbar}V_c e^{-iH_0t_c/\hbar} \tag{12.15a}$$

$$V_a(t_a)V_b(t_b)V_c(t_c) = e^{iH_0t_b/\hbar}V_a(t_a-t_b)V_b(0)V_c(t_c-t_b)e^{-iH_0t_b/\hbar}. \tag{12.15b}$$

Let us define

$$\tau_1 = t_a - t_b \tag{12.16a}$$

$$\tau_2 = t_c - t_b. \tag{12.16b}$$

Substituting eq. (12.15) and a similar expression for $V_{a'}$, $V_{b'}$, V_c into eq. (12.14) and introducing the variables of integration (12.16) we obtain

$$\langle f|S^{(3)}|i\rangle = \frac{i}{\hbar^3}\int_{-\infty}^{+\infty} dt_b e^{i(E_i-E_f)t_b/\hbar}\int_{-\infty}^{+\infty} d\tau_1 \int_{-\infty}^{+\infty} d\tau_2$$

$$\times \langle f|T\{V_a(\tau_1)V_b(0)V_c(\tau_2) + V_{a'}(\tau_1)V_{b'}(0)V_c(\tau_2)\}|i\rangle \tag{12.17}$$

where E_i and E_f are the eigenvalues of H_0 in the initial and final states, and the exponential factors are obtained by the operation of the first and last factors of eq. (12.15b) in the states $|f\rangle$ and $|i\rangle$.

Integration over t_b gives a delta function for conservation of energy,

$$\langle f|S^{(3)}|i\rangle = \frac{i}{\hbar^2}2\pi\delta(E_i - E_f)\int_{-\infty}^{+\infty} d\tau_1 \int_{-\infty}^{+\infty} d\tau_2$$

$$\times \langle f|T\{V_a(\tau_1)V_b(0)V_c(\tau_2) + V_{a'}(\tau_1)V_{b'}(0)V_c(\tau_2)\}|i\rangle. \tag{12.18}$$

Until this point, our derivation of time-dependent perturbation theory can either be used for the old 'intermediate state' approach or for the modern 'intermediate particle' approach. From here the two approaches diverge. The old approach evaluates the matrix element (12.18) by inserting complete sets of states between the various V-factors. The dependence on the time variables τ_1 and τ_2 then appears as exponential phase factors with the operators H_0 replaced by the eigenvalues in the intermediates states. Integration over these time variables then gives the energy denominators.

Rather than expanding into intermediate states, we reorder the creation and annihilation operators in the various V-factors in order to bring together operators which create and annihilate the same particle. The time integrations then give energy denominators or propagators corresponding to the virtual intermediate particles. We eliminate all the creation and annihilation operators for the initial and final particles by using the expressions (12.11) for the initial and final states to convert the matrix element (12.18) to a vacuum expectation value and then using Wick's theorem.

$$\langle f|T\{V_a(\tau_1)V_b(0)V_c(\tau_2)\}i\rangle = \langle 0|c_{j'}c_{k'}a_q T\{V_a(\tau_1)V_b(0)V_c(\tau_2)\}c_k^\dagger c_j^\dagger|0\rangle \quad (12.19a)$$

$$\langle f|T\{V_{a'}(\tau_1)V_{b'}(0)V_c(\tau_2)\}|i\rangle = \langle 0|c_{j'}c_{k'}a_q T\{V_{a'}(\tau_1)V_{b'}(0)V_c(\tau_2)\}c_k^\dagger c_j^\dagger|0\rangle. \quad (12.19b)$$

So far the operators appearing in the T-product at different times have been combinations of boson operators and pairs of fermion operators. Now that we are introducing single anticommuting fermion operators explicitly, we adopt the following convention. In the chronological reordering of a T-product, a phase of -1 is introduced whenever two fermion operators are commuted. For example,

$$T\{c_k^\dagger(\tau)c_k(0)\} = c_k^\dagger(\tau)c_k(0) \quad \text{for} \quad \tau \geqq 0, \quad (12.20a)$$

$$= -c_k(0)c_k^\dagger(\tau) \quad \text{for} \quad \tau < 0. \quad (12.20b)$$

The reason for this convention will be apparent below. To evaluate (12.19) we must commute and anticommute time dependent operators. This is done by noting that

$$c_k^\dagger(\tau) = e^{iH_0\tau/\hbar}c_k^\dagger e^{-iH_0\tau/\hbar} = e^{iE_k\tau/\hbar}c_k^\dagger \quad (12.21a)$$

$$c_j c_k^\dagger(\tau) = \delta_{jk}e^{iE_k\tau/\hbar} - c_k^\dagger(\tau)c_j, \quad (12.21b)$$

and similarly for boson commutators

$$a_q a_r^\dagger(\tau) = \delta_{qr}e^{i\omega_r\tau/\hbar} + a_r^\dagger(\tau)a_q. \quad (12.21c)$$

Thus

$$\langle 0|c_{j'}AV_c(\tau_2)A'c_j^\dagger|0\rangle = e^{i(E_{j'}-E_j)\tau_2/\hbar}\langle 0|A[a_r(\tau_2)g_{j'jr} + a_{-r}^\dagger(\tau_2)g_{jj'-r}^*]A'|0\rangle \quad (12.22a)$$

where A and A' are any operators which commute with $c_{j'}$ and $c_j^\dagger$, and similarly

$$\langle 0|c_{k'}a_q BV_a(\tau_1)B'|0\rangle = e^{i(E_{k'}+\omega_q)\tau_1/\hbar}g_{ik'q}^*\langle 0|Bc_i(\tau_1)B'|0\rangle \quad (12.22b)$$

and

$$\langle 0|a_q D V_{a'}(\tau_1) D' c_k^\dagger|0\rangle = e^{-i(E_k-\omega_q)\tau_1/\hbar} g_{kiq}^* \langle 0|D c_i^\dagger(\tau_1) D'|0\rangle \quad (12.22c)$$

where B and B' are operators commuting with $c_{k'}$ and $a_{q'}$ and D and D' are operators commuting with $c_k^\dagger$ and a_q.

The two vacuum expectation values (12.19a) and (12.19b) are seen to have the form (12.22a), after anticommuting $c_{k'}$ to the left and $c_k^\dagger$ to the right, for all time orderings of the operators in the T-products. Similarly, eq. (12.19a) has the form (12.22b) and eq. (12.19b) has the form (12.22c) for all time orderings. We can therefore simplify eqs. (12.19) to eliminate the operators with indices k, k' and q by using eqs. (12.22) to obtain

$$\langle f|T\{V_a(\tau_1)V_b(0)V_c(\tau_2)|i\rangle = e^{i[(E_{k'}+\omega_q)\tau_1+(E_{j'}-E_j)\tau_2]/\hbar} g_{ik'q}^*$$
$$\times \langle 0|T\{c_i(\tau_1)V_b(0)[a_r(\tau_2)g_{j'jr} + a_{-r}^\dagger(\tau_2)g_{jj'-r}^*]\}c_k^\dagger|0\rangle \quad (12.23a)$$

$$\langle f|T\{V_{a'}(\tau_1)V_{b'}(0)V_c(\tau_2)|i\rangle = e^{i[(\omega_q-E_k)\tau_1+(E_{j'}-E_j)\tau_2]/\hbar} g_{kiq}^*$$
$$\times \langle 0|c_{k'}T\{c_i^\dagger(\tau_1)V_{b'}(0)[a_r(\tau_2)g_{j'jr} + a_{-r}^\dagger(\tau_2)g_{jj'-r}^*]\}|0\rangle. \quad (12.23b)$$

We continue the simplification process by anticommuting $c_k^\dagger$ through V_b in eq. (12.23b) and anticommuting $c_{k'}$ through $V_{b'}$ in eq. (12.23b). Now, however, we must be careful about the time ordering because of the single fermion operators $c_i(\tau_1)$ and $c_i^\dagger(\tau_1)$ in the time ordered products. One of these single fermion operators may or may not stand between V_b and $c_k^\dagger$ in (12.23a) or between $c_{k'}$ and $V_{b'}$ in (12.23b), depending upon the time ordering. The presence of such an operator introduces a negative phase factor because of the anticommutation. This is one reason for the particular phase convention (12.20) chosen in defining the T-product. With this definition we can anticommute single fermion operators through a T-product using the ordering in which the product is written, and we do not have to worry about different orderings. The result will be a T-product in which the phase convention (12.20) automatically introduces the proper phases for the calculation of the anticommutator in the different orderings. Thus

$$\langle f|T\{V_a(\tau_1)V_b(0)V_c(\tau_2)\}|i\rangle$$
$$= e^{i[(E_{k'}+\omega_q)\tau_1+(E_{j'}-E_j)\tau_2]/\hbar} g_{ik'q}^* \langle 0|T\{c_i(\tau_1)c_i^\dagger(0)[a_r^\dagger(0)g_{kir}^* + a_{-r}(0)g_{ij-r}]$$
$$\times [a_r(\tau_2)g_{j'jr} + a_{-r}^\dagger(\tau_2)g_{jj'-r}^*]\}|0\rangle \quad (12.24a)$$

$$\langle f|T\{V_{a'}(\tau_1)V_{b'}(0)V_c(\tau_2)\}|i\rangle$$
$$= -e^{i[(\omega_q-E_k)\tau_1+(E_{j'}-E_j)\tau_2]/\hbar} g_{kiq}^* \langle 0|T\{c_i^\dagger(\tau_1)c_i(0)[a_r^\dagger(0)g_{ik'r}^* + a_{-r}(0)g_{k'i-r}]$$
$$\times [a_r(\tau_2)g_{j'jr} + a_{-r}^\dagger(\tau_2)g_{jj'-r}^*]\}|0\rangle. \quad (12.24b)$$

Eq. (12.24b) can be transformed to a form similar to eq. (12.24a) by the change of variable $\tau_1 \to -\tau_1$ and noting that

$$\langle 0|T\{c_i^\dagger(-\tau_1)c_i(0)\}|0\rangle = \langle 0|T\{c_i^\dagger(0)c_i(\tau_1)\}|0\rangle = -\langle 0|T\{c_i(\tau_1)c_i^\dagger(0)\}|0\rangle. \tag{12.24c}$$

Then

$$\langle f|T\{V_{a'}(-\tau_1)V_{b'}(0)V_c(\tau_2)\}|i\rangle = e^{i[(E_k-\omega_q)\tau_1+(E_{j'}-E_j)\tau_2]/\hbar} g_{kiq}^* \langle 0|T\{c_i(\tau_1)c_i^\dagger(0)[a_r^\dagger(0)g_{ik'r}^*+a_{-r}(0)g_{k'i'r}] \times [a_r(\tau_2)g_{j'jr} + a_{-r}^\dagger(\tau_2)g_{jj'-r}^*]\}|0\rangle. \tag{12.24d}$$

The vacuum expectation values in (12.24) can be factorized into a product of two vacuum expectation values, one for bosons and one for fermions, since the vacuum is the only intermediate state which can be connected to the vacuum on the left by fermion operators and on the right by boson operators.

The fermion vacuum expectation values in eq. (12.24) are just the Green's functions which we have already encountered in discussing the elementary excitations of many-particle systems, eq. (11.57c)

$$G(k,\tau) = -i\langle 0|T\{c_k(\tau)c_k^\dagger(0)\}|0\rangle. \tag{11.57c}$$

The Fourier transform of this expression appears in eq. (12.25)

$$G(k,E) = \int_{-\infty}^{+\infty} e^{iE\tau/\hbar} G(k,\tau)\,d\tau/\hbar. \tag{12.25}$$

The boson vacuum expectation values can also be expressed in terms of Green's functions. Let

$$D(q,\tau) = -i\langle 0|T\{\phi_q(\tau)\phi_q^\dagger(0)\}|0\rangle \tag{12.26a}$$

where the boson field operator is given by

$$\phi_q = a_q + a_{-q}^\dagger. \tag{12.26b}$$

The Fourier transform of eq. (12.26) is

$$D(q,\omega) = \int_{-\infty}^{+\infty} e^{i\omega\tau/\hbar} D(q,\tau)\,d\tau/\hbar. \tag{12.27}$$

The expression (12.25) can be written in terms of the boson Green's functions (12.27) if

$$g_{ikr} = g_{ki-r}^*. \tag{12.28}$$

This relation is generally true in practical examples. We therefore assume it and substitute the Green's functions (11.57c) and (12.26) into the matrix elements (12.24)

$$\langle f|T\{V_{\mathrm{a}}(\tau_1)V_{\mathrm{b}}(0)V_{\mathrm{c}}(\tau_2)\}|i\rangle$$
$$= g^*_{ik'q}g^*_{kir}g_{j'jr}\,\mathrm{e}^{\mathrm{i}(E_{k'}+\omega_q)\tau_1/\hbar}\,G(i,\tau_1)\,\mathrm{e}^{\mathrm{i}(E_{j'}-E_j)\tau_2/\hbar}\,D(r,\tau_2) \quad (12.28a)$$
$$\langle f|T\{V_{\mathrm{a'}}(-\tau_1)V_{\mathrm{b'}}(0)V_{\mathrm{c}}(\tau_2)\}|i\rangle$$
$$= g^*_{kiq}g^*_{ik'r}g_{j'jr}\,\mathrm{e}^{\mathrm{i}(E_k-\omega_q)\tau_1/\hbar}\,G(i,\tau_1)\,\mathrm{e}^{\mathrm{i}(E_{j'}-E_j)\tau_2/\hbar}\,D(r,\tau_2). \quad (12.28b)$$

Substituting these matrix elements into the expression for the S-matrix (12.18) gives the Fourier transforms of the Green's functions

$$\langle f|S^{(3)}|i\rangle = \mathrm{i}\,2\pi\delta(E_i-E_f)[g^*_{ik'q}g^*_{kir}g_{j'jr}G(i,E_{k'}+\omega_q)D(r,E_{j'}-E_j)$$
$$+\,g^*_{kiq}g^*_{ik'r}g_{j'jr}G(i,E_k-\omega_q)D(r,E_{j'}-E_j)]. \quad (12.29)$$

This expression contains interaction matrix elements and Green's functions, rather than the interaction matrix elements and energy denominators found in 'intermediate state perturbation theory'. We see that there is a Green's function for each virtual intermediate particle. The relation between Green's functions and energy denominators is seen from the properties of the Green's functions defined in ch. 11. The Green's function $G(k, E)$ has a simple pole at the value of E corresponding to the energy of each elementary excitation produced by adding a particle to the system in the state k. For the Hamiltonian H_0, this is just the energy E_k. Thus $G(i, E_{k'}+\omega_q)$ has a denominator $E_i-E_{k'}-\omega_q$ which gives this pole, and is just the denominator corresponding to the fermion in the state i which we have seen in the expression (12.4d).

Examination of the first term of the expression (12.29) shows that it contains the contributions arising in 'intermediate state perturbation theory' from all the diagrams of fig. 12.1 and includes as well two additional diagrams obtained from fig. 12.1d in which the boson r is emitted from the state j before and after the emission of the boson q. That the contribution from the diagram 12.1d which includes a hole or antiparticle state i is correctly described by the expression (12.29) can be seen by comparing the Green's function with the relevant energy denominator. The Green's function for a *hole* excitation has a pole at a *negative* energy $-E$ which is the negative of the elementary excitation, i.e. at $-E_k$. Thus $G(i, E_{k'}+\omega_q)$ has a denominator $E_i+E_{k'}+\omega_q$. This is just the energy denominator for the first intermediate state of fig. 12.1d. Similarly, evaluation of the boson Green's function defined by eq. (12.26) and (12.27) shows that it provides the proper function of r appearing in eq. (12.6) to give the sum of diagrams.

Examination of the second term of (12.29) shows that it gives the contributions from another set of diagrams, namely those in which the final boson q is emitted in the transition from the initial state k to the intermediate state i and the exchanged boson r is either emitted or absorbed in the transition from the intermediate state i to the final state k'.

As mentioned above, there are additional contributions from terms obtainable from eq. (12.29) by interchanging the indices k' and j' and by interchanging k and j. These are clearly obtainable directly by performing these interchanges on the result (12.29) and introducing a phase of -1 for each interchange because of the anticommutation relations. For the conventional physical bremsstrahlung process, where an electron is scattered by a nucleon or nucleus and a photon is emitted, these interchanges are not included. The labels j, j', k and k' would then include whether the particle is an electron or a nucleon, and the single interchanges would consider a transition of an electron into a nucleon with the emission of a photon. These contributions would vanish automatically, because there is no non-vanishing g-coefficient corresponding to this transition. The double exchange would describe the emission of the photon by the nucleon instead of by the electron. Although this contribution is not zero, it is much smaller than the other, because of the electron–nucleon mass difference.

Note that a summation over all possible intermediate states i and r is implied in the result (12.29). Furthermore, since the indices i and r include all degrees of freedom of the particle, a sum over spins and polarizations is also implied. For the case where the fermion is described by the Dirac equation, the Green's function is defined to include all degrees of freedom included in the Dirac spinor, including antiparticle states. For this case diagrams like 12.1d include pair production.

We can now compare the simplicity of the result (12.29) with that obtained from intermediate state perturbation theory. We see that each time-ordered product in the expression (12.18) leads to one term in the result (12.29). If, on the other hand, we proceed from eq. (12.18) by intermediate state perturbation theory and expand the matrix elements by inserting complete sets of intermediate states between each pair of V's, we see that there is a different set of intermediate states *for each ordering* of the T-product. Thus each single term in (12.29) would appear as the sum of six terms in intermediate state perturbation theory, the reduction of the number of terms for an nth-order contribution would be a factor $n!$. This is the reason that a third-order example was chosen for this presentation, rather than second order. A factor of six is already impressive, and both aspects of the reduction are illustrated: different time orderings of the same intermediate particles and treating particles and antiparticles on the same footing. The simple form of the result (12.29) and its relation to the diagrammatic representation leads to a simple set of rules, known as Feynman rules, for the execution of practical calculations. All topologically distinct diagrams are drawn for a given process and a given order in the interaction. The contribution of each diagram in eq. (12.29) is seen to include

(1) a factor $G(k, E)$ for each virtual fermion line,

(2) a factor $D(q, \omega)$ for each virtual boson line,

(3) a factor g_{ikq} for each vertex (intersection of three lines).

The indices ikq are summed over all possible values. The arguments E and ω of the Green's functions are determined by assuming energy conservation at all vertices.

For the general case, there are also phase factors related to the topological structure of the diagrams, particularly if there are 'closed loops' of virtual fermion lines.

The particular form of the Green's function depends upon the form of H_0. In the many-body problem, it could be the non-interacting Fermi gas, but it might also be the BCS approximation, with the BCS ground state as the vacuum. In relativistic quantum field theory the Dirac equation is used to define the fermion Green's function and the boson Green's function is defined by the appropriate relativistic wave equation, depending upon its spin.

Further simplifications which are very important are easily introduced in this diagrammatic approach. For example the vacuum state $|0\rangle$ is not an eigenstate of the full Hamiltonian H, but only of H_0. The exact eigenstate of H corresponding to no particles is called the 'physical vacuum', in contrast to the 'unperturbed vacuum'. For the bremsstrahlung problem the desired initial and final states are not (12.11) but should have the creation operators acting on the *physical* vacuum state. If we use the initial states (12.11) the perturbation series (12.10) contains not only the desired scattering matrix but also the corrections to the unperturbed vacuum state. It turns out that these undesired vacuum corrections can be eliminated by discarding all 'disconnected diagrams'; i.e. all diagrams which can be separated into two or more disconnected pieces without breaking any lines.

Furthermore, a state constructed by having a single particle creation operator acting on the *physical* vacuum is still not an eigenstate of H. There are correction terms which can be calculated by using this state in the perturbation series (12.8). The effect of the interaction is to change the energy of the state, i.e. to renormalize the mass, and to change the wave function into one containing not only the 'bare' one-fermion state but also states containing a 'cloud of virtual bosons'. In the diagrammatic formulation it is possible to isolate those diagrams which change the 'bare' particle to the 'dressed particle' with its cloud of virtual bosons and particle–hole pairs. The properties of the physical or dressed particle can then be used in the scattering calculation. The calculation of the masses and interactions of the dressed particles from those of the bare particles is called renormalization. In some problems, particularly in solids, these renormalization factors can be calculated and compared with experiment. The 'effective mass' of an electron moving in a solid with its cloud of virtual phonons can be calculated and measured. In other cases, particularly quantum field theory, they can neither be calculated nor measured. The 'bare mass' of an electron with its electromagnetic field turned off cannot be measured, as there is no way to turn off the field. The renormalization factor between the bare mass and the physical mass is expressed as a divergent integral in perturbation theory, as in the simple example of the kaon mass difference. Thus the approach used is to calculate physical scattering matrix elements in terms of the physical particles and their properties and to eliminate the unmeasurable properties of the bare particles from the results.

12.4 Fieldsmanship: the art of discussing field theory without really knowing anything about it

Relativistic quantum field theory treats processes like radiation where particles are emitted and absorbed. The formalism of second quantization was first derived by the quantization of classical fields. However, it is also possible to introduce the second-quantized notation directly, as in ch. 5, and describe particle emission and absorption by phenomenological interactions, like those in eqs. (4.2), (4.28), (7.12) and (12.1c). In cases like electrodynamics and gravitation, where a classical field exists and its equations are known, quantization of the field gives interesting predictions from first principles; namely the existence of the quanta of the field as particles (photons and gravitons) and the exact form of the Hamiltonian and Lagrangian, including the interactions. For electrons, nucleons and mesons, where no analogous classical field is known, field theory works backwards, and looks for field equations which will predict the existence of the observed particles when quantized. There are then no new predictions from first principles, and one might just as well start from a phenomenological second-quantized Hamiltonian like eq. (12.1).

We begin our discussion with the phenomenological Hamiltonian (12.1). (A Lagrangian rather than Hamiltonian formulation is generally preferred in quantum field theory, but we shall not consider these fine points here.) The parameters E_k, ω_q and g_{jkq} appear here as free parameters, whereas when the Hamiltonian is obtained from quantum field theory they are already determined or satisfy certain restrictions. These are of three types: 1. Restrictions following from a symmetry or invariance principle, 2. Restrictions following from the properties of the classical field, if it exists as in electrodynamics. 3. Ad hoc restrictions, such as locality for cases where there is no classical analog.

For the case where the complications of spin can be neglected, and the labels j, k and q denote momenta, the requirement of momentum conservation imposes the condition

$$\boldsymbol{k} = \boldsymbol{j} + \boldsymbol{q}. \tag{12.30}$$

The interaction (12.1c) can then be written in the simpler form

$$V = \sum_{\boldsymbol{kq}} G_{\boldsymbol{kq}} c_{\boldsymbol{k}}^{\dagger} c_{(\boldsymbol{k}+\boldsymbol{q})} a_{\boldsymbol{q}}^{\dagger} + G_{\boldsymbol{kq}}^{*} c_{(\boldsymbol{k}+\boldsymbol{q})}^{\dagger} c_{\boldsymbol{k}} a_{\boldsymbol{q}}. \tag{12.31}$$

Further simplification follows from imposing Galilean or Lorentz invariance and locality. In the examples discussed in chs. 4 and 7, symmetry and locality determine the interaction completely except for an overall constant specifying the strength of the interaction. For cases where spin is relevant, angular momentum conservation and parity conservation (if it is valid) restrict the spin couplings.

Once a Hamiltonian of the type (12.1) is obtained either by quantizing a classical field or by simply writing it down as done above, there remains the problem of solving the appropriate Schrödinger equation. There are also more esoteric open problems such as whether a particular kind of Hamiltonian has any solutions at all. This very serious problem is central in contemporary research. The principal difficulty is the appearance of divergent integrals in calculations. So far, no satisfactory method has been found for eliminating these divergences. However, methods have been developed for obtaining results which can be compared with experiment in many cases, notably that of quantum electrodynamics. These methods have been very successful in obtaining agreement between theory and experiment. We shall therefore disregard the problem of divergences and optimistically assume that solutions really exist and simply have to be found.

The Schrödinger equation defined by the Hamiltonian (12.1) cannot generally be solved exactly. On the other hand, the eigenfunctions of the free-particle Hamiltonian (12.1b) can be written down by inspection. They are simply the plane-wave states for any fixed number of bosons and fermions. One is therefore led to consider the use of perturbation theory in which the interaction (12.1c) is treated as a perturbation. Perturbation calculations of *measurable physical processes* all seem to have the following properties: 1. The calculation of any process to the lowest, non-vanishing order in perturbation theory is reasonably straightforward and gives a sensible result as in eq. (12.29). 2. Calculations of higher-order corrections are a mess. There are several aspects to this mess. First of all, there are so many possible intermediate states, that it is practically a full Ph.D. thesis just to keep track of all the terms which arise if modern methods are not used. Second, many of the contributions calculated by conventional methods lead to divergent integrals. Third, many of the terms arising in a straightforward application of conventional perturbation theory are not really relevant to the physical results desired and should be thrown away. The diagrammatic method of Feynman provides a procedure for classifying all the terms that arise in perturbation theory and representing them in a reasonable area on a piece of paper. The diagrams indicate clearly the physical nature of each contribution. The Feynman rules are applied to calculate the contribution of each diagram and to indicate the degree of divergent integral to be expected from any particular diagram. Procedures are given for eliminating and reducing diagrams to discard the large number of contributions having no physical relevance to the desired result.

The troubles are already illustrated in the simple example of a one-fermion state. This corresponds to the solution of the free particle Schrödinger equation with only one fermion present. The unperturbed energy of this state is just E_k. However, there are corrections to the energy of this state due to the interaction. Let us attempt to see how such corrections would be calculated by perturbation theory. Since the interaction (12.1c) either adds one boson to, or removes one from the system, it has no diagonal matrix element in a state having a fixed number of bosons. There is therefore no contribution to the energy in first order. In second order, there is a contribution involving an intermediate state in which the fermion has emitted a boson. This second-order matrix element includes two matrix elements of

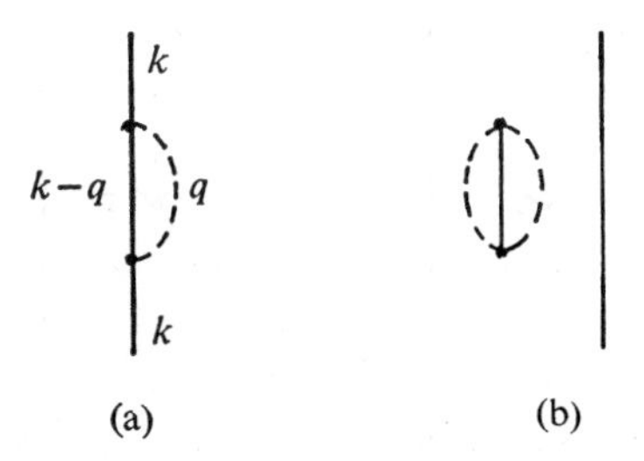

Fig. 12.2

the interaction and is represented by the diagram, fig. 12.2a, in which a boson is emitted and reabsorbed by the fermion. Another second-order process is represented by the diagram 12.2b. Here the electron initially present is not changed but a boson is created and absorbed with the creation and annihilation of a fermion pair. Further examination shows that these are the only two kinds of diagrams which can represent processes of second order.

Let us now examine higher orders. Only even-order processes can contribute since each boson that is emitted must also be absorbed in order to return the system to its initial state. We therefore next examine fourth-order processes involving the emission and absorption of two bosons. There are many different types of diagrams, as shown in fig. 12.3.

In calculating the contribution of each diagram one writes down the appropriate matrix element for the interaction (12.1c) corresponding to each vertex and propagators for each of the intermediate states. One then has to

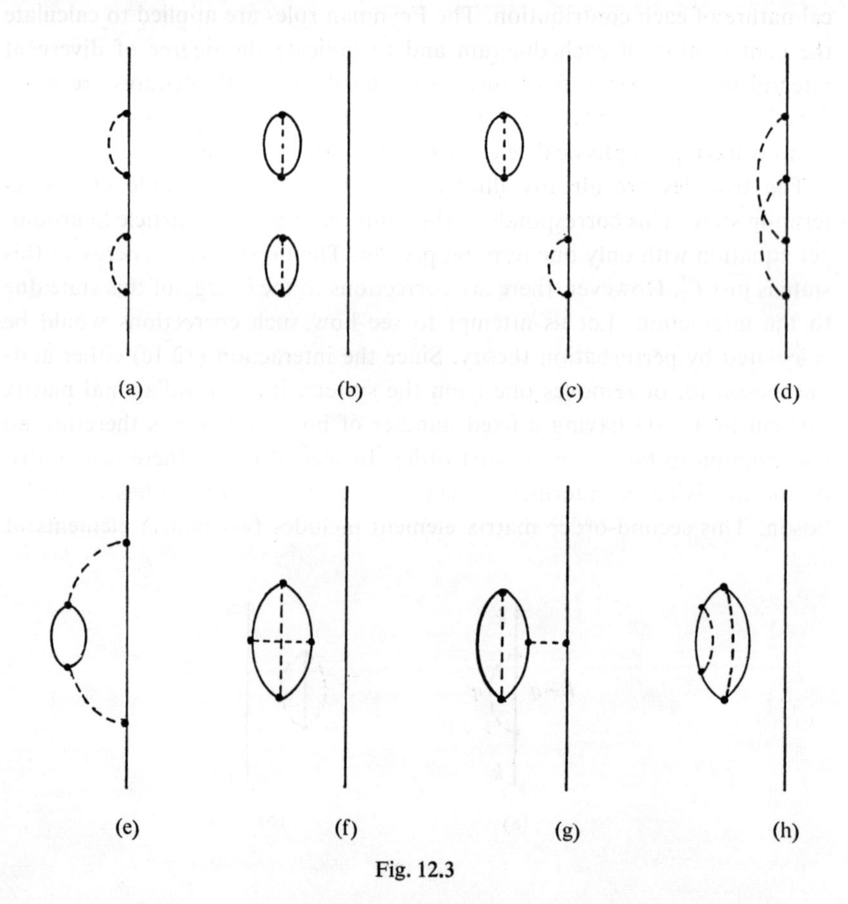

Fig. 12.3

sum over all possible intermediate states which can occur in a particular type of diagram. For example in fig. 12.2a, if the state under consideration has momentum $\hbar \boldsymbol{k}$, then the intermediate state consists of a boson having a momentum $\hbar \boldsymbol{q}$ and a fermion having a momentum $\hbar(\boldsymbol{k}-\boldsymbol{q})$. The contribution of this diagram is obtained by summing the contributions over all possible values of $\boldsymbol{q}$. This summation over $\boldsymbol{q}$ leads to the characteristic ultraviolet divergence of quantum field theory because of the large phase space available for states with large values of $\boldsymbol{q}$. By examining the properties of the interaction (12.1c) and the propagators, one can see which power of $\boldsymbol{q}$ appears in the matrix element described by the diagram, fig. 12.2a. If this power of $\boldsymbol{q}$ is too large, the integral diverges at high values of $\boldsymbol{q}$. This, in fact, happens in the calculation in quantum electrodynamics resulting from the emission and absorption of a photon.

The processes described by the diagrams of fig. 12.2 and fig. 12.3 lead to divergent results in most field theories and in particular in quantum electrodynamics. However, one can question the physical meaning of these results. The quantity calculated is the energy of the one-fermion state: i.e. its mass. This is certainly a physically meaningful quantity which can be measured experimentally. However, the contributions of the diagrams of figs. 12.2, 12.3 and all other diagrams do not give a physically measurable quantity. They give the correction to the fermion mass which is due to the presence of the interaction (12.1c). This is the difference between the physically measurable mass of the fermion and the 'bare' mass appearing in the unperturbed Hamiltonian (12.1b). This bare mass cannot be measured in any experiment. In the real world, the interaction (12.1c) always exists and it is impossible to find out what the mass of the fermion would be if it were turned off. One can talk about a bare or mechanical mass for an electron, and an electromagnetic mass representing the contribution of its interaction with the electromagnetic field. However, all that is measurable is the total mass of the electron and there is no experimental way of separating the bare and electromagnetic components. Thus one can argue that the divergences in the calculations of the diagrams of figs 12.2 and 12.3 do not correspond to any observable physical process and are irrelevant.

A similar analysis can be made for the corrections in perturbation theory to the one-boson state. These are indicated to second order in fig. 12.4. Again we can say that these calculations do not correspond to any physical processes. They give the difference in mass between the physical boson and the 'bare' boson and the latter cannot be measured.

We next consider the emission of a boson by a fermion. This is a transition from an initial state with one fermion present to a final state with a fermion

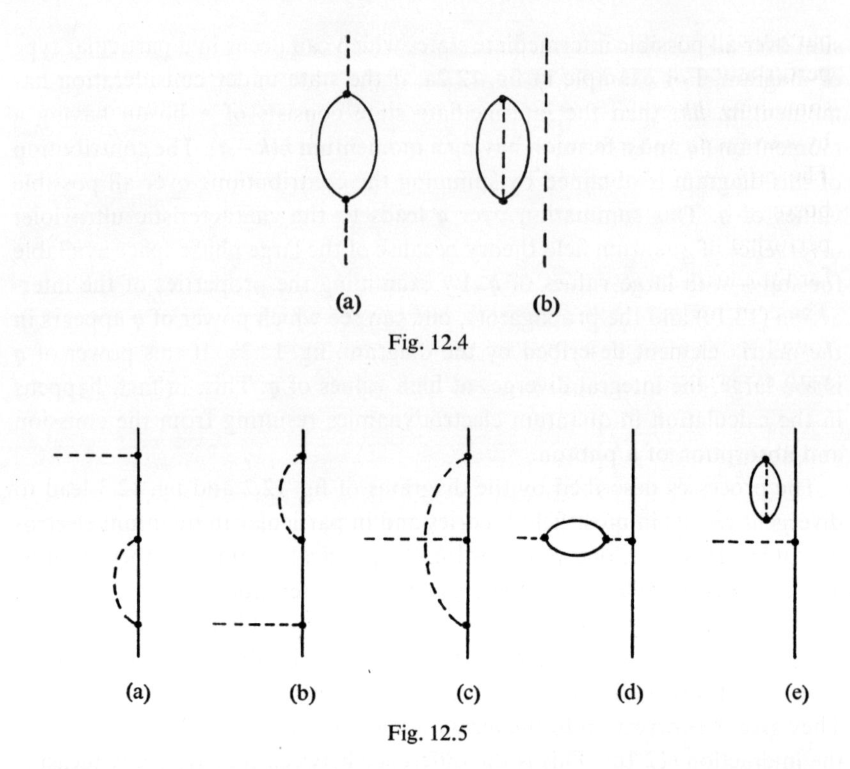

Fig. 12.4

Fig. 12.5

and a boson. In first-order perturbation theory the relevant diagram is just the simple vertex. The next higher corrections are third order, involving the emission and reabsorption of a second boson. The appropriate diagrams are shown in fig. 12.5. Here again there are divergences and one can argue that they are irrelevant. The transition probability for the emission of a boson by a fermion is proportional to the strength of the interaction or the 'coupling constant' between the bosons and fermions. Again one can define a 'bare' coupling constant which appears in the Hamiltonian (12.1c) and a 'physical' coupling constant which is actually measured in an experiment. Again there is no way of determining the bare coupling constant experimentally. The perturbation diagrams of the type shown in fig. 12.5 only describe contributions to the difference between the physical and bare coupling constants or charges. These differences are not measurable physical quantities.

So far, we have chosen the simplest possible processes, represented them by diagrams and found that they really, in the end, do not represent any physical quantities at all. They only represent the difference between physically measured quantities and unmeasurable 'bare' quantities which appear in a particular division of the real Hamiltonian into an unperturbed part and a perturbation. Let us now examine the simplest process which corresponds to something that can be measured in the laboratory, the scattering of a boson by a fermion (Compton scattering in the case of quantum electrodynamics). The process is second order, since it involves a change in the state of the boson; i.e. one emission and one absorption. The diagrams in second-order perturbation theory are shown in figs. 12.6. The calculation of these diagrams for the case of quantum electrodynamics leads to the Klein–Nishina formula. The contribution of the negative energy states (virtual pair production) was found to produce an appreciable effect and was required to give agreement with experiment.

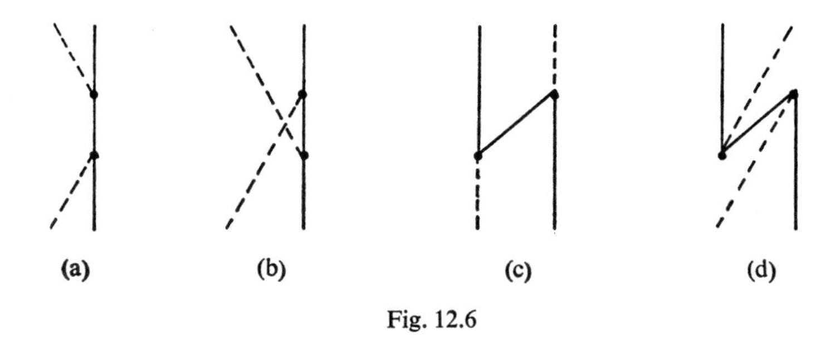

Fig. 12.6

Let us now consider the next higher-order corrections, namely fourth order, to the physical process of Compton scattering. Fig. 12.7 shows only a portion of the diagrams which enter. A conventional perturbation theory calculation in which each term is considered separately evidently leads to a real mess, to say nothing of divergences. We thus see an example of the general feature mentioned above. A process which can be measured physically in the laboratory can be treated by perturbation theory in the lowest order where a non-vanishing result is obtained and the result is sensible. However, higher-order contributions are a mess.

Let us now see how to clean up this mess. The simplest possible state of the system, the vacuum, corresponds to the unperturbed state of the free Hamiltonian (12.1b) where there are no fermions and no bosons present. If we now calculate the energy and other properties of this state by perturbation theory, we see that in second order there are contributions which are represented by the diagram shown in fig. 12.8. In higher order, there are more diagrams and all of these constitute the difference between the 'bare' vacuum and the real, physical vacuum. Again, we do not know what the bare vacuum is nor the differences between the bare and the physical vacua.

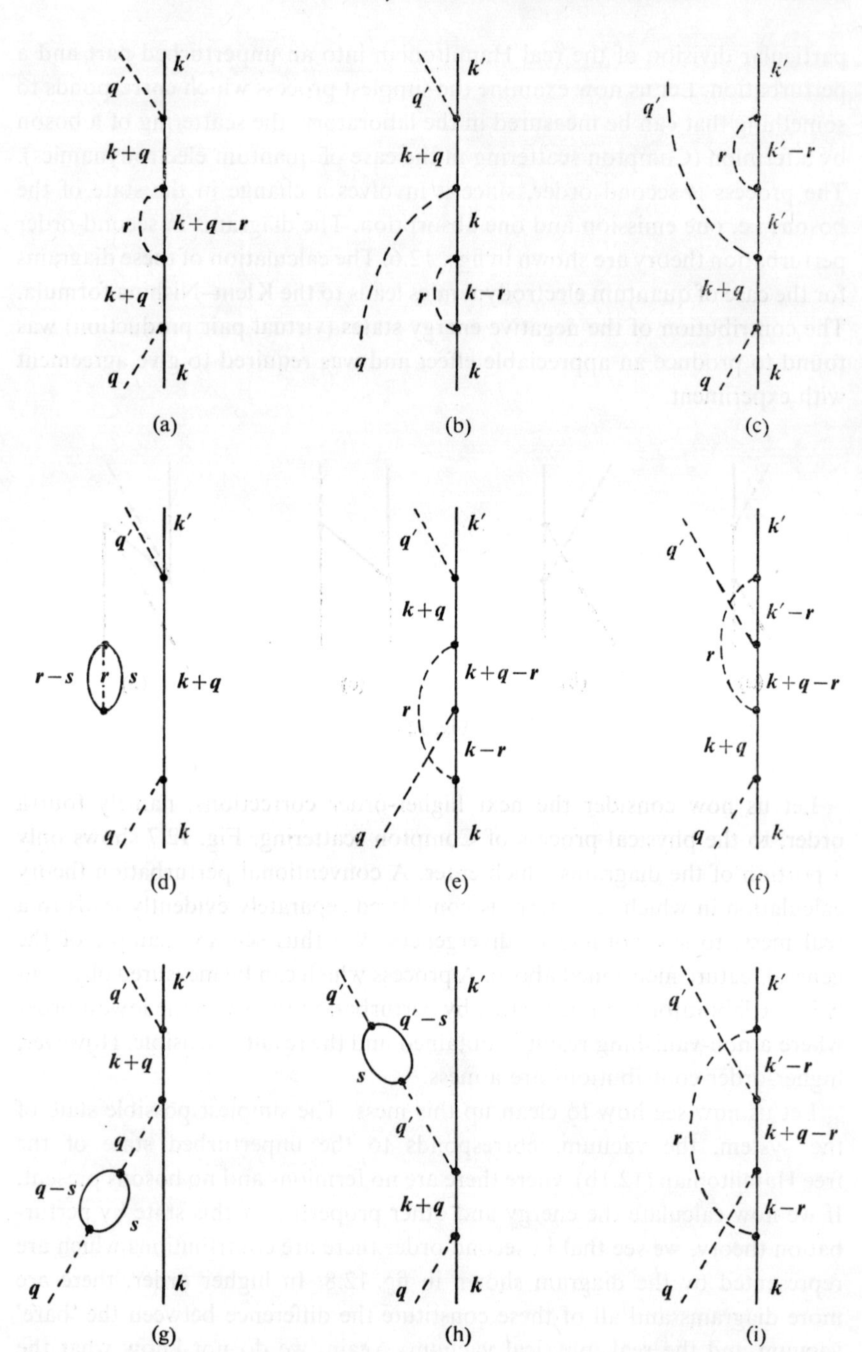
k'
q'
k+q
r
k+q−r
k+q
q
k
(a)
k'
q'
k+q
k
r
k−r
q
k
(b)
k'
q'
k'−r
r
k'
k+q
q
k
(c)
k'
q'
r−s
r
s
k+q
q
k
(d)
k'
q'
k+q
k+q−r
r
k−r
q
k
(e)
k'
q'
k'−r
r
k+q−r
k+q
q
k
(f)
k'
q'
k+q
q
q−s
s
q
k
(g)
k'
q'
q'−s
s
q
k+q
q
k
(h)
k'
q'
k'−r
r
k+q−r
k−r
q
k
(i)

Fig. 12.7

We now note that diagrams like fig. 12.8 appear in a number of the diagrams we have examined previously, such as fig. 12.2b. We can now interpret fig. 12.2b as being the unperturbed motion of a fermion with an unrelated 'vacuum fluctuation' in which a virtual fermion pair and a boson are created and then annihilated. This vacuum fluctuation diagram, fig. 12.8, contributes to the difference between the physical vacuum and the bare vacuum. It really does not belong in the properties of a physical fermion. We can therefore disregard the diagram, fig. 12.2b, as being due to the difference between the bare and physical vacua and not relevant to the process under consideration

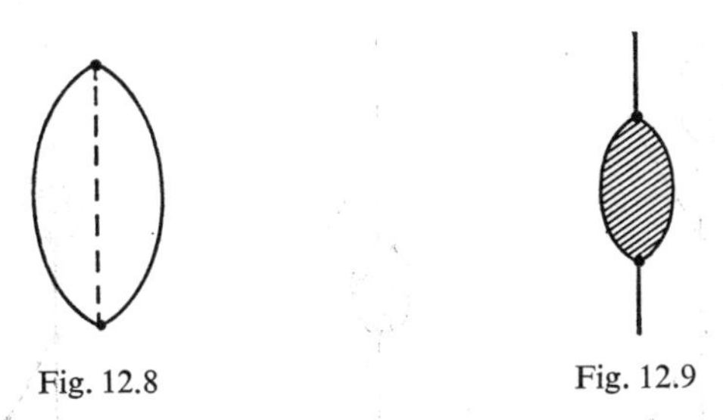

Fig. 12.8 Fig. 12.9

Similarly, we can disregard diagrams 12.3b,f,h, 12.4b, 12.5e and 12.7d. An analogous neglect of 'unlinked clusters' arises in many-body perturbation theory. Rigorous justification of this procedure is possible but belongs in field theory and not in fieldsmanship. It is analogous to the Goldstone theorem in the many-body case and is actually a precursor of the Goldstone theorem.

The next step in cleaning up the mess is to note that all diagrams not already discarded in fig. 12.2 and fig. 12.3 have the general form shown in fig. 12.9, where the shaded area represents all possible diagrams that can be inserted. Although we do not know how to calculate the contributions of all diagrams of this type we know what the answer to the sum of all such diagrams should be; namely the physical mass of the fermion. Thus, in calculating the results of physical processes, we can throw away the contributions of all diagrams of the type shown in figs. 12.2 and 12.3 including all those having the form shown in fig. 12.9. Their sum is just the difference between the physically observed mass of the fermion and the bare mass which enters into the unperturbed Hamiltonian (12.1b). We can take these diagrams into account by simply replacing all of them by a single line representing a single fermion but having the observed physical mass rather than the bare mass.

Now consider the contribution of the diagram shown in fig. 12.7b to Compton scattering. This diagram is just the fermion self-energy diagram fig. 12.2a and the second-order Compton scattering diagram, fig. 12.5a, placed end to end. We can now consider a whole series of such diagrams illustrated in fig. 12.10a in which the most general fermion self-energy diagrams are added to the simple Compton scattering diagram of fig. 12.5a. Similarly the diagrams of fig. 12.7c and fig. 12.7a are the simple Compton scattering diagram of fig. 12.6a with an additional electron self-energy diagram of the type shown in fig. 12.2a added on in the final and intermediate fermion states respectively. We can generalize these diagrams as shown in

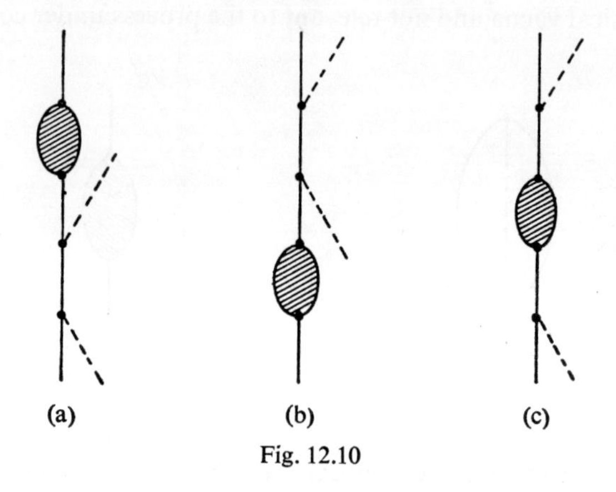

Fig. 12.10

fig. 12.10b and fig. 12.10c to include the most general type of fermion self-energy diagrams, fig. 12.9, as well as the simple second-order diagram of fig. 12.2a. These complicated diagrams shown in fig. 12.10 give the difference between the physical mass of the fermion and the bare mass in the initial, final and intermediate states. We therefore can discard all of these diagrams if we use the observed physical mass for these particles in all cases.

The above plausibility argument is not a rigorous proof that summation of all diagrams of the type shown in fig. 12.10 is equivalent to replacing the bare fermion mass by the observed physical mass. The formal proof that the summation of all the diagrams of the type shown in fig. 12.9 is equivalent to a 'renormalization' of the mass of the fermion can be carried through in detail and constitutes an important development in modern field theory.

In the renormalization approach, results are evaluated for physical quantities measurable in an experiment, without worrying about infinite values for unobservable 'bare' quantities which could only be determined if some interaction were turned off. In certain types of field theories, like quantum electrodynamics, the theory is formulated in terms of the physically observed masses and charges, and gives sensible results for all physical processes such as Compton scattering. However, the renormalization constants relating the bare and physical quantities all turn out to be infinite. The reasons for these divergences constitute one of the open, unsolved problems of quantum electrodynamics. It may be that these infinities are spurious, and are simply due to the use of an improper method of calculation; i.e. perturbation theory. On the other hand, there may be much deeper reasons for these divergences. Future developments will hopefully resolve this question.

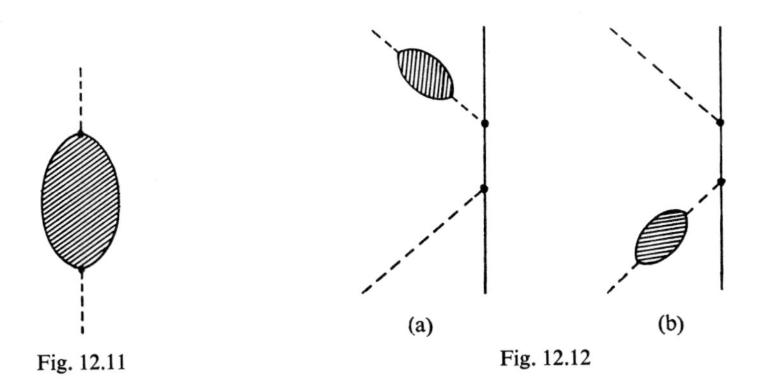

Fig. 12.11

Fig. 12.12

A similar renormalization process eliminates all diagrams having the form of fig. 12.4 or the more general form shown in fig. 12.11. These 'boson self-energy' diagrams can be replaced by using the physically measured boson mass. For the special case of quantum electrodynamics, where the boson rigorously has zero mass, gauge invariance allows many diagrams to be discarded. However, diagrams like fig. 12.11 can always be discarded simply on grounds of boson mass renormalization. Figs. 12.7g and 12.7h for Compton scattering contain such boson self-energy-diagrams. These, along with all of the more general diagrams, such as those shown in fig. 12.12 can now be discarded since they are all equivalent to the simple second-order diagram, fig. 12.6a, with self-energy corrections to the two photons.

We now examine the vertex diagrams, fig. 12.5. All of the diagrams, except fig. 12.5c, are just a simple vertex with added fermion self-energy, boson self-energy or vacuum fluctuation diagrams. However, the diagram of fig. 12.5c represents a second-order correction to the simple vertex. One can draw a general class of diagrams as shown in fig. 12.13 which are the most general vertex diagrams. Again, the sum of all these diagrams just gives the difference between the physically observed emission or absorption process of a boson by a fermion and that described in first order using the 'bare' interaction. We can therefore throw away all diagrams having complicated vertices of the type shown in fig. 12.13 in them and replace the sum of all such diagrams by a simple vertex, in which the physically observed interaction or coupling constant is used. We thus have a renormalization of the 'charge' of the fermion as well as of its mass. In the Compton scattering problem, we see that the diagrams of fig. 12.7e and fig. 12.7f can be combined with the more

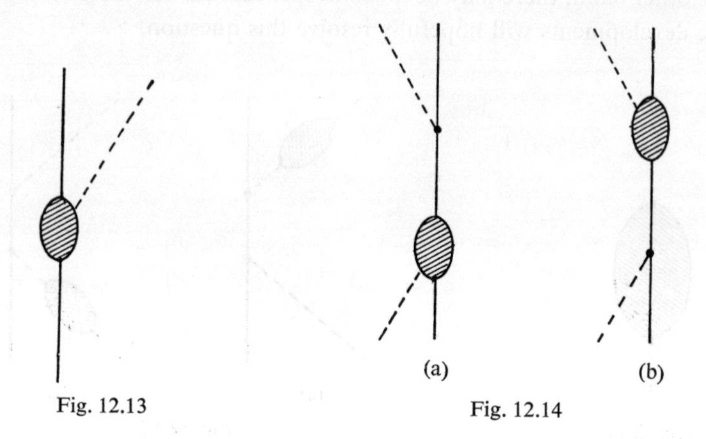

Fig. 12.13

Fig. 12.14

general class of diagrams shown in fig. 12.14 and the contributions of all of them replaced by a single second-order diagram of the type shown in fig. 12.6a with the physical coupling constant or charge used for both vertices.

Thus the renormalization method allows us to dispense with nearly all the fourth-order Compton scattering diagrams illustrated in fig. 12.7, and replace them by the inclusion of the physical masses for the fermion and the boson and the physical coupling constant in the second-order diagram of fig. 12.6a. The only diagram shown in fig. 12.7 which is not eliminated by renormalization, is the one of fig. 12.7i. This diagram turns out to give a finite and calculable result.

Inspection of the diagrams of fig. 12.7 shows that it is indeed reasonable that the one diagram, fig. 12.7i, which is not 'reducible' should have a better behavior with regard to divergences than the other reducible diagrams which contain self-energy or vertex parts. The diagrams of figs. 12.7a,e,i all consist of the simple second-order diagram, fig. 12.6a, with an additional boson emitted and reabsorbed. This boson does not appear in the initial or final states and its momentum is not determined a priori. Any value for the momentum of this boson is possible, provided that the momenta of the fermions present in the appropriate intermediate states are adjusted to give momentum conservation. It is therefore necessary to sum over all values for the momentum of this virtual boson.

The degree of divergence or convergence of the integral over the momentum of the virtual boson depends upon the highest power of the virtual boson momentum appearing in the integrand. Momentum-dependent factors appear in the phase space for the boson and may appear also in the two vertex functions if the coefficients G_{kq} in the interaction (12.31) depends upon q. These factors are the same for all three diagrams, figs. 12.7a,e,i. In addition the momentum of the virtual boson appears in the propagators for the boson and for all other particles whose momenta are related to the virtual-boson momentum by momentum conservation. The degree of divergence depends upon the relation between the powers of momentum appearing in the numerator due to phase space and the vertex function and those appearing in the denominator due to propagators. In fig. 12.7, all diagrams contain one boson propagator and three fermion propagators corresponding to the 'internal lines' in the diagrams. The momenta of these four lines are given on the diagram as functions of the initial and final momenta, $\boldsymbol{k}$, $\boldsymbol{q}$, $\boldsymbol{k}'$ and $\boldsymbol{q}'$ and additional boson and fermion momenta $\boldsymbol{r}$ and $\boldsymbol{s}$, whenever necessary, after all constraints from momentum conservation are imposed on the intermediate momenta. Each line labeled by a momentum which depends only on $\boldsymbol{k}$, $\boldsymbol{q}$, $\boldsymbol{k}'$ and $\boldsymbol{q}'$ has a propagator which is a constant and can be taken outside the integrals over intermediate momenta. All lines which have $\boldsymbol{r}$ or $\boldsymbol{s}$ appearing in their momenta have propagators which contain an integration variable in a denominator and therefore help convergence. Thus the diagram 12.7i gives the least divergent contribution of all the diagrams of fig. 12.7, as it contains only one integration variable $\boldsymbol{r}$ and all four propagators contain $\boldsymbol{r}$.

Some theories, such as quantum electrodynamics have the feature of being renormalizable. They give sensible physical results for all observable quantities although the renormalization factors are infinite. There are, however, other theories where even renormalization is not enough and diagrams exist after renormalization which still give divergent results. The treatment of such theories is still a very open question.

12.5 The polaron, a simple example from solid state of a quantum field theory

A polaron is an electron moving in a polar crystal. The dynamics of this system can be formulated by analogy with quantum field theory and offers an instructive example. We first consider the small oscillations of the ions in a harmonic crystal about their equilibrium positions, as discussed in ch. 3 and § 8.15. We denote by $\boldsymbol{q}(\boldsymbol{x})$ the coordinate describing the deviation from equilibrium of the ion whose equilibrium point in the lattice is specified by $\boldsymbol{x}$. The variable $\boldsymbol{x}$ then takes on a discrete set of values corresponding to the coordinates of the lattice points in the crystal. The variable $\boldsymbol{x}$ is therefore not a dynamical variable. It is merely a label, specifying the particular ion whose displacement is described by $\boldsymbol{q}(\boldsymbol{x})$.

We now define the transformation from the variables $\boldsymbol{q}(\boldsymbol{x})$ and their canonically conjugate momenta $\boldsymbol{p}(\boldsymbol{x})$ to the normal coordinates of the lattice vibrations, as in ch. 3 and § 8.15. As in § 8.15, we use the translational symmetry. We neglect surface and end effects which are usually handled formally by the requirement of periodic boundary conditions. Let us represent the operation of translation by a lattice vector $\boldsymbol{a}$ by $T_{\boldsymbol{a}}$. The ion coordinates and momenta transform as follows:

$$T_{\boldsymbol{a}}\boldsymbol{q}(\boldsymbol{x})T_{\boldsymbol{a}} = \boldsymbol{q}(\boldsymbol{x}+\boldsymbol{a}) \tag{12.32a}$$

$$T_{\boldsymbol{a}}\boldsymbol{p}(\boldsymbol{x})T_{\boldsymbol{a}} = \boldsymbol{p}(\boldsymbol{x}+\boldsymbol{a}). \tag{12.32b}$$

Note that the translation operation does not shift a dynamical variable by an amount $\boldsymbol{a}$ as in one-body or many-body problems. Since $\boldsymbol{x}$ here is only a label, the translation operation does not shift the value of the coordinate $\boldsymbol{q}$ of the ion, it *changes* the *label* of the ion. It moves the ion which was at the lattice point $\boldsymbol{x}$ to the lattice point $\boldsymbol{x}+\boldsymbol{a}$.

The state of the crystal is described by a complicated wave function depending on the variables $\boldsymbol{q}(\boldsymbol{x})$. The normal mode variables are eigenvectors of the translations (12.32) as in the one-dimensional case, § 8.15. In three dimensions the polarizations of the normal modes must also be specified.

$$\xi_{\boldsymbol{k}\alpha} = \frac{1}{N}\sum_{\boldsymbol{x}} \mathrm{e}^{-i\boldsymbol{k}\cdot\boldsymbol{x}} q_\alpha(\boldsymbol{x}) \tag{12.33a}$$

$$\pi_{\boldsymbol{k}\alpha} = \frac{1}{N}\sum_{\boldsymbol{x}} \mathrm{e}^{-i\boldsymbol{k}\cdot\boldsymbol{x}} p_\alpha(\boldsymbol{x}) \tag{12.33b}$$

$$T_{\boldsymbol{a}}\xi_{\boldsymbol{k}\alpha}T_{\boldsymbol{a}} = \frac{1}{N}\sum_{\boldsymbol{x}} \mathrm{e}^{-i\boldsymbol{k}\cdot\boldsymbol{x}} q_\alpha(\boldsymbol{x}+\boldsymbol{a}) = \mathrm{e}^{i\boldsymbol{k}\cdot\boldsymbol{a}}\xi_{\boldsymbol{k}\alpha} \tag{12.33c}$$

$$T_{\boldsymbol{a}}\pi_{\boldsymbol{k}\alpha}T_{\boldsymbol{a}} = \mathrm{e}^{i\boldsymbol{k}\cdot\boldsymbol{a}}\pi_{\boldsymbol{k}\alpha} \tag{12.33d}$$

where N is a normalization factor and $\alpha=1, 2, 3$ labels the three polarization directions of the normal modes. The different polarizations are not relevant to our further discussion. We therefore drop the index α.

The operators $\xi_{\boldsymbol{k}}$ and $\pi_{\boldsymbol{k}}$ defined by eq. (12.33) are not Hermitean. They satisfy the relations

$$\xi_{\boldsymbol{k}} = \xi^{\dagger}_{-\boldsymbol{k}} \tag{12.34a}$$

$$\pi_{\boldsymbol{k}} = \pi^{\dagger}_{-\boldsymbol{k}} \tag{12.34b}$$

$$[\xi_{\boldsymbol{k}}, \pi_{-\boldsymbol{k}'}] = \mathrm{i}\hbar\delta_{\boldsymbol{k}\boldsymbol{k}'}. \tag{12.34c}$$

The normal modes defined by these variables represent traveling waves. Hermitean operators describing normal modes which are *standing* waves are easily defined by choosing appropriate linear combinations* of $\xi_{\boldsymbol{k}}$ and $\xi_{-\boldsymbol{k}}$ and of $\pi_{\boldsymbol{k}}$ and $\pi_{-\boldsymbol{k}}$.

As in eq. (8.128), the normal mode transformation (12.33) separates the variables and reduces the ion Hamiltonian to that of a set of uncoupled harmonic oscillators.

$$H_{\mathrm{I}} = \sum_{\boldsymbol{k}} \frac{\pi_{\boldsymbol{k}}\pi_{-\boldsymbol{k}}}{2M} + \frac{M\omega_{\boldsymbol{k}}^2}{2}\,\xi_{\boldsymbol{k}}\xi_{-\boldsymbol{k}} \tag{12.35a}$$

$$H_{\mathrm{I}} = \tfrac{1}{2}\sum_{\boldsymbol{k}} \hbar\omega_{\boldsymbol{k}}(a_{\boldsymbol{k}}^{\dagger}a_{\boldsymbol{k}} + a_{\boldsymbol{k}}a_{\boldsymbol{k}}^{\dagger}) = \sum_{\boldsymbol{k}} \hbar\omega_{\boldsymbol{k}}(a_{\boldsymbol{k}}^{\dagger}a_{\boldsymbol{k}} + \tfrac{1}{2}) \tag{12.35b}$$

where $\omega_{\boldsymbol{k}}$ is the frequency of the normal mode $\boldsymbol{k}$, M is the mass of the ion, and $a_{\boldsymbol{k}}^{\dagger}$ and $a_{\boldsymbol{k}}$ are the oscillator creation and annihilation operators for the normal modes,

$$a_{\boldsymbol{k}}^{\dagger} = \frac{1}{(2M\hbar\omega_{\boldsymbol{k}})_0}\,(\pi_{\boldsymbol{k}} + \mathrm{i}M\omega_{\boldsymbol{k}}\xi_{\boldsymbol{k}}) \tag{12.36a}$$

$$a_{\boldsymbol{k}} = \frac{1}{(2M\hbar\omega_{\boldsymbol{k}})_0}\,(\pi_{-\boldsymbol{k}} - \mathrm{i}M\omega_{\boldsymbol{k}}\xi_{-\boldsymbol{k}}). \tag{12.36b}$$

The number of normal coordinates is equal to the total number of ions in the crystal, and is therefore finite. There are thus only a finite number of allowed values for the wave vector $\boldsymbol{k}$ of a normal mode. Since the wave amplitude is defined only at a discrete set of lattice points, a wave whose wavelength is shorter than the minimum distance between two ions in a lattice has no physical meaning. This fixes an upper limit for $\boldsymbol{k}$ or a lower limit for the wavelength.

We note the similarity between the Hamiltonian (12.34b) and the Hamil-

* The standing waves satisfy the usual equations of motion for a normal mode. In a standing wave all the coordinates $q(\boldsymbol{x})$ are proportional to a single amplitude which varies periodically with time. This is not true for a traveling wave, where the oscillations of different coordinates $q(\boldsymbol{x})$ are not in phase. Travelling waves are described formally with a single amplitude and phase as the real or imaginary part of a periodically varying complex amplitude. Thus traveling waves are described in classical mechanics by complex dynamical variables and in quantum mechanics by pairs of non-Hermitean operators satisfying commutation relations of the form (12.34).

tonian (12.16) for the free boson field. This correspondence makes many problems in solid state physics similar to those in field theory and vice versa.* Note, however, that there are no divergences in the solid state problem because of the upper limit on $\boldsymbol{k}$. In any perturbation theory involving the emission and absorption of the lattice bosons or 'phonons' as they are usually called, there is an upper limit on the wave vector of the phonon. No divergences are produced by the necessity to integrate over all values of the wave vector of a virtual phonon.

One infinity of field theory already appears in eq. (12.34b) namely the zero-point energy of the oscillations. The lowest state of the lattice has an energy of $\frac{1}{2}\hbar\omega_k$ for each normal mode. This zero point motion is observable as fluctuations in the position and the momentum of the ion and is measurable experimentally in many ways; e.g. X-ray diffraction or Doppler shift of nuclear gamma rays emitted by a vibrating nucleus. The total energy of this zero point motion is finite and causes no difficulty, since there are a finite number of values of $\boldsymbol{k}$. For a field having an infinite number of degrees of freedom, representing the system by a set of harmonic oscillators leads to an infinite zero-point energy for the lowest state of the system; i.e. the vacuum. This can be eliminated without difficulty by simply throwing it away as in eq. (12.16). This is sometimes called writing second-quantized operators in 'normal order' with annihilation operators on the right so that they give zero when acting on the vacuum.

Although the energy of the zero-point motion has been thrown away in eq. (12.16), other physical effects of this motion really exist and cannot be eliminated. They are associated with the uncertainty principle and the impossibility of preparing a state in which the values of two canonically conjugate variables are known both to be precisely zero. There are thus vacuum fluctuations in the magnitudes of physical fields such as the electromagnetic field. These fluctuations have physically observable consequences.

* In treating classical solids; e.g. in problems of elasticity, thermal conductivity, electrical conductivity, etc, the discrete variable $\boldsymbol{x}$ is often treated as a continuous variable and the displacement $q(\boldsymbol{x})$ as a function of $\boldsymbol{x}$. Thus the equations of motion become partial differential equations, rather than sets of ordinary differential equations and are much easier to handle. The continuous variable $\boldsymbol{x}$ is not a dynamical variable, but a coordinate, labeling the space point specifying the particular dynamical variable $q(\boldsymbol{x})$. In classical field theory, an analogous situation arises, in which the continuum formulation is exact and the dynamical variables are the field variables, e.g. the electric and magnetic fields $\boldsymbol{E}(\boldsymbol{x})$ and $\boldsymbol{B}(\boldsymbol{x})$. Here also the continuous variable $\boldsymbol{x}$ is just a label and not a dynamical variable. In quantum field theory, where dynamical variables are represented by operators which satisfy commutation relations, this can be confusing to one who is accustomed to seeing $\boldsymbol{x}$ as an operator. In field theory $\boldsymbol{x}$ is a label, not an operator.

Let us now introduce an electron into our crystal, which we assume to be a polar crystal like NaCl with alternating positive and negative ions. An electron held at a particular point polarizes the crystal, attracting the positive ions and repelling the negative ions. This polarization produces an electrostatic potential in the neighborhood of the electron which is positive since there is a net motion of positive charge *toward* the electron. A fixed electron in the polar crystal therefore produces an attractive potential well around the electron. If the electron is free to move in the crystal, the motion depends on the coupling between the electron and the lattice. The electron tends to carry the potential well along with it. If the inertia of the ions is large, the well cannot follow the rapid motion of the electron, which can oscillate back and forth rapidly in the potential well. This is called the strong coupling limit. The potential is determined by a time average of the electron's position. It would be that produced by a smeared-out charge distribution obtained from integrating the electron charge density over a time characterized by the ability of the lattice to follow the motion. If the coupling is weak, the well is not deep enough to bind the electron. The electron then moves more or less freely through the lattice, and its interaction can excite waves or phonons. The basic process is scattering of the electron with emission or absorption of phonons. This can be described formally as in § 8.15 by adding terms to the lattice Hamiltonian describing the motion of the electron and its interaction with the lattice. Thus if $\boldsymbol{r}$ is the coordinate of the electron, and $\boldsymbol{p}$ its momentum, we have

$$H_{\mathrm{e}} = \frac{p^2}{2m} \tag{12.37a}$$

$$H_{\mathrm{int}} = \sum_{\boldsymbol{x}} V[\boldsymbol{r} - \boldsymbol{x} - \boldsymbol{q}(\boldsymbol{x})] \tag{12.37b}$$

$$= \sum_{\boldsymbol{x}} \sum_{\boldsymbol{k}} V_{\boldsymbol{k}} \mathrm{e}^{-\mathrm{i}\boldsymbol{k}\cdot[\boldsymbol{r}-\boldsymbol{x}-\boldsymbol{q}(\boldsymbol{x})]} \tag{12.37c}$$

where m is the mass of the electron. The interaction V depends only on the distance between the electron and the ion, whose coordinate is $\boldsymbol{x}+\boldsymbol{q}(\boldsymbol{x})$. This interaction is then summed over all the ions and conveniently written, as in § 8.15, in terms of its Fourier transform (12.37c). This leads naturally to the expression of the interaction in terms of normal coordinates or phonon variables.

In general the displacement $\boldsymbol{q}(\boldsymbol{x})$ is small compared to the phonon wavelength. We can thus expand

$$\mathrm{e}^{\mathrm{i}\boldsymbol{k}\cdot\boldsymbol{q}(\boldsymbol{x})} \approx 1 + \mathrm{i}\boldsymbol{k}\cdot\boldsymbol{q}(\boldsymbol{x}). \tag{12.38}$$

Substituting the approximation (12.38) into the interaction (12.37c) and noting that $\sum_x e^{i\boldsymbol{k}\cdot\boldsymbol{x}}=0$, we obtain

$$H_{\text{int}} \approx \sum_{\boldsymbol{x}}\sum_{\boldsymbol{k}} V_{\boldsymbol{k}} e^{-i\boldsymbol{k}\cdot\boldsymbol{r}} e^{i\boldsymbol{k}\cdot\boldsymbol{x}} [i\boldsymbol{k}\cdot\boldsymbol{q}(\boldsymbol{x})]. \tag{12.39}$$

The interaction (12.39) is easily expressed in terms of the normal mode variables (12.33)

$$H_{\text{int}} \approx \sum_{\boldsymbol{k}} iNV_{\boldsymbol{k}} e^{-i\boldsymbol{k}\cdot\boldsymbol{r}} \xi_{\boldsymbol{k}}. \tag{12.40}$$

Let us introduce a second-quantized description for the electron and the boson operators (12.36) for the ions. Then

$$H_{\text{int}} \approx \sum_{\boldsymbol{k}\boldsymbol{k}'} NV_{\boldsymbol{k}} \sqrt{\frac{\hbar}{m\omega_{\boldsymbol{k}}}}\, c^{\dagger}_{\boldsymbol{k}'} c_{\boldsymbol{k}'+\boldsymbol{k}} (a^{\dagger}_{\boldsymbol{k}} - a_{-\boldsymbol{k}}). \tag{12.41}$$

The interaction (12.41) has the same form as the interaction (12.1c) describing the emission and absorption of a boson by a fermion. The polaron problem is thus formally identical to that of the interacting system of fermions and bosons described by the Hamiltonian (12.31). There is here only one fermion, the electron, and the bosons are phonons. The same perturbation methods using Feynman diagrams can be applied to the polaron problem, but there are important differences. There are no divergences, because of the natural cut-off at short phonon wavelengths. Renormalization effects are thus finite. They are also measurable, since it is possible to take an electron out of the crystal and to measure its 'bare' mass and charge in the absence of the coupling to the 'phonon field'. The 'renormalized mass' of the polaron is usually called the 'effective mass'.

12.6 The strong-coupling and classical limits and the 'unquantization' of a field

In the standard treatments, a classical field theory is quantized to produce a theory of quanta or particles. In this section we shall do it backwards and 'unquantize' the particles to produce a field. For simplicity we consider the one-dimensional case.

Let us examine again the Hamiltonian (12.31) and look for possible approximation methods for treating it. The perturbation theory described above using Feynman diagrams can be expected to be useful if the interaction (12.1c) in the Hamiltonian (12.31) is sufficiently weak to be treated as a perturbation. If the coupling is not weak, then perturbation theory is not satisfactory for the treatment of the polaron problem.

Another approximate method for treating the Hamiltonian (12.31) is analogous to the Hartree–Fock method for the many-body problem. This corresponds in the polaron problem to the case where the electron produces a potential well by polarizing the lattice. Since the motion of the electron is determined by the potential which is in turn produced by the electron, the problem is one of a self-consistent field. Such a solution can be expected to be valid in the *strong*-coupling limit, where the interaction can produce a potential strong enough to bind the electron.

Let us assume 'independent-particle motion' of the bosons and fermions and look for the best trial wave function of this type from the variational principle. There is an additional complication because the number of bosons is not conserved by the Hamiltonian (12.31), and we wish to consider the emission and absorption of bosons within this approximation. We therefore choose a trial function of the form

$$|\psi\rangle = f(c^\dagger)g(a^\dagger)|0\rangle, \tag{12.42}$$

where f and g are arbitrary functions of the fermion and boson operators respectively, and need not create a fixed number of particles. It is convenient to define the states

$$|f\rangle = f(c^\dagger)|0\rangle \tag{12.43a}$$

$$|g\rangle = g(a^\dagger)|0\rangle \tag{12.43b}$$

and to normalize the functions f and g so that the states (12.43) are both normalized. The states (12.43) contain only fermions and only bosons respectively. The expectation value of the Hamiltonian (12.31) with the trial wave function (12.42) can be expressed in terms of the states (12.43) since boson and fermion operators commute

$$\begin{aligned}\langle\psi|H|\psi\rangle &= \langle f|\sum E_k c_k^\dagger c_k|f\rangle + \langle g|\sum_q \omega_q a_q^\dagger a_q|g\rangle \\ &\quad + \sum_{kq} G_{kq}\langle f|c_k^\dagger c_{(k+q)}|f\rangle\langle g|a_q^\dagger|g\rangle + \sum_{kq} G_{kq}^*\langle f|c_{(k+q)}^\dagger c_k|f\rangle\langle g|a_q|g\rangle.\end{aligned} \tag{12.44}$$

The variational principle applied to (12.44) varying f and g independently leads to two coupled equations which look like Schrödinger equations for the states (12.43)

$$H_F|f\rangle = E_F|f\rangle \tag{12.45a}$$

$$H_B|g\rangle = E_B|g\rangle \tag{12.45b}$$

where

$$H_F = \sum_k E_k c_k^\dagger c_k + \sum_{kq} \{G_{kq} c_k^\dagger c_{(k+q)} \langle g|a_q^\dagger|g\rangle + G_{kq}^* c_{(k+q)}^\dagger c_k \langle g|a_q|g\rangle\} \tag{12.46a}$$

$$H_B = \sum_q \omega_q a_q^\dagger a_q + \sum_{kq} \{G_{kq} \langle f|c_k^\dagger c_{(k+q)}|f\rangle a_q^\dagger + G_{kq}^* \langle f|c_{(k+q)}^\dagger c_k|f\rangle a_q\}. \tag{12.46b}$$

The fermion Hamiltonian H_F describes independent fermion motion in a one-particle potential which depends upon the expectation values of the boson variables. The boson Hamiltonian H_B describes an assembly of uncoupled independent harmonic oscillators, with a 'forcing term' for each oscillator which depends upon expectation values of the fermion operators. The situation is thus directly analogous to the many-body Hartree–Fock equations. The fermions move independently in the 'average field' of the bosons; the boson oscillators move under the influence of the average force of the fermions. These average fields must then be determined in a self-consistent way.

The boson equation (12.45b) can be easily solved exactly. It is just a set of harmonic oscillators, each having a linear forcing term. The Hamiltonian H_B can be transformed to a simple set of oscillators by changing the origin of each oscillator. Let us define new boson operators

$$\alpha_q^\dagger = a_q^\dagger + \frac{1}{\omega_q} \sum_k G_{kq}^* \langle f|c_{(k+q)}^\dagger c_k|f\rangle \tag{12.47a}$$

$$\alpha_q = a_q + \frac{1}{\omega_q} \sum_k G_{kq} \langle f|c_k^\dagger c_{(k+q)}|f\rangle. \tag{12.47b}$$

The operators $\alpha_q^\dagger$ and α_q satisfy the usual boson commutation rules

$$[\alpha_q, \alpha_r] = [\alpha_q^\dagger, \alpha_r^\dagger] = 0 \tag{12.48a}$$

$$[\alpha_q, \alpha_r^\dagger] = \delta_{qr}. \tag{12.48b}$$

The boson Hamiltonian H_B can now be written

$$H_B = \sum_q \omega_q \alpha_q^\dagger \alpha_q - \sum_q \left| \frac{1}{\omega_q} \sum_k G_{kq} \langle f|c_k^\dagger c_{(k+q)}|f\rangle \right|^2. \tag{12.49}$$

This is just a set of uncoupled oscillators plus a constant term.

The lowest eigenfunction of the Hamiltonian (12.49) is the vacuum for the α-operators, which satisfies the relation

$$\alpha_q|g\rangle = \left(a_q + \frac{1}{\omega_q} \sum_k G_{kq} \langle f|c_k^\dagger c_{(k+q)}|f\rangle \right)|g\rangle = 0. \tag{12.50}$$

The mean number of bosons in this state (12.50) can be calculated from eq. (12.50) and its Hermitean conjugate expression

$$\langle g|n_q|g\rangle = \langle g|a_q^\dagger a_q|g\rangle = \left|\sum_k G_{kq}\langle f|c_k^\dagger c_{(k+q)}|f\rangle\right|^2/\omega_q^2 \tag{12.51a}$$

$$= |S(q)/\omega_q|^2 \tag{12.51b}$$

where

$$S(q) = \sum_k G_{kq}\langle f|c_k^\dagger c_{(k+q)}|f\rangle. \tag{12.51c}$$

The mean square number of bosons is given by

$$\langle g|n_q^2|g\rangle = \langle g|a_q^\dagger a_q a_q^\dagger a_q|g\rangle = \langle g|a_q a_q^\dagger|g\rangle|S(q)/\omega_q|^2 \tag{12.52a}$$

$$= \langle g|n_q + 1|g\rangle\langle g|n_q|g\rangle \tag{15.22b}$$

or

$$\langle g|n_q^2|g\rangle - |\langle g|n_q|g\rangle|^2 = \langle g|n_q|g\rangle. \tag{12.52c}$$

This is the usual statistical result for fluctuations in the number of bosons, varying as the square root of n. Thus if the number of bosons is large, the fluctuations are relatively small.

Now consider the case where the mean number of bosons (12.50) is very large. This corresponds to 'large quantum numbers' for the individual boson operators and thus to the classical limit of the corresponding oscillators. The fluctuation in the number of bosons is small and we can consider n_q as a classical variable whose value is given by the expectation value (12.51). We can also consider the variables a_q and $a_q^\dagger$ as classical variables in this limit. The fermion Hamiltonian H_F thus describes the motion of independent fermions in a classical field. Let us 'unsecond-quantize' H_F and write it as a Hamiltonian in the Schrödinger representation for the fermions. The first term is the total kinetic energy of all the fermions. The second term involves the momentum shift operators $c_{(k+q)}^\dagger c_k$ which become $e^{i\boldsymbol{q}\cdot\boldsymbol{r}}$ in the Schrödinger representation. We thus have

$$H_F = \sum_i \{T_i + \sum_{\boldsymbol{q}} G(\boldsymbol{p}_i, \boldsymbol{q})e^{-i\boldsymbol{q}\cdot\boldsymbol{r}_i}a_{\boldsymbol{q}}^\dagger + \sum_{\boldsymbol{q}} e^{i\boldsymbol{q}\cdot\boldsymbol{r}_i}G^*(\boldsymbol{p}_i, \boldsymbol{q})a_q\}, \tag{12.53}$$

where the dependence of the coefficient $G_{\boldsymbol{kq}}$ on the wave number $\boldsymbol{k}$ now appears as a dependence upon the momentum operator $\boldsymbol{p}_i$. If we wish to assume a local interaction; i.e. independent of velocity, then $G(\boldsymbol{p}, \boldsymbol{q})$ is independent of $\boldsymbol{p}$. Since locality is commonly assumed in field theory, we make this assumption here.

We now note that the variables $a_{\boldsymbol{q}}^\dagger$ and $a_{\boldsymbol{q}}$ define the potential in which the fermions move. If we write this potential as some function $V(\boldsymbol{r}_i)$ in configuration space we see that the classical variables $a_q^\dagger$ and a_q are proportional to Fourier components of this potential. The two equations (12.45) thus represent the motion of a system of particles in a classical external field, and a classical equation determining the field.

This classical limit is simply interpreted in the polaron problem. It corresponds to the case where the polarization of the lattice produced by the electron is so large that the quantum fluctuations can be neglected. This means that the displacements $q(x)$ of ions near the electron are so large that the fluctuations in $q(x)$ required by the uncertainty principle produce a negligible effect on the potential as seen by the electron. This is a classical limit only on the field or phonon variables, which can be considered as a classical field. The fermion variables (the electron in the polaron) are not necessarily classical in this limit. In general, the fermions behave like quantum-mechanical particles obeying the Schrödinger equation in a classical external potential.

The limit in which the 'quantum fluctuations of the fermion field' are neglected is sometimes called the case of a fixed source. If the fermions have an 'infinite mass', their kinetic energy can be neglected. In this case the coordinates $\boldsymbol{r}_i$ in eq. (12.53) become classical variables, since no variable appears in the Hamiltonian (12.53) which does not commute with $\boldsymbol{r}_i$. This holds for the exact Hamiltonian (12.31) as well as for the 'Hartree–Fock' Hamiltonians (12.46). We therefore are not restricted to the classical limit for the boson variables. In this limit, which can be specified in the Hamiltonian (12.31) by setting $E_k = 0$, the factorization into fermion and boson factors (12.42) is exact. Since the Hamiltonian now commutes with the operators $\boldsymbol{r}_i$, the eigenfunctions can be simultaneous eigenfunctions of H and of $\boldsymbol{r}_i$; i.e. their dependence on $\boldsymbol{r}_i$ is given by $\delta(\boldsymbol{r}_i - \boldsymbol{R}_i)$, where $\boldsymbol{R}_i$ is any number. This corresponds to the fermions being *fixed* at the points $\boldsymbol{R}_i$. This fixed position is possible, because the corresponding infinite uncertainty in the fermion momentum has no dynamical effect in the limit of infinite mass. The boson equation can be solved exactly by the transformation (12.47), leading to an exact solution in the static limit. The functions $S(\boldsymbol{q})$ represents a 'source' for the boson field which is fixed in space in this approximation. If the field is assumed to be local, then singularities arise as a result of the delta functions of the point sources. These are avoided by the use of an arbitrary 'form factor' giving a finite extension to the source.

In the limit where both the bosons and the fermions are treated as classical fields, the problem reduces to that of the static boson field produced by an assembly of fixed sources. One would expect an equation analogous to Poisson's equation to arise for this case. To see this let us define field variables in configuration space analogous to the displacement variables $q(\boldsymbol{x})$ for the crystal. These are related to the boson creation and annihilation operators by the same expressions, eqs. (12.36) and (12.33) where $\boldsymbol{x}$ is considered to be a continuous variable and all sums are replaced by integrals.

The free boson Hamiltonian then becomes

$$H_{\text{free boson}} = \sum_s \omega_s a_s^\dagger a_s = \tfrac{1}{2} \sum_s \pi_s \pi_{-s} + \omega_s^2 \xi_s \xi_{-s} \tag{12.54a}$$

$$= \tfrac{1}{2}\left[\int p^2(x)\,\mathrm{d}x + \frac{1}{N^2}\int \mathrm{d}x\,\mathrm{d}x' \sum_s q(x)q(x')\mathrm{e}^{\mathrm{i}s\cdot x}\mathrm{e}^{-\mathrm{i}s\cdot x'}\omega_s^2\right]. \tag{12.54b}$$

If ω_s is the energy of a boson of mass m and momentum $\hbar s$

$$\omega_s^2 = (\hbar c s)^2 + m^2 c^4. \tag{12.55a}$$

Then

$$\sum_s \omega_s^2 \mathrm{e}^{\mathrm{i}s\cdot x} = \sum_s \{-(\hbar c)^2 \Delta + m^2 c^4\} \mathrm{e}^{\mathrm{i}s\cdot x} \tag{12.55b}$$

and by partial integration

$$\int \mathrm{d}\boldsymbol{x}\, q(\boldsymbol{x})\omega_s^2 \mathrm{e}^{\mathrm{i}s\cdot x} = \int \mathrm{d}\boldsymbol{x}\, \mathrm{e}^{\mathrm{i}s\cdot x}\{-(\hbar c)^2 \Delta + m^2 c^4\} q(\boldsymbol{x}). \tag{12.55c}$$

Substituting eq. (12.55c) into eq. (12.54b) and performing the sum on $\boldsymbol{s}$ and the trivial integration over $\boldsymbol{x}'$, we obtain

$$H_{\text{free boson}} = \tfrac{1}{2}[\int p^2(\boldsymbol{x})\,\mathrm{d}\boldsymbol{x} + \int q(\boldsymbol{x})\{-(\hbar c)^2 \Delta + m^2 c^4\} q(\boldsymbol{x})\,\mathrm{d}\boldsymbol{x}]. \tag{12.56}$$

In the static limit, the variables $q(\boldsymbol{x})$ are treated as classical variables. This means that the contributions of the non-commuting variables $p(\boldsymbol{x})$ to the Hamiltonian (12.56) are neglected. The interaction (12.1c) can also be expressed in terms of the variables $q(\boldsymbol{x})$ and is obviously linear in $q(\boldsymbol{x})$

$$H_{\text{int}} = \int s(\boldsymbol{x}) q(\boldsymbol{x})\,\mathrm{d}\boldsymbol{x} \tag{12.57}$$

where $s(\boldsymbol{x})$ is the source function expressed in configuration space, related to $s(\boldsymbol{q})$ by a Fourier transformation.

Thus in the static limit we have

$$H_{\text{static}} = \tfrac{1}{2}\int q(\boldsymbol{x})[-(\hbar c)^2 \Delta + m^2 c^4] q(\boldsymbol{x}) + \int s(\boldsymbol{x}) q(\boldsymbol{x})\,\mathrm{d}\boldsymbol{x}. \tag{12.58}$$

The solution of the 'Schrödinger equation 'for the lowest state of the system is obtained by choosing the function $q(\boldsymbol{x})$ which minimizes the expression (12.58). Requiring that the value of (12.58) be stationary with respect to an arbitrary variation

$$q(\boldsymbol{x}) \to q(\boldsymbol{x}) + \delta(\boldsymbol{x}) \tag{12.59}$$

leads directly to the equation

$$\{-(\hbar c)^2 \Delta + m^2 c^4\} q(x) + s(x) = 0. \tag{12.60}$$

This has the form of a Poisson equation with the additional mass term $m^2 c^4 q(x)$.

The solution of this equation for a point source at the origin, $s(x) = \delta(x)$ in three dimensions is just

$$q(x) \propto \frac{e^{-\mu r}}{r} \tag{12.16a}$$

where

$$\mu = mc/\hbar. \tag{12.61b}$$

Thus we see that a point source of a boson field leads to a potential of the Yukawa type, which reduces to the case of a Coulomb potential for a zero-mass boson. The parameter μ which characterizes the range of the field around the source is related to the mass of the boson. The range is the order of the boson Compton wavelength.

SYMMETRIES, INVARIANCE AND RELATIVISTIC QUANTUM MECHANICS FOR PEDESTRIANS

Chapter 13 introduces the Dirac equation from the extreme relativistic limit with a two-component wave equation for a spin-$\frac{1}{2}$ particle. Many features of the Dirac equation are present in this simple limit, such as negative-energy states and their interpretation in terms of holes and antiparticles, the role of helicity and the decoupling of states having opposite helicities in this limit, motion in external fields and the existence of the Dirac magnetic moment. The Dirac equation is then presented as a generalization of this extreme-relativistic wave equation which reduces to the non-relativistic Schrödinger equation in the appropriate limits. The Dirac equation for a free particle is solved and the roles of orbital angular momentum, spin and parity are discussed. The scattering by a spin-independent potential in Born approximation shows the tendency of the helicity to remain conserved at relativistic velocities.

Chapter 14 first discusses two types of transformations. 1) Those which describe the *same* physical state in different ways are useful for finding convenient coordinates in a particular problem; e.g. the transformation from cartesian to polar coordinates. 2) Those which describe *different* physical states in the *same* coordinate system create families of states having related properties (multiplets). Symmetry transformations are next considered. These commute with the Hamiltonian and define conservation laws. The eigenvalues of conserved quantities define new quantum numbers. Linear operators are conveniently classified by their behavior under the transformation (irreducible tensors) and selection rules and relations between operator matrix elements are obtained (Wigner–Eckart theorem). Examples of symmetry transformations begin with parity and translations. The combination of parity and translations as two non-commuting symmetry transformations provides an elementary example of the irreducible representations of a Lie group and of the degeneracy and multiplet structure of the eigenfunctions of a Hamiltonian invariant under a set of non-commuting transformations. Time reversal and antilinear operators are briefly presented.

Gauge transformations change redundant variables not directly measurable like the potentials in electrodynamics, without changing the physically measurable variables like the electric and magnetic fields. They provide *formally different* descriptions of the *same* physical system. The simple example of a particle in an external vector potential is discussed and the canonical momentum and de Broglie wavelength of a particle are shown not to be gauge invariant. They are not uniquely defined in the presence of magnetic fields.

Symmetry transformations on spin and internal degrees of freedom and on Dirac spinors show the roles of intrinsic parity and intrinsic angular momentum, the ambiguity in phase resulting from rotations by 2π of a spin-$\frac{1}{2}$ particle, more peculiarities of antilinear time reversal and Kramers degeneracy. The detailed charge conjugation properties of a Dirac electron are introduced by a paradox usually ignored in conventional treatments. The formal charge conjugation operator acts only in the internal spinor space and not in configuration space. It therefore transforms the electron wave function for the ground state of a hydrogen atom into another wave function with the same spatial properties, i.e. a bound state wave function. Yet the transformed wave function must be a solution of the Dirac equation for a positron moving in the repulsive field of a proton where no bound state exists. The paradox is resolved by showing that this solution represents a negative energy positron; it is simply another representation of the original hydrogen atom in a formulation where positrons are particles and electrons are holes. This transformation shows the equivalence of particles and holes and is not directly related to charge conjugation invariance. Charge conjugation invariance of the Coulomb interaction in the hydrogen atom is tested by changing the proton into an antiproton at the same time that the electron is changed into a positron.

Chapter 15 presents the Lorentz group as a particular case of symmetry transformations which depend explicitly on the time. Galilean and Lorentz transformations depend linearly on time. The symmetry requires the coefficient of the time, momentum, to be a constant of the motion and determines the relations between energy, momentum and helicity. Section 15.2 considers Lorentz transformations in a 2+1-dimensional space. A time-independent rotation operator and two time-dependent Lorentz transformation operators are defined which satisfy angular momentum commutation rules. The momenta in the two directions are required to be constants of the motion.

The quantum numbers used to label states are not analogous to J, M for ordinary angular momentum because the 'total angular momentum' operator formed by taking the sum of the squares of the three operators does not commute with the momentum or with the Hamiltonian. The 'inhomogeneous' algebra larger than ordinary angular momentum must be considered which includes the energy–momentum vector as well as the three 'angular momentum' operators, i.e. translations as well as rotations. The multiplets (irreducible representations) containing states which go into one another under these transformations are labeled by new quantum numbers obtained from two new independent operators which commute with all operators in the larger algebra. Each multiplet consists of all the possible states of a particle having a given mass and a given spin, and the new quantum numbers are just the mass and spin.

The behavior of these states under Lorentz transformations is examined. The product of two Lorentz transformations in different directions is shown to be equal to a Lorentz transformation in a third direction with an additional spin rotation (Wigner rotation).

Section 15.4 considers the full $3+1$-dimensional Lorentz group. The irreducible representations (multiplets) of the ordinary rotation group in four dimensions are first considered. The Lorentz group algebra is constructed by analogy with the three-dimensional case by giving the three operators which rotate into the fourth (time) dimension a linear time dependence. The irreducible representations are again shown to be characterized by the mass and spin of the set of states, i.e. the energy and angular momentum in the rest frame, and to consist of all possible momentum states of such a system. The 'little group' is defined and the Wigner rotations produced by successive Lorentz transformations are shown.

The Dirac spinor notation for the representations of the Lorentz group is introduced by a search for a representation which can be used to describe local interactions in configuration space, such as that of an electron with an external electromagnetic field. This requires the construction of densities like the four-vector charge–current density. These have a four-vector index and a dependence on the four-vector space–time variable which both transform independently under Lorentz transformations. The Dirac representation is obtained by requiring a representation of Lorentz transformations which factorizes into two transformations acting separately in configuration space and in another space associated with spin. In the $2+1$-dimensional case two-component spinors are obtained with the spin degree of freedom, described by operators which look formally like Pauli spin matrices, but which in reality act in a space of positive and negative energy states having the same spin. The condition that a wave function is an eigenfunction of the operators corresponding to mass and spin with spin $\frac{1}{2}$ is just the $2+1$-dimensional analog of the Dirac equation. The extension to the full Lorentz group and the conventional Dirac equation has one additional complication because of space inversion.

CHAPTER 13

INTRODUCTION TO RELATIVISTIC QUANTUM MECHANICS

Prologue

Any formulation of relativistic quantum mechanics begins with a discussion of the motion of free particles and then extends the treatment to include interactions between the particles. The discussion of free particles is always relatively easy, regardless of the particular formulation used. The troubles arise when interactions are considered. There are two kinds of troubles, old troubles and new troubles.

The old troubles caused serious concern to those physicists who first developed relativistic quantum theory, but they are now well understood. Today we all know that positrons exist and that electrons and positrons are created and annihilated in pairs, as a result of the interaction of electrons with the electromagnetic field. If a high-energy electron is scattered by a Coulomb field, there is a certain probability of pair production. This must be taken into account, in any correct description of the interaction of a relativistic electron with a static external field.

There is therefore difficulty in extending the one-particle Schrödinger equation to the relativistic domain; one particle can suddenly turn into three particles; an electron and a pair. The probability density and probability current must be redefined or reinterpreted in order to take into account the creation and annihilation of particles. The number of electrons is no longer conserved. It is the electric charge which is still conserved, and the charge density and current density satisfy an equation of continuity. These are equivalent to the probability density and probability current for systems of nonrelativistic electrons, but not for the relativistic case. Furthermore, the charge density is no longer positive definite when both electrons and positrons can be present.

When the Dirac and Klein–Gordon equations were first written down, the positron had not yet been discovered (its existence was later predicted by Dirac), and the concept of antiparticle was unknown. The 'old troubles' were just the difficulty of describing in a relativistic theory a world without positrons, antiparticles and pair production. Now that we know that we need not describe such a world, these old troubles do not bother us.

The new troubles arise from our lack of understanding of phenomena occurring at very small distances, and involving particle states of very short wavelength and high energy. If an electron is a point particle, the energy in the Coulomb field around it is infinite, because the field increases very rapidly at small distances. If the electron can emit and absorb photons, the contribution to its energy; i.e. its mass, from virtual emission and absorption processes can be calculated by perturbation theory. Here again an infinite result is obtained. The sum over all possible values of the virtual photon momentum diverges because of the contribution from high momenta (short wavelengths). Other 'ultraviolet divergences' occur elsewhere in the theory because of the necessity to include contributions from high momenta.

It is easy to say that these divergences are unphysical, because the electron is really not a point particle but has a finite size. This will introduce form factors into all calculations which will reduce the contributions from high momentum transfers (like the Debye–Waller factor in the Mössbauer effect). It is not so easy to define these form factors in a convincing and relativistically consistent way. Furthermore, any suggestion of a finite size for the electron requires the introduction of a new physical constant with the dimensions of a length, which specifies the *size* of the electron. So far, there is no experimental indication of such a finite size for the electron. The predictions of quantum electrodynamics are continually tested by experiment to the highest precision and at the highest energies available in the laboratory. There is still no evidence for a deviation from a point structure for the electron. Thus, even if the electron is found to have a finite size when experimental probes reach shorter distances, it will be a very short distance, corresponding to a very high energy, much higher than the electron rest energy. The 'ultraviolet divergences' will disappear, but the integrals will still be very large, although not infinite. The electron size could then define an absolute energy scale with a unit several orders of magnitude larger than the electron mass.

The present formulation of relativistic quantum electrodynamics uses perturbation theory successfully to compute observable properties of the electron and electromagnetic field. Many unobservable properties turn out to be infinite, e.g. what the electron mass would be if the interaction between the electron and the electromagnetic field could be switched off. Introducing a cut-off or finite size for the electron would make this unobservable mass finite and very large, and give no reason for the value of the physical mass observed for the electron, or the high ratio between the unobservable 'bare' mass and the small physical mass. Thus until there is experimental evidence for a particular size for the electron, or a 'breakdown of quantum electrodynamics at small distances', all theories including cut-offs or finite sizes have an element of arbitrariness which is not very satisfactory.

We shall present relativistic quantum mechanics from today's point of view, where we understand the old troubles but not the new ones, rather than giving a historical approach. Since processes which involve the creation and destruction of particles are best described in the formulation of second quantization, which we already know, we use it freely, without first developing the formal apparatus of quantum field theory.

13.1 Introduction

The non-relativistic Schrödinger equation, written in the form

$$\left(H - i\hbar \frac{\partial}{\partial t}\right)\psi = 0, \tag{13.1}$$

is easily seen to be relativistic as well. Since the operator H is an energy, both H and $\partial/\partial t$ are time-components of four-vectors and can be combined linearly without violating any principles of relativity. The transformation properties of ψ under Lorentz transformations remain to be defined. However, once they are, an equation is obtained whose transformation properties in special relativity are well defined and contain no internal contradictions.

The Hamiltonian for a system of free particles is easily defined in the second-quantized notation. Let $a_k^\dagger$ represent a creation operator for a particle in an eigenstate of the momentum $\hbar k$ and energy ε_k. Then

$$H = \sum_k \varepsilon_k a_k^\dagger a_k. \tag{13.2}$$

A relativistically invariant theory is obtained by using the relativistic relation between energy and momentum for a particle of mass m.

$$\varepsilon_{km} = [(\hbar k)^2 c^2 + m^2 c^4]^{\frac{1}{2}}. \tag{13.3}$$

The eigenfunctions of the Hamiltonian (13.2) describe states of several free particles. These states are seen to transform properly under Lorentz transformation to a moving coordinate system. The energies and momenta of the particles are defined by eq. (13.3) to transform in just the right way.

The addition of relativistic interactions to the problem poses no fundamental difficulty. A typical two-body interaction would have the form

$$V = \sum_{k_1} \sum_{k_2} \sum_{k_3} \sum_{k_4} V(k_1 k_2 k_3 k_4) a_{k_1}^\dagger a_{k_2}^\dagger a_{k_3} a_{k_4}. \tag{13.4}$$

This interaction can be made relativistic, by requiring the dependence of the function $V(k_1 k_2 k_3 k_4)$ on its argument to give it the proper behavior under Lorentz transformations.

Addition to the Hamiltonian of several different kinds of particles and of emission and absorption interactions also cause no great difficulty. The inclusion of spin requires the examination of how spin transforms under Lorentz transformations. Once this is established, interactions involving particles with spin can also be written with appropriate restrictions on the forms of the interactions to make them consistent with relativistic invariance and with all known conservation laws and invariance principles.

As long as everything is written in momentum space, everything is easy. However, when wave functions are required to have simple properties in configuration space, difficulties arise. We would not like a theory with a two-body interaction of the form (13.4) if it showed two particles disappearing in Rehovoth and suddenly reappearing in Timbuctoo without ever having been in between. (This should not be confused with the tunnel effect in quantum mechanics where a particle can escape through a potential barrier; in the wave description, the wave passes continuously through the barrier.) To check that such phenomena do not occur, the creation operators should be expressed in configuration space, rather than in momentum space. This is easily achieved by a Fourier transformation. One then can require an interaction to be local; i.e. that for the example (13.4) each of the two particles should be created at the same point where one particle disappears (or perhaps at least in the same small region of space). The particle creation and annihilation operators in configuration space are called quantum field operators. The theory of relativistic fields is beset by a number of difficulties, the principal ones being that nobody knows how to solve, even approximately, the equations analogous to the Schrödinger equation for systems of interacting particles. Perturbation theories have divergence difficulties because of the large number of intermediate states of high momenta (ultraviolet divergences).

The problem of describing a number of interacting particles relativistically in quantum mechanics leads eventually to quantum field theory and to the difficulties which still remain to be resolved. However, a large number of physical problems can be treated using methods which are now well established, although they must still be considered incomplete in some sense. Many problems can be handled by a method resembling the non-relativistic Schrödinger equation for a single particle. A naive extension of the non-relativistic Schrödinger theory immediately encounters difficulties which once worried eminent physicists, but which are now reasonably well understood. Rather than follow the historical path, we point out the difficulty immediately, since the relevant phenomena are familiar to everyone and no longer mysterious.

The basic difficulty is the impossibility of defining a system having a fixed number of particles. A fast charged particle moving in the electric field of a nucleus can produce an electron–positron pair with a corresponding energy loss. Thus a particle moving in an external electromagnetic field can suddenly change into three particles. Any description of the motion of a charged particle in an external electromagnetic field must include the possibility of pair production if the kinetic energy of the particle is sufficiently high (greater than the rest energy of the pair). Any attempt to describe the motion of a charged particle in an external field as a pure one-body problem, without the possibility of pair production is therefore incorrect and contradictions can be expected.

13.2 The relativistic free particle

Let us now extend the Schrödinger picture of a single-particle wave function to treat the relativistic case and see the difficulties as they arise. For a free particle, any state can be expressed as a linear combination of plane-wave states

$$\psi(\boldsymbol{x}, t) = \int \mathrm{d}^3\boldsymbol{k}\; g(\boldsymbol{k}) \mathrm{e}^{\mathrm{i}[\boldsymbol{k}\cdot\boldsymbol{x} - \omega(\boldsymbol{k})t]}. \tag{13.5}$$

If the relation between the frequency and wave number, $\omega(\boldsymbol{k})$, is known for all $\boldsymbol{k}$, the wave function (13.5) is defined for all times t if it is known at the time $t=0$. The value of the wave function at time $t=0$ is specified either by $\psi(\boldsymbol{x}, 0)$ or by its Fourier transform $g(\boldsymbol{k})$.

The frequency and wave number of a plane wave are related to the energy and momentum by the following well-known quantum conditions:

$$E = \hbar\omega \tag{13.6a}$$

$$\boldsymbol{p} = \hbar\boldsymbol{k}. \tag{13.6b}$$

The relations between frequency and wave number are obtained directly from the conditions (13.6) and the relation between energy and momentum. A natural generalization of the Schrödinger picture is to take the relativistic relation between energy and momentum. Thus

$$E = p^2/2m \Rightarrow \omega = \hbar k^2/2m \qquad \text{(non-relativistic)} \tag{13.7a}$$

$$E = (p^2c^2 + m^2c^4)^{\frac{1}{2}} \Rightarrow \omega = (k^2c^2 + m^2c^4/\hbar^2)^{\frac{1}{2}} \qquad \text{(relativistic).} \tag{13.7b}$$

We obtain a perfectly good relativistic description of free particles in quantum mechanics from eqs. (13.5), (13.6) and (13.7).

One point appears peculiar. The relativistic description should reduce to the non-relativistic case for small momenta. Expanding (13.7b) for small k gives

$$\omega \approx (mc^2/\hbar) + \hbar k^2/2m. \tag{13.8}$$

This is not the same as the non-relativistic value (13.7a) because of the rest energy. This difference between the non-relativistic and relativistic definitions of energy causes no difficulty, since absolute energy is not measurable and a difference in the zero level for energy does not have any physical consequences. However the discrepancy between (13.8) and (13.7a) can also be expressed as a shift of the zero of the frequency scale. This seems peculiar, since anyone who has dealt with radio waves knowns that the absolute frequency is measurable, that dc is different from ac, and that shifting the frequency zero is not allowed. This difficulty does not arise in the Schrödinger waves, because the absolute *frequency of these waves is not measurable and has no physical significance.* All physically measurable quantities contain products $\psi_i^* \psi_j$, which are unchanged by multiplying every state vector by the same periodically varying phase factor $e^{i\omega t}$.

The description of relativistic free particles given by eqs. (13.5), (13.6) and (13.7b) is complete even though no wave equation has been introduced. These equations determine the time behavior of the wave function for all times, if the value of the function is known at any particular time, in the same way as does the solution of the Schrödinger wave equation. Similar results could be obtained by writing down a relativistic wave equation whose solution would be given by eqs. (13.5), (13.6) and (13.7b). The appropriate wave equation is just the time-dependent Schrödinger equation (13.1) with the Hamiltonian defined by the relativistic relation

$$H = (p^2c^2 + m^2c^4)^{\frac{1}{2}}. \tag{13.9}$$

This wave equation reduces to the ordinary non-relativistic Schrödinger equation for small momenta, and has as its solutions the functions (13.5) with the frequency defined by (13.7b).

A relativistic treatment of free particles can begin with plane-wave momentum eigenstates and their time behavior, rather than with a wave equation. For interacting particles, the time development of a state cannot be written directly, and the dynamics are usually formulated with a wave equation which must be solved. The simplest interaction problem is the motion of a charged particle in an external electromagnetic field. The electromagnetic interaction is introduced into the free-particle wave equation by replacing the momentum $\boldsymbol{p}$ with $\boldsymbol{p}-e\boldsymbol{A}/c$ where $\boldsymbol{A}$ is the vector potential and e is the charge of the particle.

Since the vector potential $\boldsymbol{A}$ is a function of the position of the electron $\boldsymbol{x}$, it does not commute with the electron momentum $\boldsymbol{p}$. The substitution of $\boldsymbol{p}-e\boldsymbol{A}/c$ into the Hamiltonian (13.9) thus places linear combinations of non-commuting operators under the square root and leads to an equation which cannot be handled except in a few special cases.

13.3 The extreme-relativistic limit

Instead of looking for another relativistic equation, we note that the Hamiltonian reduces to a simpler form not only in the non-relativistic limit, but also in the extreme-relativistic limit, where $pc \gg mc^2$. For this case we have

$$H = |pc| \qquad \text{(E.R.)} \tag{13.10}$$

where E.R. denotes the extreme-relativistic limit. Let us now look for a satisfactory description in the simpler extreme-relativistic limit, rather than over the whole range of momenta. The presence of the absolute value in (13.10) introduces just as much difficulty as the square root, when $\boldsymbol{p}$ is replaced by $\boldsymbol{p}-e\boldsymbol{A}/c$. However, a tractable equation almost equivalent to eq. (13.10) can be written for the case of a particle with spin.

Consider a particle of spin $\frac{1}{2}$, whose state is described by a two-component Pauli spinor. The dynamical variables describing the spin are just the Pauli spin matrices σ_x, σ_y and σ_z. Since these behave like the components of a vector, we can form a rotationally invariant Hamiltonian by writing

$$H' = (\boldsymbol{\sigma}\cdot\boldsymbol{p})c \qquad \text{(E.R.).} \tag{13.11}$$

The eigenfunctions and eigenvalues of H' are easily obtained. For a plane wave in the z-direction, $p_z=p$, $p_x=p_y=0$, and $\boldsymbol{\sigma}\cdot\boldsymbol{p}=\sigma_z|p|$. The eigenvalues of σ_z are ± 1, corresponding to the spin 'parallel' or 'antiparallel' to the z-axis. We denote these two states by 'right-handed' $|R\rangle$, and 'left-handed' $|L\rangle$, respectively, since the spin and momentum are related like those for right and left-handed screws.

Because of the difference in sign of the two eigenvalues of σ_z, the eigenvalues of H' for the two states have opposite signs:

$$H'|R\rangle = pc|R\rangle \tag{13.12a}$$

$$H'|L\rangle = -pc|L\rangle. \tag{13.12b}$$

Since the Hamiltonian H' is invariant under rotations, these results do not depend upon our choice of the positive z-axis as the direction of the momentum. For any plane-wave state, there are two eigenfunctions of H' with eigenvalues $\pm pc$ corresponding to right and left-handed spin orientations relative to the momentum.

The same result can also be obtained by choosing an arbitrary direction for the momentum, or a state which is an eigenfunction of p^2 and not necessarily of $\boldsymbol{p}$, such as a spherical wave with a well-defined angular momentum. Let us choose the usual matrix representation for the Pauli spin matrices:

$$\sigma_x = \begin{pmatrix} 0 & 1 \\ 1 & 0 \end{pmatrix} \quad \sigma_y = \begin{pmatrix} 0 & -\mathrm{i} \\ \mathrm{i} & 0 \end{pmatrix} \quad \sigma_z = \begin{pmatrix} 1 & 0 \\ 0 & -1 \end{pmatrix} \tag{13.13a}$$

$$\boldsymbol{\sigma}\cdot\boldsymbol{p} = \sigma_x p_x + \sigma_y p_y + \sigma_z p_z = \begin{pmatrix} p_z & p_x - \mathrm{i}p_y \\ p_x + \mathrm{i}p_y & -p_z \end{pmatrix}. \tag{13.13b}$$

Since $\boldsymbol{\sigma}\cdot\boldsymbol{p}$ is a 2×2 matrix with zero trace, its two eigenvalues are equal and opposite. The product of the two eigenvalues is the determinant

$$\det(\boldsymbol{\sigma}\cdot\boldsymbol{p}) = -p_z^2 - p_x^2 - p_y^2 = -p^2. \tag{13.14}$$

The eigenvalues of $\boldsymbol{\sigma}\cdot\boldsymbol{p}$ are thus $\pm p$, as expected. This result depends only upon p and not on the individual components. It is therefore not restricted to plane waves and valid for any state which is an eigenfunction of p^2, e.g. a spherical wave.

The eigenfunctions of the Hamiltonian H' are highly degenerate, as expected for a free particle, since the energy depends only upon the magnitude of the momentum and not on its direction. Two particular representations are of physical interest, corresponding to plane and spherical waves. For both cases four quantum numbers are required to specify the states. There are the usual three spatial quantum numbers (p_x, p_y, p_z) or (p, j, m) for the plane and spherical cases. The spin direction is specified as either R or L. This designation of spin direction relative to the momentum direction is called 'helicity'.

For the Hamiltonian (13.11) the angular momentum used to specify spherical waves must be the *total* angular momentum. The orbital angular momentum cannot be specified since the orbit and spin are coupled by the term $\boldsymbol{\sigma}\cdot\boldsymbol{p}$, which does not commute with the orbital angular momentum. The eigenfunctions of (13.11) with a definite total angular momentum j are thus mixtures of two values of orbital angular momentum, $l=j+\frac{1}{2}$ and $l=j-\frac{1}{2}$. The spin direction also cannot be specified independently of the momentum, because of the spin–momentum coupling. Only the helicity eigenstates R and L are allowed.

The mixture of different orbital angular momenta and the distinction between 'left' and 'right'-handed states indicate that the eigenfunctions of (13.11) are *not* states having a well-defined parity. This is expected since the Hamiltonian (13.11) is the scalar product of an axial vector and a polar vector. It is therefore a pseudoscalar rather than a scalar, changes sign under space inversion and does not conserve parity. In the early days this would have provided sufficient grounds for rejecting the Hamiltonian H'. At present, after a Nobel prize has been awarded for the discovery that parity is *not conserved* in certain processes in nature; namely, the weak interactions, we are not disturbed by a Hamiltonian which does not conserve parity and we do not throw it away.

13.4 Negative energies, holes and antiparticles

The most peculiar feature of the Hamiltonian H' is its spectrum of negative eigenvalues, extending to $-\infty$. These negative energies constitute a characteristic difficulty of attempts to construct a one-particle relativistic theory. Negative-energy states appear and lead to unphysical results. These states cannot simply be discarded. An interaction, such as the electromagnetic interaction, has finite matrix elements connecting positive and negative-energy states. These allow a particle in a positive-energy state to make a transition to one of negative energy, and to cascade downward to lower and lower negative-energy states while radiating unlimited quantities of energy.

One method for eliminating these unphysical processes while still retaining the framework of a Schrödinger-type theory is to assume that the negative-energy states are all initially occupied. Then all transitions from positive to negative-energy states are forbidden by the Pauli principle. The presence of this 'infinite sea' of particles in negative-energy states is not observable, as long as the particles remain in these states. However, the possibility exists for a transition from a negative-energy state to a positive-energy state. Such a transition costs energy instead of liberating it and has no absurd physical consequence. Rather, the vacant negative-energy state appears as a 'hole' in the same sense as we have considered for the many-body problem.

Although the physical interpretation of 'holes in a sea of unobservable negative-energy states' was rather mysterious when first proposed by Dirac, we now interpret them like the hole states in the many-body problem: a convenient shorthand for an approximate description of a more complicated phenomenon. In both cases, the processes of physical interest involve elementary excitations of a complicated system possessing many degrees of freedom. By calling the physical ground state of the system the 'vacuum' and discussing the excitation of a small number of particles and holes from this vacuum, we avoid considering the complicated structure of the physical ground state and treat only simple excitations which are directly related to certain experimentally measured quantities.

The justification of the 'particle–hole' picture and the detailed investigation of the properties of the ground state require many-body theory for the many-body case and quantum field theory for the case of relativistic particles. These treatments show indeed that the vacuum possesses a complicated structure. It has quantum fluctuations and can be 'polarized' as if an infinite sea of negative-energy particles really exist. This sea is unobservable when undisturbed, but can give observable results when acted upon by external fields.

We now use the 'hole theory' to interpret the simple Hamiltonian (13.11) and its eigenfunctions (13.12) .The excitation of a particle from a state of negative energy $-E_1$ to a state of positive energy $+E_2$ requires an energy E_1+E_2, and creates a particle of energy $+E_2$ and a hole. The hole thus has a positive energy $+E_1$ as expected. If the negative-energy state has momentum $\boldsymbol{p}_1$ and the positive-energy state momentum $\boldsymbol{p}_2$, the momentum required to cause the transition is $\boldsymbol{p}_2-\boldsymbol{p}_1$; i.e. the momentum of the hole is $-\boldsymbol{p}_1$. Similarly for angular momentum, so that the energy, momentum and angular momentum of a hole are all equal in magnitude and *opposite* in sign to those of the corresponding negative-energy state. Note, however, that reversing the signs of *both* the momentum and angular momentum leaves the *helicity* unchanged. Thus the *particles* defined by the Hamiltonian (13.11) are all right-handed and the *holes* are all left-handed.

Now let the particles under consideration have an electric charge e and be placed in a uniform scalar potential ϕ. The Hamiltonian becomes

$$H' = (\boldsymbol{\sigma}\cdot\boldsymbol{p})c + e\phi \qquad \text{(E.R.)}. \tag{13.15}$$

The energies of the right-handed particle state E_{pR} and the left-handed hole state E_{hL} are given by

$$E_{\text{pR}} = pc + e\phi \tag{13.16a}$$

$$E_{\text{hL}} = -E_{\text{pL}} = -(-pc + e\phi) = pc - e\phi. \tag{13.16b}$$

The energy of the particle state is shifted by the amount expected for a particle of charge e in a potential ϕ. The energy of the hole state is also shifted by the appropriate amount, but in the *opposite* direction. A hole state behaves like a particle with the *opposite sign* of electric charge. This is not surprising, since electric charge is conserved and should also be conserved in the process of creating a particle–hole pair.

We thus encounter another characteristic feature of relativistic theories of charged particles. They must appear in pairs with equal and opposite charges, thus allowing them to be produced in pairs and conserve charge. These pairs of states, corresponding to particle and hole in our description, are called particle and antiparticle. It makes no physical difference which of the two states is called the 'particle' and which is the 'antiparticle'. The pair of corresponding particle and antiparticle states must have equal and opposite eigenvalues of all conserved quantities.

13.5 Relation between charge and helicity

We now find a very peculiar feature of the eigenfunctions of H' (13.15) if the particles are charged. The left-handed and right-handed states have *opposite* charge. Depending upon how one chooses the sign of e, we have a theory which can describe left-handed positrons and right-handed negative electrons or vice versa, but not all four possibilities together. We can add perturbations and interactions to the Hamiltonian (13.15) which cause transitions between different free-particle states or which change the eigenfunctions to hydrogen atom-type wave functions or anything else. However, these interactions cannot change the *helicity* of the particle. The interaction may have matrix elements between states of opposite helicity. However, the transitions produced by such matrix elements correspond to the transition between a particle and a hole state; i.e. to the physical processes of pair creation and annihilation. There are no transitions possible in this theory between particle states of opposite helicity.

It is also possible to define an alternative Hamiltonian by reversing the sign of the Hamiltonian (13.11)

$$H'' = -(\boldsymbol{\sigma}\cdot\boldsymbol{p})c \qquad \text{(E.R.).} \tag{13.17}$$

This Hamiltonian has left-handed particle states and right-handed hole states. One might consider that both Hamiltonians (13.11) and (13.17) are necessary for a description of extremely relativistic electrons. One describes right-handed positrons and left-handed electrons while the other describes left-handed positrons and right-handed electrons. However, these two pairs are completely decoupled. Interactions depending upon space and spin variables can create or annihilate electron–positron pairs with opposite helicity, but the helicity of a given particle cannot be changed, and electron–positron pairs of the same helicity cannot be created or annihilated.

Looking back at the non-relativistic limit, one finds a similar decoupling of two pairs of states, but they are different pairs. The non-relativistic description of electrons using Pauli spinors allows for reversal of the helicities of the particles, but the two charges are completely disconnected from one another. A completely different Hamiltonian is used to describe positive and negative particles, and it is perfectly possible in this description for electrons to exist and positrons not to exist and vice versa. Pair creation is not possible at non-relativistic energies and transitions involving both signs of charge are not found.

We might expect that a complete description of spin $\frac{1}{2}$ particles, valid both at the non-relativistic and extreme-relativistic limits should require a four-component wave function, which splits into two decoupled two-component spinors in the two limits. This is usually the case, but is not necessary. The nature of the splitting is different in the two limits, as indicated above. The decoupling of the different charge states in the non-relativistic limit seems quite reasonable, since we are used to this limit. The peculiar decoupling in the extreme-relativistic limit of right-handed electrons and left-handed positrons from left-handed electrons and right-handed positrons comes as a surprise. However, this peculiar decoupling is more fundamental than the non-relativistic decoupling of different charges.

A non-relativistic particle can always be made highly relativistic by moving into a different Lorentz frame. The decoupling of different charge states can only be an approximation valid in a particular set of Lorentz frames but cannot be true in general. Similar statements can generally be made for the extreme-relativistic limit; a particle can be made non-relativistic by moving into a different Lorentz frame, one close to the rest frame of the particle. In the rest frame helicity is undefined. In any frame a particular state can have its helicity reversed by a Lorentz transformation with a velocity high enough to reverse the particle momentum. However, there is one case where this cannot be done. A particle of zero rest mass is always extremely relativistic and has no rest frame. The extreme-relativistic approximation is always exact for a zero-mass particle.

The neutrino is just such a particle, and the decoupling of the different helicity states is now a well established experimental fact. In fact, for the neutrino, only one of the two Hamiltonians (13.11) or (13.17) seems to exist. (Which of the two depends upon sign conventions.) Although neutrinos have no electric charge, they have a different 'charge' or internal quantum number called the lepton number. Corresponding particles and antiparticles have opposite lepton numbers. It turns out that the neutrino having the same lepton number as the electron is always *left*-handed and that the one having the same lepton number as the positron is always *right*-handed.

13.6 Motion in a magnetic field. The magnetic moment

Consider the classical motion of a charged particle with spin in electromagnetic fields in the extreme relativistic limit. There are forces which change the direction of the momentum. If the particle has a magnetic moment, there are forces which change the direction of the spin. These forces also change the helicity, unless they are so related that they rotate the spin and the momentum by exactly the same amount. This turns out to be the case in the extreme-relativistic limit only if the charge and magnetic moment are related by a g-factor which is exactly equal to 2. By the correspondence principle one would expect to find cases of relativistic motion describable in terms of classical relativistic trajectories; e.g. the motion of electrons in a high-energy synchrotron. If the helicity decoupling of the extreme relativistic particles described by the Hamiltonian (13.11) is to be consistent with the correspondence principle, then charged particles of spin $\frac{1}{2}$ must have magnetic moments whose value is given by a g-factor of 2. We shall see that this *is* the case for the Hamiltonian (13.17).

Let us now investigate the behavior of the Hamiltonian (13.11) if an external magnetic field, described by a vector potential $\boldsymbol{A}$, is added in the conventional manner.

$$H' = \boldsymbol{\sigma}\cdot(\boldsymbol{p}c - e\boldsymbol{A}). \tag{13.18}$$

The vector potential $\boldsymbol{A}$ is a function of position, and does not commute with $\boldsymbol{p}$, or with $\boldsymbol{\sigma}\cdot\boldsymbol{p}$. One might be led to the rash and wrong conclusion that the helicity is no longer a constant of the motion because $\boldsymbol{\sigma}\cdot\boldsymbol{p}$ does not commute with the Hamiltonian. However, $\boldsymbol{p}$ is no longer the kinetic momentum $m\boldsymbol{v}$ in the presence of the vector potential. The helicity is no longer proportional to $\boldsymbol{\sigma}\cdot\boldsymbol{p}$, but to $\boldsymbol{\sigma}\cdot(\boldsymbol{p}-e\boldsymbol{A}/c)$, and is still conserved.

The properties of the eigenfunctions of (13.18) can be seen by some simple manipulation of the eigenvalue equation

$$\{\boldsymbol{\sigma}\cdot(\boldsymbol{p}c - e\boldsymbol{A}) - E\}\psi = 0. \tag{13.19a}$$

Operating on both sides with $\boldsymbol{\sigma}\cdot(\boldsymbol{p}c-e\boldsymbol{A})+E$ we obtain

$$\{[\boldsymbol{\sigma}\cdot(\boldsymbol{p}c - e\boldsymbol{A})]^2 - E^2\}\psi = 0. \tag{13.19b}$$

This expression is simplified by use of commutation relations to give

$$\{(\boldsymbol{p}c - e\boldsymbol{A})^2 - ec\hbar\boldsymbol{\sigma}\cdot\boldsymbol{\nabla} \times \boldsymbol{A} - E^2\}\psi = 0. \tag{13.19c}$$

The first term is just the square of the energy of a spinless relativistic charged particle in an electromagnetic field. The second term is a coupling between the spin of the particle $\boldsymbol{\sigma}$ and the magnetic field $\boldsymbol{H}=\boldsymbol{\nabla}\times\boldsymbol{A}$. Let us assume that the first term is very large compared to the second (again an extreme relativistic assumption) and define the kinetic momentum

$$\boldsymbol{p}_{\mathrm{kin}} = (\boldsymbol{p}c - e\boldsymbol{A}). \tag{13.20}$$

Then eq. (13.19c) can rewritten, neglecting terms of order $1/E^2$,

$$\left[(p_{\mathrm{kin}}c)^2 - \left(E + \frac{e\hbar c\boldsymbol{\sigma}\cdot\boldsymbol{H}}{2E}\right)^2\right]\psi \approx 0 \qquad \text{(E.R.)} \tag{13.21a}$$

or

$$\left[\pm|p_{\mathrm{kin}}c| - \frac{e\hbar c\boldsymbol{\sigma}\cdot\boldsymbol{H}}{2E}\right]\psi \approx E\psi \qquad \text{(E.R.)} \tag{13.21b}$$

The energy E of the state is thus given by the normal extreme-relativistic energy $p_{\mathrm{kin}}c$ with an added term corresponding to the interaction of the external magnetic field with a magnetic moment

$$\boldsymbol{\mu} = e\hbar c\boldsymbol{\sigma}/2E. \tag{13.22}$$

This expression has the form used in defining the magnetic g-factor for a relativistic particle. In the non-relativistic limit $E\approx mc^2$ this reduces to the more familiar non-relativistic expression

$$\boldsymbol{\mu} = e\hbar\boldsymbol{\sigma}/2mc. \tag{13.23}$$

Since $\hbar\boldsymbol{\sigma}$ is twice the spin angular momentum ($s=\frac{1}{2}$, and σ_z has eigenvalues ± 1), we obtain a g-factor of 2. This is just the value required to give helicity conservation in a classical trajectory.

13.7 The Dirac equation

Let us now attempt to generalize the Hamiltonians (13.11) and (13.17) to apply throughout the entire range of momenta, to give the proper relation (13.17b) between energy and momentum, to reduce to the appropriate simpler forms in the non-relativistic and extreme-relativistic limits. We first note that the two Hamiltonians required for the extreme-relativistic description

(13.11) and (13.17) can be combined into a single Hamiltonian as follows

$$H_{\text{ER}} = \varrho_1(\boldsymbol{\sigma}\cdot\boldsymbol{p}), \tag{13.24}$$

where ϱ_1 is an operator whose eigenvalues are ± 1 and which commutes with all the operators describing the space and spin degrees of freedom. The eigenfunctions of (13.24) are the same as those of (13.11) for the eigenvalue $+1$ of ϱ_1 and are the same as those of (13.17) for the eigenvalue -1 of ϱ_1. The operator ϱ_1 defines an additional degree of freedom, having two eigenvalues, like the spin. It is thus possible to represent the operator ϱ_1 by a 2×2 matrix and its eigenvectors by two-component spinors. The Hilbert space defined by the Hamiltonian (13.24) is thus the direct product of the spaces of two two-component spinors, one operating in the space defined by the operator ϱ_1 and the other in the space of the Pauli spin matrices.By analogy with the Pauli spin matrices, it is possible to define a set of three matrices in the ϱ-space, call them ϱ_1, ϱ_2 and ϱ_3, and have them satisfy similar commutation and anticommutation relations:

$$\varrho_1\varrho_2 + \varrho_2\varrho_1 = \varrho_2\varrho_3 + \varrho_3\varrho_2 = \varrho_3\varrho_1 + \varrho_1\varrho_3 = 0 \tag{13.25a}$$

$$\varrho_1\varrho_2 = \mathrm{i}\varrho_3;\ \varrho_2\varrho_3 = \mathrm{i}\varrho_1;\ \varrho_3\varrho_1 = \mathrm{i}\varrho_2 \tag{13.25b}$$

$$\varrho_1^2 = \varrho_2^2 = \varrho_3^2 = 1 \tag{13.25c}$$

$$[\varrho_i, \sigma_k] = 0. \tag{13.25d}$$

It is also possible to represent the wave function by a four-component spinor, which is the direct product of the two-component spinors in the ϱ and σ-spaces. The ϱ and σ-matrices then become 4×4 matrices which satisfy the relations (13.25) as well as the standard relations for the 2×2 Pauli σ-matrices.

So far the introduction of the operator ϱ_1 to combine the two two-component Hamiltonians (13.11) and (13.17) into a single framework is purely formal and does not change the complete decoupling of the states of the same charge and opposite helicity. Although ϱ_1 has been defined formally as an operator it is effectively just a number, since no operators which do not commute with ϱ_1 have yet been introduced into the dynamics. One way to mix the two helicity states is to add a term to (13.24) which is proportional to ϱ_3:

$$H = \varrho_1(\boldsymbol{\sigma}\cdot\boldsymbol{p})c + K\varrho_3 \tag{13.26}$$

where K is a constant. In the extreme-relativistic limit, $p \gg K$, and eq. (13.25) reduces to eq. (13.24). However, at lower momenta, ϱ_1 is no longer diagonal, because of the added term which does not commute with it, and the opposite helicity states are mixed.

We can examine the properties of the eigenfunctions of (13.26) in the same way as for eq. (13.18); namely by multiplying the eigenvalue equation by an operator which gets rid of the ϱ and σ-matrices through the anticommutation rules. Thus

$$[\varrho_1(\boldsymbol{\sigma}\cdot\boldsymbol{p})c + K\varrho_3 - E]\psi = 0 \tag{13.27a}$$

$$[\varrho_1(\boldsymbol{\sigma}\cdot\boldsymbol{p})c + K\varrho_3 + E][\varrho_1(\boldsymbol{\sigma}\cdot\boldsymbol{p})c + K\varrho_3 - E]\psi = 0 \tag{13.27b}$$

$$[p^2c^2 + K^2 - E^2]\psi = 0. \tag{13.27c}$$

Here all the matrices drop out by the anticommutation relations. The result obtained (13.27c) is just the relation (13.27b) between energy and momentum if $K=mc^2$. The substitution $K=mc^2$ gives the equation originally proposed by Dirac and known as the Dirac equation.

The Dirac equation is often written in terms of operators $\boldsymbol{\alpha}$ and β defined by

$$\boldsymbol{\alpha} = \varrho_1\boldsymbol{\sigma} \tag{13.28a}$$

$$\beta = \varrho_3. \tag{13.28b}$$

The Dirac Hamiltonian is then written

$$H_{\mathrm{D}} = \varrho_1(\boldsymbol{\sigma}\cdot\boldsymbol{p})c + \varrho_3 mc^2 \tag{13.29a}$$

$$H_{\mathrm{D}} = (\boldsymbol{\alpha}\cdot\boldsymbol{p})c + \beta mc^2 \tag{13.29b}$$

or in the presence of an electromagnetic field

$$H_{\mathrm{D}} = \boldsymbol{\alpha}\cdot(\boldsymbol{p}c - e\boldsymbol{A}) + \beta mc^2 + e\phi. \tag{13.29c}$$

Another notation commonly used for the Dirac matrices is based on transformation properties under Lorentz transformations. Unfortunately the exact definition of these gamma matrices depends on the choice of metric, and there are two different conventions in common use. For the present we shall use the ϱ and σ-matrices or the α and β-matrices which are uniquely defined. In terms of the γ-matrices, $\varrho_3=\beta$ is γ_0 or γ_4 and ϱ_1 is γ_5, with phase factors depending upon the metric.

13.8 Solution of the Dirac equation

The Dirac Hamiltonian (13.29) describes a particle having two degrees of freedom in addition to the usual three degrees of freedom of motion in configuration space. Both of the additional degrees of freedom involve operators having only two possible eigenvalues and represented by 2×2

matrices. These two 'internal' degrees of freedom are respectively the orientation of the spin and a dynamical variable associated with the operators ϱ_1 and ϱ_3. To specify a wave function five quantum numbers are needed, three for the spatial degrees of freedom and one each for the two internal degrees of freedom.

To help choose appropriate quantum numbers, we look for operators which commute with the Hamiltonian. Consider first the Hamiltonian (13.29a) with no external field. The momentum operator $\boldsymbol{p}$ commutes with H_D, the orbital angular momentum $\boldsymbol{l}=\boldsymbol{r}\times\boldsymbol{p}$ clearly does not and neither does the spin $\boldsymbol{\sigma}$. The total angular momentum $\boldsymbol{j}=\boldsymbol{l}+\frac{1}{2}\boldsymbol{\sigma}$ commutes with H_D and p but not with $\boldsymbol{p}$. The helicity $h=\boldsymbol{\sigma}\cdot\boldsymbol{p}/|p|$ commutes with H_D and also with $\boldsymbol{p}$ and $\boldsymbol{j}$.

$$[H_\mathrm{D}, \boldsymbol{p}] = [H_\mathrm{D}, \boldsymbol{j}] = [H_\mathrm{D}, h] = [\boldsymbol{p}, h] = [\boldsymbol{j}, h] = 0 \tag{13.30a}$$

$$[H_\mathrm{D}, \boldsymbol{l}] \neq 0; \qquad [H_\mathrm{D}, \boldsymbol{\sigma}] \neq 0; \qquad [\boldsymbol{j}, \boldsymbol{p}] \neq 0. \tag{13.30b}$$

We can choose either $\boldsymbol{p}$ or $\boldsymbol{j}$ and p^2 to define three quantum numbers. The helicity commutes with both $\boldsymbol{j}$ and $\boldsymbol{p}$, and can be added to either triplet to make two sets of four commuting operators which all commute with H_D. One set (p_x, p_y, p_z, h) corresponds to plane waves and one (p, j^2, j_z, h) corresponds to spherical waves.

The parity operator in configuration space is defined by the relation

$$P\psi(x, y, z) = \psi(-x, -y, -z). \tag{13.31}$$

This operator does not commute with the Hamiltonian H_D. However, the operator $\varrho_3 P$ is seen to commute with H_D, since both factors anticommute individually with the first term. In the notation of the Dirac α and β-matrices, (13.28), $\varrho_3 P=\beta P$.

$$[H, P] \neq 0 \tag{13.32a}$$

$$[H, \varrho_3 P] = [H, \beta P] = 0. \tag{13.32b}$$

The operator βP can be considered as the 'relativistic parity' or 'intrinsic parity' implying that the operation of space inversion in configuration space should be accompanied by an operation in the space of the internal degree of freedom associated with the ϱ_1-variable. The parity operator does not commute with the helicity, but it does commute with the angular momentum variables. Thus another set of four commuting operators $(p, j^2, j_z, \beta P)$ can be used to specify spherical waves.

In defining a complete set of eigenfunctions of H_D, there is a choice between plane and spherical waves, and also a choice for the latter case between eigenfunctions of the helicity and those of the relativistic parity. With the addition of an external field, many of these quantum numbers are no longer 'good quantum numbers'. For an external spherically symmetric potential (e.g. the hydrogen atom), the operators j^2, j_z and βP still commute with the Hamiltonian, while the others do not, and can be used to classify the eigenfunctions. For a scattering problem, the Hamiltonian in the asymptotic region is that of the free particle, and the desired solution has the asymptotic form of an incident plane wave and an outgoing spherical wave. The helicity eigenfunctions are better suited for this problem.

The fifth degree of freedom is not simply described. Both ϱ_1 and ϱ_3 do not commute with H_D, and there is no linear combination of them which commutes and which is simpler than the total Hamiltonian itself. Since both of these operators have only two eigenvalues, an eigenfunction of H_D can be written as a linear combination of two eigenfunctions of ϱ_1 with eigenvalues ± 1, or the same for ϱ_3. Once we choose the eigenvalues of the four space-spin quantum numbers, we need only determine the values of the coefficients appearing in the expansion in eigenfunctions of ϱ_1 or ϱ_3. Either ϱ_1 or ϱ_3 can be chosen to specify the states, and the results must be equivalent. Since ϱ_1 commutes with H_D in the extreme relativistic limit, while ϱ_3 commutes with H_D in the non-relativistic limit, the use of ϱ_1 to specify the states should lead to simpler calculations in the extreme-relativistic limit and similarly for ϱ_3 in the non-relativistic limit. We consider both cases.

Let us investigate the eigenfunctions of H_D having momentum $\boldsymbol{p}'$ and helicity h'. We first write the eigenfunctions as a linear combination of two eigenfunctions of ϱ_1 with eigenvalues $\pm\varrho_1'$, where $\varrho_1' = \pm 1$.

$$|\psi\rangle = A|\boldsymbol{p}', h', \varrho_1'\rangle + B|\boldsymbol{p}', h', -\varrho_1'\rangle, \tag{13.33}$$

where A and B are constants to be determined, normalized so that

$$A^2 + B^2 = 1. \tag{13.34}$$

The constants A and B and the eigenvalue E can be determined by substituting in the eigenvalue equation for H_D. We must define the action of the operator ϱ_3 on the eigenfunctions of ϱ_1. Since the two operators are unitary, anticommute with one another and commute with all the space and spin variables, ϱ_3 changes the eigenvalue of ϱ_1 without changing any of the other eigenvalues and preserves the norm of a state. We can thus write

$$\varrho_3|\boldsymbol{p}', h', \varrho_1'\rangle = |\boldsymbol{p}', h', -\varrho_1'\rangle, \tag{13.35}$$

where we have chosen a phase convention to eliminate an arbitrary phase factor which could be present. This phase can be changed by a change in the phase of the basic wave functions and a corresponding change in the phases of A and B. With our choice (13.35), A and B come out positive and real.

Substituting eq. (13.33) and (13.35) into the eigenvalue equation for H_D gives

$$\begin{aligned} H_D|\psi\rangle &= A\varrho_1' h' p' c|\boldsymbol{p}', h', \varrho_1'\rangle - B\varrho_1' h' p' c|\boldsymbol{p}', h', -\varrho_1'\rangle \\ &\quad + Bmc^2|\boldsymbol{p}', h', \varrho_1'\rangle + Amc^2|\boldsymbol{p}', h', -\varrho_1'\rangle \\ &= EA|\boldsymbol{p}', h', \varrho_1'\rangle + EB|\boldsymbol{p}', h', -\varrho_1'\rangle. \end{aligned} \tag{13.36}$$

Since the two eigenfunctions of ϱ_1 are linearly independent, we can equate the individual coefficients to obtain

$$A\varrho_1' h' p' c + Bmc^2 = EA \tag{13.37a}$$

$$-B\varrho_1' h' p' c + Amc^2 = EB. \tag{13.37b}$$

Solving each of these equations for B/A gives

$$\frac{B}{A} = \frac{E - \varrho_1' h' p' c}{mc^2} = \frac{E + \varrho_1' h' p' c}{mc^2}. \tag{13.38}$$

Eq. (13.38) can be solved for E to give the expected energy eigenvalue

$$E^2 = m^2c^4 + p'^2c^2. \tag{13.39}$$

Multiplying the two expressions (13.38) for B/A gives a convenient expression for B^2/A^2 which together with the normalization condition (13.34) gives values for A and B.

$$\frac{B^2}{A^2} = \frac{E - \varrho_1' h' p' c}{E + \varrho_1' h' p' c} \tag{13.40a}$$

$$1 + \frac{B^2}{A^2} = \frac{1}{A^2} = \frac{2E}{E + \varrho_1' h' p' c} \tag{13.40b}$$

$$A^2 = \frac{E + \varrho_1' h' p' c}{2E}; \qquad B^2 = \frac{E - \varrho_1' h' p' c}{2E}. \tag{13.40c}$$

In taking the square root, we determine the relative phase of B and A from eq. (13.38):

$$A = \left(\frac{E + \varrho_1' h' p' c}{2E}\right)^{\frac{1}{2}}; \qquad B = \frac{E}{|E|}\left(\frac{E - \varrho_1' h' p' c}{2E}\right)^{\frac{1}{2}}. \tag{13.40d}$$

The result (13.40) has the expected simplified form in the extreme-relativistic limit, where $pc \to E$, and $A \to 0$ if $\varrho_1' h = -1$, while $B \to 0$ if $\varrho_1' h = +1$. Near this limit, one of the two terms of the state (13.33) is of order unity, while the other is small. Since each of the terms is a two-component Pauli spin wave function, the state (13.33) has two large components and two small components near the extreme-relativistic limit. (The use of the terms 'large components and small components' in the literature does *not* refer to this classification but rather to the non-relativistic limit discussed below).

The results (13.40) have been obtained using plane-wave eigenfunctions, but are independent of the direction of the momentum. One can therefore form linear combinations of plane waves having the form (13.33) with A and B independent of the direction. Thus the same treatment holds for spherical waves which are eigenfunctions of the helicity. However, the coefficients A and B (13.40d) depend upon the helicity h', and the treatment cannot be extended to eigenfunctions of the *relativistic parity* which are mixtures of the helicity states. This agrees with our previous observation that states of opposite helicity are decoupled in the extreme-relativistic limit and helicity eigenstates are necessarily of mixed parity.

In the non-relativistic limit $p=0$, $A^2 = B^2 = \frac{1}{2}$, and all components are of the same order.

We now solve the Dirac equation using eigenfunctions of ϱ_3, rather than those of ϱ_1 used in eq. (13.33). We can try an eigenfunction having the form

$$|\psi\rangle = C|\boldsymbol{p}', h', \varrho_3'\rangle + D|\boldsymbol{p}', h', -\varrho_3'\rangle \tag{13.41}$$

where C and D are to be determined. We now define the operation of ϱ_1 on these eigenfunctions of ϱ_3. By analogy with eq. (13.35) we set

$$\varrho_1|\boldsymbol{p}', h', \varrho_3'\rangle = |\boldsymbol{p}', h', -\varrho_3'\rangle. \tag{13.42}$$

Substituting in the Dirac equation and equating coefficients leads to relations directly analogous to eqs. (13.36)–(13.40).

$$\begin{aligned} H_{\mathrm{D}}|\psi\rangle &= Ch'p'c|\boldsymbol{p}', h', -\varrho_3'\rangle + Dh'p'c|\boldsymbol{p}', h', \varrho_3'\rangle \\ &\quad - D\varrho_3' mc^2|\boldsymbol{p}', h', -\varrho_3'\rangle + C\varrho_3' mc^2|\boldsymbol{p}', h', \varrho_3'\rangle \\ &= DE|\boldsymbol{p}', h', -\varrho_3'\rangle + CE|\boldsymbol{p}', h', -\varrho_3'\rangle \end{aligned} \tag{13.43}$$

$$Ch'p'c - D\varrho_3' mc^2 = DE \tag{13.44a}$$

$$Dh'p'c + C\varrho_3' mc^2 = CE \tag{13.44b}$$

$$\frac{D}{C} = \frac{h'p'c}{E + \varrho_3' mc^2} = \frac{E - \varrho_3' mc^2}{h'p'c} \tag{13.45}$$

$$\frac{D^2}{C^2} = \frac{E - \varrho_3' mc^2}{E + \varrho_3' mc^2} \tag{13.46a}$$

$$1 + \frac{D^2}{C^2} = \frac{1}{C^2} = \frac{2E}{E + \varrho_3' mc^2} \tag{13.46b}$$

$$C^2 = \frac{E + \varrho_3' mc^2}{2E}; \quad D^2 = \frac{E - \varrho_3' mc^2}{2E} \tag{13.46c}$$

$$C = \left(\frac{E + \varrho_3' mc^2}{2E}\right)^{\frac{1}{2}}; \quad D = h' \frac{E}{|E|}\left(\frac{E - \varrho_3' mc^2}{2E}\right)^{\frac{1}{2}}. \tag{13.46d}$$

The result (13.46) has the expected simplified form in the non-relativistic limit, where $E = mc^2$, and $C = 0$ if $\varrho_3' = -1$ and $D = 0$ if $\varrho_3' = +1$. Near this limit one of the two terms in (13.41) is of order unity and the other is small. The designation 'large components' and 'small components' commonly found in the literature refers to *this* classification in the non-relativistic limit.

The results (13.46) are clearly also valid for the use of spherical waves, since the coefficients C and D turn out to be independent of the direction of $\boldsymbol{p}'$. Unlike the previous case, C and D are also independent of the helicity h' except for an overall phase and the treatment (13.41)–(13.46) can also be applied to states which are linear combinations of different helicities; e.g. the eigenfunctions of the relativistic parity $\varrho_3 P$. One can therefore define spherical wave eigenfunctions which are also parity eigenfunctions

$$|p', j', m', h'\rangle = C|p', j', m', h', \varrho'\rangle + D|p', j', m', h', \varrho_3'\rangle \tag{13.47a}$$

$$|p', j', m', \varrho_3 P\rangle = C|p', j', m', P', \varrho_3'\rangle \pm D|p', j', m', -P', -\varrho_3'\rangle \tag{13.47b}$$

where the phase depends on the convention used to define the parity states.

For convenience, the state (13.47) has been written as a linear combination of eigenfunctions of P and ϱ_3 separately, with both signs reversed in going from one term to the other. The state is thus an eigenfunction of $\varrho_3 P$. A state which is an eigenfunction of j, with spin $\frac{1}{2}$ is a linear combination of eigenfunctions of the orbital angular momentum l with two eigenvalues $j \pm \frac{1}{2}$. Since these two adjacent l-values have opposite parity in configuration space, individual terms in (13.47) which each have a definite parity are also eigenfunctions of the orbital angular momentum l. Thus in the non-relativistic limit where either C or D vanishes, the state (13.47) approaches an eigenfunction of the orbital angular momentum. This is in accord with the non-relativistic Pauli spin wave functions, for which l can be a good quantum number. The relativistic parity βP is thus equivalent to the non-relativistic parity P in the non-relativistic limit. The helicity states, however, have neither a good parity nor a good orbital angular momentum even in the non-relativistic limit.

The coefficients C and D are of the same order in the extreme-relativistic limit $E \gg mc^2$, as is expected. If helicity eigenstates are used, the particular linear combination (13.41) approaches just that state which is an eigenfunction of ϱ_1. However, if parity eigenstates are used, the state is an equal mixture of opposite helicities, and therefore also an equal mixture of eigenfunctions of ϱ_1 with opposite eigenvalues. Thus the helicity eigenfunctions have a simple form in both the extreme-relativistic and non-relativistic limits, with two large components and two small components in the appropriate representation (13.33) or (13.41). The parity eigenfunctions have a simple form in the non-relativistic limit in the representation (13.41) with two large components and two small components. However, they do not have a simple form in the extreme-relativistic limit, having four components of approximately equal magnitude in both representations (13.33) and (13.41).

The obvious symmetry between the treatments using the wave functions (13.33) and (13.41) in the two opposite limits should be noted, as well as the differences between the helicity and parity properties. Since both solutions are correct, either type of wave function can be used in a given calculation. However, results are often obtained much more easily with one than with the other, depending upon the conditions of the problem.

13.9 Scattering by a spin-independent potential in Born approximation

Consider the scattering of a particle of spin $\frac{1}{2}$ by a potential which is independent of the spin orientation of the particle. The cross section for the scattering of a particle from an initial state $|i\rangle$ into a final state $|f\rangle$ is proportional in Born approximation to the square of the transition matrix element.

$$\sigma_{i \to f} \propto \left|\langle f|V|i\rangle\right|^2. \tag{13.48}$$

In the absence of spin and neglecting relativity, the transition matrix element for scattering from a plane-wave state $\boldsymbol{p}_i$ into a plane-wave state $\boldsymbol{p}_f$ is

$$\langle \boldsymbol{p}_f|V|\boldsymbol{p}_i\rangle = M(\boldsymbol{p}_i, \boldsymbol{p}_f). \tag{13.49}$$

Let us now consider non-relativistic scattering using Pauli spin wave functions $|\boldsymbol{p}_i, \sigma_i\rangle$ and $|\boldsymbol{p}_f, \sigma_f\rangle$ specifying the momenta and spin orientations of the initial and final states. Since V is independent of spin, and the wave function is the direct product of a space wave function $|\boldsymbol{p}\rangle$ and a spinor $|\sigma\rangle$,

$$\langle \boldsymbol{p}_f, \sigma_f | V | \boldsymbol{p}_i, \sigma_i \rangle = \langle \boldsymbol{p}_f | V | \boldsymbol{p}_i \rangle \langle \sigma_f | \sigma_i \rangle = M(\boldsymbol{p}_i, \boldsymbol{p}_f) \langle \sigma_f | \sigma_i \rangle . \tag{13.50}$$

This result has a very simple interpretation. The potential V does not affect the spin and scatters the particle exactly like a spinless particle. The transition matrix element is thus the product of the matrix element for spinless scattering multiplied by a spin overlap factor. The latter is unity if the spin orientation of the final state is the same as that of the initial state and zero if the two orientations are opposite (i.e. a spin-independent potential cannot produce 'spin-flip' scattering).

The result (13.50) can also be written for particle states specified by the helicity, rather than the orientation of the spin with respect to fixed axes. For a scattering angle θ, the angle between the orientations of σ_f and σ_i is θ or $\pi - \theta$ if the initial and final helicities are the same or opposite respectively. Thus

$$\langle \boldsymbol{p}_f, h_f | V | \boldsymbol{p}_i, h_i \rangle_{\mathrm{NR}} = M(\boldsymbol{p}_i, \boldsymbol{p}_f) \cos \tfrac{1}{2}\theta \quad \text{if} \quad h_i = h_f \tag{13.51a}$$

$$= M(\boldsymbol{p}_i, \boldsymbol{p}_f) \sin \tfrac{1}{2}\theta \quad \text{if} \quad h_i = -h_f \tag{13.51b}$$

where the factors $\cos \frac{1}{2}\theta$ and $\sin \frac{1}{2}\theta$ are just the spin overlaps $\langle \sigma_f | \sigma_i \rangle$ between states differing by a rotation of θ and $\pi - \theta$ respectively.

The relativistic extension of this result (13.51) is straightforward. Instead of using Pauli wave functions, we substitute the eigenfunctions (13.33) of the Dirac Hamiltonian into eq. (13.48). The initial and final states are

$$|\boldsymbol{p}_i, h_i\rangle = A_i |\boldsymbol{p}_i, h_i, \varrho_1'\rangle + B_i |\boldsymbol{p}_i, h_i, -\varrho_1'\rangle \tag{13.52a}$$

$$|\boldsymbol{p}_f, h_f\rangle = A_f |\boldsymbol{p}_f, h_f, \varrho_1'\rangle + B_f |\boldsymbol{p}_f, h_f, -\varrho_1'\rangle \tag{13.52b}$$

where the coefficients A_i, B_i, A_f and B_f are given by eq. (13.40d). The transition matrix element of the potential V between the two states (13.52) is easily calculated because V commutes with ϱ_1 and has no matrix elements between states having different eigenvalues of ϱ_1. The transition matrix element is just the sum of the contributions of the first terms and of the second terms of (13.52) with no cross terms. The individual contributions are just matrix elements between different helicity states described by Pauli spinors and given by eq. (13.51). Thus

$$\langle \boldsymbol{p}_f, h_f | V | \boldsymbol{p}_i, h_i \rangle_{\mathrm{Rel}} = [A_f A_i + B_f B_i] \langle \boldsymbol{p}_f, h_f | V | \boldsymbol{p}_i, h_i \rangle_{\mathrm{NR}} . \tag{13.53}$$

From eqs. (13.4) we obtain

$$A_f A_i + B_f B_i$$

$$= \frac{1}{2E} \{\sqrt{(E + h_i pc)(E + h_f pc)} + \sqrt{(E - h_i pc)(E - h_f pc)}\} \qquad (13.54a)$$

$$= 1 \qquad (h_i = h_f) \qquad (13.54b)$$

$$= \frac{mc^2}{E} (h_i = -h_f). \qquad (13.54c)$$

Substituting the results (13.54) and the non-relativistic result (13.51) into eq. (13.53) we obtain

$$\langle \boldsymbol{p}_f h_f | V | \boldsymbol{p}_i h_i \rangle_{\text{Rel}} = M(\boldsymbol{p}_i, \boldsymbol{p}_f) \cos \tfrac{1}{2}\theta \qquad (h_i = h_f) \qquad (13.55)$$

$$= M(\boldsymbol{p}_i, \boldsymbol{p}_f) \sin \tfrac{1}{2}\theta \qquad (h_i = -h_f).$$

The relativistic result is the *same* as the non-relativistic result for scattering *with no helicity change*, but the scattering with helicity reversal is reduced by the factor mc^2/E. This factor is unity in the non-relativistic limit, and becomes very small in the extreme relativistic limit, as expected, since the two helicity states are completely decoupled in the extreme-relativistic limit.

We can now ask how a potential which is independent of spin can give spin-dependent results in a relativistic treatment (13.55) when they are absent in a non-relativistic treatment (13.51). How can relativistic effects cause a force which does not act on the spin to change the direction of the spin? The answer is clarified by the example of scattering of a physical electron by an electric field, say of a nucleus. If the electron is moving with relativistic velocity through an electric field, it sees a magnetic field as well. A field which is a pure electric field in the laboratory system contains both electric and magnetic fields in the rest frame of the electron. This magnetic field seen by the electron rotates the spin.

The Dirac equation automatically introduces the effects of the magnetic moment of the electron in its interaction with the electromagnetic field. Thus in the relativistic Dirac theory of the potential V, considered as an electrostatic potential in the laboratory system, the effects of the magnetic field seen in the rest frame of the electron and its interaction with the magnetic moment of the electron are automatically included.

We can also ask what would happen if the potential V were not electrostatic, but were some other kind of field, say a scalar meson field which would not give rise to magnetic effects in a moving coordinate system. Although we have apparently not specified the nature of the potential V in our problem, we have simply added it to the Hamiltonian; i.e. to the time component of a four-vector. We have therefore implicity assumed the potential V to be the time component of a four-vector, like the electromagnetic scalar potential. It automatically has 'magnetic-type' effects in moving coordinate systems where the other three components of the four-vector no longer vanish. The effect is relativistic since the intensity of the magnetic field is proportional to v/c.

For the case of a relativistic scalar potential, one might add $\varrho_3 V$ to the Hamiltonian, rather than just V. This would be adding V to the term mc^2, which is a relativistic scalar, rather than to H which is a four-vector.

The total scattering cross section (assumed over all polarizations of final states) is proportional to the sum of the squares of the two matrix elements (13.55a) and (13.55b).

$$\begin{aligned}\sigma_{\text{Rel}}(\boldsymbol{p}_i, h_i \to \boldsymbol{p}_f) &\propto \sum_{h_f} \left|\langle \boldsymbol{p}_f h_f | V | \boldsymbol{p}_i h_i \rangle\right|^2 \\ &= |M(\boldsymbol{p}_i, \boldsymbol{p}_f)|^2 \left(\cos^2 \tfrac{1}{2}\theta + \frac{m^2c^4}{E^2} \sin^2 \tfrac{1}{2}\theta\right) \\ &= |M(\boldsymbol{p}_i, \boldsymbol{p}_f)|^2 [1 - (v^2/c^2) \sin^2 \tfrac{1}{2}\theta]. \end{aligned} \tag{13.56}$$

To relate the cross section for the corresponding relativistic and non-relativistic processes, we must also consider the relativistic corrections to the proportionality factors multiplying the square of the matrix element. Using the derivation of the Born approximation from the 'golden rule' of time-dependent perturbation theory,

$$\frac{\mathrm{d}\sigma}{\mathrm{d}\Omega} = \frac{2\pi}{\hbar v} |\langle f|V|i\rangle|^2 \varrho_E. \tag{13.57}$$

Relativistically, the factor ϱ_E/v becomes

$$\frac{\varrho_E}{v} \propto \frac{p^2}{v} \frac{\mathrm{d}p}{\mathrm{d}E} = \frac{p^2 E}{vcp^2} = \frac{pE}{vc^2} = \frac{m^2}{1 - v^2/c^2}. \tag{13.59}$$

The factor $(1 - v^2/c^2)^{-1}$ is unity in the non-relativistic limit and represents a relativistic correction. Inserting this factor into eq. (13.56) we obtain

$$\sigma_{\text{Rel}}(\boldsymbol{p}_i, h_i \to \boldsymbol{p}_f) = \sigma_{\text{NR}}(\boldsymbol{p}_i \to \boldsymbol{p}_f) \frac{1 - (v^2/c^2) \sin^2 \frac{1}{2}\theta}{1 - v^2/c^2}. \tag{13.59}$$

This result (13.59) is general and independent of the form of the potential V, as long as V is a scalar component of a four-vector potential and independent of spin. The result holds for either helicity of the incident beam, and is independent of the helicity. It therefore holds as well for an unpolarized beam, which is an incoherent mixture of the two helicities.

A particular example of physical interest where eq. (13.59) is applicable is the scattering of relativistic electrons by the Coulomb field of a light nucleus. For this case the non-relativistic result is just the Rutherford formula and eq. (13.59) gives the relativistic correction. The Born approximation is expected to hold for light nuclei, where $Z/137 \ll 1$, but to break down for heavy nuclei where $Z/137 \approx 1$. The accidental result that the Rutherford formula is obtained in the non-relativistic Born approximation is no longer true relativistically. The higher order Born corrections are not zero, and are appreciable for heavy nuclei.

CHAPTER 14

INVARIANCE, SYMMETRY TRANSFORMATIONS AND CONSERVATION LAWS

14.1 Types of transformations

Many quantum-mechanical problems are simplified by examining the behavior of states and dynamical variables under certain transformations. We present different aspects of these transformations by considering a few examples in detail.

The simplest transformations are coordinate transformations like the transformation from cartesian to spherical coordinates or the canonical transformations which mix coordinates and momenta. These transformations can always be made on any system and are useful if they convert the Hamiltonian or the equations of motion to a simpler form. The Hamiltonian always changes its form under such transformations and the wave functions before and after the transformation look quite different. The line $x=\text{constant}$, for example, has completely different physical properties from the line $r=\text{constant}$. The former is a straight line which can be a possible path for motion in the absence of forces, the latter is a circle which cannot. The wave function e^{ikx} is a plane wave which is a possible state for a free particle. The wave function e^{ikr} (*not* e^{ikr}/r) is a peculiar spherical wave which is not a possible state for a free particle. The line $x=\text{constant}$ and the plane-wave state e^{ikx} can of course be expressed in spherical coordinates. However, these expressions have a very different form and involve all the polar variables r, θ and ϕ. This difference in *form* of the states before and after the transformation is the essential difference between this type of transformation and the symmetry transformations discussed below. (We are not considering here the possibility of gravitational fields and curvature in the metric as in the general theory of relativity.)

These coordinate and canonical transformations are associated with different descriptions of the same physical states and the same dynamical variables. One does not change the states or the dynamical variables by making these transformations; one simply changes the language used in describing them.

Canonical transformations can always be expressed by the action of a unitary operator e^{iS}. The wave functions undergo the transformation

$$\psi \to e^{iS}\psi. \tag{14.1a}$$

Operators representing observables undergo the transformation

$$A \to e^{iS} A e^{-iS}. \tag{14.1b}$$

When a canonical transformation is performed, *both the states and the operators* are transformed. Thus, although the states and operators change their form, all matrix elements remain invariant.

An important special class of coordinate or canonical transformations are those which correspond to a symmetry of the system. These include translations, rotations, space inversion and time reversal. Under a symmetry transformation which does not depend explicitly on time, the *form* of the Hamiltonian and of the equations of motion remain invariant. In contrast to the case of the transformation to spherical coordinates discussed above, both the lines $x = \text{constant}$ and $y = \text{constant}$ represent possible paths of motion in the absence of forces and both the wave functions e^{ikx} and e^{iky} represent plane waves which are possible states of motion for a free particle. Thus the 90° rotation about the z-axis which transforms x into y does not change the *form* of the equations of motion of a free particle, its possible paths of motion or the eigenfunctions of the Hamiltonian.

These transformations can be interpreted in three ways, as transformations of the coordinate system without changing the physical system, as transformations on the physical system without changing the coordinate system, or as changing both. In discussing the behavior of the hydrogen atom under translations one can examine the change in the description of a hydrogen atom when the origin of the coordinate system is moved from Rehovoth to Timbuctoo without doing anything to the hydrogen atom. One can also examine what happens when the hydrogen atom itself is moved from Rehovoth to Timbuctoo without any change in the origin of the coordinate system. One can also move *both* the hydrogen atom *and* the coordinate system from Rehovoth to Timbuctoo.

In the first case the translational invariance of the dynamics of a hydrogen atom is expressed by the invariance of the Hamiltonian and of its equations of motion when the variables are changed by moving the origin of the coordinate system. In the second case the translational invariance is indicated by the existence of two different hydrogen atoms, one in Rehovoth and one in Timbuctoo which *both* satisfy the *same* equations of motion. In quantum mechanics such a transformation can be applied to an eigenfunction of the

Hamiltonian, in order to construct another eigenfunction of the Hamiltonian with the same eigenvalue. We therefore learn something about the classification and degeneracies of the eigenfunctions of the Hamiltonian from these transformations. In the third case, nothing is really changed and all results must be the same. No symmetry or dynamics is tested. It is just an ordinary canonical transformation.

The three ways of interpreting these transformations are expressed formally using the transformations (14.1).

A. *The physical system is changed without changing the coordinate system.* The wave function changes according to eq. (14.1a), but the operators are expressed in terms of same coordinate system and do not change. Matrix elements of operators then undergo the transformation

$$\langle\psi'|A|\psi\rangle \rightarrow \langle\psi' \mathrm{e}^{-\mathrm{i}S}|A|\mathrm{e}^{\mathrm{i}S}\psi\rangle = \langle\psi'|\mathrm{e}^{-\mathrm{i}S} A \mathrm{e}^{\mathrm{i}S}|\psi\rangle. \tag{14.2a}$$

In calculating the matrix element, the unitary transformation can always be shifted from the wave functions to the operators. We see that matrix elements are changed by the transformation unless the operator A commutes with the transformation S.

B. *The physical system is unchanged, but the coordinate system is changed.* The wave function remains the same, but the operators are expressed in terms of the new coordinate system and change according to eq. (14.1b). Matrix elements then undergo the transformation

$$\langle\psi'|A|\psi\rangle \rightarrow \langle\psi'|\mathrm{e}^{\mathrm{i}S} A \mathrm{e}^{-\mathrm{i}S}|\psi\rangle. \tag{14.2b}$$

Again the matrix elements are changed by the transformation unless the operator commutes with the transformation S. The difference in the sign of the exponents between (14.2a) and (14.2b) is easily understood. If the transformation is a clockwise rotation, for example, (14.2a) describes a clockwise rotation of the physical system, keeping the coordinate system fixed. Eq. (14.2b) describes a clockwise rotation of the coordinate system, keeping the physical system fixed; i.e. a *counter-clockwise* of the physical system relative to the coordinate system.

C. *Both the physical system and the coordinate system are transformed.* Both the wave functions and the operators are transformed by eqs. (14.1). The matrix elements are then unchanged,

$$\langle\psi'|A|\psi\rangle \rightarrow \langle\psi' \mathrm{e}^{-\mathrm{i}S}|\mathrm{e}^{\mathrm{i}S} A \mathrm{e}^{-\mathrm{i}S}|\mathrm{e}^{\mathrm{i}S}\psi\rangle = \langle\psi'|A|\psi\rangle. \tag{14.2c}$$

This is just an ordinary canonical transformation, like to one from cartesian to spherical coordinates, which simply allows calculations of the *same* results in different ways.

The relations (14.2a) and (14.2b) describe transformations which *change* the matrix elements of some observables. It is meaningful to ask 'how particular observables transform under the transformation'. Under rotations for example, matrix elements of certain operators can be invariant or can transform like components of a vector or tensor. If the Hamiltonian is invariant under the particular transformation, then the Schrödinger equation is invariant, and the transformation describes a *symmetry* of the system. The operator S is then a constant of the motion and defines a conservation law.

In between these two types of transformations lies the gauge transformation. Like the coordinate or canonical transformation and unlike the symmetry transformations of translations and rotations a gauge transformation can only be interpreted as a change in the description of a given physical state. It cannot be interpreted as a transformation of the physical state itself into another state. Yet, although one does not expect any invariance under the canonical transformations which merely change the coordinate system, gauge invariance is well defined and has interesting implications as discussed in section 14.6 below.

A special class of symmetry transformations with particularly interesting properties are those which depend explicitly on the time, like Galilean or Lorentz transformations. These time-dependent translations can either be interpreted as a change between two coordinate systems moving relative to one another or a change of the physical system by giving it a velocity. Eqs. (14.1) and (14.2) all apply to this case, but the operator S depends explicitly on the time. Invariance with respect to these transformations means in the first interpretation that the laws of nature are the same for all observers moving relative to one another with constant velocities. The second interpretation tells us that given any possible state of motion for the system, we can construct other possible states of motion by giving the system as a whole an additional uniform velocity. In quantum mechanics the latter interpretation tells us how to transform a state which is an eigenfunction of the Hamiltonian to another state which is also an eigenfunction of the Hamiltonian but with a different eigenvalue. We therefore learn something about the eigenvalue spectrum of the Hamiltonian from these transformations.

We now consider some individual transformations in detail. We first consider the case of a single spinless particle and generalize the treatment later.

14.2 Parity and space inversion

The parity or space inversion operation P is defined by the relation

$$P\psi(x, y, z) = \psi(-x, -y, -z). \tag{14.3}$$

For any state ψ of a quantum-mechanical system, one can define another state by the parity operation (14.3). The two states are 'mirror images' of one another about the origin. For example, if one state describes a particle moving in a 'right-handed' helical orbit, the other state describes the corresponding 'left-handed' helical orbit, progressing in the opposite direction with the same sense of angular momentum.

The parity operator has a particular significance in a physical problem if it is a constant of the motion; i.e. if it commutes with the Hamiltonian of the system

$$[H, P] = 0. \tag{14.4}$$

In this case parity is said to be conserved and the Hamiltonian invariant under space inversion. If any state ψ is a solution of the Schrödinger equation, either time-independent or time-dependent, the state $P\psi$ is also a solution. The operator P is linear and Hermitean, and satisfies the equation

$$P^2 = 1. \tag{14.5}$$

It is therefore unitary and its eigenvalues are ± 1, called even and odd.

Any state ψ which is not an eigenfunction of P can be separated into even and odd parts

$$\psi_{\pm} = (1 \pm P)\psi \tag{14.6a}$$

$$P\psi_{\pm} = \pm(1 \pm P)\psi. \tag{14.6b}$$

Parity conservation helps in solving the Schrödinger equation, because we can look for simultaneous eigenfunctions of H and P. If we choose as a basis of functions for solving the Schrödinger equation a set which are already eigenfunctions of P, we cut our work in half, because the Hamiltonian cannot mix even and odd states. We have separated our Hilbert space in two pieces which are decoupled from one another. As long as $[H, P] = 0$ for the *total* Hamiltonian including all perturbations, there can be no transitions between even and odd states; i.e. parity is conserved in all transitions.

We can also classify operators as even or odd under parity, according to whether they commute or anticommute with P. Any operator A can be written as the sum of an even part A_+ and an odd part A_-:

$$A = A_+ + A_- \tag{14.7a}$$

$$A_+ = \tfrac{1}{2}(A + PAP) \tag{14.7b}$$

$$A_- = \tfrac{1}{2}(A - PAP). \tag{14.7c}$$

The even and odd operators satisfy the relations

$$PA_+P = A_+ \tag{14.8a}$$

$$PA_-P = -A_- \tag{14.8b}$$

Even and odd operators satisfy simple selections rules. Even operators have non-vanishing matrix elements only between states of the same parity; odd operators have non-vanishing matrix elements only between states of opposite parity. This is seen formally by considering the matrix elements between two states of parity P' and P''

$$\begin{aligned}\langle P'|A_+|P''\rangle &= \langle P'|PA_+P|P''\rangle \\ &= P'P''\langle P'|A_+|P''\rangle = 0 \quad \text{if} \quad P' = -P''\end{aligned} \tag{14.9a}$$

$$\begin{aligned}\langle P'|A_-|P''\rangle &= -\langle P'|PA_-P|P''\rangle \\ &= -P'P''\langle P'|A_-|P''\rangle = 0 \quad \text{if} \quad P' = P''.\end{aligned} \tag{14.9b}$$

This treatment is easily generalized to more complicated systems. There are simple rules for combining parities of different parts of a system. The parity of a complex system is just the product of the parities of its component parts.

The transformation (14.3) can be interpreted in three ways as discussed above. The two states on the left and right-hand sides of eq. (14.3) can either be considered as two different physical states of the system, or as the same physical state seen from two coordinate systems. In the latter case, one coordinate system is a left-handed coordinate system, the other is a right-handed coordinate system. Confusion can arise if it is not understood explicitly which of the two interpretations is being used.

Consider for example the expectation value of the odd operator x in a particular state ψ.

$$\langle\psi|x|\psi\rangle = \int \psi^*(x, y, z)x\psi(x, y, z)\,\mathrm{d}x\,\mathrm{d}y\,\mathrm{d}z. \tag{14.10}$$

Suppose we wish to find out 'how the expectation value behaves under the parity transformation' (14.3). There are three interpretations of this statement as discussed above.

A. The transformation transforms the physical ψ into another physical state according to the transformation (14.3) but does not change the operator x, as in eq. (14.2a)

$$\begin{aligned}\langle\psi|x|\psi\rangle \to \langle P\psi|x|P\psi\rangle &= \int \psi^*(-x, -y, -z)x\psi(-x, -y, -z)\,\mathrm{d}x\,\mathrm{d}y\,\mathrm{d}z \\ &= -\langle\psi|x|\psi\rangle.\end{aligned} \tag{14.11a}$$

B. The transformation leaves the physical state alone, but changes the coordinate system from a right-handed one to a left-handed one. This can be expressed by leaving the wave function alone and transforming the operator, as in eq. (14.2b),

$$\langle\psi|x|\psi\rangle \rightarrow \langle\psi|PxP|\psi\rangle = -\langle\psi|x|\psi\rangle. \tag{14.11b}$$

C. The transformation is applied both to the wave function *and* to the operators, as in eq. (14.2c)

$$\langle\psi|x|\psi\rangle \rightarrow \langle P\psi|PxP|P\psi\rangle = \langle\psi|PPxPP|\psi\rangle = \langle\psi|x|\psi\rangle. \tag{14.11c}$$

The first two interpretations lead to the same result, while the negative sign is missing in the third case. Interpretation A is the one most easily understood. A different physical state is defined with no change of coordinate system and no change in the meaning of any observables. The new state has an expectation value of x opposite in sign from the old one. Interpretation C is simply a particular canonical transformation applied to both the state vector and the operator, and cannot change the value of the result. It is the same physical quantity calculated in two coordinate systems.

Interpretation B is the most confusing. The same state is described in a different coordinate system. In the change of coordinate system, the *physical meaning of the operator x is also changed.* It is the coordinate in the x-direction, but the x-direction is now opposite to what it was before the transformation. The result therefore has the opposite sign.

Let us now examine the distinction between interpretations B and C, which both change the coordinate system. In the interpretation C the expectation value (14.11c) is considered to be a physical quantity independent of the coordinate system. Suppose that the x-axis originally points in the direction of the star Sirius. Then the expectation value (14.11c) is the mean distance of the particle from the origin *in the direction of Sirius.* This is unaffected by our labeling the coordinate x or $-x$, as indicated by the result (14.11c). In interpretation B, the expectation value is defined as the mean distance of the particle from the origin *in the direction which we choose to call the positive x-axis.* If we change the direction of our coordinate axes, we change this x-direction, even though we do not change anything in the physical system. This accounts for the difference between the results (14.11b) and (14.11c).

Interpretation A is the clearest. The coordinate axes are not changed, and two different physical states are considered. We therefore recommend that symmetry transformations be considered as taking place on the physical system, with no change in the operators or the coordinate system, whenever this is feasible.

This simple example exhibits a number of features common to all transformations *which can be interpreted as changing the physical state*:

1. If the operation commutes with the Hamiltonian,
   (a) A new constant of the motion (conserved quantity) is defined.
   (b) Applying the transformation to any solution of the Schrödinger equation gives a state which is also a solution.
   (c) The new constant of the motion defines a new quantum number which can be used to classify the eigenfunctions of the Hamiltonian.

2. Linear operators can be expressed in terms of certain operators transforming in a simple way under the operation (in this case even or odd; in more general cases, so-called 'irreducible tensors').

3. Relations between the matrix elements of the simple operators (14.5) follow from the transformation properties. These may simply be selection rules as in this case, eqs. (14.9), or relations between different matrix elements (Wigner–Eckart theorem).

4. The transformation can also be interpreted as leaving the physical state unchanged and changing the coordinate system. For the case where the operation commutes with the Hamiltonian, the form of the Hamiltonian and of the equations of motion do not change in the transformation to the new coordinate system. This interpretation often causes confusion.

14.3 Translations and momentum conservation

Consider the operator T_a producing a translation by the finite amount a in the x-direction

$$T_a \psi(x, y, z) = \psi(x + a, y, z). \tag{14.12}$$

For any state of a physical system, this transformation can produce another state of the system. The operation has a particular significance if it commutes with the Hamiltonian of the system

$$[H, T_a] = 0. \tag{14.13}$$

Unlike the parity operator, the square of T_a is not one, but corresponds to a translation by a distance $2a$. In general

$$(T_a)^n = T_{na}, \tag{14.14}$$

for all integral values of n, positive and negative. The infinite set of operators (14.14) define a group of transformations. If T_a commutes with the Hamiltonian, all the operators (14.14) commute as well. The operator T_a is linear and unitary, but is not Hermitean. Its eigenvalues therefore include all complex numbers of modulus unity.

If T_a commutes with the Hamiltonian, solutions of the Schrödinger equation can be constructed by operating on any solution ψ with the translation operators (14.14). A complete set of simultaneous eigenfunctions of H and of T_a can be found. A state having the form

$$\psi(x) = \mathrm{e}^{ikx} u(x), \tag{14.15}$$

where $u(x)$ is a periodic function of x with period a, is clearly an eigenfunction of T_a with eigenvalue e^{ika}. Since the function $\mathrm{e}^{i2\pi x/a}$ is a periodic function of x with period a, any integral multiple of $2\pi/a$ can be added to k in a given function (14.11) with an appropriate change of the periodic function $u(x)$, but no change in the eigenvalue e^{ika}. Thus k is defined only modulo $2\pi/a$.

Let us now consider the case where the Hamiltonian commutes with any translation operator; i.e. eq. (14.13) holds for all values of a. It is then convenient to define the infinitesimal translation T_ε

$$T_\varepsilon \psi(x) = \psi(x + \varepsilon) = \psi(x) + \varepsilon \frac{\partial \psi(x)}{\partial x} \tag{14.16a}$$

$$T_\varepsilon = (1 + \varepsilon \partial/\partial x) = (1 + \mathrm{i}\varepsilon p_x/\hbar), \tag{14.16b}$$

where p_x is the x-component of the momentum operator.

If T_ε commutes with the Hamiltonian, then p_x commutes with the Hamiltonian, and the x-component of the momentum is an integral of the motion and is conserved.

The finite translation operators T_a are given by

$$T_a = \exp(\mathrm{i}ap_x/\hbar). \tag{14.17}$$

The complete set of all translation operators constitute a continuous or Lie group. The momentum operator p_x is called the infinitesimal operator or generator of the group, since it can be used to generate any translation in the x-direction (14.17). The invariance of the Hamiltonian under translations thus leads to the *conservation of momentum*.

In the same way that operators can be divided into two types, corresponding to their behavior under space inversion, they can be divided into a continuous infinity of types corresponding to their behavior under translations. These 'eigenoperators' go into themselves under the transformation, multiplied by an eigenvalue.

$$T_a A_k (T_a)^{-1} = \mathrm{e}^{\mathrm{i}ap_x/\hbar} A_k \mathrm{e}^{-\mathrm{i}ap_x/\hbar} = \mathrm{e}^{ika} A_k. \tag{14.18a}$$

The expansion of an arbitrary operator A into a continuous set of operators A_k, analogous to the parity expansion is just a Fourier expansion:

$$A = \int A_k \mathrm{d}k \tag{14.18b}$$

where

$$A_k = \frac{1}{2\pi} \int_{-\infty}^{+\infty} \mathrm{e}^{-\mathrm{i}ka} T_a A (T_a)^{-1} \mathrm{d}a$$

$$= \frac{1}{2\pi} \int_{\infty}^{\infty} \mathrm{e}^{-\mathrm{i}ka} \mathrm{e}^{\mathrm{i}p_x a/\hbar} A \, \mathrm{e}^{-\mathrm{i}p_x a/\hbar} \mathrm{d}a. \tag{14.18c}$$

The eigenoperator relation (14.18a) can also be written in a form resembling an eigenvalue equation

$$[p_x, A_k] = \hbar k A_k. \tag{14.18d}$$

The operators A_k have the property of adding a momentum $\hbar k$ to a state. Simple examples of such operators are $\mathrm{e}^{\mathrm{i}kx_j}$, $p_i \mathrm{e}^{\mathrm{i}kx_j}$ and $(\mathrm{e}^{\mathrm{i}qx_i})(\mathrm{e}^{\mathrm{i}(k-q)x_j})$ where x_i, x_j, p_i and p_j are coordinates and momenta of two particles in the many-particle system and q is arbitrary. These momentum eigenoperators satisfy momentum conservation selection rules, analogous to the parity selection rules satisfied by the parity eigenoperators. The matrix elements of the operators $A_{k'}$ between two momentum eigenstates $|k''\rangle$ and $|k'''\rangle$ vanish unless momentum is conserved. This can also be seen from the formal properties (14.18) of the operators:

$$\hbar k' \langle k'''|A_{k'}|k''\rangle = \langle k'''|[p_x, A_{k'}]|k''\rangle$$
$$= \hbar(k''' - k'')\langle k'''|A_{k'}|k''\rangle = 0 \quad \text{unless} \quad k' = k''' - k''. \tag{14.19}$$

The above treatment is easily extended to include translations in all directions in three-dimensional space. All translation operators commute with one another; the translation group is an *Abelian* group, and all translation operators can be simultaneously diagonalized. The generators of the infinitesimal translations in the three directions are just the momentum operators, and the commutation of the Hamiltonian with the translation operators implies momentum conservation.

There are two distinct cases, one in which the Hamiltonian commutes with the whole group of continuous translations; the other when it commutes only with a discrete set of translations. In the first case, there is complete momentum conservation, and the simultaneous eigenfunctions of the Hamiltonian and the translation operators are just plane waves. In the latter case, which corresponds to a system with the translational symmetry of a crystal lattice, momentum vectors can be defined 'modulo a reciprocal lattice' and its conservation is required allowing the addition of an arbitrary multiple of a reciprocal lattice vector. The eigenfunctions of the set of translation operators have a form analogous to (14.15) generalized to three dimensions, with a function $u(x, y, z)$ periodic with a period of the unit cell of the lattice.

The results for a one-particle system are easily generalized to a many-particle system, where the translation operation shifts the position of all particles by an equal amount and leads to conservation laws for the total momentum of the system.

The four features summarized in the discussion of parity clearly apply as well to translations. However, the translations are a stronger symmetry and impose stronger restrictions on the solutions of the Schrödinger equation. This is described more precisely as follows.

Space inversion is a single discrete transformation defined by an operator with only two eigenvalues. If the Hamiltonian is invariant under space inversion, the space of eigenfunctions is separated into two decoupled spaces, and the Hamiltonian can be diagonalized separately in each subspace. If we consider the Schrödinger equation as a differential equation, the parity symmetry allows us to consider even and odd functions separately. For each we have only to solve the differential equation in the interval $0<x<\infty$, and the symmetry gives us the value of the function for negative x. Thus in both the algebraic and the differential-equation approach, the parity symmetry allows us to separate the problem into two decoupled problems, each of 'half the complexity of the original one'.

Discrete translations constitute a discrete group of transformations defined by an infinite discrete set of commuting operators, having a continuous bounded spectrum of eigenvalues. A Hamiltonian invariant under these translations can be diagonalized separately in an infinite number of decoupled subspaces, each labeled by a different eigenvalue of the translation operator. The Schrödinger differential equation need be solved only in the finite interval from $-\frac{1}{2}a$ to $+\frac{1}{2}a$, and the values elsewhere are given by the symmetry through the relation (14.15). Thus the discrete translation symmetry separates the problem into an infinite number of decoupled problems, each 'infinitely simpler' than the original.

The full continuous translation group is defined by a Hermitean operator which is the generator of the infinitesimal translation and has a continuous unbounded spectrum. If the Hamiltonian is invariant under these translations, it is completely diagonalized for one degree of freedom by this symmetry. For a problem with only one degree of freedom, like motion of a single particle in one dimension, translation invariance completely determines the eigenfunctions, namely plane waves. For a many-particle system, translation invariance for each dimension separates one degree of freedom from the rest, namely center-of-mass motion and the Schrödinger equation needs to be solved only for the remaining degrees of freedom.

14.4 Parity and translations, rotations and Lie groups

Consider a Hamiltonian invariant under both space inversion and translations; i.e.

$$[H, P] = [H, T_a] = 0. \tag{14.20}$$

We now can define a new group of transformations including both the translations, space inversion, and the operators $P(T_a)^n$ which are combinations of a space inversion and a translation. Following our previous example, we can look for simultaneous eigenfunctions of H and both P and T. However, we now have a new feature in this group of transformations. It is no longer Abelian.

$$T_a P = P T_{-a} \tag{14.21a}$$

$$[P, T_a] \neq 0. \tag{14.21b}$$

Momentum and parity do not commute. A momentum eigenstate does not have a definite parity, and a parity eigenstate does not have a definite momentum (except for the trivial case of zero momentum). Thus we can find solutions of the Schrödinger equation for this system which have either a definite momentum or a definite parity, but not both. The solutions with a definite momentum are those in which the center-of-mass motion is described by a traveling plane wave, e^{ikx}. The solutions with a definite parity are those in which the center-of-mass motion is described by a standing plane wave, $\cos kx$ or $\sin kx$. The parity and momentum eigenstates having the same energy eigenvalue are clearly related by a simple linear transformation. The eigenvalue spectrum of the Hamiltonian is characterized by a two-fold degeneracy. For each momentum eigenvalue $k>0$, there are two degenerate eigenfunctions of H having the form $e^{ikx}\phi_\alpha$ and $e^{-ikx}\phi_\alpha$ where ϕ_α describes all the other degrees of the system except center-of-mass motion. We have chosen the momentum eigenstates. The corresponding parity eigenstates are $\phi_\alpha \sin kx$ and $\phi_\alpha \cos kx$.

From this example we see that when a Hamiltonian commutes with two operators which do not commute with one another, the eigenvalue spectrum of the Hamiltonian consists of degenerate multiplets. Furthermore, we can determine the characteristics of these multiplets without knowing anything more about the details of the Hamiltonian and the dynamics of the system. If the Hamiltonian is invariant under symmetry operations, such as space inversion or translations, then an eigenfunction of the Hamiltonian is transformed into another eigenfunction with the same eigenvalue by these transformations. If there are two non-commuting symmetry operations, like space inversion and translation, both operations cannot leave a state invariant (except for special cases like zero momentum), and therefore the successive operation with these transformations creates new states which constitute a multiplet of degenerate eigenfunctions of the Hamiltonian. The properties of the multiplets which can arise are determined by the relations between the different non-commuting symmetry operations, and are independent of further properties of the Hamiltonian.

Let us now examine our particular example in more detail and show formally how many properties of the eigenfunctions of H follow from the interplay of the translation and space inversion transformations.

Consider an eigenfunction of H which is an eigenfunction of T_a with the eigenvalue e^{ika},

$$H|k\rangle = E|k\rangle \tag{14.22a}$$

$$T_a|k\rangle = e^{ika}|k\rangle. \tag{14.22b}$$

Then

$$HP|k\rangle = EP|k\rangle \tag{14.22c}$$

$$T_a P|k\rangle = PT_{-a}|k\rangle = e^{-ika}P|k\rangle. \tag{14.22d}$$

The spectrum of eigenfunctions of H thus consists of degenerate doublets, which are eigenfunctions of T_a with eigenvalues $\exp(\pm ika)$. For the case of continuous translations, this corresponds to states having equal and opposite momenta. The two states go into one another under space inversion.

The one exception to the rule that all eigenvalues are doubly degenerate is the case $k=0$ where $e^{ika}=1=e^{-ika}$. The states (14.22b) and (14.22d) are the same and are eigenfunctions of P as well as of T_a. This occurs, even though P and T_a do not commute, because the commutator $[P, T_a]=0$, when acting on a $k=0$ state. The state $|k=0\rangle$ is an eigenfunction of $[P, T_a]$ with eigenvalue zero.

The eigenfunctions of H separate into degenerate multiplets each characterized by a number $k \geqq 0$. The multiplets are doublets if $k \neq 0$ and singlets if $k=0$. In the representation we have chosen, the operator p_x is diagonal, while the operator P is 'almost diagonal'; i.e. P has non-vanishing matrix elements only between states within the same multiplet. We could have chosen a representation in which P would be diagonal; then p_x would be 'almost diagonal'.

Let us specify the eigenfunctions of H by the following quantum numbers: the magnitude of the momentum $\hbar k'$, the sign of the momentum, $\sigma' = \pm 1$, and a set of quantum numbers α' which specify the other degrees of freedom of the system. The quantum number k' specifies the kind of multiplet containing the state. The quantum number σ' specifies the particular state within the multiplet. The matrix elements of the operators p_x and P are completely specified in this representation:

$$p_x|k'. \sigma', \alpha'\rangle = \hbar\sigma' k'|k', \sigma', \alpha'\rangle \tag{14.23a}$$

$$P|k', \sigma', \alpha'\rangle = |k', -\sigma', \alpha'\rangle. \tag{14.23b}$$

The operator p_x is diagonal with the eigenvalue $\hbar\sigma' k'$. The operator P is 'almost diagonal', with matrix elements of magnitude unity between states of the same multiplet.

Instead of choosing the sign of the momentum to specify the state within the multiplet, we can choose the parity. For this case,

$$p_x|k', P', \alpha'\rangle = \hbar k'|k', -P'. \alpha'\rangle \tag{14.24a}$$

$$P|k', P', \alpha'\rangle = P'|k', P', \alpha'\rangle. \tag{14.24b}$$

Here P is diagonal and p_x is almost diagonal.

The parity and momentum 'eigenoperators' also form multiplets when we consider parity and momentum together. For each operator $A_{k'}$ satisfying eq. (14.18) we can define a companion $A_{-k'}$.

$$A_{-k'} = PA_{k'}P \tag{14.25a}$$

$$[p_x, A_{-k'}] = [p_x, PA_{k'}P] = -\hbar k' A_{-k'}. \tag{14.25b}$$

The operator multiplets $(A_{k'}, A_{-k'})$ have a structure resembling the wave function multiplets.

We now find a very important relation between matrix elements of operator multiplets between sets of states belonging to multiplets. Consider the matrix element

$$\langle k', \sigma', \alpha'|A_{k'''\sigma'''}|k'', \sigma'', \alpha''\rangle. \tag{14.26}$$

If we consider all the matrix elements of the two components of the operator multiplet $A_{k'''}$ between states of the multiplets (k', α') and (k'', α'') there are a total of eight independent matrix elements. We shall see that these are all proportional to a single quantity depending upon the properties of the system, with proportionality factors depending only upon the algebra of the operators P and p_x. First we note that no more than two of the eight matrix elements can differ from zero, because the momentum conservation relation (14.19) requires that $k'\sigma' = k''\sigma'' + k'''\sigma'''$. The two non-vanishing matrix elements are equal since

$$\langle k', \sigma', \alpha' | A_{k'''\sigma'''} | k'', \sigma'', \alpha'' \rangle = \langle k', \sigma', \alpha' | P(PA_{k'''\sigma'''}P)P | k'', \sigma'', \alpha'' \rangle$$
$$= \langle k', -\sigma', \alpha' | A_{k''', -\sigma'''} | k'', -\sigma'', \alpha'' \rangle. \qquad (14.27)$$

We can thus write

$$\langle k'\sigma'\alpha' | A_{k'''\sigma'''} | k''\sigma''\alpha'' \rangle = V(k', \sigma', k'', \sigma'', k''', \sigma''') \langle k'\alpha' \| A_{k'''} \| k''\alpha'' \rangle \qquad (14.28)$$

where $V(k', \sigma', k'', \sigma'', k''', \sigma''')$ is a coefficient depending only upon the parity and momentum quantum numbers $(k', k'', k''', \sigma', \sigma'', \sigma''')$ and independent of the other quantum numbers (α', α'') and the particular nature of the operator A. The double-barred 'reduced matrix element' $\langle k'\alpha' \| A_{k'''} \| k''\alpha'' \rangle$ depends only upon the multiplets, but is independent of the quantum numbers which specify the particular members of the multiplets. In this case all the coefficients V vanish except for the two corresponding to values of the argument which satisfy momentum conservation, and $V=1$ for these cases. This result (14.28) is a simplified version of the Wigner–Eckart theorem.

In this simple example we have seen how certain symmetry properties of a Hamiltonian lead to many useful results. This can be summarized in a form which has a more general validity:

Whenever the Hamiltonian of a physical system is invariant under two or more transformations which do not commute with one another, one can define a group of non-commuting transformations (a set of non-commuting operators) under which the Hamiltonian is invariant. In this case:

1. The eigenvalue spectrum of the Hamiltonian consists of degenerate multiplets.

2. The structure of the possible multiplets (singlet and doublets in the parity–momentum example, $2j+1$-plets for angular momentum) is determined completely by the relations of the transformations among themselves and is independent of the detailed properties of the Hamiltonian. Singlets must be eigenfunctions of all commutators with the eigenvalue zero.

3. The Hamiltonian can be diagonalized in a representation in which all the operators of the group are either diagonal or 'almost diagonal', they have non-zero matrix elements only between states which are members of the same degenerate multiplet. The matrix elements of these operators are determined completely by the algebra of the operators and are independent of the specific details of the Hamiltonian.

4. 'Operator multiplets', generally called 'irreducible tensor operators' can be defined by analogy with the wave function multiplets. These have the same structure as the wave function multiplets.

5. The matrix elements of members of a given irreducible tensor operator between states of two multiplets are all proportional to one (in some special cases more than one) reduced matrix element which is independent of the quantum numbers specifying the particular member of the multiplet. The coefficient is independent of the details of the wave functions and operators and depends only on the quantum numbers associated with the symmetry group. These are called Wigner coefficients or generalized Clebsch–Gordan coefficients.

6. There are simple rules for combining multiplets which depend only upon the algebra of the group.

The well-known results of the continuous non-commutative rotation group follow directly from the above considerations. If the Hamiltonian commutes with rotation operators, then angular momentum is conserved, as the angular momentum operators generate infinitesimal rotations. The energy spectrum of a rotationally invariant Hamiltonian consists of multiplets, with the well-known structure of $2j+1$ states, characterized by the quantum numbers j and m. The $j=0$ multiplet is a singlet and is an eigenfunction of all the commutators with eigenvalue zero. The 'operator multiplets' are the irreducible tensor operators which add a given amount of angular momentum to a state in a manner determined by angular momentum coupling rules. These coupling rules are used to combine multiplets. The relations between matrix elements are just the triangular inequality selection rules and the Wigner–Eckart theorem.

Similar results are obtained for other continuous groups of transformations (Lie groups). The algebra of the generators can be used to define the allowed multiplet structure and generalized irreducible tensor operators can be defined whose matrix elements satisfy a generalized Wigner–Eckart theorem. Some examples are isospin, SU(3) in particle physics, the SU(3) symmetry of the three-dimensional harmonic oscillator and the O(4) symmetry of the hydrogen atom.

14.5 Complex conjugation and time reversal

Let ψ be a solution of the time-dependent Schrödinger equation

$$H\psi(x, t) = i\hbar \frac{\partial \psi}{\partial t}. \tag{14.29a}$$

Then if H is real, the complex conjugate of (14.29a) is

$$H\psi^*(x, t) = -i\hbar \frac{\partial \psi^*}{\partial t} = i\hbar \frac{\partial \psi^*}{\partial(-t)}. \tag{14.29b}$$

Thus $\psi^*(x, -t)$ is also a solution of the Schrödinger equation and corresponds to a 'time-reversed' motion. The two wave functions are equal, except for a phase, at time $t=0$. At any time t, the value of one wave function t is equal to the value of the other at the time $-t$, except for a phase. Thus if one state represents a traveling wave packet, the other represents a wave packet traveling in the opposite direction.

There are complications in the formal description of this transformation because complex conjugation is not a linear operation. If we define an operator K_0 by the relation

$$K_0\psi = \psi^* \tag{14.30a}$$

then

$$K_0(a_1\psi_1 + a_2\psi_2) = a_1^*\psi_1^* + a_2^*\psi_2^* = a_1^*K_0\psi_1 + a_2^*K_0\psi_2. \tag{14.30b}$$

The operator K_0 is not a linear operator and does not commute with c-numbers unless they are real.

$$K_0 a = a^* K_0 \tag{14.31}$$

for any number a. Such an operator is called an *antilinear* operator. Repeating complex conjugation gives the identity

$$K_0^2 = 1. \tag{14.32}$$

The eigenvalues of K_0 are ± 1 for functions which are pure real or pure imaginary. The phase of the eigenvalue has no physical meaning since it can be changed by multiplying the eigenfunction by i. An antilinear operator with eigenvalue of modulus unity is called antiunitary.

The behavior of operators under complex conjugation is given by

$$K_0 A K_0 = A^* \tag{14.33}$$

where A^* means *complex conjugate*, not Hermitean conjugate. Thus the momentum operator $\boldsymbol{p} = -i\hbar\boldsymbol{\nabla}$ is imaginary in the Schrödinger representation, although it is Hermitean in all representations. The coordinate operator $\boldsymbol{x}$ is both real and Hermitean. Thus

$$K_0 \boldsymbol{p} K_0 = -\boldsymbol{p} \tag{14.34a}$$

$$K_0 \boldsymbol{x} K_0 = +\boldsymbol{x}. \tag{14.34b}$$

This agrees with the intuitive picture of time reversal, under which momentum is reversed, but position is not.

Whether an operator is real or imaginary depends upon the representation. In the momentum space representation, where $x = \mathrm{i}\hbar\,\partial/\partial p$, momentum would be real and the coordinate imaginary. Thus the formal representation of time reversal depends on the representation. In the Schrödinger configuration space representation, where x is real and p is imaginary, the operator K_0 is the time reversal operator.

If a Hamiltonian is real, it is invariant under time reversal (14.33), and a complete set of real eigenfunctions can be found. This can be shown as follows: For any eigenfunction ψ of H,

$$H\psi = E\psi, \tag{14.35a}$$

then

$$H\psi^* = E\psi^*. \tag{14.35b}$$

If ψ is not real, then the real combinations $\psi + \psi^*$ and $\mathrm{i}(\psi - \psi^*)$ can be chosen for the complete set.

The existence of a real basis has only been shown (14.35) for the case of spinless particles. The situation is somewhat different when spin is included, as shown below.

14.6 Gauge transformations

Consider the Hamiltonian for a free particle in a magnetic field

$$H = (\boldsymbol{p} - e\boldsymbol{A}/c)^2/2m. \tag{14.36}$$

If the vector potential $\boldsymbol{A}$ is a constant, independent of position, then there is no electromagnetic field, since all derivatives of $\boldsymbol{A}$ vanish. For this case, the Hamiltonian (14.36) must be equivalent to that of a free particle; i.e. it should be obtainable from the free particle Hamiltonian by a gauge transformation.

The eigenfunctions of (14.36) with constant $\boldsymbol{A}$ are clearly plane waves

$$\psi_{\boldsymbol{k}} = \mathrm{e}^{\mathrm{i}\boldsymbol{k}\cdot\boldsymbol{x}} \tag{14.37a}$$

$$H\psi_{\boldsymbol{k}} = E\psi_{\boldsymbol{k}} = [(\hbar\boldsymbol{k} - e\boldsymbol{A}/c)^2/2m]\psi_{\boldsymbol{k}}. \tag{14.37b}$$

The eigenfunctions are the same as those for a free particle, but the relation between the energy and the wave number $\boldsymbol{k}$ is not the same. This can be understood by remarking that the kinetic momentum is not the same as the canonical momentum $\boldsymbol{p}$, in the presence of a vector potential $\boldsymbol{A}$. For this case, the velocity operator is

$$\boldsymbol{v} = \mathrm{i}[H, \boldsymbol{x}]/\hbar = (\boldsymbol{p} - e\boldsymbol{A}/c)/m \tag{14.38a}$$

and

$$E = \tfrac{1}{2}mv^2. \tag{14.38b}$$

The energy thus has the usual form when expressed in terms of the velocity $\boldsymbol{v}$, but *not* when expressed in terms of the wave vector $\boldsymbol{k}$. The energy can be made to have the usual form when expressed in terms of the wave vector by performing the following canonical transformation:

$$\psi \to \mathrm{e}^{\mathrm{i}S}\psi \tag{14.39a}$$

where

$$S = -e\boldsymbol{A}\cdot\boldsymbol{x}/\hbar c. \tag{14.39b}$$

The state (14.37a) then transforms as follows:

$$\mathrm{e}^{\mathrm{i}\boldsymbol{k}\cdot\boldsymbol{x}} \to \mathrm{e}^{\mathrm{i}\boldsymbol{k}'\cdot\boldsymbol{x}}, \tag{14.40a}$$

where

$$\boldsymbol{k}' = \boldsymbol{k} - e\boldsymbol{A}/\hbar c = m\boldsymbol{v}/\hbar. \tag{14.40b}$$

The new wave number $\boldsymbol{k}'$ now has the usual relation to the velocity. The transformation (14.39) transforms the Hamiltonian (14.36) into the usual form for a free particle

$$\mathrm{e}^{\mathrm{i}S}H\mathrm{e}^{-\mathrm{i}S} = p^2/2m. \tag{14.41}$$

The canonical transformation (14.39) is thus just the gauge transformation required to eliminate from the Hamiltonian the non-physical vector potential which produces no observable fields.

The gauge transformation (14.39b) is a special case of a canonical transformation which represents two ways of describing the same physical state, *not two different states of the same system*. The Hamiltonian is not invariant under this transformation, and the eigenfunctions are not generally invariant either, although in this trivial case they remain plane waves. The gauge transformation differs from the general canonical transformation in that *all physically measurable quantities remain invariant in form* under the gauge transformation. Thus the concept of gauge invariance has a meaning, whereas nothing remains invariant under the transformation from cartesian to polar coordinates.

Consider the general gauge transformation (14.39a), where

$$S(x) = -e\boldsymbol{\nabla}\phi\cdot\boldsymbol{x}/\hbar c, \tag{14.42}$$

and ϕ is any scalar function of $\boldsymbol{x}$. Such a transformation multiplies the wave function by a phase factor, and therefore does not change the probability density. It also does not change the form of any function of $\boldsymbol{x}$. The form of the Hamiltonian is changed, but only in a way which indicates a change in the vector potential with no change in the electromagnetic fields. Gauge transformations are thus transformations which change *unobservable* dynamical variables. They add gradients of scalars which do not change fields to electrodynamic potentials, change absolute phases of wave functions, and change canonical (as opposed to kinetic) momenta. Although it may seem surprising to refer to momentum as not being observable, it is the kinetic momentum, $m\dot{x}$ which is observable. The canonical momentum p is observable only in the conventional representation where $p=m\dot{x}$. As long as no magnetic fields are present, a gauge can always be chosen in which the kinetic and canonical momenta are equal. In the presence of magnetic fields, this is no longer possible.

The wave vector $\boldsymbol{k}$ and the De Broglie wavelength of a particle are *not gauge-invariant quantities*. Intuitive pictures of interfering De Broglie waves must therefore be accepted with reservations when there are magnetic fields present. With no field, the gauge with no vector potential is preferable physically to other gauges, such as that of the Hamiltonian (14.36) which introduce a non-physical potential. For this case the De Broglie wavelength can be defined unambiguously. If a field is present, there is no a priori choice of gauge, and thus no unambiguous value for the De Broglie wavelength!

14.7 Spin and internal degrees of freedom

We now examine the behavior of spin and other internal degrees of freedom under the above transformations. For a composite particle with a well-defined internal structure described by a wave function, e.g. an alpha particle, the transformation can be applied directly to the internal wave function. The transformation of a state of a composite particle is the product of two partial transformations, an 'orbital' part which transforms the center-of-mass motion of the particle and an 'intrinsic' part which transforms the internal structure relative to the center-of-mass. For an 'elementary' particle, or one whose structure is not known, similar separation into orbital and intrinsic parts is assumed, but the intrinsic transformations are not defined a priori and must be considered in each individual case.

The parity of a state of a composite particle is the product of the intrinsic parity and the orbital parity. If the internal structure is not known, the intrinsic parity may be determined from experiment if the particle is produced or absorbed in processes where parity is conserved. The neutral pion, for example, has been shown experimentally to have odd intrinsic parity.

The intrinsic parity of a particle which is produced or absorbed only with the emission or absorption of other particles cannot be measured experimentally. The intrinsic parity of an electron cannot be measured, since electrons are only produced or annihilated in pairs. The total intrinsic parity of an electron–positron pair can be measured, and is found experimentally to be odd. The intrinsic parity of a charged pion cannot be measured absolutely, since it cannot be emitted or absorbed without a change in charge elsewhere in the system. A proton can emit a π^+, for example, and change into a neutron. Experiment shows that the pion has odd intrinsic parity, if the neutron and proton have the *same* intrinsic parity, whereas the pion has even intrinsic parity, if the neutron and proton have opposite intrinsic parity. Experiment cannot distinguish between these two possibilities. By convention, odd parity is chosen for π^+ and π^-, since this gives them the same parity as the π^0 (which can be directly measured) and also gives the neutron and proton the same parity.

A particle need not have a well-defined intrinsic parity. A simple classical example of such a particle is an organic molecule with a complicated internal helical structure. A space inversion transforms this 'particle' into another particle with left and right-handed helices interchanged. If the Hamiltonian of the system is *invariant under space inversion*, these two 'mirror image' states must have the same energy (mass). 'Mass doublets' of this type must occur whenever a particle exists which is not an eigenstate of an intrinsic parity operation which commutes with the Hamiltonian.

The angular momentum of a particle is the sum of the orbital angular momentum and the intrinsic angular momentum or spin. The spin transforms under space inversion like any angular momentum. Since P commutes with orbital angular momentum, it commutes with all spin operators as well.

The spin operators also transform under rotation like any angular momentum. The operators generating infinitesimal rotations of the system are thus the operators of the *total* angular momentum of the system, the sum of the spin and orbital angular momenta. The transformation under rotations of wave functions for particles with spin is then described completely by the action of these finite rotation operators on the wave functions.

A peculiar ambiguity arises for particles with half-integral spin. The phase of the rotated state is not single-valued. The operator generating infinitesimal rotations about the x-axis for example is the component in the x-direction of the total angular momentum, J_x. A rotation of any state ψ by an angle θ is then given by

$$\psi \rightarrow e^{iJ_x\theta}\psi. \tag{14.43}$$

Consider the case $\theta=2\pi$. A state of a system with integral spin is a linear combination of eigenfunctions of J_x with integral eigenvalues. Thus $e^{2\pi i J_x}$ acting on such a state is $+1$ and leaves the state invariant, as expected for a rotation by 2π. However, a state of a system with *half*-integral spin is a linear combination of eigenfunctions of J_x with *half*-integral eigenvalues. Here $e^{2\pi i J_x}$ acting on such a state is -1 and reverses its phase. A rotation by 2π should not change the state but multiplies it by a phase factor -1. Rotations by angles θ, $\theta-2\pi$ and $\theta+2\pi$ are equivalent for any physical quantity but produce different transformed wave functions which can differ by a phase factor of -1.

This ambiguity does not affect measurable physical quantities. Matrix elements of operators between two states always depend quadratically on wave functions in processes where no particles are created or absorbed. Since particles of half-integral spin are always created and absorbed in pairs, an even number of wave functions for such particles always appear in transition matrix elements for creation and absorption processes. Thus multiplying all wave functions of particles with half-integral spins by factors of -1 does not change physical results. The transformation of particles of half-integral spin under rotation is therefore expressed as a double-valued operation on the wave functions which does not lead to ambiguities in observable quantities.

Since rotations by 2π can change the phases of wave functions without affecting physical results, one can ask whether similar phase changes can occur under other transformations which are physically equivalent to the identity. The square of the parity operator P^2 brings the system back to its original state and could be assumed for some spin $\frac{1}{2}$ particles to be equivalent to a rotation by 2π. The sign of the wave function would reverse under P^2, for such particles, thus giving them an 'imaginary intrinsic parity' of $+i$. Such possibilities have been considered but have not led to any useful results. The square of the time reversal operation can also be defined to reverse phases of certain states without changing physical results, as discussed below.

The intrinsic angular momentum is a vector and must transform under rotations like any other vector. However the orientation of the angular momentum vector can be defined either with respect to an external coordinate system or to an 'intrinsic' coordinate system fixed in the system. Such a choice arises in the description of the motion of a top in quantum mechanics and in molecules and deformed nuclei. The components of the angular momentum of the top can be defined either in the directions of the laboratory axes x, y and z, or in the directions of 'intrinsic' axes fixed in the top, e.g. the principal axes of inertia, and moving relative to the coordinate system.

The former depend upon the direction of the coordinate system and therefore transform under rotations, the latter do not depend upon the direction of the coordinate system and do not change under rotations.

The spin transforms under time reversal like any angular momentum. It changes sign under time reversal. The formal definition of the time reversal operator (14.30) must be generalized to be the product of the complex conjugation operator K_0 and an appropriate spin operator. Since K_0 is anti-unitary its operation depends upon the choice of representation and some representation must be chosen for the spin operators before an explicit form can be chosen for the time reversal operator. The common choice makes S_x and S_z real and S_y imaginary (cf. the Pauli spin matrices).

For this representation

$$K_0 S_x K_0 = S_x \tag{14.44a}$$

$$K_0 S_y K_0 = -S_y \tag{14.44b}$$

$$K_0 S_z K_0 = S_z. \tag{14.44c}$$

Complex conjugation reverses S_y but not S_x nor S_z. The generalized time reversal operator K must therefore be the product of K_0 and an operator which reverses S_x and S_z without reversing S_y; i.e. rotation by π about the y-axis.

$$K = e^{i\pi S_y} K_0. \tag{14.45a}$$

Then

$$KSK^{-1} = -S \tag{14.45b}$$

as desired. For spin one-half eq. (14.45a) reduces to

$$K = i\sigma_y K_0. \tag{14.45c}$$

Since iS_y is real, K_0 commutes with iS_y and with $e^{i\pi S_y}$. Thus since $K_0^2 = 1$,

$$K^2 = e^{2\pi i S_y} = \pm 1. \tag{14.46}$$

If S is integral, $K^2 = 1$, as for spinless particles. However, if S is half-integral, then $K^2 = -1$. Since S is the *total spin* for the case of a system of many particles, time reversal has a different character for systems containing even or odd numbers of particles of half-integral spin.

For integral S, the argument based on eqs. (14.35) can be generalized to construct a complete set of *real* eigenfunctions of any Hamiltonian invariant under time reversal. Instead of eq. (14.35)

$$H\psi = E\psi \tag{14.47a}$$

$$H(K\psi) = E(K\psi). \tag{14.47b}$$

The states $(1+K)\psi$ and $i(1-K)\psi$ are thus simultaneous eigenfunctions of H and of K, and are both *real*; i.e. they have positive eigenvalue of K.

For half-integral S, $K^2=-1$, and we again encounter an operation which does not change the physical system but changes the phase of a half-integral spin state. Because K is antilinear, *no eigenfunctions of K exist*. To prove this, suppose ψ is an eigenfunction of K with eigenvalue c. Then

$$K^2\psi = KK\psi = Kc\psi = c^*K\psi = c^*c\psi \neq -\psi. \qquad (14.48)$$

For any eigenfunction of a Hamiltonian invariant under time reversal one can construct another eigenfunction $K\psi$, eqs. (14.47a) and (14.47b). Since ψ cannot be an eigenfunction of K for half-integral S the two states (14.47a) and (14.47b) must be different (they are in fact orthogonal) and the eigenvalue has a two-fold degeneracy. All eigenvalues have this two-fold degeneracy in a system of half-integral total spin invariant under time reversal. This is called Kramers degeneracy.

Thus for systems of integral spin, a complete set of simultaneous eigenfunctions of H and K can be found (a real basis). For systems of half-integral spin, no such real basis exists, since K has no eigenfunctions, but all states are members of degenerate Kramers doublets which go into one another under time reversal.

14.8 Transformations of Dirac spinors

The Dirac electron has internal degrees of freedom which are described theoretically, but which have no classical analog. There is therefore no way a priori to define the actions of transformations like space inversion, rotation and time reversal on these degrees of freedom. The intrinsic properties of the Dirac electron are determined by requiring that known conservation laws hold in processes involving electrons, including electron pair creation and annihilation. The intrinsic part of the transformation is then defined to make the Dirac Hamiltonian invariant under the total transformation.

Let us now examine transformations of the Dirac Hamiltonian. We use the Dirac representation, with ϱ and σ-matrices

$$H_{\mathrm{D}} = \varrho_1(\boldsymbol{\sigma}\cdot\boldsymbol{p})c + \varrho_3 mc^2. \qquad (14.49)$$

By inspection, H_{D} commutes with $\rho_3 P$ but not with the orbital parity operator P alone. If we define ρ_3 as the intrinsic parity operator and $\rho_3 P$ as the total parity, H_{D} is invariant under the 'total space inversion' operation.

H_{D} is manifestly invariant under rotations if $\boldsymbol{\sigma}$ is a vector and ρ_1 and ρ_3 are invariant under rotations.

The definition of time reversal depends on the representation. We choose one in which σ_x and σ_z are real, σ_y imaginary, ρ_1 and ρ_3 and ρ_2 imaginary. Then with the definition (14.45b) the Dirac Hamiltonian is invariant under time reversal.

$$KH_D K^{-1} = H_D. \tag{14.50}$$

The addition of an external electromagnetic field appears to introduce complications in time reversal invariance. Both the non-relativistic and Dirac Hamiltonians seem to lose time reversal invariance when an external vector potential is added.

$$(\boldsymbol{p} - e\boldsymbol{A}/c)^2 \rightarrow (\boldsymbol{p} + e\boldsymbol{A}/c)^2 \tag{14.51a}$$

$$\varrho_1[\boldsymbol{\sigma}\cdot(\boldsymbol{p} - e\boldsymbol{A}/c)] \rightarrow \varrho_1[\boldsymbol{\sigma}\cdot(\boldsymbol{p} + e\boldsymbol{A}/c)]. \tag{14.51b}$$

Time reversal invariance is restored if physical magnetic fields are produced by currents, since time reversal reverses these currents and therefore all magnetic fields. We can take this into account by defining an operator K_A which reverses the vector potential

$$K_A \boldsymbol{A} K_A = -\boldsymbol{A}. \tag{14.52a}$$

For the case where all magnetic fields are produced by currents, we define the 'total time reversal operator'

$$K^{(t)} = K_A K. \tag{14.52b}$$

Then

$$K^{(t)} H_D [K^{(t)}]^{-1} = H_D \tag{14.52c}$$

even in the presence of external fields.

The existence of an elementary magnetic monopole would violate time reversal invariance in this sense (or would require that either the elementary electric charge or magnetic pole be odd under time reversal).

14.9 Charge conjugation

Consider the time-dependent Dirac equation in the presence of an external electromagnetic field.

$$\{\rho_1[\boldsymbol{\sigma}\cdot(\boldsymbol{p}c - e\boldsymbol{A})] + \rho_3 mc^2 + e\phi\}\,\psi = \mathrm{i}\hbar\frac{\partial\psi}{\partial t}. \tag{14.53}$$

The operator ρ_2 is defined to anticommute with both ρ_1 and ρ_3. Let us define

$$K_C = \rho_2 K = \mathrm{i}\rho_2 \sigma_y K_0 \tag{14.54}$$

where K is the time reversal operator (14.45b) for any state ψ satisfying the Dirac equation (14.53),

$$\{\rho_1[\boldsymbol{\sigma}\cdot(\boldsymbol{p}c + e\boldsymbol{A})] + \rho_3 mc^2 - e\phi\}\, K_C\psi = i\hbar \frac{\partial}{\partial t} K_C\psi. \tag{14.55}$$

The state $K_C\psi$ thus satisfies a Dirac equation for a particle with the *opposite sign of electric charge.* The operator K_C is sometimes called the 'charge conjugation' operator.

That an operator having the simple form (14.54) should produce charge conjugation may seem surprising. Consider, for example the case where $A=0$ and $\phi=-e/r$ as in a hydrogen atom. The operator K_C contains only spinor transformations and complex conjugation and does not change the probability density in space. How can this operator change a hydrogen atom wave function into one describing a *positron* in the external field of a proton? Such a wave function describes a positron *bound* to the proton with a probability density which is a maximum at the proton and decreases exponentially at large distances. Try to understand this paradox before reading further!

In a non-relativistic theory, the electron and the positron are described by completely different Hamiltonians and wave equations. An electron is attracted by a positively charged proton; a positron is repelled. There is no solution of the non-relativistic Schrödinger equation for an electron in a given external field which can be transformed in a simple way to get the corresponding solution for a positron. Yet, in the relativistic case, the simple transformation (14.54) changes a solution of the Dirac electron equation into a solution of the Dirac positron equation.

However, this apparent one-to-one correspondence between electron and positron states is not a real correspondence between physical states. The solutions of the Dirac electron equation include the set of negative-energy states which cannot be interpreted as states of a physical electron. These formal solutions can only be given a physical interpretation by use of the hole theory or quantum field theory. Examination of the transformation generated by the operator K_C (14.54) shows that for any eigenfunction ψ of the Dirac Hamiltonian, the state $K_C\psi$ is indeed an eigenfunction of the corresponding Hamiltonian with the electric charge reversed, but the eigenvalue of the Hamiltonian has the opposite sign.

$$H_D(e)\psi = E\psi \tag{14.56a}$$

$$H_D(-e)K_C\psi = -EK_C\psi. \tag{14.56b}$$

The transformation K_C thus transforms a *positive energy* solution of the *electron* Dirac equation into a *negative energy* solution of the *positron* Dirac equation, and vice versa. It either transforms a physical state of an electron into a formal solution of the positron equation which has negative energy, or transforms a formal solution of the electron equation with negative energy into a solution of the positron equation which can represent a physical state. It is always a transformation between two formal solutions of Dirac equations, only *one* of which is *directly* interpreted as a physical state. This transformation is thus qualitatively different from those of translations, rotations, parity and time reversal as discussed above. It does not transform a physical state into another different physical state. The negative-energy electron state can only be interpreted as describing in some way a physical positron state, and this is just the state obtained from it by the transformation (14.54). The transformation is thus only between two different descriptions of the same physical state which appear as different formal solutions of different Dirac equations.

The relation with the non-relativistic theory is now clear. In the non-relativistic limit the positive and negative-energy states are completely decoupled in the Dirac equation. The result is a combination of two theories, one for electrons and one for positrons. The two theories become coupled in the relativistic region where pair production can take place. The transformation (14.53)–(14.56) merely shows that the Dirac equation describes both positrons and electrons and *does not indicate any invariance principle.*

On the other hand, one can also formulate a charge conjugation transformation between two different physical states. Suppose that the electron and the proton in a hydrogen atom were transformed into a positron and an antiproton. This 'antihydrogen atom' should have very similar properties to the hydrogen atom, since the electrodynamic forces are identical to those in a hydrogen atom. This is a transformation between two different physical states which are both solutions of the same dynamical equations of quantum mechanics. However, this transformation is not the same as the transformation (14.55) which considers the electron as moving in a given external field unchanged by the transformation. If the system under consideration is a hydrogen atom, the transformation (14.55) changes the electron into a positron while leaving the field in which it moves to be that of a proton. The two systems related by this transformation have completely different dynamics, and there is no one-to-one correspondence between their eigenfunctions. The hydrogen atom has an infinite number of bound states while the positron–proton system has none.

The formal transformation (14.45)–(14.56) can be modified to obtain a charge conjugation transformation which agrees with the above intuitive picture. We follow a procedure analogous to that used for the time reversal transformation of external magnetic fields. If all electromagnetic fields are produced by charges, a complete charge conjugation transformation of all particles in the universe also reverses all electromagnetic potentials, $\boldsymbol{A}$ and ϕ. We can then define a total charge conjugation operator

$$K_{\mathrm{C}}^{(\mathrm{t})} = K_{\mathrm{E}} K_{\mathrm{C}} \tag{14.57}$$

where K_{E} is the electromagnetic charge conjugation operator which reverses the electromagnetic potentials

$$K_{\mathrm{E}} \boldsymbol{A} K_{\mathrm{E}} = -\boldsymbol{A} \tag{14.58a}$$

$$K_{\mathrm{E}} \phi K_{\mathrm{E}} = -\phi. \tag{14.58b}$$

Then

$$K_{\mathrm{C}}^{(\mathrm{t})} H_{\mathrm{D}} [K_{\mathrm{C}}^{(\mathrm{t})}]^{-1} = -H_{\mathrm{D}}. \tag{14.59}$$

Thus for any state ψ which is an eigenfunction of the Dirac Hamiltonian H_{D}

$$H_{\mathrm{D}} \psi = E \psi \tag{14.60a}$$

$$H_{\mathrm{D}} K_{\mathrm{C}}^{(\mathrm{t})} \psi = -E K_{\mathrm{C}}^{(\mathrm{t})} \psi. \tag{14.60b}$$

Thus the total charge conjugation operation transforms an eigenfunction of H_{D} into another eigenfunction with the opposite sign of eigenvalue.

Note the difference between eqs. (14.60) and the similar eqs. (14.56). In eqs. (14.60) both equations refer to the *same* Hamiltonian, and two different eigenfunctions of the same Hamiltonian must refer to different physical states. In eqs. (14.56) the two equations refer to *different* Hamiltonians, and two eigenfunctions are merely different descriptions of the same physical state. In eqs. (14.56) the electromagnetic potentials appear as external parameters in the Hamiltonian, unchanged by the charge conjugation transformation. In eqs. (14.60) they are dynamical variables, reversing their sign under charge conjugation.

Suppose the state ψ in eqs. (14.50a) and (14.60a) represents the ground state of a hydrogen atom. Then the state described by eq. (14.56b) is a state of a *positron of negative energy* moving in the field of the proton. If this state is interpreted as a 'hole state', it is a hole in a sea of normally filled negative energy *positron* states since it is a solution of the Dirac equation for positrons. The state represents an antipositron; i.e. a negative electron, moving in the field of the proton in the hydrogen atom ground state. On the other hand, the state described by eq. (14.60b) is a state of a *negative energy electron* moving in the field of opposite sign from that of a proton; i.e. that which would be produced by an antiproton. If this state is interpreted as a 'hole state' it is a hole in a sea of negative electrons, since it is a solution of the Dirac equation for electrons, not for positrons. It represents a positron, moving in the field of an antiproton in the ground state of the 'antihydrogen atom'. Thus the state described by eq. (14.56b) is exactly the same hydrogen atom state described by eq. (14.56a), while the state described by eq. (14.60b) is the physically charge conjugate antihydrogen atom state.

The transformations of parity, time reversal and charge conjugation are considered together in relativistic quantum field theory. It is possible to prove on very general grounds that such a theory must be invariant under the product of all three of these transformations. This is called the *CPT* theorem. However, it is not required that the theory be invariant separately under *C*, *P* and *T* individually. The strong and electromagnetic interactions seem to be invariant under all three, whereas the weak interactions associated with nuclear beta decay are now known not to be invariant under *C* and *P* individually, and also not invariant under the product *CP*.

PROBLEMS

1. Let K be an *antiunitary* operator satisfying the relation $K^4 = 1$. Discuss the classification of quantum-mechanical states according to eigenfunctions of this operator, the classification of operators, relations between matrix elements of operators. If the Hamiltonian of a system commutes with K, what can be said about its eigenfunctions?

2. Let A, B, and C be three Hermitean operators satisfying the commutation rules:

$$[A, B] = -\mathrm{i}C; \qquad [C, A] = \mathrm{i}B; \qquad [B, C] = \mathrm{i}A.$$

Note that these differ from ordinary angular momentum commutation rules by a sign in the first commutator. Consider a system whose Hamiltonian H commutes with both A and B; i.e.

$$[H, A] = [H, B] = 0,$$

A. If ψ is a *non-degenerate* eigenfunction of H, what conditions must ψ satisfy.

B. Show that any eigenfunction of H which does not satisfy the conditions given in part A must have an infinite degeneracy.

3. A particle with electric charge e and mass m is constrained to move in a thin cylinder of radius R and thickness δr. There is a magnetic flux through the center of the cylinder, but no magnetic field in the region where the electron moves. This is described by a constant vector potential $\boldsymbol{A}$ which is everywhere tangent to the surface of the cylinder and normal to the axis; i.e. it is in the θ-direction in cylindrical coordinates.

A. Write the Hamiltonian for this system in cylindrical coordinates.

B. Calculate the ground state wave function and the ground state energy as a function of the strength of the vector potential $\boldsymbol{A}$. Calculate the electric current flowing around the cylinder.

C. Since there is no magnetic field at the electron, changing the strength of the vector potential does not change the forces on the electron. Yet it changes the ground state energy. Is this energy change observable? Explain your answer.

4. Consider a system of identical bosons of mass m and charge e in a cylinder with inner radius R_1 and outer radius R_2. Assume that all the bosons are in the lowest state corresponding to a given vector potential $A(r)$, which is normal to r and z; i.e. in the θ-direction, and depends only on r. Assume that the radial and axial parts of the wave function are constants, independent of r and z, and that the wave function depends only on θ. Let the number of particles per unit volume be n, assumed to be constant over the cylinder.

A. Calculate the ground state energy and the electric current density in the cylinder as a function of r.

B. Assume that there is no external magnetic field, and that all of the magnetic field in the cylinder is produced by the circulating current due to the bosons. Use Maxwell's equations to relate the vector potential to the current density and find self-consistent solutions in part A, where the vector potential $A(\boldsymbol{r})$ is just that produced by the bosons themselves.

C. What are the physical implications of the results of part B.

5. Consider the rotation of an asymmetric body or system about the center-of-mass (e.g. an ellipsoidal top). Let $\boldsymbol{V}_1$, $\boldsymbol{V}_2$ and $\boldsymbol{V}_3$ be three orthogonal unit vectors fixed in the system (e.g. the directions of the principal axes of the ellipsoid), whose orientations in space are dynamical variables depending upon the coordinate specifying the motion of the system. Let $\boldsymbol{J}$ be the total angular momentum of the system.

A. Using the fact that $\boldsymbol{V}_1$, $\boldsymbol{V}_2$ and $\boldsymbol{V}_3$ are vectors, write the commutation relations between the three components of the total angular momentum J_x, J_y and J_z and the three components of each vector V_{1x}, V_{1y}, V_{1z}, V_{2x}, V_{2y} V_{2z}, V_{3x}, V_{3y} and V_{3z}.

B. Show that the projections of the angular momentum on the directions of these vectors,

$$J_1 = \boldsymbol{J}\cdot\boldsymbol{V}_1, \quad J_2 = \boldsymbol{J}\cdot\boldsymbol{V}_2 \quad \text{and} \quad J_3 = \boldsymbol{J}\cdot\boldsymbol{V}_3,$$

all commute with the angular momentum operators J_x, J_y and J_z.

C. Show that the operators J_1, J_2 and J_3 satisfy commutation rules like angular momenta with opposite signs.

D. Suppose that the operators J_1, J_2 and J_3 are considered as generators of infinitesimal rotations. What is the difference between the rotations generated by these operators and those generated by J_x, J_y and J_z. How is it that J_1 and J_x commute, although they are components of the angular momentum in different directions? Hint: consider the commutativity of the corresponding rotations.

CHAPTER 15

THE LORENTZ GROUP

15.1 A symmetry generator with explicit time dependence

Lorentz transformations are symmetry transformations which mix space and time. To consider these we first consider the special properties of a symmetry generator which depends explicitly on the time. Let $\Psi(t)$ be a solution of the time-dependent Schrödinger equation

$$H\Psi(t) = \mathrm{i}\hbar \frac{\partial \Psi}{\partial t} \tag{15.1a}$$

for a quantum-mechanical system having a Hamiltonian H. Let G be an operator that is a generator of a symmetry transformation. We then require that an infinitesimal transformation generated by the operator G transforms the state Ψ into a state that is also a solution of the Schrödinger equation,

$$\begin{aligned} H(1+\mathrm{i}vG/\hbar)\Psi(t) &= \mathrm{i}\hbar \frac{\partial}{\partial t}(1+\mathrm{i}vG/\hbar)\Psi \\ &= -v\frac{\partial G}{\partial t}\Psi + \mathrm{i}\hbar(1+\mathrm{i}vG/\hbar)\frac{\partial \Psi}{\partial t} \end{aligned} \tag{15.1b}$$

where v is an infinitesimal constant, and the partial time derivative $\partial G/\partial t$ refers to the explicit dependence of G on t as a parameter. Multiplying eq. (15.1a) on the left by $1+\mathrm{i}vG/\hbar$ and subtracting it from eq. (15.1b), we obtain

$$[H, G]\Psi(t) = \mathrm{i}\hbar \frac{\partial G}{\partial t}\Psi(t). \tag{15.2a}$$

If this condition is to hold for any state Ψ, the condition on G becomes the operator relation*

* More precisely, the condition (15.2a) must hold at all times t for any state $\Psi(t)$ which satisfies the time-dependent Schrödinger equation (15.1a). For such states all time derivatives acting on Ψ in eq. (15.2a) can be eliminated by substituting eq. (15.1a) and replacing

$$[H, G] = \mathrm{i}\hbar \frac{\partial G}{\partial t}. \tag{15.2b}$$

This condition can also be written

$$\mathrm{e}^{\mathrm{i}\alpha G}\left[H - \mathrm{i}\hbar \frac{\partial}{\partial t}\right]\mathrm{e}^{-\mathrm{i}\alpha G} = H - \mathrm{i}\hbar \frac{\partial}{\partial t} \tag{15.2c}$$

or

$$\dot{G} \equiv \frac{\partial G}{\partial t} + \frac{\mathrm{i}}{\hbar}[H, G] = 0, \tag{15.2d}$$

where $\dot{G}$ denotes the total time derivative in the Heisenberg representation.

Each of the equivalent equations (15.2) expresses the condition to be satisfied by an operator G if it generates transformations which connect solutions of the time-dependent Schrödinger equation only to other states that are also solutions. In the form (15.2c), *the Schrödinger equation of motion is invariant under the transformations generated by the operator G*. In the form (15.2d), *the total time derivative of G is zero* – i.e., it is a constant of the motion even though it may depend explicitly on time. In all future discussion we use the Schrödinger representation in which the state vectors depend on time and operators are time-dependent unless they depend explicitly on t as a paramer.

From eq. (15.2b) an operator G that satisfies eqs. (15.1) and does not depend explicitly on the time must commute with the Hamiltonian. This reduces to the case of time-independent symmetry transformations considered in ch. 14. If G depends explicitly on the time, eq. (15.2b) shows that G cannot commute with H. If Ψ is an eigenfunction of H, $(1+\mathrm{i}vG/\hbar)\Psi$ *cannot be an eigenfunction of H with the same eigenvalue*. A time-dependent operator G satisfying eqs. (15.1) must therefore generate sets of states having different energies and provide information about the spectrum of H.

The case of linear time dependence is of particular interest, since the generators of Galilean and Lorentz transformations all have this form*

$$G = \hbar J - Pt, \tag{15.3a}$$

$\partial\Psi/\partial t$ by $H\Psi/\mathrm{i}\hbar$. In this form we can require eq. (15.2a) to hold at some fixed time $t=t_1$ for a complete set of time-independent functions $\Psi(t_1)$, since all these functions can be continued in time to construct a solution of the Schrödinger equation (15.1a). Eq. (15.2b) then follows.

* This is seen as follows: (1) *In a non-relativistic many-body problem*, the coordinates of the particles are dynamical variables that do not depend explicitly on the time. Under infinitesimal Galilean or Lorentz transformations, they are transformed into linear functions of

where J and P are both time-independent Hermitean operators. It is common to set $\hbar=1$ in treatments of this type. We resist this temptation, in order to keep track of dimensions. Instead we introduce $\hbar$ in expressions like eq. (15.3a) to give the operator G the dimensions of action (or angular momentum) while J is dimensionless and P has dimensions of momentum. Substituting (15.3a) into the condition (15.2a) we obtain

$$[H, G] = \hbar[H, J] - [H, P] = -i\hbar P. \tag{15.3b}$$

Since eq. (15.3b) must hold for all values of t, we can equate the constant and linear terms separately. Thus

$$[H, P] = 0 \tag{15.4a}$$

$$[H, G] = \hbar[H, J] = -i\hbar P. \tag{15.4b}$$

Invariance of the equations of motion (15.1) under the transformations generated by an operator G that depends linearly on the time necessarily requires the coefficient of the time to be an operator that commutes with the Hamiltonian. For example, if G is the generator of Galilean or Lorentz transformations in the x-direction, the operator P is proportional to the momentum in the x-direction. Momentum conservation is *automatically* implied by the requirement of Galilean or Lorentz invariance, and translations *must enter* into physical applications of the homogeneous Lorentz group. *Lorentz invariance without translational invariance is impossible.*

In order to proceed further we need to know the commutation relations between J and P. Let us define the commutator

$$\hbar[J, P] = [G, P] \equiv i\hbar C. \tag{15.5a}$$

In any special case, where J and P are given, C can be calculated. However, for all cases, eqs. (15.4) show that

$$[H, C] = [[H, G], P] + [G, [H, P]] = 0. \tag{15.5b}$$

Thus C is a constant of the motion and is either a new operator or some function of H and P. In the cases of practical interest, Lorentz and Galilean

the time. Since time appears in this description as a parameter and not as a dynamical variable, the generators of the transformations must be linear functions of the time.

(2) *In a relativistic quantum field theory*, the total energy and total momentum are constants of the motion and transform like a four-vector under Lorentz transformations. The commutator of a Lorentz generator with the Hamiltonian, which is a component of this four-vector, is thus equal to a linear combination of the other components of the four-vector which are all independent of time. Substituting a time-independent commutator into eq. (15.2b) shows that the generator must be a linear function of time.

transformations, C commutes with P. We therefore assume

$$[C, P] = 0. \tag{15.5c}$$

The operators H, P and C can therefore be diagonalized simultaneously to give a basis $|E', P', C', \alpha'\rangle$ of simultaneous eigenfunctions,

$$H|E', P', C', \alpha'\rangle = E'|E', P', C', \alpha'\rangle \tag{15.6a}$$

$$P|E', P', C', \alpha'\rangle = P'|E', P', C', \alpha'\rangle \tag{15.6b}$$

$$C|E', P', C', \alpha'\rangle = C'|E', P', C', \alpha'\rangle, \tag{15.6c}$$

where α' denotes the eigenvalues of other operators which commute with G and which are necessary for a complete specification of the state.

We now examine the infinitesimal transformation (15.1b) on these states. To first order in v,

$$\begin{aligned} H(1 + ivG/\hbar)|E', P', C', \alpha'\rangle &= [(1 + ivG/\hbar)H + iv[H, G]/\hbar]|E', P', C', \alpha'\rangle \\ &= (1 + ivG/\hbar)(H + iv[H, G]/\hbar)|E'P', C', \alpha'\rangle, \end{aligned} \tag{15.7a}$$

where we have simplified the expression by adding a second order term in v to enable the removal of the factor $1+ivG/\hbar$. Similarly

$$\begin{aligned} &P(1 + ivG/\hbar)|E', P', C', \alpha'\rangle \\ &\quad = (1 + ivG/\hbar)(P + iv[P, G]/\hbar)|E', P', C', \alpha'\rangle. \end{aligned} \tag{15.7b}$$

Substituting eqs. (15.4) and (15.5),

$$H(1 + ivG/\hbar)|E', P', C', \alpha'\rangle = (1 + ivG/\hbar)(E' + vP')|E', P', C', \alpha'\rangle \tag{15.8a}$$

$$P(1 + ivG/\hbar)|E', P', C', \alpha'\rangle = (1 + ivG/\hbar)(P' + vC')|E', P', C', \alpha'\rangle \tag{15.8b}$$

The state $(1+ivG/\hbar)|E', P', C', \alpha'\rangle$ is thus a simultaneous eigenfunction of H and P with eigenvalues differing from E' and P' by amounts proportional to the continuous infinitesimal parameter v. The infinitesimal transformation (15.7) thus generates a continuous spectrum of simultaneous eigenfunctions of H and P. The changes in the eigenvalues given by eqs. (15.8) can be expressed by the differential equations,

$$\frac{\partial E'}{\partial v} = P' \tag{15.9a}$$

$$\frac{\partial P'}{\partial v} = C'. \tag{15.9b}$$

Multiplying eqs. (15.9a) and (15.9b) we obtain

$$P' \frac{\partial P'}{\partial v} = C' \frac{\partial E'}{\partial v}. \tag{15.9c}$$

So far we have not specified the commutator $[C, J]$. In the two cases of direct physical interest, (1) Galilean and (2) Lorentz transformations, the operator C turns out respectively to be (1) a c-number and (2) an operator proportional to the Hamiltonian H. We therefore consider these two cases in detail, defining the constants of proportionality to give the result in the conventional form.

Case 1: $C = C' = M$. We can solve the differential equations (15.9) to obtain

$$P' = M(v - v_0) \tag{15.10a}$$

$$E' = \tfrac{1}{2}M(v - v_0)^2 + E_0 \tag{15.10b}$$

where v_0 and E_0 are integration constants. Thus

$$E' = E_0 + P'^2/2M \tag{15.10c}$$

This has the familiar form of the non-relativistic relations between energy, velocity, and momentum.

Case 2: $C = H/c^2$, $C' = E'/c^2$. For this case eq. (15.9c) gives

$$P'^2 + M^2c^2 = E'^2/c^2 \tag{15.11}$$

where M^2c^2 is an integration constant. This is just the relativistic relation between energy and momentum.

Thus the existence of an operator G depending linearly on the time and satisfying the commutation relation (15.2b) leads to an operator P that commutes with the Hamiltonian, to a continuous spectrum for both H and P, and to a continuous 'multiplet' of states obtained from one another by the transformations generated from the operator G. The eigenvalues of H and P in this continuous multiplet are related to one another in a manner determined by the algebra generated by the commutators of the operators H, J and P and give the relations between energy, momentum and velocity for Galilean and Lorentz transformations.

15.2 *G*-spin. Lorentz transformations in 2+1 dimensions

We now consider transformations in a (2+1)-dimensional space-time; i.e. Lorentz transformations in the x and y-directions and rotations in the xy-plane. These are described by a time-dependent quasi-spin whose transformations leave the Schrödinger equation invariant; i.e., its generators satisfy eq. (15.2b). We consider three operators G_{12}, G_{20} and G_{01}, which satisfy angular momentum commutation rules among themselves, and which each satisfy eq. (5.2b).

$$[G_{12}, G_{20}] = i\hbar G_{01}, \tag{15.12a}$$

$$[G_{20}, G_{01}] = i\hbar G_{12}, \tag{15.12b}$$

$$[G_{01}, G_{12}] = i\hbar G_{20}. \tag{15.12c}$$

The indices 1 and 2 refer to the spatial dimensions, the index 0 to time. For convenience, we choose the signs of the commutators to correspond to the ordinary angular momentum algebra; i.e. to rotations in a space with a positive definite metric. Our treatment applies to the Lorentz metric if G_{20} and G_{01} are anti-Hermitean; i.e. the Hermitean Lorentz generators are $\mathrm{i}G_{20}$ and $\mathrm{i}G_{01}$.

The two Lorentz generators depend linearly on the time, the rotation in the 12-plane is time-independent. Thus

$$G_{12} = \hbar J_{12}, \tag{15.13a}$$

$$\mathrm{i}G_{20} = \hbar J_{20} + P_2 t, \tag{15.13b}$$

$$\mathrm{i}G_{01} = \hbar J_{01} - P_1 t. \tag{15.13c}$$

The phase factors in eqs. (15.13b) and (15.13c) are chosen to provide a notation easily generalized to the physical four-dimensional Lorentz group, as shown below, and to make the operators J_{ij} and P_i all Hermitean.

Since the commutation relations (15.12) are those of ordinary angular momentum, these operators, which we call G-spin, can be manipulated in the conventional manner to obtain all results normally obtained from angular momentum commutation rules. However, since the operators are not Hermitean we must be careful to use only results following from the commutation rules *without assuming hermeticity*. We define the operators,

$$G^2 = G_{12}^2 + G_{20}^2 + G_{01}^2 \tag{15.14a}$$

$$G_+ = G_{20} + \mathrm{i}G_{01} = \hbar(J_{01} - \mathrm{i}J_{20}) - (P_1 + \mathrm{i}P_2) \tag{15.14b}$$

$$G_- = G_{20} - \mathrm{i}G_{01} = -\hbar(J_{01} + \mathrm{i}J_{20}) + (P_1 - \mathrm{i}P_2) \tag{15.14c}$$

$$G_0 = G_{12} = \hbar J_{12}. \tag{15.14d}$$

Operators can be classified according to their transformation properties in 'G-spin space'. These are determined by the commutation relations of the operators with the G-spin operators (15.14).

The commutation relations of the Hamiltonian with the G-spin operators are determined by the requirement that the Schrödinger equation is invariant under the G-spin transformations. Eqs. (15.3) and (15.4) then apply to all the G-spin operators. We thus obtain

$$[H, P_2] = [H, P_1] = [H, G_{12}] = [H, J_{12}] = 0, \tag{15.15a}$$

$$\mathrm{i}[H, G_{20}] = \hbar[H, J_{20}] = \mathrm{i}\hbar P_2, \tag{15.15b}$$

$$\mathrm{i}[H, G_{01}] = \hbar[H, J_{01}] = -\mathrm{i}\hbar P_1. \tag{15.15c}$$

Combining eqs. (15.14) and (15.15) we obtain

$$[G_+, H] = +\mathrm{i}\hbar(P_1 + \mathrm{i}P_2), \tag{15.16a}$$

$$[G_-, H] = -\mathrm{i}\hbar(P_1 - \mathrm{i}P_2), \tag{15.16b}$$

$$[G_0, H] = 0. \tag{15.16c}$$

These commutation relations show that H is not a G-spin scalar but could be the zero component of a G-spin vector. Relations (15.16a) and (15.16b) show that the other components of the vector are proportional to P_1 and P_2. This is just the energy–momentum vector for the particular example of the Lorentz group. Let us define the three components of this G-spin vector with the proper phases indicated by the commutators (15.16a) and (15.16b).

$$V_0 = H, \tag{15.17a}$$

$$V_1 = -\mathrm{i}P_1, \tag{15.17b}$$

$$V_2 = -\mathrm{i}P_2. \tag{15.17c}$$

Commutation relations between the J and P-operators are obtained by substitution into the commutation rules (15.12) and equating the coefficients of equal powers of the time,

$$[J_{12}, J_{20}] = \mathrm{i}J_{01} \qquad [J_{12}, J_{01}] = -\mathrm{i}J_{20}, \tag{15.18a}$$

$$[J_{20}, J_{01}] = -\mathrm{i}J_{12}, \tag{15.18b}$$

$$[J_{12}, P_2] = -\mathrm{i}P_1 \qquad [J_{12}, P_1] = \mathrm{i}P_2, \tag{15.18c}$$

$$[P_1, P_2] = 0, \tag{15.18d}$$

$$-\hbar[J_{20}, P_1] = \hbar[J_{01}, P_2] = -\mathrm{i}[G_{20}, P_1] = \mathrm{i}[G_{01}, P_2]. \tag{15.18e}$$

The three operators J_{12}, J_{20} and J_{01} satisfy the commutation relations of dimensionless angular momentum operators except for the sign in eq. (15.18b). This sign difference expresses the Lorentz metric rather than the usual possitive definite metric. We also define commutators, by analogy with eq. (15.5) between the G and P-variables. These are determined by the requirement that H, P_1, and P_2 are the components of the vector, the energy–momentum vector, and do not contain components of higher rank. Thus

$$[G_+, (P_1 + iP_2)] \equiv 0 \tag{15.19a}$$

$$[G_-, (P_1 - iP_2)] \equiv 0 \tag{15.19b}$$

$$i[G_+, (P_1 - iP_2)] = i[G_-, (P_1 + iP_2)] = 2\hbar H. \tag{15.19c}$$

From eqs. (15.15a) and (15.17c) the invariance of the Schrödinger equation under the transformations generated by the time-dependent operators (15.18) requires the existence of two operators P_1 and P_2 that commute with one another and with the Hamiltonian. These are just the momenta in the 1 and 2 directions which generate the corresponding translations. Thus an Abelian group of translations is very intimately associated with any time-dependent Lie algebra. One might think that it would be useful to find the simultaneous eigenfunctions of G^2 and G_{12} by analogy with ordinary angular momentum. However, G^2 does not commute with the Hamiltonian,

$$[H, G^2] = \hbar[-P_1G_{01} - G_{01}P_1 + G_{20}P_2 + P_2G_{20}] \neq 0. \tag{15.20}$$

Thus, H and G^2 cannot be simultaneously diagonalized, and the eigenfunctions of G^2 have no simple physical meaning. The G-spin multiplets cannot be used for classifying the eigenfunctions of H. This explains why the multiplets or irreducible representations of the *homogeneous* Lorentz group are never used to classify states of particles or quantum-mechanical systems. One must always take into account the translations and consider the *inhomogeneous Lorentz group*, also called the *Poincaré group*. We must consider the simultaneous eigenfunctions of H, P_1, and P_2 and examine the multiplets generated by the 'inhomogeneous algebra' that includes the translations P_1 and P_2 as well as the homogeneous algebra (15.12).

Two new G-spin scalars can be constructed from scalar products of the vector V with itself and with the G-spin generators,

$$V^2 = V_0^2 + V_1^2 + V_2^2 = H^2 - (P_1^2 + P_2^2), \tag{15.21a}$$

$$\begin{aligned}(\boldsymbol{G}\cdot \boldsymbol{V}) &= G_{12}V_0 + G_{20}V_1 + G_{01}V_2 \\ &= G_{12}H - i(G_{20}P_1 + G_{01}P_2).\end{aligned} \tag{15.21b}$$

From eqs. (15.15) we see that both scalars (15.21) commute with the Hamiltonian H, and with P_1 and P_2.

$$[H, V^2] = [H, (\boldsymbol{G}\cdot \boldsymbol{V})] = 0 = [P_i, V^2) = [P_i, (\boldsymbol{G}\cdot \boldsymbol{V})]. \tag{15.21c}$$

15.3 The structure of G-spin multiplets

We now consider the detailed structure of the multiplets generated by our 'inhomogeneous' G-spin algebra which includes the 'translations' P_1 and P_2. These consist of the set of all simultaneous eigenfunctions of H, P_1 and P_2 which can be generated from a particular eigenfunction by G-spin transformations. The operators V^2 and $\boldsymbol{G}\cdot \boldsymbol{V}$ can be diagonalized simultaneously with H, P_1 and P_2. These operators are invariant under 'G-spin rotations'. Their eigenvalues remain constant for all states in a multiplet and can therefore be used to characterize the whole multiplet in the same way that the eigenvalues of J^2 are used to characterize an angular momentum multiplet.

To evaluate the eigenvalues of V^2 and $\boldsymbol{G}\cdot \boldsymbol{V}$, which are the same for all states in a multiplet, we can choose any convenient state. The most convenient state is the one with the zero eigenvalue for P_1 and P_2. Such a state always exists, since the operators P_1 and P_2 have been shown to have unbounded continuous spectra. For this state,

$$V^2 = H^2 \qquad \text{if} \quad P_1 = P_2 = 0 \tag{15.22a}$$

$$\boldsymbol{G}\cdot \boldsymbol{V} = G_{12}H = \hbar J_{12}H \quad \text{if} \quad P_1 = P_2 = 0. \tag{15.22b}$$

The physical significance of the multiplets and the eigenvalues (15.22) is easily seen. The state $P_1 = P_2 = 0$ corresponds to the rest frame. The eigenvalue of V^2 is the square of the energy of the zero-momentum state, i.e. the square of the rest energy or mass. We denote it by M^2. The operator G_{12} is the angular momentum operator in the 12-plane. Its eigenvalue for the zero-momentum 'rest' state is the angular momentum at rest, usually called the intrinsic 'spin' of the state. The eigenvalue of the operator $\boldsymbol{G}\cdot \boldsymbol{V}$, which we denote by $MS\hbar$ is the product of the mass and spin of the state.

The states of these multiplets are classified by giving the eigenvalues of the set of operators $(V^2, \boldsymbol{G}\cdot \boldsymbol{V}; P_1, P_2)$. We denote these states as $|M, S; P_1, P_2\rangle$. The eigenvalues of V^2 and $\boldsymbol{G}\cdot \boldsymbol{V}$ are the same for all states in the multiplet and are functions of the mass and spin of the state. For simplicity we use the labels M, S, rather than the eigenvalues M^2, $MS\hbar$, in the same way that the label j denotes the eigenvalue $j(j+1)\hbar^2$. The operators P_1 and P_2 label the individual states within a multiplet, like the magnetic quantum number in an angular momentum multiplet. The multiplet thus consists of a zero-momentum rest state, with well-defined mass and spin, and all states of finite momentum obtained by 'boosting' this state to any arbitrary value of P_1 and P_2. The inhomogeneous G-spin multiplets are generalizations of angular momentum multiplets in that they have two labels for the multiplet, M and S instead of one label J and two labels to specify the state in the multiplet, P_1 and P_2 instead of one label M. Also P_1 and P_2 have unbounded continuous spectra.

The states of finite momentum can be constructed from the rest by operating with the unitary operator corresponding to the appropriate boost transformation,

$$\exp[\alpha(G_{01}\cos\theta - G_{20}\sin\theta)]|M, S; 0,0\rangle$$
$$= |M, S, M\cos\theta\sinh(\alpha\hbar), M\sin\theta\sinh(\alpha\hbar)\rangle \qquad (15.23)$$

where the eigenvalues of P_1 and P_2 as functions of the parameters α and θ of the transformation are obtained directly from the commutation relations (15.15).

Eq. (15.23) shows how to construct all the states of a multiplet specified by given eigenvalues of V^2 and $\boldsymbol{G}\cdot\boldsymbol{V}$ (or equivalently labelled by the quantum numbers M and S). The matrix elements of all of the G-spin operators between any two states of the multiplet are obtainable by straightforward algebra using the commutation relations. The set of all these matrix elements constitute a representation of the algebra; i.e., a set of matrices which satisfy the desired commutation rules. Since G^2 is not diagonal and many eigenvalues of G^2 are included, the multiplet contains many G-spin multiplets; i.e. it is a *reducible* representation of the homogeneous algebra (15.12). However, it is an *irreducible representation* of the inhomogeneous algebra (including the momentum operators P_1 and P_2) and cannot be broken up into several smaller multiplets. Each set of eigenvalues of P_1 and P_2 appears only once, and all these states must be in the same multiplet, since any pair of eigenvalues can be transfromed into any other pair by a G-spin transformation.

In this simplified version of the Lorentz group, there is only a single angular momentum operator and only a single state in each multiplet for a given value of the momenta. For the physical four-dimensional Lorentz group there are three independent angular momenta and the multiplets may contain sets of states having the same eigenvalues of all momenta and differing from one another by rotations of the spin. In the present simple example the 'spin rotation' operator is just $(\boldsymbol{G}\cdot\boldsymbol{V})/M$ and reduces to G_{12} for the case at rest. This operator generates a trivial group of transformations which simply multiply any angular momentum eigenstate by a phase factor. This is the analog of the 'little group', in the Lorentz case, of transformations which leave the momentum invariant.

The states within a multiplet can be specified by the eigenvalues of the generator G_{12} and of $P^2 = P_1^2 + P_2^2$. This is simply the use of polar variables (cylindrical waves) instead of cartesian variables (plane waves). The eigenvalues of P^2 are continuous while those of $G_{12}=\hbar J_{12}$, denoted by $J\hbar$, are discrete. The latter follows from the observation that all states are obtained by operating on the rest state with functions of G-spin operators and these can only change the eigenvalue of J_{12} by integral units. We have not yet required the eigenvalues of the spin to be integral or half-integral, and a continuous spin spectrum is in principle possible. However, once the spin value is chosen for a given multiplet, the allowed values for the total angular momentum in the multiplet can differ from the spin value only by integral amounts.

The polar eigenstates are constructed directly from the cartesian states (15.25). We first note the behavior of the state (15.25) under a finite rotation in the 12-plane, remembering that the zero momentum state is an eigenfunction of J_{12} with the eigenvalue S. For convenience we use the dimensionless operator J_{12} rather than G_{12}

$$
\begin{aligned}
&\mathrm{e}^{-\mathrm{i}\beta J_{12}}|M, S, M\cos\theta\sinh(\alpha\hbar), M\sin\theta\sinh(\alpha\hbar)\rangle \\
&\quad = \mathrm{e}^{-\mathrm{i}\beta J_{12}}[\exp\alpha(G_{01}\cos\theta - G_{20}\sin\theta)]\mathrm{e}^{+\mathrm{i}\beta J_{12}}\mathrm{e}^{-\mathrm{i}\beta J_{12}}|M, S, 0, 0\rangle \\
&\quad = [\exp\alpha\{G_{01}\cos(\theta+\beta) - G_{20}\sin(\theta+\beta)\}]\mathrm{e}^{-\mathrm{i}\beta S}|M, S, 0, 0\rangle. \qquad (15.24)
\end{aligned}
$$

The rotation thus separates into an 'orbital' part which rotates the momentum and a 'spin' part which rotates the spin. We can now construct the angular momentum eigenfunction,

$$|M, S, P, J\rangle = N\int_0^{2\pi} \mathrm{d}\theta\, \mathrm{e}^{\mathrm{i}L\theta}|M, S, P_1 = P\cos\theta, P_2 = P\sin\theta\rangle \qquad (15.25)$$

where N is a normalization factor. The total angular momentum J is the sum of the orbital angular momentum L and the spin S, as is seen explicitly by operating on the state (15.25) with the operator J_{12},

$$J_{12}|M, S, P, J\rangle = (L+S)|M, S, P, J\rangle. \qquad (15.26)$$

Thus

$$J = L + S. \qquad (15.27)$$

This neat separation of orbital and spin angular momentum may seem surprising to anyone familiar with the Dirac equation. This point is clarified by examining the operation of a Lorentz transformation (a G-spin transformation generated by the operators G_{01} or G_{20}) on the state (15.23). For simplicity, we set $\theta=0$, let α be small, and consider an infinitesimal transformation generated by G_{20},

$$\mathrm{e}^{\gamma G_{20}}|M, S, M\alpha/\hbar, 0\rangle = \mathrm{e}^{\gamma G_{20}}\mathrm{e}^{\alpha G_{01}}|M, S, 0, 0\rangle. \qquad (15.28)$$

Expanding the exponential and keeping terms only to second order in the infinitesimals γ and α, we obtain

$$\begin{aligned} e^{\gamma G_{20}} e^{\alpha G_{01}} &\approx (1 + \gamma G_{20} + \tfrac{1}{2}\gamma^2 G_{20}^2)(1 + \alpha G_{01} + \tfrac{1}{2}\alpha^2 G_{01}^2) \\ &\approx 1 + (\gamma G_{20} + \alpha G_{01}) + \tfrac{1}{2}(\gamma G_{20} + \alpha G_{01})^2 + \tfrac{1}{2}\gamma\alpha [G_{20}, G_{01}] \\ &\approx e^{(\gamma G_{20} + \alpha G_{01})} e^{\frac{1}{2}\gamma\alpha\hbar G_{12}}. \end{aligned} \tag{15.29}$$

The second factor on the right-hand side of eq. (15.29) produces a spin rotation when acting on the zero-momentum state. The magnitude of this spin rotation depends on both parameters, γ and α. Substituting eq. (15.29) into eq. (15.28) we obtain

$$e^{\gamma G_{20}} |M, S, M\alpha\hbar, 0\rangle \approx e^{\frac{1}{2}i\gamma\alpha\hbar S} |M, S, M\alpha\hbar, M\gamma\hbar\rangle. \tag{15.30}$$

Thus, a Lorentz transformation on a state of the form (15.15) produces not only the expected transformation of the momentum, but also a *spin rotation whose magnitude depends both on the parameters of the Lorentz transformation and on the momentum of the state* to which the transformation is applied. Thus, although momentum and spin variables are neatly separated when we consider the effect of a rotation on the state (15.25), momentum and spin are mixed together in a complicated way by the action of the Lorentz transformation. This is just the opposite of the case of the Dirac equation where the action of the Lorentz transformation on space and spin variables is neatly separated but they are mixed together by the action of rotations. We shall return to this point after generalizing the results obtained thus far to the full Lorentz group.

15.4 The representations of the Lorentz group

We now consider the full Lorentz group in four-dimensional space-time. We first develop the algebra of operators generating the infinitesimal rotations in a Euclidean four-dimensional space with a positive definite metric. We denote the dimensions by 0, 1, 2 and 3. There are six generators corresponding to rotations in all possible planes defined by two axes. It is convenient to group these as follows.

$$G_{12} \quad G_{23} \quad G_{31} \tag{15.31a}$$

$$G_{03} \quad G_{01} \quad G_{02} \tag{15.31b}$$

A rotation in the $\mu\nu$-plane can be written either $G_{\mu\nu}$ or $G_{\nu\mu}$. Since these would be rotations in opposite directions, the generators are antisymmetric in the two indices,

$$G_{\mu\nu} = -G_{\nu\mu}. \tag{15.32a}$$

Their commutation relations are easily obtained, since the generators in any three-dimensional subspace satisfy angular momentum commutation rules

$$[G_{\mu\nu}, G_{\nu\varrho}] = i\hbar G_{\varrho\mu}. \tag{15.32b}$$

Rotations in two completely independent planes (no common index) must commute.

$$[G_{\mu\nu}, G_{\varrho\sigma}] = 0 \quad \text{if} \quad \varrho \neq \mu,\ \varrho \neq \nu,\ \sigma \neq \mu \text{ and } \sigma \neq \nu. \tag{15.32c}$$

Eqs. (15.32b) and (15.32c) specify all the commutators.

The three generators (15.31a) are the three angular momentum operators in the three-dimensional subspace labelled by indices 1, 2, and 3. The three operators (15.31b) can be considered as three components of a vector in the subspace, since they satisfy commutation relations with the operators (15.31a) like those of a vector.

It is convenient to define the sums and differences of the corresponding operators in the sets (15.31a) and (15.31b)

$$G_{ij}^{1} = \tfrac{1}{2}[G_{ij} + \varepsilon_{ijk}G_{0k}] \qquad (i = 1, 2, 3) \tag{15.33a}$$

$$G_{ij}^{2} = \tfrac{1}{2}[G_{ij} - \varepsilon_{ijk}G_{0k}] \qquad (i = 1, 2, 3). \tag{15.33b}$$

The two sets of operators $G^{(1)}$ and $G^{(2)}$ are seen to satisfy angular momentum commutation rules among themselves and to commute with one another.

$$[G_{ij}^{\alpha}, G_{jk}^{\alpha}] = i\hbar G_{ki}^{\alpha} \qquad (\alpha = 1, 2) \tag{15.34a}$$

$$[G_{ij}^{1}, G_{rs}^{2}] = 0 \qquad \text{for all } i, j, r, s. \tag{15.34b}$$

The operators (15.33a) and (15.33b) thus behave like two independent angular momenta, whose sum is just the total angular momentum in the 1, 2, 3-space.

$$G_{ij}^{1} + G_{ij}^{2} = G_{ij}. \tag{15.35}$$

…n thus define the squares of these two angular momenta:

$$(G^{\alpha})^2 = (G_{12}^{\alpha})^2 + (G_{23}^{\alpha})^2 + (G_{31}^{\alpha})^2; \qquad \alpha = 1, 2. \tag{15.36a}$$

…rs commute with all the operators (15.33) and therefore …1).

$$\ldots, G_{ij}^{\beta}] = 0 = [(G^{\alpha})^2, G_{\mu\nu}] \tag{15.36b}$$

where we use the convention that Greek indices on a generator G run through the values 0, 1, 2 and 3, while Latin indices take on the values 1, 2 and 3.

We thus see that the multiplets or irreducible representations defined by the four-dimensional rotation group can be labelled by the quantum numbers $(G^1G^2; M^1M^2)$ of the two independent angular momenta. Each multiplet is labelled by two numbers G^1 and G^2 and consists of $(2G^1+1)\times(2G^2+1)$ states. The states within the multiplet can be labelled by quantum numbers M^1 and M^2 representing the eigenvalues of one of the G^1 and one of the G^2-operators. They can also be labelled by the quantum numbers $(G^1, G^2; GM)$ obtained coupling the two angular momenta to a total angular momentum G, with projection M. The matrix elements of all the operators (15.31) between states in a multiplet are now completely determined by standard angular momentum algebra. Note that here, as in the 'inhomogeneous G-spin algebra' the multiplets require two quantum numbers to specify the multiplet and two quantum numbers to label the states within the multiplet. Algebras whose multiplets (irreducible representations) have this property are called rank-two algebras. The angular momentum algebra is rank one.

The separation of the algebra of the generators of the four-dimensional rotation group into the direct product of two angular momentum algebras is not a general feature of rotation groups. In the case of the five-dimensional rotation group, for example, this does not occur and more complicated algebraic manipulations are necesary to determine the structure of the multiplets.

We now consider the case of the Lorentz group. From our analysis of the corresponding group in three-dimensional space-time, we know how to treat any three-dimensional subspace of our four-dimensional space. The operators (15.31a) which describe spatial rotations are all independent of time, Hermitean and commute with the Hamiltonian. The operators (15.31b) which generate Lorentz transformations depend linearly on the time, are anti-Hermitean and satisfy a particular commutation relation (15.15) with the Hamiltonian. Eqs. (15.13) and (15.15) are easily generalized to give

$$G_{ij} = \hbar J_{ij} \tag{15.37a}$$

$$[H, G_{ij}] = 0 \qquad (1$$

$$\mathrm{i}G_{0j} = \hbar J_{0j} - P_j t$$

$$[H, G_{0j}] = -\hbar P_j$$

$$[H, P_j] = 0.$$

The operators P_j together with the Hamiltonian H, satisfy the commutation relations of a vector with the generators (15.31). In the four-dimensional case, this is a four-vector, whose components are defined with the phases

$$V_0 = H \tag{15.39a}$$

$$V_j = -\mathrm{i}P_j. \tag{15.39b}$$

The magnitude of this vector is again a scalar which commutes with all the generators (15.31) and also with the Hamiltonian H and is interpreted physically as a square of the mass.

$$M^2 = \sum_\mu V_\mu^2 = H^2 - (P_1^2 + P_2^2 + P_3^2). \tag{15.40}$$

We now consider the detailed structure of the multiplets or irreducible representations of the inhomogeneous Lorentz group. Again, as in eq. (15.20) the operators (15.36a) which label the multiplets of the homogeneous group do not commute with the Hamiltonian and cannot be used to classify states in the multiplet.

In the four-dimensional case we cannot form another invariant by taking the scalar product of the energy–momentum vector with the generators of the group because the generators are components of a second rank antisymmetric tensor which is different from a vector in four dimensions. The third-rank antisymmetric tensor which turned out to be the spin in the three-dimensional case is the axial vector

$$W_\mu = \tfrac{1}{2}\,\varepsilon_{\mu\nu\varrho\sigma} G_{\nu\varrho} V_\sigma. \tag{15.41}$$

This operator is seen to commute with the Hamiltonian

$$\begin{aligned}[H, W_\mu] &= \tfrac{1}{2}\{\varepsilon_{\mu 0\varrho\sigma}[H, G_{0\varrho}] + \varepsilon_{\mu\nu 0\sigma}[H, G_{\nu 0}]\}V_\sigma \\ &= \tfrac{1}{2}\mathrm{i}\hbar\,[\varepsilon_{\mu 0\varrho\sigma} P_\varrho P_\sigma + \varepsilon_{\mu\nu 0\sigma} P_\nu P_\sigma] = 0.\end{aligned} \tag{15.42a}$$

The square of the operator (15.41) is therefore a scalar invariant which commutes with all of the generators (15.31) as well as with the Hamiltonian H. Let us define

$$\sum_\mu W_\mu W_\mu \equiv \hbar^2 M^2 S^2. \tag{15.42b}$$

Then

$$[H, S^2] = [G_{\mu\nu}, S^2] = 0. \tag{15.42c}$$

The eigenvalues of the operator S^2 can then be used to label the multiplets.

The physical significance of the operators W_μ and S^2 can be seen by examining the rest states in which the operators P_1, P_2 and P_3 all have the eigenvalue zero. For this case we have:

$$W_0 = 0 \qquad (P_1 = P_2 = P_3 = 0) \quad (15.43a)$$

$$W_i = \tfrac{1}{2}\varepsilon_{ijk} G_{jk} V_0 = G_{jk} H = G_{jk} M \qquad (P_1 = P_2 = P_3 = 0). \quad (15.43b)$$

The space components of the operator W_μ are thus proportional to the angular momentum generators for the zero-momentum states. This is a natural generalization of the three-dimensional result (15.22b). The three operators W_i/M are thus the generators for the spin rotations of a state at rest and the operator S^2 defined by eq. (15.42b) is the total angular momentum of the zero-momentum state and is just the spin of the state.

Thus the irreducible representations of the Lorentz group consist of states having continuous spectra of the momenta P_1, P_2 and P_3. The multiplet as a whole is characterized by its mass and by its spin. The value of the mass determines the relation between the energy and the momenta, and the value of the spin determines the multiplicity of states at a given value of the momenta and also the behavior under rotations. The group of transformations generated by the operators W_i, namely the group of transformations corresponding to a fixed set of values of all momenta, is called the little group. For the case of the Lorentz group the little group is the ordinary three-dimensional rotation group. For the simpler case of Lorentz transformations in a three-dimensional space-time the little group is just that of rotation about a single axis.

Once, we have chosen the values of the mass and the spin, we can construct the whole irreducible representation corresponding to these values. We begin with the $2S+1$ zero-momentum states which go into one another under rotations. Lorentz transformations acting on these states change the momenta in a manner corresponding to the Lorentz transformation. We can determine the effect of these Lorentz transformations on the spin by using the algebra of the operators, exactly as in the three-dimensional case.

We can construct the representation in spherical coordinates by a generalization of the method used in the three-dimensional case. Since angular momenta are vectors, they must be combined with vector coupling techniques and are not simply additive, as in eq. (15.27). However, the result is essentially the same. The total angular momentum J is the vector sum of the orbital angular momentum L and the spin S and the state is an eigenfunction of L, in contrast to the impression received from the Dirac equation.

The combining of two successive Lorentz transformations is formally the same as in the three-dimensional case, as given by eq. (15.29). The product of two Lorentz transformations in two different directions cannot be expressed as a single Lorentz transformation in a direction at an appropriate angle between the two directions but is the product of such a Lorentz transformation and a rotation. However, the rotation is no longer a trivial phase factor but a real spin rotation. If one applies this Lorentz transformation to a zero-momentum state and then transforms the state back to zero momentum with a third Lorentz transformation in the appropriate direction, the rotation remains. This rotation is called a Wigner rotation.

15.5 *G*-spin multiplets in the Dirac spinor notation

The representations of the Lorentz group which we have developed are suitable for the description of free particles in momentum space. However, difficulties arise if we wish to describe interacting particles in configuration space. The simplest interaction is that of a single electrically charged particle with an external electromagnetic field. To calculate this interaction we need an expression for the charge and current densities carried by this particle. The charge-current density $j_\mu(x_\nu)$ has the form of a four-vector field. Under Lorentz transformations the index μ and the argument x transform separately and independently. The index transformation makes one component go into a linear combination of all four components, with coefficients independent of x, and the argument transforms like a four-vector independent of the indices. The same would be true for the Fourier transform of the current $j_\mu(p_\nu)$. Thus the Lorentz transformation of the charge-current density is expressed as the direct product of two independent transformations, one in space-time and the other in the 'intrinsic' index space.

The representations which we have obtained have the peculiar feature of eq. (15.29) that the space and intrinsic spin variables are mixed together in Lorentz transformations; the spin is rotated by an angle depending upon the momenta and not only on the parameters of the Lorentz transformation. The Dirac spinor representation has the feature of separating the spatial and spin degrees of freedom and allowing covariant local densities, like the charge-current density to be easily constructed.

We now develop the Dirac representation. We begin with the three-dimensional case and study the multiplet structure and the matrix elements of the *G*-spin operators in a notation where the wave function factorizes into a plane wave in the spatial variables, and a spinor which acts in another space $|\xi\rangle$ associated with spin. In the Dirac notation we write the state (15.23) as

$$|M, S, P_1, P_2\rangle^{\mathrm{D}} = |\xi\rangle |M, 0, P_1, P_2\rangle. \tag{15.44a}$$

The first factor on the right-hand side represents the spinor. The second factor is the plane wave. The momentum operators P_1 and P_2 act only on the second factor and not on the first, while the second factor has no spin; i.e. it is an eigenfunction of the operator $\boldsymbol{G}\cdot\boldsymbol{V}$ with the eigenvalue zero. The action of any G-spin transformation on this state factorizes into transformations on the two factors

$$e^{i\alpha G}|M, S, P_1, P_2\rangle^{D} = (e^{i\alpha G(s)}|\xi\rangle)(e^{i\alpha G(p)}|M, 0, P_1, P_2\rangle) \quad (15.44b)$$

where

$$(\boldsymbol{G}^{(p)}\cdot \boldsymbol{V})|M, 0, P_1, P_2\rangle = 0 \quad (15.44c)$$

and $G^{(s)}$ and $G^{(p)}$ act in the spin and momentum spaces respectively and are represented individually by matrices satisfying the G-spin commutation rules. The matrices representing $G^{(p)}$ are just the conventional G-spin matrix representation for spin zero states. In the auxiliary spin space we can choose any set of matrices which satisfy the G-spin commutation rules.

The state (15.44a) is an eigenfunction of all components of the vector $\boldsymbol{V}$

$$\boldsymbol{V}|M, S, P_1, P_2\rangle^{D} = \boldsymbol{V}_{PM}|M, S, P_1, P_2\rangle^{D} \quad (15.44d)$$

where $\boldsymbol{V}_{PM}$ represents the eigenvalues of the vector operator $\boldsymbol{V}$ for a state labelled by the quantum numbers P_1, P_2 and M. Thus using eqs. (15.44)

$$(\boldsymbol{G}\cdot \boldsymbol{V})|M, S, P_1, P_2\rangle^{D} = \{(\boldsymbol{G}^{(s)}\cdot \boldsymbol{V}_{PM})|\xi\rangle\}|M, 0, P_1, P_2\rangle. \quad (15.45a)$$

Since this state (15.44) must be an eigenfunction of $\boldsymbol{G}\cdot\boldsymbol{V}$ with the eigenvalue $MS\hbar$

$$(\boldsymbol{G}^{(s)}\cdot \boldsymbol{V}_{PM})|\xi\rangle = \hbar MS|\xi\rangle. \quad (15.45b)$$

This equation must be satisfied by the spinor $|\xi\rangle$.

The simplest non-trivial example of matrices satisfying the angular momentum algebra is the Pauli spin algebra. Let us set the three G-spin generators proportional to the corresponding Pauli matrices.

$$G_{12}^{(s)} = \tfrac{1}{2}\hbar\sigma_A \quad (15.46a)$$

$$G_{20}^{(s)} = \tfrac{1}{2}\hbar\sigma_B \quad (15.46b)$$

$$G_{01}^{(s)} = \tfrac{1}{2}\hbar\sigma_C. \quad (15.46c)$$

Substituting these into eq. (15.45b) we obtain

$$\tfrac{1}{2}\{\sigma_A E - i\sigma_B P_1 - i\sigma_C P_2\}|\xi\rangle = \tfrac{1}{2}M|\xi\rangle \quad (15.47a)$$

where E is the eigenvalue of H and we have chosen the eigenvalue $\frac{1}{2}$ for S, as is required for this case. Multiplying eq. (15.47a) by σ_A and rearranging terms we obtain

$$E|\xi\rangle = \{i(\sigma_A\sigma_B)P_1 + i(\sigma_A\sigma_C)P_2 + \sigma_A M\}|\xi\rangle$$
$$= \{-\sigma_C P_1 + \sigma_B P_2 + \sigma_A M\}|\xi\rangle. \tag{15.47b}$$

This can be written in a more familiar form by relabelling the Pauli matrices σ_A, σ_B and σ_C(as β, α_2, and $-\alpha_1$

$$E|\xi\rangle = (\alpha_1 P_1 + \alpha_2 P_2 + \beta M)|\xi\rangle. \tag{15.47c}$$

Thus the equation which the spinor must satisfy is just the Dirac equation.

We have represented the G-spin generators in the spin space (15.33) by Hermitean matrices, in apparent contradiction with the requirement that the operators G_{20} and G_{01} be anti-Hermitean. This is permissible because the matrices $G^{(s)}$ are not defined as operators in Hilbert space. A G-spin transformation on a physical state defined by eq. (15.44b) requires two simultaneous transformations, one in the space of the momentum variables and one in the auxiliary spin space. A formal transformation generated by the matrix $G_{20}^{(s)}$ acting only in the spin space without the corresponding transformation in momentum space is not a transformation in Hilbert space. It gives a state vector which has the form (15.44a) but is not a physical state, since it does not satisfy the condition (15.44b) or the equivalent Dirac equation (15.47) for the case of spin $\frac{1}{2}$. Since $G^{(s)}$ is not an operator in Hilbert space, the requirements of hermiticity for the full G-spin operator do not apply to the matrices $G^{(s)}$.

For given values of P_1, P_2 and M there are two solutions of eq. (15.47c), one with a positive and one with a negative eigenvalue. However at this stage we need not concern ourselves with the physical meaning of the negative-energy state since the two states are not connected to one another by G-spin transformations. It is perfectly consistent to choose the positive solution and throw away the negative one. The properties of the other solution are discussed below.

The use of the Dirac notation achieves a formal separation between the space and spin variables and allows Lorentz generators to be defined to act on each of these variables separately and independently. This is in apparent contradiction with the result (15.30) indicating that a Lorentz transformation also involves a Wigner spin rotation whose magnitude depends both on the parameters of the Lorentz transformation and on the momentum of the state. This coupling between spin and space is not lost, but now appears in the Dirac equation. In effect one has introduced an additional degree of freedom to describe the spin and restricted the dependence of the wave function on this variable by a subsidiary condition (15.45) or (15.46) which depends upon the momentum. Thus the dependence of the Lorentz transformation on the momentum appears in the momentum dependence of the spinor describing the spin state.

The Dirac representation does not exhibit the simple separation of orbital and spin angular momentum. This is seen by constructing the polar eigenfunctions.

$$
\begin{aligned}
|M, S, J, M\rangle^{\mathrm{D}} &= N \int_0^{2\pi} \mathrm{d}\theta\, \mathrm{e}^{\mathrm{i}L\theta} |M, S, P\cos\theta, P\sin\theta\rangle^{\mathrm{D}} \\
&= N \int_0^{2\pi} \mathrm{d}\theta\, \mathrm{e}^{\mathrm{i}L\theta} [\exp \alpha(G_{01}\cos\theta - G_{20}\sin\theta)] |M, S, 0, 0\rangle^{\mathrm{D}} \\
&= N \int_0^{2\pi} \mathrm{d}\theta\, \mathrm{e}^{\mathrm{i}L\theta} [\exp \alpha(G_{01}^{(s)}\cos\theta - G_{20}^{(s)}\sin\theta)] |\xi\rangle_0 \\
&\quad \times |M, 0, P\cos\theta, P\sin\theta\rangle
\end{aligned}
\tag{15.48}
$$

where $|\xi\rangle_0$ denotes the spinor for the state $P_1 = P_2 = 0$. Since the variable of integration θ appears *both* in the *space* and *spin* factors, the angular momentum eigenfunction does not separate into space and spin parts.

The mixing of different values of L is easily seen explicitly in the simple case where P and α are small. Then to first order in α,

$$
\exp[\alpha(G_{01}^{(s)}\cos\theta - G_{20}^{(s)}\sin\theta)] \approx 1 + \tfrac{1}{2}\alpha \mathrm{e}^{\mathrm{i}\theta}[G_{01}^{(s)} + \mathrm{i}G_{20}^{(s)}] + \tfrac{1}{2}\alpha \mathrm{e}^{-\mathrm{i}\theta}[G_{01}^{(s)} - \mathrm{i}G_{20}^{(s)}] \tag{15.49a}
$$

$$
\begin{aligned}
|M, S, J, M\rangle \approx N \int_0^{2\pi} \mathrm{d}\theta \{ &\mathrm{e}^{\mathrm{i}L\theta} + \tfrac{1}{2}\alpha \mathrm{e}^{\mathrm{i}(L+1)\theta}[G_{01}^{(s)} + \mathrm{i}G_{20}^{(s)}] \\
&+ \tfrac{1}{2}\alpha \mathrm{e}^{\mathrm{i}(L-1)\theta}[G_{01}^{(s)} - \mathrm{i}G_{20}^{(s)}]\} |\xi\rangle_0 |M, 0, P\cos\theta, P\sin\theta\rangle.
\end{aligned}
\tag{15.49b}
$$

The leading term on the right-hand side of the relation (15.49b) has the separated form of an orbital factor with angular momentum L and a spinor with angular momentum S. However, the additional terms which are linear in α ('small components') have the form appropriate to orbital angular momenta $L+1$ and $L-1$, with the appropriate G-spin step operators in the spinor space to change the angular momentum of the spinor part of the wave function by one unit to keep the same total angular momentum J. The state (15.49b) thus has the appearance of a state with a well-defined total angular momentum J but mixtures of orbital angular momenta L, $L+1$ and $L-1$. For the particular case of spin $\frac{1}{2}$ where the G-spin generators are Pauli matrices, one of the two linear terms in α vanish because $|\xi\rangle$ is a two-component spinor and is annihilated either by the raising operator or the lowering operator.

15.6 The Dirac notation for the Lorentz group. The Dirac equation

We now extend the Dirac spinor notation to the four-dimensional case; let us write

$$|M, S, M_s, P_1, P_2, P_3\rangle^{\mathrm{D}} = |\xi\rangle \times |M, 0, 0, P_1, P_2, P_3\rangle \quad (15.50a)$$

where the additional quantum number M_s gives the projection of the spin on a quantization axis. As in eq. (15.44), the generators are expressed formally as the sum of a term acting only in the spin space and one acting on the momenta.

$$G_{\mu\nu} = G_{\mu\nu}^{(s)} + G_{\mu\nu}^{(p)}. \quad (15.50b)$$

The second term of eq. (15.50b) acts on the second factor in the wave function (15.50a) and is determined completely by the mass M and the action of Lorentz transformations on momenta. The set of six matrices $G_{\mu\nu}^{(s)}$ must satisfy the commutation relations (15.32). They therefore must correspond to the well-known representations of two independent angular momentum variables (15.33) discussed in section 15.4.

The simplest non-trivial case, that of spin $\frac{1}{2}$, can be constructed in two ways. We can choose $(G^1, G^2) = (\frac{1}{2}, 0)$ or $(0, \frac{1}{2})$. In either case the $G_{\mu\nu}^{(s)}$ are 2×2 matrices and can be represented by some combination of Pauli spin matrices. Either of these representations is suitable for the description of the neutrino.

In other cases of physical interest parity must be considered. A space inversion reverses the direction of the three spatial coordinates but does not reverse the direction of time. Space inversion does not change any of the angular momentum operators (15.31a) which are axial vectors in ordinary three-dimensional space, but reverses all of the Lorentz generators (15.31b) which are polar vectors in three-dimensional space. Thus

$$\Pi G_{ij} \Pi = G_{ij} \quad (15.51a)$$

$$\Pi G_{0j} \Pi = -G_{0j} \quad (15.51b)$$

where Π is the space inversion operator. In the simpler three-dimensional case the reflections pose no extra problem because the reversal of both spatial directions in a plane is equivalent to a rotation of 180° about an axis perpendicular to the plane. This operation is already included in the algebra and need not be added. When there are three spatial dimensions, a reflection is not a rotation and represents a transformation outside the proper Lorentz group. The improper Lorentz group is defined by including the discrete transformation Π together with the transformations generated by the six generators (15.51).

We now apply space inversion to the operators (15.33)

$$\Pi G^1_{ij} \Pi = G^2_{ij} \tag{15.52a}$$

$$\Pi G^2_{ij} \Pi = G^1_{ij}. \tag{15.52b}$$

Space inversion thus interchanges the two independent angular momenta (15.33). To define an irreducible representation of the improper Lorentz group including space inversion, we must combine pairs of representations of the proper Lorentz group, defined by the eigenvalues $(G^1, G^2) = (m, n)$ and (n, m). For the particular case of spin $\frac{1}{2}$, this means combining the representations $(\frac{1}{2}, 0)$ and $(0, \frac{1}{2})$. The generators (15.31) are therefore 4×4 matrices acting on four-component spinors for the spin $\frac{1}{2}$ representation of the *improper* Lorentz group.

A space inversion operator which satisfies the relation (15.51b) must have a factor which acts in the spin space as well as one which reverses the directions of all momenta. Thus we write

$$\Pi = \Pi^{(s)} \Pi^{(p)}. \tag{15.53}$$

For the case of spin $\frac{1}{2}$, $\Pi^{(s)}$ is a 4×4 matrix. We denote this matrix by γ_0. It must satisfy the relations:

$$\gamma_0 G^{(s)}_{ij} \gamma_0 = G^{(s)}_{ij} \tag{15.54a}$$

$$\gamma_0 G^{(s)}_{0j} \gamma_0 = -G^{(s)}_{0j}. \tag{15.54b}$$

The existence of a matrix γ_0 satisfying the relations (15.54) picks out the zero direction as a preferred direction. Since the commutation relations (15.32b) are completely symmetric in the indices 0, 1, 2 and 3, there must also be three other matrices γ_k which pick out the other three directions in the same manner. Thus, there are four matrices γ_μ satisfying the relations:

$$\gamma_\mu G_{\varrho\sigma} \gamma_\mu = \pm G_{\varrho\sigma} \qquad (+ \text{ if } \varrho \neq \mu \text{ and } \sigma \neq \mu; \; - \text{ if } \varrho = \mu \text{ or } \sigma = \mu). \tag{15.55a}$$

This can also be written

$$\gamma_\mu G_{\varrho\sigma} \gamma_\mu = G_{\varrho\sigma}[1 - 2\delta_{\mu\varrho} - 2\delta_{\mu\sigma}]. \tag{15.55b}$$

There are only 16 independent Hermitean 4×4 matrices. Of these we need to find a set of ten matrices $G_{\mu\nu}$ and γ_ϱ satisfying the commutation relations (15.32b) and (15.54). A simple way to construct the matrices is to choose four Hermitean matrices which anticommute with one another and whose square is unity.

$$\gamma_\mu \gamma_\nu + \gamma_\nu \gamma_\mu = 2\delta_{\mu\nu} \tag{15.56a}$$

$$\gamma_\mu = \gamma_\mu^\dagger. \tag{15.56b}$$

We now define

$$G_{\mu\nu}^{(s)} = -\tfrac{1}{2}\mathrm{i}\gamma_\mu \gamma_\nu. \tag{15.57}$$

It follows immediately from the anticommutation relation (15.56a) that the matrices (15.57) satisfy the commutation relations (15.32b) of generators of the four-dimensional rotation group and that the matrices γ_μ satisfy the commutation relations of a four-vector with the G-matrices (15.57) as well as the relations (15.54). We cannot proceed as in the three-dimensional case with the analog of eq. (15.45b), since the invariant $\boldsymbol{G}\cdot\boldsymbol{V}$ does not exist. The invariant $W_\mu W_\mu$ (15.42b) is too complicated. Instead we can use the new four-vector γ_μ to construct an invariant useful for this case.

The 'scalar product' of the vector γ_μ and the vector V_μ defined by eq. (15.39) commutes by construction with the generators (15.50b) expressed as the sums of generators acting in momentum space and matrices acting in spin space.

$$[(\gamma_\mu V_\mu), (G_{\mu\nu}^{(s)} + G_{\mu\nu}^{(p)})] = 0. \tag{15.58a}$$

As in the previous case, we can choose the zero-momentum state $P_1 = P_2 = P_3 = 0$ to evaluate the invariants. For a zero-momentum state the operator $\gamma_\mu V_\mu$ has the simple form

$$(\gamma_\mu V_\mu)|M, S, M_s, 0, 0, 0\rangle = M\gamma_0|M, S, M_s, 0, 0, 0\rangle. \tag{15.58b}$$

The zero-momentum state is chosen to be an eigenfunction of the total spin and its projection on some axis. The total space inversion operator $\gamma_0 \Pi$ commutes with the spin generators at zero momentum by eq. (15.54). It also commutes with the Hamiltonian, if the Hamiltonian is invariant under space inversion. Then $\gamma_0 \Pi$ can be diagonalized simultaneously with the desired spin operators and with the Hamiltonian. Since Π is the identity on a zero-momentum state, except for a possible phase factor, the zero-momentum state can be chosen to be an eigenfunction of γ_0 and therefore, by eq. (15.58b) also of the operator $\gamma_\mu V_\mu$. By eq. (15.58a) this operator is invariant under Lorentz transformations, thus all states which can be generated from the zero-momentum state by Lorentz transformations are also eigenfunctions of $\gamma_\mu V_\mu$ with the same eigenvalue.

Since γ_0 has the eigenvalues ± 1,

$$(\gamma_\mu V_\mu)|M, S, M_s, P_1, P_2, P_3\rangle = \pm M|M, S, M_s, P_1, P_2, P_3\rangle. \tag{15.59a}$$

Multiplying through by γ_0, substituting for V from eq. (15.39) and rearranging terms we obtain

$$H|M, S, M_s, P_1, P_2, P_3\rangle = (i\gamma_0\gamma_i P_i \pm \gamma_0 M)|M, S, M_s, P_1, P_2, P_3\rangle. \quad (15.59b)$$

This can be rewritten

$$H|M, S, M_s, P_1, P_2, P_3\rangle = (\alpha_i P_i + \beta M)|M, S, M_s, P_1, P_2, P_3\rangle. \quad (15.59c)$$

where we have set

$$\alpha_i = i\gamma_0\gamma_i \quad (15.59d)$$

$$\beta = \gamma_0. \quad (15.59e)$$

The matrices α_i and β are Hermitean and satisfy the relations

$$\alpha_i\alpha_j + \alpha_j\alpha_i = 2\delta_{ij} \quad (15.60a)$$

$$\alpha_i\beta + \beta\alpha_i = 0 \quad (15.60b)$$

$$\beta^2 = 1. \quad (15.60c)$$

The equation (15.59c) was originally derived with a completely different approach by Dirac for the relativistic description of an electron. When applied to a wave function defined to have the form (15.50a) this equation (15.59c) has as solutions the four-component spinors ψ which can represent physical states having given values of the momenta, P_i. It can be written in configuration space by a Fourier transformation, with the operator P_i replaced by $(\hbar/i)(\partial/\partial x_i)$. The probability and current densities can be defined with proper Lorentz transformation properties, and interactions can be constructed.

BIBLIOGRAPHY

Detailed references are not given in this book. Instead a list of useful books is presented here.

1 Mathematical background

The mathematical background necessary for this book is all found in J.D. Jackson, Mathematics for Quantum Mechanics (W.A. Benjamin, New York 1962).

2 General references

The following quantum mechanics textbooks are useful as general references:

Gordon Baym, Lectures on Quantum Mechanics (W.A. Benjamin, New York 1969).
H.A. Bethe, Intermediate Quantum Mechanics (2nd edition) (W.A. Benjamin, New York 1966).
D. Bohm, Quantum Theory (Prentice-Hall, New York 1951).
P.A.M. Dirac, Quantum Mechanics (3rd edition) (Oxford University Press, New York 1947).
K. Gottfried, Quantum Mechanics (W.A. Benjamin, New York 1966).
E. Merzbacher, Quantum Mechanics (John Wiley, New York 1961).
A. Messiah, Quantum Mechanics (North-Holland Publ. Co., Amsterdam 1961).
L.I. Schiff, Quantum Mechanics (McGraw-Hill, New York 1955).

3 References for specific topics

The following books present more detailed and extensive treatment of the topics covered in the individual 'monographs' of this book.

A. *Mössbauer effect*, Chapters 2, 3 and 4

Hans Frauenfelder, The Mössbauer Effect (W.A. Benjamin, New York 1962).

B. *Many-body problems*, Chapters 5, 6, 9, 10, 11

J.M. Blatt and V.F. Weisskopf, Theoretical Nuclear Physics (John Wiley, New York 1952).

G.E. Brown, Many-Body Problems (North-Holland Publ. Co., Amsterdam 1972).

E.U. Condon and G.H. Shortley, Theory of Atomic Spectra (Cambridge University Press, New York 1935).

A. de-Shalit and I. Talmi, Nuclear Shell Theory (Academic Press, New York, 1963).

J. Robert Schrieffer, Theory of Superconductivity (W.A. Benjamin, New York 1964).

C. *Kaon decay*, Chapter 7

W. Heitler, The Quantum Theory of Radiation (3rd edition) (Oxford University Press, New York 1954).

D. *Scattering theory*, Chapter 8

M.L. Goldberger and K.M. Watson, Collision Theory (John Wiley, New York 1964).

N.F. Mott and H.S.W. Massey, The Theory of Atomic Collisions (2nd edition) (Oxford University Press, New York 1949).

E. *Feynman diagrams*, Chapter 12

J.D. Bjorken and S.D. Drell, Relativistic Quantum Fields (McGraw-Hill, New York 1965).

J.J. Sakurai, Advanced Quantum Mechanics (Addison-Wesley, Reading, Mass. 1967).

F. *Symmetries, invariance and relativistic quantum mechanics*, Chapters 13, 14, 15

F.J. Dyson, Symmetry Groups (W.A. Benjamin, New York 1965).

M. Hamermesh, Group Theory (Addison-Wesley, Reading, Mass. 1962).

H.J. Lipkin, Lie Groups for Pedestrians (North-Holland Publ. Co., Amsterdam 1965).

E.P. Wigner, Group Theory and Its Applications to Quantum Mechanics (Academic Press, New York 1959).

SUBJECT INDEX